IET ENERGY ENGINEERING SERIES 166

Advanced characterization of thin film solar cells

Other volumes in this series

Volume 1 **Power Circuit Breaker Theory and Design** C.H. Flurscheim (Editor)
Volume 4 **Industrial Microwave Heating** A.C. Metaxas and R.J. Meredith
Volume 7 **Insulators for High Voltages** J.S.T. Looms
Volume 8 **Variable Frequency AC Motor Drive Systems** D. Finney
Volume 10 **SF_6 Switchgear** H.M. Ryan and G.R. Jones
Volume 11 **Conduction and Induction Heating** E.J. Davies
Volume 13 **Statistical Techniques for High Voltage Engineering** W. Hauschild and W. Mosch
Volume 14 **Uninterruptible Power Supplies** J. Platts and J.D. St Aubyn (Editors)
Volume 15 **Digital Protection for Power Systems** A.T. Johns and S.K. Salman
Volume 16 **Electricity Economics and Planning** T.W. Berrie
Volume 18 **Vacuum Switchgear** A. Greenwood
Volume 19 **Electrical Safety: A guide to causes and prevention of hazards** J. Maxwell Adams
Volume 21 **Electricity Distribution Network Design, 2nd Edition** E. Lakervi and E.J. Holmes
Volume 22 **Artificial Intelligence Techniques in Power Systems** K. Warwick, A.O. Ekwue and R. Aggarwal (Editors)
Volume 24 **Power System Commissioning and Maintenance Practice** K. Harker
Volume 25 **Engineers' Handbook of Industrial Microwave Heating** R.J. Meredith
Volume 26 **Small Electric Motors** H. Moczala *et al.*
Volume 27 **AC-DC Power System Analysis** J. Arrillaga and B.C. Smith
Volume 29 **High Voltage Direct Current Transmission, 2nd Edition** J. Arrillaga
Volume 30 **Flexible AC Transmission Systems (FACTS)** Y.-H. Song (Editor)
Volume 31 **Embedded generation** N. Jenkins *et al.*
Volume 32 **High Voltage Engineering and Testing, 2nd Edition** H.M. Ryan (Editor)
Volume 33 **Overvoltage Protection of Low-Voltage Systems, Revised Edition** P. Hasse
Volume 36 **Voltage Quality in Electrical Power Systems** J. Schlabbach *et al.*
Volume 37 **Electrical Steels for Rotating Machines** P. Beckley
Volume 38 **The Electric Car: Development and future of battery, hybrid and fuel-cell cars** M. Westbrook
Volume 39 **Power Systems Electromagnetic Transients Simulation** J. Arrillaga and N. Watson
Volume 40 **Advances in High Voltage Engineering** M. Haddad and D. Warne
Volume 41 **Electrical Operation of Electrostatic Precipitators** K. Parker
Volume 43 **Thermal Power Plant Simulation and Control** D. Flynn
Volume 44 **Economic Evaluation of Projects in the Electricity Supply Industry** H. Khatib
Volume 45 **Propulsion Systems for Hybrid Vehicles** J. Miller
Volume 46 **Distribution Switchgear** S. Stewart
Volume 47 **Protection of Electricity Distribution Networks, 2nd Edition** J. Gers and E. Holmes
Volume 48 **Wood Pole Overhead Lines** B. Wareing
Volume 49 **Electric Fuses, 3rd Edition** A. Wright and G. Newbery
Volume 50 **Wind Power Integration: Connection and system operational aspects** B. Fox *et al.*
Volume 51 **Short Circuit Currents** J. Schlabbach
Volume 52 **Nuclear Power** J. Wood
Volume 53 **Condition Assessment of High Voltage Insulation in Power System Equipment** R.E. James and Q. Su
Volume 55 **Local Energy: Distributed generation of heat and power** J. Wood
Volume 56 **Condition Monitoring of Rotating Electrical Machines** P. Tavner, L. Ran, J. Penman and H. Sedding
Volume 57 **The Control Techniques Drives and Controls Handbook, 2nd Edition** B. Drury
Volume 58 **Lightning Protection** V. Cooray (Editor)
Volume 59 **Ultracapacitor Applications** J.M. Miller

Volume 62 **Lightning Electromagnetics** V. Cooray
Volume 63 **Energy Storage for Power Systems, 2nd Edition** A. Ter-Gazarian
Volume 65 **Protection of Electricity Distribution Networks, 3rd Edition** J. Gers
Volume 66 **High Voltage Engineering Testing, 3rd Edition** H. Ryan (Editor)
Volume 67 **Multicore Simulation of Power System Transients** F.M. Uriate
Volume 68 **Distribution System Analysis and Automation** J. Gers
Volume 69 **The Lightening Flash, 2nd Edition** V. Cooray (Editor)
Volume 70 **Economic Evaluation of Projects in the Electricity Supply Industry, 3rd Edition** H. Khatib
Volume 72 **Control Circuits in Power Electronics: Practical issues in design and implementation** M. Castilla (Editor)
Volume 73 **Wide Area Monitoring, Protection and Control Systems: The enabler for Smarter Grids** A. Vaccaro and A. Zobaa (Editors)
Volume 74 **Power Electronic Converters and Systems: Frontiers and applications** A.M. Trzynadlowski (Editor)
Volume 75 **Power Distribution Automation** B. Das (Editor)
Volume 76 **Power System Stability: Modelling, analysis and control** B. Om P. Malik
Volume 78 **Numerical Analysis of Power System Transients and Dynamics** A. Ametani (Editor)
Volume 79 **Vehicle-to-Grid: Linking electric vehicles to the smart grid** J. Lu and J. Hossain (Editors)
Volume 81 **Cyber-Physical-Social Systems and Constructs in Electric Power Engineering** S. Suryanarayanan, R. Roche and T.M. Hansen (Editors)
Volume 82 **Periodic Control of Power Electronic Converters** F. Blaabjerg, K. Zhou, D. Wang and Y. Yang
Volume 86 **Advances in Power System Modelling, Control and Stability Analysis** F. Milano (Editor)
Volume 87 **Cogeneration: Technologies, Optimisation and Implementation** C.A. Frangopoulos (Editor)
Volume 88 **Smarter Energy: From smart metering to the smart grid** H. Sun, N. Hatziargyriou, H.V. Poor, L. Carpanini and M.A. Sánchez Fornié (Editors)
Volume 89 **Hydrogen Production, Separation and Purification for Energy** A. Basile, F. Dalena, J. Tong and T.N. Veziroğlu (Editors)
Volume 90 **Clean Energy Microgrids** S. Obara and J. Morel (Editors)
Volume 91 **Fuzzy Logic Control in Energy Systems with Design Applications in MATLAB®/Simulink®** İ.H. Altaş
Volume 92 **Power Quality in Future Electrical Power Systems** A.F. Zobaa and S.H.E.A. Aleem (Editors)
Volume 93 **Cogeneration and District Energy Systems: Modelling, analysis and optimization** M.A. Rosen and S. Koohi-Fayegh
Volume 94 **Introduction to the Smart Grid: Concepts, technologies and evolution** S.K. Salman
Volume 95 **Communication, Control and Security Challenges for the Smart Grid** S.M. Muyeen and S. Rahman (Editors)
Volume 96 **Industrial Power Systems with Distributed and Embedded Generation** R Belu
Volume 97 **Synchronized Phasor Measurements for Smart Grids** M.J.B. Reddy and D.K. Mohanta (Editors)
Volume 98 **Large Scale Grid Integration of Renewable Energy Sources** A. Moreno-Munoz (Editor)
Volume 100 **Modeling and Dynamic Behaviour of Hydropower Plants** N. Kishor and J. Fraile-Ardanuy (Editors)
Volume 101 **Methane and Hydrogen for Energy Storage** R. Carriveau and D.S.-K. Ting
Volume 104 **Power Transformer Condition Monitoring and Diagnosis** A. Abu-Siada (Editor)
Volume 106 **Surface Passivation of Industrial Crystalline Silicon Solar Cells** J. John (Editor)
Volume 107 **Bifacial Photovoltaics: Technology, applications and economics** J. Libal and R. Kopecek (Editors)

Volume 108 **Fault Diagnosis of Induction Motors** J. Faiz, V. Ghorbanian and G. Joksimović
Volume 110 **High Voltage Power Network Construction** K. Harker
Volume 111 **Energy Storage at Different Voltage Levels: Technology, integration, and market aspects** A.F. Zobaa, P.F. Ribeiro, S.H.A. Aleem and S.N. Afifi (Editors)
Volume 112 **Wireless Power Transfer: Theory, technology and application** N. Shinohara
Volume 115 **DC Distribution Systems and Microgrids** T. Dragičević, F.Blaabjerg and P. Wheeler
Volume 117 **Structural Control and Fault Detection of Wind Turbine Systems** H.R. Karimi
Volume 119 **Thermal Power Plant Control and Instrumentation: The control of boilers and HRSGs, 2nd Edition** D. Lindsley, J. Grist and D. Parker
Volume 120 **Fault Diagnosis for Robust Inverter Power Drives** A. Ginart (Editor)
Volume 121 **Monitoring and Control Using Synchrophasors in Power Systems with Renewables** I. Kamwa and C. Lu (Editors)
Volume 123 **Power Systems Electromagnetic Transients Simulation, 2nd Edition** N. Watson and J. Arrillaga
Volume 124 **Power Market Transformation** B. Murray
Volume 125 **Wind Energy Modeling and Simulation, Volume 1: Atmosphere and plant** P. Veers (Editor)
Volume 126 **Diagnosis and Fault Tolerance of Electrical Machines, Power Electronics and Drives** A.J.M. Cardoso
Volume 128 **Characterization of Wide Bandgap Power Semiconductor Devices** F. Wang, Z. Zhang and E.A. Jones
Volume 129 **Renewable Energy from the Oceans: From wave, tidal and gradient systems to offshore wind and solar** D. Coiro and T. Sant (Editors)
Volume 130 **Wind and Solar Based Energy Systems for Communities** R. Carriveau and D.S.-K. Ting (Editors)
Volume 131 **Metaheuristic Optimization in Power Engineering** J. Radosavljević
Volume 132 **Power Line Communication Systems for Smart Grids** I.R.S. Casella and A. Anpalagan
Volume 139 **Variability, Scalability and Stability of Microgrids** S.M. Muyeen, S.M. Islam and F. Blaabjerg (Editors)
Volume 145 **Condition Monitoring of Rotating Electrical Machines** P. Tavner, L. Ran, C. Crabtree
Volume 146 **Energy Storage for Power Systems, 3rd Edition** A.G. Ter-Gazarian
Volume 147 **Distribution Systems Analysis and Automation, 2nd Edition** J. Gers
Volume 155 **Energy Generation and Efficiency Technologies for Green Residential Buildings** D. Ting and R. Carriveau (Editors)
Volume 157 **Electrical Steels, 2 volumes** A. Moses, K. Jenkins, Philip Anderson and H. Stanbury
Volume 158 **Advanced Dielectric Materials for Electrostatic Capacitors** Q Li (Editor)
Volume 160 **Microgrids for Rural Areas: Research and case studies** R.K. Chauhan, K. Chauhan and S.N. Singh (Editors)
Volume 172 **Lighting Interaction with Power Systems, 2 volumes** A. Piantini (Editor)
Volume 905 **Power system Protection, 4 volumes**

Advanced characterization of thin film solar cells

Edited by
Mowafak Al-Jassim and Nancy Haegel

The Institution of Engineering and Technology

Published by The Institution of Engineering and Technology, London, United Kingdom

The Institution of Engineering and Technology is registered as a Charity in England & Wales (no. 211014) and Scotland (no. SC038698).

First published 2020

The Institution of Engineering and Technology
Michael Faraday House
Six Hills Way, Stevenage
Herts, SG1 2AY, United Kingdom

www.theiet.org

British Library Cataloguing in Publication Data
A catalogue record for this product is available from the British Library

ISBN 978-1-83953-023-4 (hardback)
ISBN 978-1-83953-024-1 (PDF)

Typeset in India by MPS Limited

Cover image designed by Alfred Hicks. Source file courtesy of First Solar, Inc

Contents

About the editors

Mowafak Al-Jassim is a senior research fellow at the NREL and manages the Analytical Microscopy and Imaging Science group. His research interests include multiscale characterization of the structural, chemical, interfacial, electro-optical, and electrical properties of bulk and thin-film semiconductors by (scanning) transmission electron microscopy, cathodoluminescence, electron-beam-induced current, atomic force microscopy, and other techniques. His work has involved developing Si, III–V, and thin-film solar cells. He has published more than 500 articles in peer-reviewed journals and conference proceedings.

Nancy Haegel is the director of the Materials Science Center in the Materials and Chemical Science and Technology Directorate at the National Renewable Energy Laboratory (NREL), USA. Her research interests include imaging of electronic and energy transport and characterization of semiconductor materials and devices. Her awards include the NPS Schieffelin Award for Teaching Excellence, the 2004 APS Prize for Research at an Undergraduate Institution, a Humboldt Research Fellowship in Germany and the U.S. Fulbright Senior Scholar at Hebrew University. She has authored some 125 publications.

Chapter 1

Introduction—Motivation of polycrystalline thin-film solar cells

Wyatt Metzger[1]

The future is bright for solar energy. Solar electricity costs increasingly compare favorably with every other generation source [1]. The annual installation rate of solar electricity has been doubling about every 2–3 years in the past decade, and solar energy is abundant [2]. More than 173,000 terawatts (TW) of solar energy strike the Earth continuously. This is approximately 10,000 times more than current global energy demand, and it will last for billions of years [3]. Forecasts are that solar energy may provide 30%–60% of electricity generation by 2050, and certain regions such as California, Hawaii, Massachusetts, and Vermont already derive 10%–20% of their electricity from solar energy [4,5]. Despite these successes, photovoltaics (PV) is still an emerging and fiercely competitive market, and there is considerable potential for further improvements. This situation is similar to the computer industry in the 1980s, where early products and market penetration were only a small harbinger of the performance and impact which would come with time and further research and development (R&D).

Semiconductor materials such as CdTe, $CuInGaSe_2$ (CIGS), and perovskites generally have an absorption depth of less than 1 μm in the spectral region of most interest for capturing solar energy. Consequently, the primary semiconductor layers in these solar cell technologies are generally in the order of hundreds of nanometers to several microns. This is less than one-tenth diameter of a human hair—hence, the name "thin-film technology." Depositing thin films, rather than processing bulk materials, opens up new paths to make solar panels with distinct processes, capital equipment, and cost structure.

In current mainstream thin-film PV production, sheets of coated glass enter a factory and exit in several hours as completed solar panels. Figure 1.1 illustrates a commercial method called vapor-transport deposition used in CdTe solar technology [6]. A powder source is heated and turned into a vapor stream that travels into a tube and exits very small perforations and a slit at the bottom. This process creates a vapor stream that is directed and condensed onto panels moving on a conveyor belt underneath the vapor stream. The time required to deposit the film across a

[1]National Renewable Energy Laboratory, Golden, CO, USA

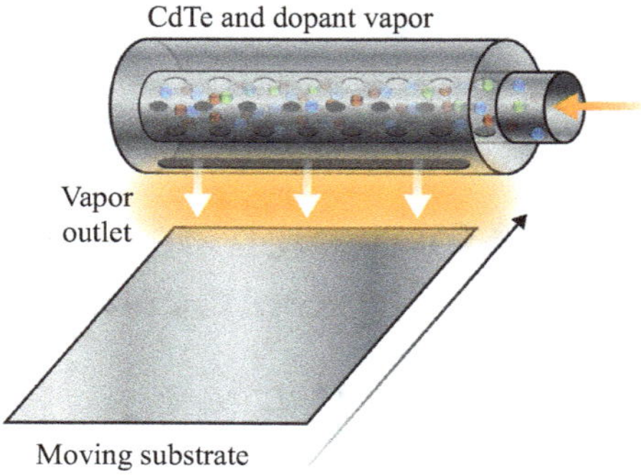

Figure 1.1. Illustration of a high-throughput process to deposit thin films on a moving substrate [7]

10–25 sq ft panel is less than a minute, allowing for very high throughput. A number of similarly fast deposition approaches can be implemented for CdTe and other thin-film materials.

After the semiconductor layers are deposited, laser scribing delineates and connects cells within a panel to achieve the desired voltage and current output. Packaging is then completed. A common approach is to enclose the semiconductor layers between two glass sheets. Metal foils or polymers may also be used in place of glass for flexible and lightweight products. Sealants are applied to the packaging to hold the module together, prevent moisture ingress, and isolate the electrically active regions within the panel and away from handlers. Two cables are connected through the packaging to provide the module electrical output.

The processing and cost structure of thin-film technology is widely different than silicon PV, allowing distinct pathways to lower the costs of electricity for consumers and the upfront capital costs for investment in large-scale deployment. For example, First Solar has constructed a new thin-film solar panel factory in Ohio with annual production of 1.2 gigawatts (GW). The factory was scheduled to take 12–15 months to build, with upfront capital costs of US$400 million [8]. In a period of 30 years, this single factory can manufacture enough panels to match the equivalent output of about 6–8 nuclear plants. In comparison, it would be expected to take 7–10 years to construct one nuclear plant with costs of 6–10 billion dollars [9–10]. Furthermore, this PV factory can be replicated many times over. Currently, the levelized cost of electricity of CdTe PV is competing directly with Si and is already less than fossil fuels in most regions across the world [11,12].

CIGS solar panels have reached sales exceeding 1 GW annually. Companies have deposited CIGS using a wide range of methods including co-evaporation, sputtering, printable inks, and roll-to-roll processing, and they have implemented distinct substrates such as glass, metal foils, and polymers. Consequently, this material is used in a number of diverse applications ranging from rooftop installations to ultralightweight defense applications. At the same time, to date, CIGS companies have struggled to keep pace with the declining costs of other utility-scale terrestrial technologies, although significant cost reductions may still be possible with continuing research and increasing deployment [13].

Perovskite absorber materials represent a wide range of compounds and are labeled based on their underlying perovskite crystal structure. They have received immense attention from the PV research community due, in large part, to a meteoric rise in record laboratory cell efficiencies in the past decade, which have reached 25% at present. By varying the chemical compositions, the bandgap of perovskite films can be tuned over a wide range of the solar spectrum for multi-junction applications. Perovskite solar cells have not been implemented yet in large-scale manufacturing, and they can degrade quickly due to moisture, heat, or light. Consequently, scaling and stability issues remain critical R&D challenges, and perovskite solar cells represent an emerging thin-film technology.

Figure 1.2 displays a common thin-film solar cell configuration [14]. A number of different materials can be used for the absorber, buffer, transparent conducting oxide, and contact layers, providing a rich area for materials research. The primary absorbers discussed in this book are CdTe, CIGS, and perovskites. Yet, the characterization techniques can be applied to other thin-film materials such as CuZnSnS alloys.

The approaches to deposit the different layers can vary for different technologies and within a single solar cell and they include sputtering, electrodeposition, physical evaporation, and solution processing. The substrates may be formed with glass, metal foils, and polymers, enabling a number of applications such as utility-scale solar farms, building-integrated PV, aerospace power, portable electronics, and military applications. For example, thin-film technology has been used to create solar panels that can be rolled up, stored in a backpack, and then deployed to provide remote power (see Figure 1.3).

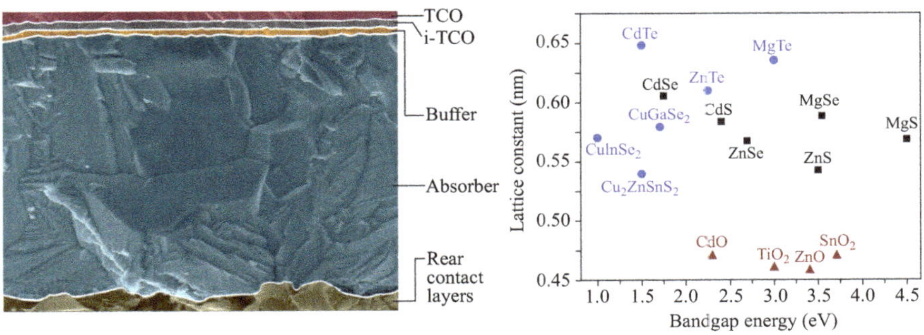

Figure 1.2. A cross-section of a typical thin-film stack from Ref. [14]. TCO refers to a transparent conducting oxide, and the buffer or window layer can help passivate the interface while transmitting most sunlight. Differing band energies can alter the optical absorption, and different lattice constants can introduce strain and defects.

(a) (b) (c)

Figure 1.3. Examples of thin-film solar cell applications: (a) utility-scale solar farm, (b) building-integrated roof shingles, and (c) flexible lightweight remote power

The distinctive features of thin-film technology also introduce key technology challenges. Placing different semiconductors onto glass creates stress and defects between the different layers, as well as optical and band-energy mismatch between each layer, as illustrated in Figure 1.2 [14]. In addition, the transition from the substrate to the semiconductor, coupled with high-throughput manufacturing, leads to interruptions in the periodic array of atoms of the semiconductor, thus leading to grain boundaries and extended defects. All of these phenomena can have deleterious effects on electro-optical properties and performance. Impurities introduced during deposition or by the glass can migrate and alter material properties, and materials or methods that may be advantageous for performance may have trade-offs in cost, product lifetime, and other attributes. In addition, novel substrates such as metal foils and polymers generally cost more than glass and allow more moisture ingress, which can affect stability.

To further increase performance and stability of thin-film technologies, it is critical to understand how different growth processes and materials affect electrical and optical properties. The number of techniques is vast, and the particulars of their capabilities and limitations are important to understand. The spatial scale can range from the dimensions of atoms to modules. For example, high-resolution transmission electron microscopy can examine structural defects in atomic layering across interfaces at the subnanometer scale. Laser spectroscopy and cathodoluminescence can characterize defects such as impurities on crystal lattice sites throughout a thin-film absorber layer on the scale of tens of nanometers, and they can also analyze electron–hole transport on the scale of microns. At the macroscopic scale, infrared and electroluminescence imaging can identify defects such as degraded cells or hot spots caused when specks of dirt block laser light from fully completing scribes during module completion.

Thin-film solar cell development may be conceptualized as improving processes to lower costs and improve performance, with characterization as a guide along the way. Yet, pushing the boundaries of characterization is as much at the core of thin-film solar cell development as material synthesis. For example, although there has been some progress, the thin-film solar community still cannot quickly parse front interface, grain boundary, bulk, and back-surface recombination to pinpoint problems and allocate resources accordingly. Where there is a lack of quantitative characterization, guesswork and empiricism are often substituted, and scientific understanding and technical progress can be hindered. For companies trying to scale a technology in an ever-increasingly competitive landscape, with limited investor dollars and high burn rates, the lack of quantitative characterization and understanding slows development and can contribute to the failure of companies and technologies.

As PV continues to mature, there is an endless drive to improve performance to compete. Hence, PV technology is striving for increasing levels of perfection, requiring a corresponding sophistication in characterization and understanding. Consequently, the topic of this book—advanced characterization—is timely and critical to develop the potential of thin-film solar technology.

References

[1] *Lazard's Levelized Cost of Energy Analysis—Version 13.0* [online]. Available from https://www.lazard.com/media/451086/lazards-levelized-cost-of-energy-version-130-vf.pdf [accessed June 10, 2020].

[2] P. Mints, "Photovoltaic Manufacturer Capacity, Shipments, Price, and Revenues 2017/2018." SPV Market Research, Report SPV-Supply6, April 2018.

[3] National Oceanic and Atmospheric Administration. *Science On a Sphere* [online]. Available from https://sos.noaa.gov/datasets/energy-on-a-sphere/ [accessed June 10, 2020].

[4] W. Cole, B. Frew, P. Gagnon, J. Richards, Y. Sun, J. Zuboy, and R. Margolis, *SunShot 2030 for Photovoltaics (PV): Envisioning a Low-Cost PV Future, NREL/TP-6A20-68105*. Available from https://www.nrel.gov/docs/fy17osti/ 68105.pdf [accessed June 10, 2020].

[5] Solar supplies more than 10% of electricity in 5 states [online]. Available from https://www.pv-magazine.com/2018/08/28/solar-supplies-morethan-10-of-electricity-in-five-us-states/ [accessed June 10, 2010].

[6] A. Luque and S. Hegedus (Eds.), *Handbook of Photovoltaic Science and Engineering*, 2nd Edition, West Sussex, United Kingdom: Wiley, 2011.

[7] B.E. McCandless, W.A. Buchanan, C.P. Thompson, *et al.*, Overcoming carrier concentration limits in polycrystalline CdTe thin films with *in-situ* doping, *Scientific Reports* 2018, **8**, 14519.

[8] *First Solar Announces New Manufacturing Plant* [online]. Available from https://investor.firstsolar.com/news/press-release-details/2018/First-Solar-Announces-New-US-Manufacturing-Plant/default.aspx [accessed June 10, 2010].

[9] International Atomic Energy Agency, Reference Data Series No. 2, 2017 edition. *Nuclear Power Reactors in the World* [online]. Available from https://www.iaea.org/publications/12237/nuclear-power-reactors-in-the-world [accessed June 10, 2020].

[10] U.S. Energy Information Administration | Cost and Performance Characteristics of New Generating Technologies, Annual Energy Outlook 2019.

[11] C. Kost, S. Shammugam, V. Jūlch, H. Nuygen, and T. Schlegl, *Levelized Cost of Electricity Renewable Energy Technologies* [online]. Available from http://www.ise.fraunhofer.de/content/dam/ise/en/documents/publications/studies/EN2018_Fraunhofer-ISE_LCOE_Renewable_Energy_Technologies.pdf [accessed June 10, 2020].

[12] GTM Research, *Global Solar Demand Monitor Q4 2017*.

[13] M.O. Reese, S. Glynn, M.D. Kempe, *et al.*, Increasing markets and decreasing package weight for high-specific-power photovoltaics, *Nat. Energy* 2018, **3**, 1002.

[14] J. Burst, J. Duenow, D. Albin, *et al.*, CdTe solar cells with open-circuit voltage breaking the 1 V barrier, *Nat. Energy* 2016, **1**, 16015.

Chapter 2
Patterns in the control of CdTe solar cell performance*

Ken Durose[1]

2.1 Introduction

Thin-film polycrystalline CdTe solar photovoltaic (PV) cells are the most successful thin-film PV technology in history and currently represent the largest single challenger to the mass-produced wafer silicon products that dominate the market. In 2016 combined sales of single- and multi-crystalline wafer silicon PV were estimated as ~300 GW_p, while those of CdTe were ~3 GW_p. Set against the ongoing expansion of silicon PV module manufacture and tumbling production costs, the challenge for thin-film CdTe is to maintain its competitiveness and market share. To this end, the manufacturers and the research community are faced with a continual challenge of decreasing the costs of producing thin-film modules and increasing their photon conversion efficiency (PCE).

Presently, the record one-of-a-kind CdTe thin-film solar cell performance record for a lab device is 21.0% while that for a module is 18.6%. World production of modules is dominated by First Solar Inc. (USA), whose production is estimated at 5–6 GW_p per annum at the time of writing (2019). Other companies with an interest include Advanced Solar Power (Hangzhou) Inc. (China), Calyxo (Germany), CT Solar (China/Germany), and Reel Solar (USA).

The approach of this review is to set the PV-relevant aspects of CdTe devices, materials, and processing into the context of the wider body of knowledge for both materials science in general and CdTe in particular. It is not intended to be a data review, and for that the reader is referred to the existing book-length reviews on CdTe, the most recent of which is the comprehensive work by Triboulet and Siffert [1]. Here the focus is on how the devices are designed and performed, how the materials are processed, and the evidence from advanced characterisation. It begins with a summary of the features of the main device architectures, and their PV performance as it has developed over time. Section 2.3 outlines the important

**Dedicated to Dieter Bonnet 1937–2016. Inventor of the modern cadmium telluride solar cell.*

[1]Stephenson Institute for Renewable Energy/Department of Physics, University of Liverpool, Liverpool, UK

deposition methods and the resulting materials, and this is followed by a longer discussion (Section 2.4) of the effects of the chloride processing which is vital for the formation of viable CdTe solar cell devices. Sections 2.5–2.7 focus on the partner layers for the CdTe absorber, i.e. the back contact (which has a fundamental limitation) and the n-CdS and latterly graded Cd(Se,Te) that are used at the front of the cell. Issues related to elemental cadmium and tellurium are summarised in Section 2.8. Section 2.9 concludes with a summary and an opinion on the outlook for thin-film CdTe PV.

2.2 Device architectures and key innovations in the development of CdTe solar cells

Although it is rarely mentioned in such terms nowadays, the vision for thin-film solar PV in the 1970s was to use cheap, impure polycrystalline materials in a p-n heterojunction configuration without intentional doping: given the correct combination of materials, it was hoped that heating or low-cost processing would create photoactive junctions at low production cost. To this end, the combination of CdTe and CdS was attractive – heating CdTe introduces cadmium vacancies (acceptors) whereas CdS has a natural population of sulphur vacancies (donors). Formation of a p-n heterojunction with CdS/CdTe – and with other materials such as CdS/Cu_2S and $CuInSe_2$ – appeared to be feasible by following this 'thin-film paradigm'. Direct gap semiconductors also require less material than indirect ones for the same optical absorption. Modern thin-film solar cells have absorbers <2.5-μm thick compared with wafer silicon at ~100 μm. At the limits of mass production, the costs of thin-film solar modules are predicted to fall sufficiently low as to be limited by the substrate, which is usually glass.

The remainder of this section highlights the key steps in the development of the CdTe solar cell and it should be read with reference to the graph of technological developments shown in Figure 2.1.

2.2.1 Superstrate CdS/CdTe solar cell

The modern CdTe solar cell (Figure 2.2(a)) was proposed in the late 1960s by Bonnet and Rabenhorst from the Battelle Institute in Germany – and publications began to appear in the IEEE Photovoltaic Specialists Conference in the early 1970s [2]. This became the most important design for CdTe PV devices in the period 1970–2010, as shown in Figures 2.2 and 2.3. It comprises the following features:

(i) *Glass and front transparent electrode.* Mass production of float line soda lime glass is compatible with in-situ chemical vapour deposition of SnO_2:F (usually as a multilayer) to provide the 'substrate' for growth of the other films. The transparent conducting oxide (TCO) is often augmented by a resistive oxide top layer that acts to increase the device voltage. Other TCOs have been used, and one of a kind world records have often been achieved with borosilicate or similar glass that withstands higher processing temperatures.

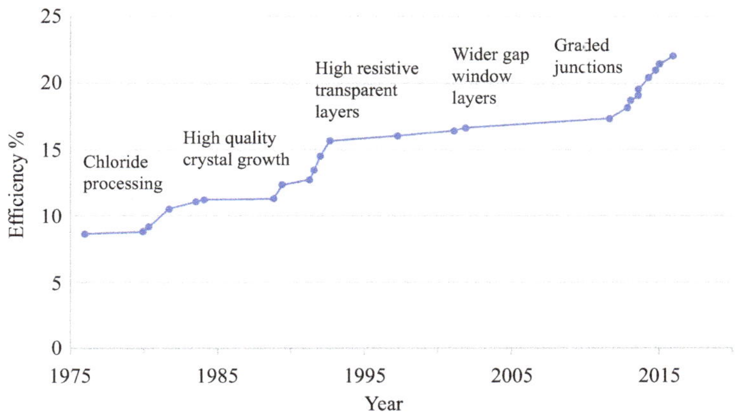

Figure 2.1 Technological development of the CdTe thin-film solar cell showing the key processing innovations responsible for efficiency improvements. See Sections 2.2.1–2.2.4 for brief descriptions of the advances. Data redrawn from NREL: 'Best Research-Cell Efficiency Chart' [10]

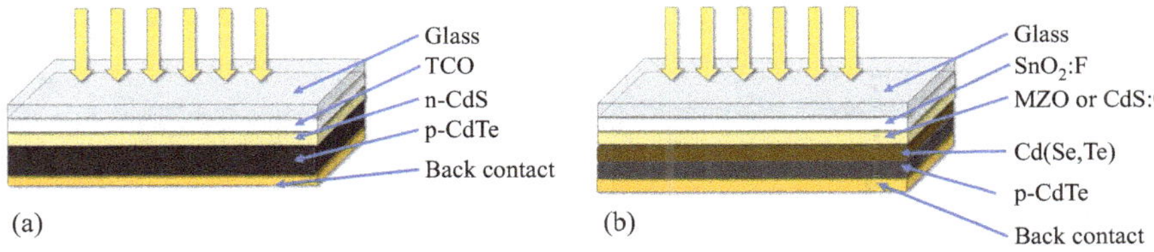

Figure 2.2 CdTe solar cell device architectures: (a) the 'superstrate' design [2] in which the active layers are deposited on glass, which is inverted in service such that light enters the PV absorber through the glass. The heterojunction partner layer is CdS which is the cause of parasitic absorption, (b) the graded junction design using a low-gap Cd(Se,Te) solid solution near to the front of the CdTe absorber. In place of CdS, this design uses either a wider gap partner layer and therefore has an optical advantage over that in (a). Published variations of the partner layer include (MgZn)O (MZO) and CdS:O, although it is likely that others will work well too

(ii) *n-CdS heterojunction partner layer or 'window layer'.* CdS is naturally n-type and forms a heterojunction field with p-CdTe that is free from band alignment discontinuities. Using such a 'window layer' is preferable to a homojunction since it avoids serious surface recombination loss from the CdTe. However, absorption in the CdS does not contribute to the photocurrent and it is a source of parasitic loss of light having energy greater than the bandgap of 2.45 eV (see Section 2.6).

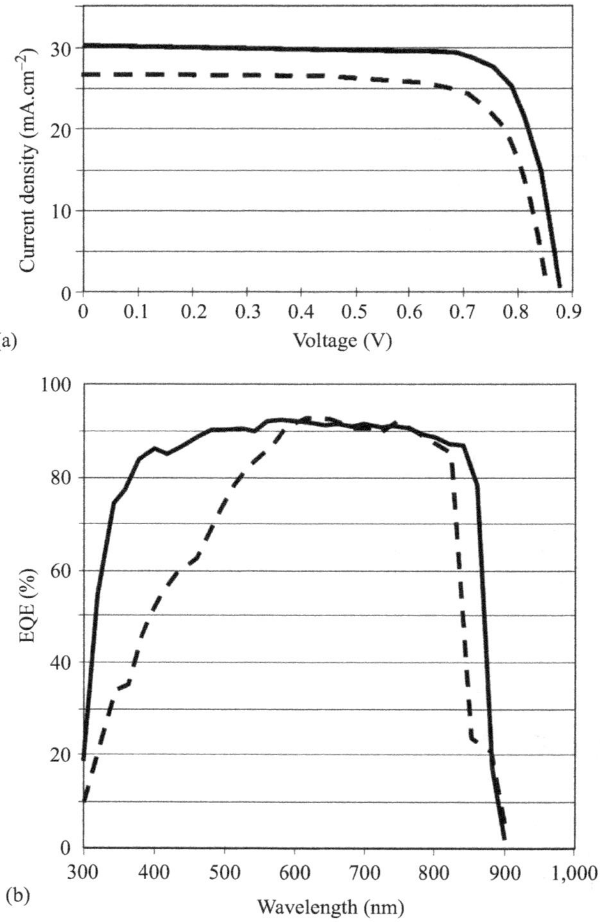

Figure 2.3 CdTe solar cell device evolution: AM1.5 (a) and EQE (b) curves for CdTe solar cells showing the change in practice between Britt's 1993 (15.8%, dashed lines) and First Solar's 2014 (21.0%, full lines) records. Britt's device used CdS which caused parasitic absorption at short wavelengths, leading to the deficiency in EQE at the left-hand side of the spectrum. First Solar's device [11] corrects for this by omitting the CdS and replacing it with an unspecified wider gap partner layer. This device also uses a graded junction with low gap Cd (Se,Te) near to the heterointerface: this leads to the extension of the EQE on the long wavelength side of the spectrum. Most of the gains in recent years have therefore arisen from increased current harvesting rather than voltage increases. Data redrawn from Britt and Ferekides [3] and Wu [14]

(iii) *p-CdTe absorber layer.* This is the site of the active absorption and while between 2 and 3 μm is more than sufficient optically, higher thicknesses have sometimes been used to ensure electrical continuity (e.g. to avoid pinholes). CdTe thin films require chloride treatment to become active in PV devices, and this has both optoelectronic and metallurgical effects. The origin of the actual p-doping was controversial for many years (see Section 2.3).

(iv) *'Back contact' (dark side contact) to the CdTe.* Since p-CdTe has an unusually high electron affinity, it is not possible to make Ohmic contacts with it (there is no metal having a high enough work function). Instead, to form low resistance contacts, narrow highly doped regions are applied to encourage tunnelling. It is an advantage of the 'superstrate' design that the contact is the last layer to be deposited and that it is therefore not disrupted by further high temperature processing (see Section 2.5).

A retrospective view is that apart from the introduction of essential processing and growth innovations in the early 1980s and 1990s (see 2.22 and 2.2.3 below), the research community was fixated and stuck with this design of CdTe PV device for too long. Despite the record device from Britt and Ferekides [3] and some prescient innovations from Wu [4], during the period 1993–2012 new innovations and records failed to translate into further increases in solar cell efficiency, and the record performance stagnated at about 16%, as shown in Figure 2.1.

2.2.2 Chloride treatment of CdTe

Annealing in the presence of chloride ions is an essential processing step for the formation of high efficiency junctions in thin-film polycrystalline CdTe solar cells. Without this, they do not work. Their origins were in the processing of electrochemically deposited material in the early 1980s using $CdCl_2$ [5,6]. However, chlorides were known as a sintering aid (flux) for II–VI semiconductors much earlier ($ZnCl_2$ for ZnS phosphors, see, e.g. [7–9]), and recently other sources of chloride have been found to be effective. Chloride treatment of CdTe is a topic of major importance in solar cell manufacturing and is treated in depth in Section 2.4.

2.2.3 Implementation of growth methods that work best for CdTe and CdS

(i) *CdS growth methods and conditions.* Growth of CdS at substrate temperatures of 180–200 °C gives the best outcomes providing that the films can be grown thin enough and without pinhole defects. Physical vapour deposition (evaporation) is prone to source spitting which forms pinholes and it is not suitable. Sputtering is capable of achieving the necessary thickness control and film integrity. Chemical bath deposition was popular on account of the uniformity achievable, but the highest efficiency devices and production environments use vacuum methods nowadays.

(ii) *CdTe growth methods and conditions.* The best outcomes for CdTe are considered to be associated with high growth temperatures, e.g. >500 °C.

Both close space sublimation (CSS) and blowing powderised feedstock onto heated substrates have both been used in production.

2.2.4 Graded junction Cd(Se,Te) solar cell

Figure 2.2(b) shows a more recent innovation adopted by First Solar: it is a variation of the superstrate design in which part of the absorber film is a low-gap solid solution of (CdSe,Te). It is formed by first depositing CdSe, and then CdTe, after which interdiffusion takes place. Another important feature of this design is that the 'window layer' is no longer CdS but is a wider gap alternative. Both (Zn,Mg)O and CdS:O have been reported to work well but have issues with stability. Replacement of the window layer acts to improve the short wavelength response, while the narrow gap region of the graded absorber increases the long wavelength response. The J-V and external quantum efficiency (EQE) performances of the 'traditional' and graded gap devices are compared in Figure 2.3(a) and (b) while the highest performances for both cells and modules are shown in Table 2.1

2.3 CdTe deposition and the As-deposited films

While many methods have been used to deposit thin-film CdTe, those that produce the highest efficiency solar cells generally use higher temperatures, typically above 500 °C. These materials have larger grain size, and it is also likely that the point defect population in the material is more optimised.

An early success in growth method innovation was the use of CSS, a physical vapour transport that is characterised by transport under a reduced pressure of an inert gas e.g. 100 Torr of N_2 or Ar. It uses a large area source tray of CdTe held just a few millimetres from the substrate. Both Ferekides's 1993 [3] and Wu's 2002 [4] longstanding records used this method of fabrication, and other labs reported similarly high efficiencies [15]. CSS was developed industrially by Antec GmbH (now CT Solar) in the 1990s. An alternative method based on blowing powderised feedstock through a high temperature slotted dispenser tube onto a hot substrate has been industrialised by First Solar Inc. and has the advantage of high throughput rates.

The influence of growth temperature on the crystallinity of as-grown polycrystalline deposits is best understood with respect to the materials science

Table 2.1 Recent CdTe record solar cell and module efficiencies, both held by First Solar

Efficiency (%)	Area (cm^2)	V_{oc} (V)	J_{sc} (mA/cm^2)	Fill factor (%)	Ref.
One of a kind cell					
21.0 ± 0.4	1.0623	0.8759	30.25	79.4	2014 [11]
Module					
18.6 ± 0.5	7038.8	110.6	*1.533	74.2	2015 [12]

Source: Data as reported in Green *et al.* [13].

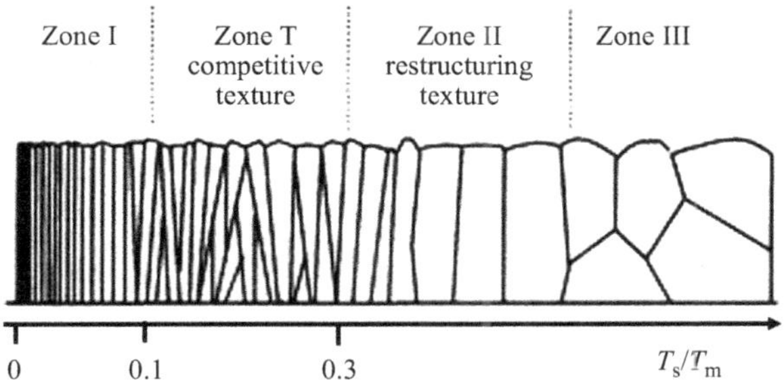

Figure 2.4 The 'Structure Zone Model' which accounts for the relationship between the grain morphologies in thin polycrystalline films and the substrate temperature expressed as a fraction of the material's melting temperature (T_m/T_s). Reprinted from Barna and Adamik [17] with permission from Elsevier

framework known as the 'Structure Zone Model' (see Figure 2.4). This systematises the grain morphologies generated by growth at temperatures expressed as a fraction of the melting temperature of the material (T_m/T_s). For example, in 'Zone 1' ($T_m/T_s > 0.1$) materials species arriving at the surface stick to (or close to) the grains where they arrive, and the temperature is not high enough for other processes to be influential. Hence a columnar, needle-like grain structure is generated and preserved. On the other hand, at high temperatures, both the nucleated and subsequent grain structures are substantially modified by both surface migration and solid-state diffusion phenomena. The result is that the grain structure is substantially altered by these processes, as seen in 'Zone III'.

Luschitz *et al.* [16] made a systematic study of CSS growth in the range 240–520 °C, as shown in Figure 2.5. Growth at 240 °C shows distinct needle-like grains characteristic of Zone I. Increasing the temperature increased the grain size systematically as expected. Gradually the cross-section grain shapes move from left to right through the Structure Zone model types, although in the cleaved cross-section images in Figure 2.5 they are difficult to be visible. Figure 2.6 shows a clearer focused ion beam scanning electron microscopic (SEM) cross section of a CdTe film grown by CSS at 550 °C, in which the grain structure clearly conforms to that expected for Zone III.

From both the Structure Zone Model schematic in Figure 2.4 [17] and the cross-section image in Figure 2.6, it may be seen that the grain size in the films increases with distance from the first growth interface. This is seen experimentally more widely, e.g. by Cousins and Durose [18]. Cousins and Durose also explored this computationally with an atom-by-atom crystal growth model [18,19]. Atoms arrived on a polycrystalline substrate and were allowed to leave or else stick then migrate and bond a favourable site. Each atom that came to rest was assigned to the grain to which it was most strongly bonded. This yielded grain structures similar to those in 'Zone T' in which some grains grew at the expense of others. In the model, and in experimental studies of close space sublimated CdTe films, the grain size

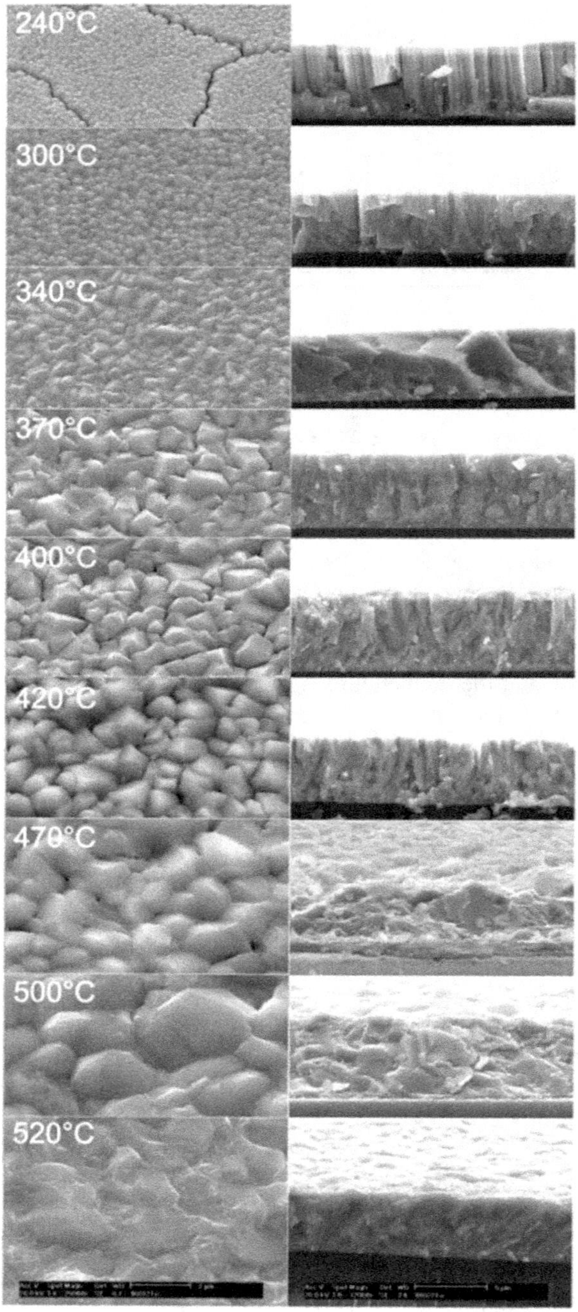

Figure 2.5 Variation in the grain structure of thin-film CdTe as a function of the substrate temperature during growth. There is a good correlation with the grain morphologies expected from the 'Structure Zone Model' as shown in Figure 2.4. Reproduced from Luschitz et al. [16] with permission from Elsevier

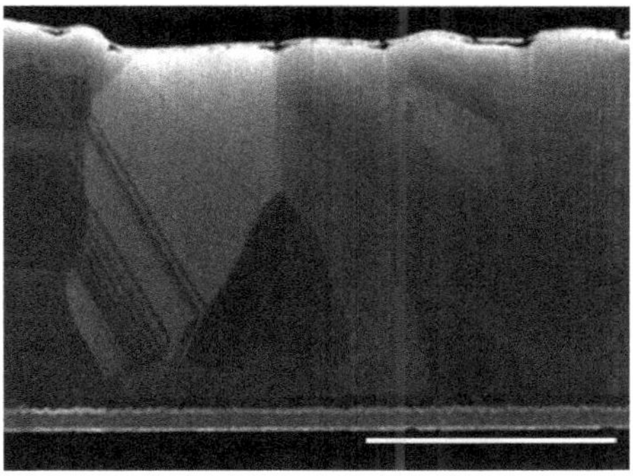

Figure 2.6 Secondary electron SEM image of a focused ion beam cut cross section of a CdTe solar cell structure for which the CdTe was grown at 550 °C by CSS. Its grain structure is as expected for Zone III of the Structure Zone Model, as shown in Figure 2.4 and for which the growth temperature is sufficiently high to encourage metallurgical change by diffusive processes, including horizontal grain size increases. Device quality material often has this kind of grain structure

varied with distance h from the interface as $R = R_0 + C \times h^k$, where R_0 and C are constants for a particular film and $k \approx 0.5$.

The general outline of the growth temperature–grain size relationship also gives some insight into the grain sizes achievable from different methods of thin-film CdTe growth. Lower temperature growth methods such as electrochemical, sputtered and MOCVD CdTe tend to have submicron grains (smallest to largest in that order). Physical vapour deposition (evaporation) uses a large source–substrate distance allowing the temperature of the substrate to be much lower than that of the source: small grains can result. CSS tends to be done at higher temperatures, and as a result the grains tend to be >1–2 μm in size, and under certain conditions can reach 10 μm or greater.

Overall, the growth method, growth conditions and substrate can give CdTe having considerable differences in the grain size, both the vertical and horizontal grain size distributions and the amount of residual strain in the thin films. These factors have an influence on the strain state of the material and hence on the metallurgical effects of the subsequent processing that are discussed in the next section.

2.4 Processing the CdTe and its effects on device performance, microstructure and defect chemistry

2.4.1 Effects of chlorides on device performance

Thin-film CdTe solar cells only work after the CdTe has been treated by annealing so as to introduce chloride ions. This processing affects every aspect of the material and device performance and appears to be vital for the formation of an active p-n junction in polycrystalline thin films made by all methods.

Chlorides have long been known as a 'flux' for II–VI semiconductors, and both $CdTe–CdCl_2$ [20,21] and $CdS–CdCl_2$ [22] have eutectic form phase diagrams, i.e. there are compositions having a low melting temperature. However, the use of chlorides in solar cell processing appear to have developed independently and may be traced back to the work of Basol on electrochemically deposited CdTe films [5]. Follow-on work by Al-Allak *et al.* [23] serves to illustrate the effects on device performance in early CSS devices: as-grown material had PCEs of ~1%, heat treatment alone gave ~3%, while heating with $CdCl_2$ in air increased the efficiency to ~10%. The effects depend on the duration and temperature of the treatment, with overtreatment past the optimum peak causing a decline in performance.

All present-day thin-film CdTe solar cells use chloride processing in one form or another: it is considered essential to form the p-n junction and to control its position, and it has a beneficial effect on grain growth and is considered by some to influence interfacial diffusion.

An example of the effect of junction position is shown in Figure 2.7 [24]. An optimum level of treatment gave the highest PCE, and this correlated with the electrical (p-n) junction being close to the metallurgical (CdS–CdTe) junction where the photoexcitation is the highest. Without chloride treatment, the p-n junction is buried deep in the CdTe and photoexcited carriers are lost.

2.4.2 Doping in CdTe and the origins of impurities in solar cells

(i) *Origins of the field in solar cells and the impact of doping density and carrier lifetime on V_{oc}.* Doping is of critical importance to solar cells since the field in a p-n junction is created by the fixed ionised dopant atoms or centres that are exposed in the region that is depleted of mobile carriers (the depletion region). It is this field that imparts usable energy to the photo-excited charge carriers. If a high built-in voltage is engineered using high doping, and the carriers have long lifetimes, then high V_{oc} values may be expected. Sites have simulated the effects of both parameters on V_{oc}, as shown in Figure 2.8 (it is highlighted here that special precautions must be taken to ensure that time-resolved luminescence measurements of carrier lifetime give reliable results that are not influenced by surfaces [25]).

(ii) *The thin-film doping paradigm and overall doping mechanisms in CdTe.* In the case of thin-film (polycrystalline) CdTe solar cells, the origin of doping has only recently come to be understood from the combined effects of copper (an acceptor dopant) and chlorine, which is essential to electrically passivate the grain boundaries [27]. In the earliest days of the n-CdS–p-CdTe solar cell, it was assumed that the doping was natural, i.e. it arose from the native populations of V_S and V_{Cd} that could be encouraged by heating (V_{Cd} is a deeper defect than anticipated). Since the early 1980s, processing with chlorides has been considered essential for high performance, and its

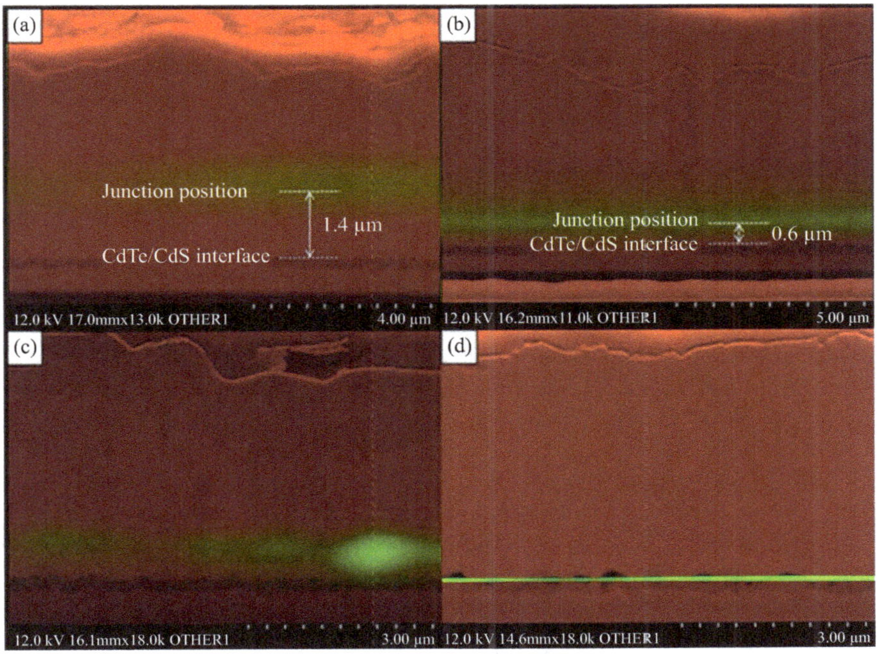

Figure 2.7 Junction position as influenced by the chloride treatment on identical CdTe solar cell structures deposited by CSS and $CdCl_2$-treated for different lengths of time. The junction position, shown in green, is obtained from focused ion beam cross sections using the electron beam-induced current (EBIC) method in the SEM. The EBIC amplifier filter setting was selected so as to give smooth out local variation and to give the average junction position. (a) As-grown (no chloride), (b) undertreated, (c) optimised and (d) overtreated, as determined from efficiency measurements. Reproduced from Progress in PVs [24] (Creative Commons)

effects have been widely studied. Copper is an endemic accidental impurity in II–VI semiconductors and may have been responsible for doping even when not intentionally introduced. It could also get into the bulk CdTe from the contacts, where it is used to make a p^+ surface layer. The situation with CdTe PVs is further complicated by the fact of 'solar grade' CdTe having a purity of 99.999% ('5N'), which implies a concentration of 10 ppm of potentially and electrically damaging impurities capable of causing unwanted recombination.

(iii) *Doping mechanisms in CdTe*. Understanding the factors controlling conductivity in CdTe benefits from an extensive literature on single crystal work from the radiation and (Hg,Cd)Te infrared detector communities. There are book-length reviews [1] and data collections [28].

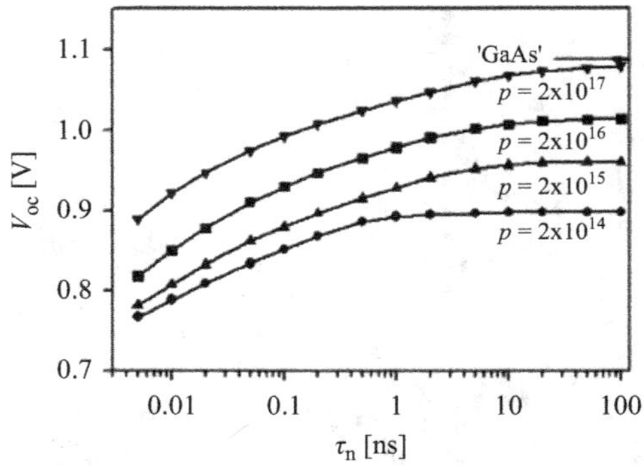

Figure 2.8 Simulation of the combined effects of carrier density and lifetime on the V_{oc} achievable for thin-film CdTe solar cells. The full V_{oc} expected from the band gap value is never achieved for any polycrystalline thin-film PV device technology, and the shortfalls for both CdTe and CIGS are comparable at about 40%. For epitaxial GaAs, the shortfall is about 20%. Fundamental limits to the doping densities for CdTe from self-compensation are described in the text. Reproduced from Sites 2007 [26] with permission from Elsevier

Conductivity in many II–VI semiconductors is considered to be controlled by a combination of the effects of native defects and impurities. In the case of CdTe, these include the vacancies (e.g. V_{Cd} – double acceptor; V_{Te} – double donor) and substitutional impurities (e.g. Ag_{Cd} and Cu_{Cd} – acceptors on the Cd site; In_{Cd} – donor on the Cd site; As_{Te} and As_{Te} – acceptors on the Te site and I_{Te} and Cl_{Te} – donors on the Te site). Fuller lists may be found in Triboulet and others [1,28]. Combinations of defects in 'complexes' may also be present, with the 'A centre' ($[V_{Cd} – Cl_{Te}]$ – a single acceptor being one of the few to be studied in detail in single crystals [29]. Where these point defects comprise easily ionisable shallow levels, they may contribute to conductivity necessary for the formation of the p-n junction. On the contrary, where they form deep levels, they may promote recombination and are deleterious to the operation of solar cells.

(iv) *Accidental impurities in CdTe solar cells.* Understanding the role of electrically active centres in CdTe devices is not supported by the fact that the production standard purity for CdTe is 99.999%, ('5N'), i.e. the feedstock contains 10 ppm of unidentified impurities, any of which could be in principle electrically active. Since 1 ppm corresponds to ~10^{16} cm^{-3} dopant atoms, this level of impurities is more than enough to disrupt the control of a p-n junction. Moreover, the purity standard for $CdCl_2$ often used in research labs is 99.9% ('3N') and the potential for contamination is significant.

Emziane and colleagues [30–38] conducted a series of elimination experiments using quantitative SIMS to identify the origins and identities of impurities in CdTe solar cells fabricated by CSS. These experiments included the fabrication of cells using 4, 5, 6 and 7N CdTe, 2, 3 and 4N $CdCl_2$ and a comparison of the effects of both sapphire and soda lime glass substrates. Copper was found to be present at high levels (~10^{18} cm^{-3}) irrespective of the source purity (it arose from the growth apparatus or back-contact doping); sodium originated from both the glass and as an impurity in the $CdCl_2$ while the impurities present and their levels in $CdCl_2$ powder appeared to bear little relation to the indicative analyses from manufacturers or the declared purity levels.

(v) *Role of chlorine in CdTe solar cells and segregation to grain boundaries.* The apparently concrete framework for the understanding of electrically active centres in CdTe throws up a conundrum: Cl_{Te} would be expected to be a donor, and yet the inclusion of chlorine is essential for the formation of a p-n junction in which the bulk of the CdTe is p-type. Chlorine clearly plays some more complex role.

Both direct [39] and electrical measurements [40] indicated that chlorine electrically passivates grain boundaries in CdTe and it appeared inevitable that chlorine concentrates at the boundaries [41] in accordance with classical metallurgical expectations [42]. In recent years, advances in spatially resolved analytical methods have confirmed this directly. Various methods concur, including high resolution *z*-contrast scanning transmission electron microscopy (STEM) [43] atom probe tomography [43] and imaging time of flight secondary ion microscopy (TOF-SIMS) [44] all confirm the segregation of chlorine at the grain boundaries. Moreover, substitution of chlorine on the tellurium sites is consistent with local n-doping [43].

Theoretical models (DFT) have indicated that substitutional Cl_{Te} will act so as to electrically passivate in-gap states that would otherwise introduce fields near to the grain boundaries that would negate p-doping. However, Cu remains uniformly distributed. It is therefore current opinion that the origin of p-type conductivity in CdTe PV devices is from Cu_{Cd}. The function of the chlorine is to electrically passivate the grain boundaries so as to make the copper doping effective [27]. The controversial topic of grain boundaries in CdTe solar cells is developed later in this section.

(vi) *Substitutional doping with group V elements*. Substitutional group V doping in CdTe is expected to generate p-type conductivity from P_{Te} or As_{Te}. This suggests an alternative doping paradigm for CdTe PV with intentional group V p-type doping rather than with Cu.

Examples relevant to solar work are as follows:

Phosphorous: Alnajjar *et al.* [45] achieved 5.5×10^{17} cm^{-3} p-type conductivity by post growth doping of CdTe by heating in orthophosphoric acid vapour, but not without experimental hazards.

Burst *et al.* used phosphorous doping for in-situ doping of Cd-rich melt in Bridgman growth to achieve stable doping densities of up to 10^{16} cm^{-3} combined with lifetimes of 30 ns [46]. Further doping soon leads to a fall in dopant activation caused by self-compensation [47] as shown in Figure 2.9 (see also (vii)).

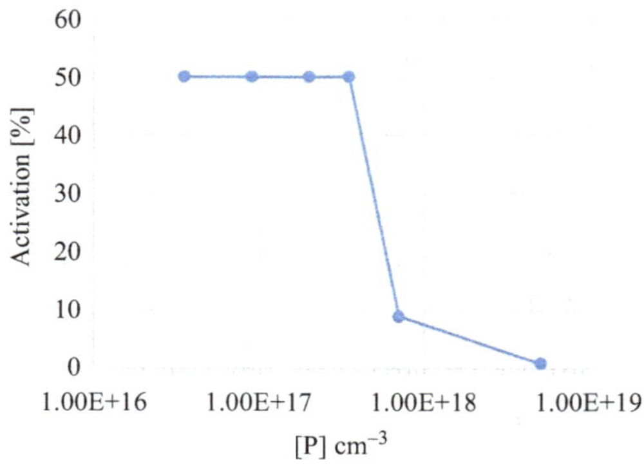

Figure 2.9 Compensation in phosphorous doping of Bridgman-grown bulk single crystal CdTe. Above mid 10^{17} atoms/cm^3, the activation ratio crashes due to the onset of self-compensation effects. p-Type doping of CdTe will always have an upper limit due to self-compensation. Data extracted from Figure 2(a) by Burst et al. [47]

Device results achieved on single crystals using CdTe:P – including the impressive result of $V_{oc} > 1$ V – were mentioned in Section 2.4.5 (vii).

Arsenic: Early ideas on doping CdTe with As for use in solar cells using MOCVD were reported in 1998 [48] and was later combined with the use of Cl in the CdTe, presumably in a passivating role [49]. Deep levels in this CdTe:As are reported from thermal admittance spectroscopy [50] and deep level transient spectroscopy [51]. Activation ratios for the MOCVD-grown As have been reported in the range 100–1,000 indicating that there is significant compensation. It is speculated here that this is either due to overdoping or perhaps interference from hydrogen. Nevertheless, efficiencies of up to 13.3% have been reported [51].

For the case of molecular beam epitaxy (MBE), Farrell *et al.* [52] reported $p = 5 \times 10^{16}$ cm^{-3} in CdTe on (211)Si, and incorporation is enhanced under an excess flux of Cd, as expected.

(vii) *Self-compensation and doping limits in CdTe*. p-Type doping in CdTe is known to be especially susceptible to self-compensation, i.e. when the doping density exceeds a certain limit, it becomes more energetically favourable for native defects to neutralise existing doping and to prevent further increases in extrinsic conductivity. There is an important implication for solar cells: there will be a fundamental limit to carrier density in CdTe, and depending on the doping density required for high efficiency this may be low enough to cap V_{oc}.

The systematic description of self-compensation phenomena has been reported in detail in the 1980s and beforehand, but it is rarely referred to in the mainstream PV literature (see Marfaing [53] for details). The 'AX' centre has been identified as a possible cause of the compensation in group V-doped CdTe [54]. Above a certain limit, it becomes energetically favourable for the acceptor, e.g. P_{Te}, to change its environment to form a complex which involves three of the four Te sites surrounding the central Cd atom. Two of the Te atoms move together to form a dimer (breaking two bonds in the process) while P_{Te} continues to occupy the third site. This centre now acts as a (compensating) donor. The effect of this is shown in Figure 2.9 for CdTe:P – above ~3×10^{17} cm^{-3} there is a very sharp drop-off in the activation ratio.

More recently, Yang *et al.* [55] made a theoretical study of P and As in CdTe and concluded that even if high doping densities (~10^{18} cm^{-3}) are achieved experimentally, then they will be unstable. Further findings from first principles are reviewed by Yang *et al.* [56].

2.4.3 Grain boundaries, extended defects, metallurgical changes and electrical passivation

Extended defects in CdTe, notably grain boundaries, are of particular importance to solar cells since they may influence performance and stability by acting: (i) as recombination centres, (ii) as resistive barriers, (iii) as charge separators and conducting channels for current transport and (iv) acting as pathways for enhanced interdiffusion of the cell's layers. Thermal processing (always used in CdTe cell-making) may be expected to promote metallurgical changes in the material. While the majority of these factors may be expected to be deleterious to solar cell operation, there is nevertheless a debate in the literatures as to whether grain boundaries harm PV performance (the 'traditional' view) or whether they contribute positively to it. All aspects of extended defects are reviewed later, including defect crystallography, the systematic effects of processing, evidence for and against grain boundary recombination and charge collection together with evidence from modelling. It is concluded that grain boundaries are always harmful to PV operation and that chlorine minimises the harm.

Understanding of extended defects in semiconductors is a mature topic that is the subject of a book-length review by Holt and Yacobi [57] to which the reader is referred for a detailed overview. Extended defects in CdTe in particular are reviewed in [58] and the main points of relevance are:

(i) CdTe adopts the sphalerite (zinc blende) cubic structure, having non-centrosymmetric space group $F\bar{4}3m$. It therefore exhibits crystallographic polarity, e.g. the Cd(111)A and Te$(\overline{111})$B faces being Cd- and Te-terminated, respectively. Care should be taken in identifying polar surfaces experimentally [59].
(ii) Random grain boundaries will most usually contain wrong bonds which will be likely to carry electrical charge and contain recombination centres. These are the most common and most important types of grain boundary in CdTe thin films and were shown to comprise 40%–45% of the grain boundaries in CSS material [60].
(iii) Special (highly symmetric) grain boundaries have been widely studied since they occur reproducibly and are prevalent in CdTe. While their study

provides some lessons for CdTe PV devices, the most common forms are of little consequence. These boundaries are 'coincidence site lattice' types for which the most common type, the first-order twin, has a Friedel index of $\Sigma = 3$ (lattice site overlap ratio). The boundary plane separating the grains can adopt four different sets of planes. Of these, the lowest energy and most common is the long straight twin boundary lying on $\{111\}$–$\{111\}$ – it has no wrong bonds and is not electrically active when clean. The other four types ('lateral' or 'incoherent' boundaries) cannot be constructed without wrong bonds [61,62] and are expected to be electrically active – this is seen experimentally.

(iv) Dislocations are of the 60° ½$\{110\}$ type but these often dissociate into two $1/6$ $\{211\}$ types having Cd and Te terminations [63] and being separated by a ribbon of stacking fault.

(v) CdTe is stabilised in its zinc blende form, for which the stacking sequence is $A\alpha B\beta C\gamma$ .... However, while the wurtzite form is virtually unknown, the stabilisation energy is sufficiently weak to allow planar defects. The thinnest of these, the stacking fault, comprises a perfectly bonded interface terminated by partial dislocations (see earlier). The planar fault itself is not significantly electrically active, but the terminating partials are known recombination centres [63]. It should be recognised that the limiting case of a thin twin band is a stacking fault and that the 'lateral' twin boundaries and their electrical activity are analogous to that of partial dislocations. However, both dislocations and lateral twin boundaries have low incidence in device grade CdTe thin films and will have an insignificant effect compared with random grain boundaries.

It is the conventional view that the wrong bonds associated with grain boundaries and dislocations in semiconductors occupy states in the bandgap that promote harmful recombination (alternative postulates follow later).

2.4.4 Metallurgical changes in CdTe upon processing with chlorides

Given the importance of chloride processing, there has been significant effort in the study of its influence on the evolution of grain structure on CdTe films. These may be systematised within the framework of the theories of recovery, recrystallisation and grain growth which were first developed for cold worked metals and which were later applied to semiconductors. A summary of the main effects is described as follows, but the reader is again referred to a book-length review for a detailed account [64].

(i) *Recovery*. Cold worked metals have highly disrupted crystal lattices with high densities of tangled dislocations having a high-volume strain. Individual grains are not present, but they emerge during the first stages of heating, i.e. 'recovery'. This is driven by the volume strain.

(ii) *Recrystallisation*. The recovered grains still contain high levels of volume strain which drives the process of recrystallisation: new, unstrained, grains form at the interstices of the original ones and this new grain structure takes over the film.
(iii) *Grain growth*. The recrystallised grains grow to reduce the grain boundary area energy.

It is expected that the processes may be encouraged by heat, time of exposure and 'flux', i.e. chemical agents that promote the metallurgical change, e.g. by decreasing the melting temperature and facilitating the mobility of atoms. For the case of CdTe, $CdCl_2$ may be considered a fluxing agent. Hiie and colleagues have studied the possibilities in detail and considered the role of the low melting point phases containing Cd, Te, Cl and O [65,66] which remain otherwise under-investigated given their importance to CdTe solar cell production.

The three processes of recovery, grain growth and recrystallisation may overlap. Their operation is also dependent on the initial starting form of the material. For example, the processes may proceed during growth itself. Also, grain growth may initiate at from the bottom or the top of a film, depending on the grain size distribution. Some examples from the CdTe literature follows:

(i) *Drivers for recrystallisation*. Moutinho *et al.* [67] reported some elegant studies in which the growth temperature of PVD CdTe films was used to vary the grain size, and hence to demonstrate that grain growth was a function of the grain size rather than the growth method. The studies also used Nelson–Riley precision lattice parameter evaluation to demonstrate that the onset of recrystallisation of CdTe films caused a second population of grains to emerge, and having a relaxed (unstrained) lattice parameter. These grains were also identified directly by atomic force microscopy.
(ii) *Development of grains during growth*. Cousins measured the gain size distribution as a function of distance from the interface in CSS films for which the as-grown material comprised small grains near to the interface and larger gains as the thickness increased. While the gains in the bulk of the film were stable to any level of $CdCl_2$ treatment, those near to the interface (in the buried part of the film) were seen to grow [18].
(iii) *Diffusion-limited transformation*. Qi *et al.* examined the kinetics of grain growth phenomena in electrochemically deposited material. Arrhenius plots yielded an activation energy for grain growth processes of 2.5 ± 0.3 eV, which the authors associated with the activation energy for Cd diffusion in CdTe [68].
(iv) *Use of electron backscattered electron diffraction (EBSD) to monitor changes and intergranular relationships*. Quinones *et al.* [69] gave an early account of the use of EBSD to study grains in as-grown films whereas Moutinho *et al.* [70] applied the method to studying recrystallisation phenomena. Stechmann *et al.* combined focused ion beam milling with EBSD to give three-dimensional orientation maps before and after processing [71].

Figure 2.10 shows multimode SEM correlating cathodoluminescence (CL) microscopy to a texture map of a CdTe film [72].

2.4.5 Role of grain boundaries in solar cell operation

The conventional view of grain boundaries in solar cells is that they reduce the PCE. Deep levels at the boundary interfaces promote minority carrier recombination, and this reduces both voltage and current. On the contrary, for CdTe devices there are some remarkable claims in high profile papers that the grain boundaries increase the performance of the solar cells by virtue of their having electric fields which aid carrier separation and by their being conductive which helps current flow. Overall it may be demonstrated that this view is not correct and that grain boundaries are always deleterious to PV device operation. A brief review and critical evaluation of the (seemingly contradictory) evidence follows:

(i) *Evidence for grain boundary recombination*. There is compelling evidence from EBIC and CL that recombination is strong at grain boundaries

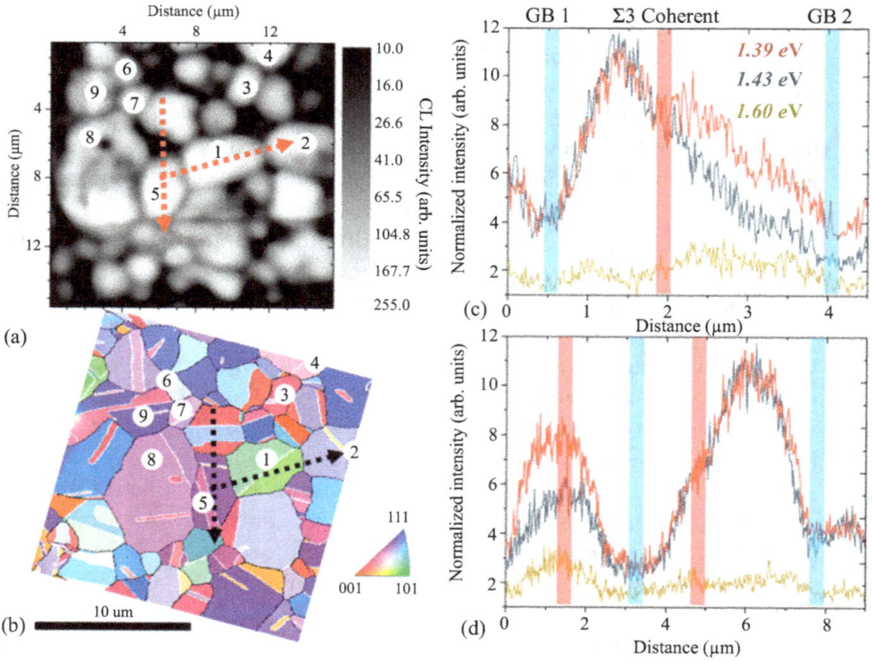

Figure 2.10 Correlation of luminescence with grain orientations in a CdTe film (plan view). (a) Panchromatic CL image, (b) EBSD map showing the colour-coded standard triangle. (c) and (d) show line-scans from (a) and (b). Reproduced from Stechmann et al. [72] with permission from Elsevier

(including in bulk material [73]), and that it is dependent on the grain boundary type [60] as expected. For example, EBIC shows that the disrupted 'lateral' or 'incoherent' $\Sigma = 3$ grain boundaries in CdTe have stronger recombination than the more perfectly bonded 'coherent' $\{111\}$ type [62]. CL contrast at grain boundaries is dark and Mendis *et al.* interpreted the line shape using a recombination model that estimated the recombination velocity in CdTe to be ~500–750 cm/s [74], this being much lower than the surface recombination value. Nevertheless, high-resolution spectrally and time-resolved CL reveals very different behaviour at the grain boundaries compared with that of the grains [75], as shown in Figure 2.11. Recombination at the boundaries is 250–1,000 times faster than in the grains.

Twinning is also endemic in CdTe, but it is not too influential on PV devices: Figure 2.6 shows the presence of long straight twin boundaries in the CdTe, twins being grains having a special symmetric relationship to their host lattices. Al-Jassim has shown that the incidence of twins in CdTe decreases with increasing growth temperature [76], a finding in common

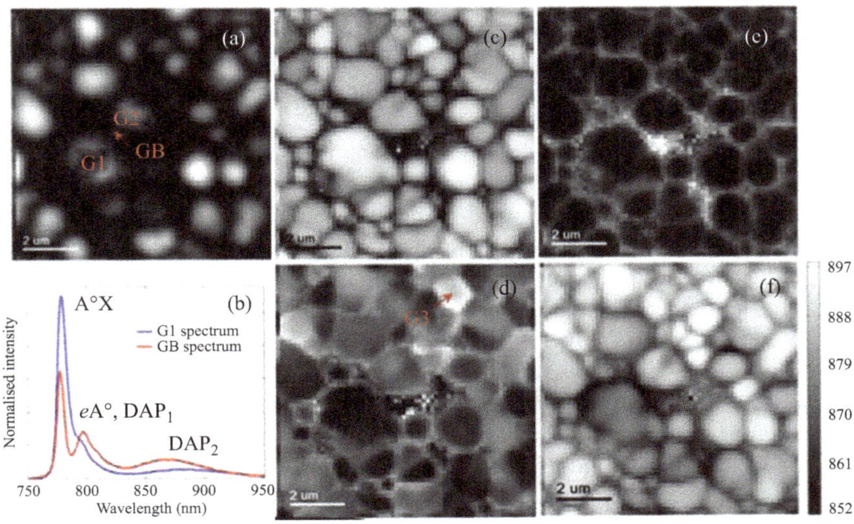

Figure 2.11 12K CL microscopy and spectroscopy of polycrystalline CdTe solar cell material. (a) Panchromatic CL image showing grains marked 'G1' and 'G2' and a grain boundary at 'GB'. (b) Normalised CL spectra from the grain and boundary features shown in (a). (c) A^0X monochromatic image showing that shallow defect luminescence dominates the grains. (d, e) eA^0/DAP_1 and DAP_2 monochromatic images showing the presence of deeper levels at the grain boundaries. (f) The peak wavelength distribution for the DAP_2 peak. From [75] with permission of the American Physical Society

with the result for CdTe bulk crystals. However, these long straight Friedel index $\Sigma = 3$ twin boundaries are not electrically influential [62]. Moreover, both twins and the related extrinsic stacking faults extend to grain boundaries where they terminate [76,77], meaning that they introduce few additional harmful defects.

(ii) *Evidence for band bending and fields at grain boundaries.* Kelvin probe microscopy has been used to profile the fields at grain boundaries in thin-film CdTe solar cells. It indicates that there is downward band bending near the grain boundaries, i.e. that they are locally n-type in the p-type host grains. This would cause holes and electrons to be separated, leading to the hypothesis that the grain boundaries assist the PV functionality of the devices [78].

(iii) *Evidence for grain boundary conductivity.* Conductive mode AFM indicates that conductivity at grain boundaries is higher than in the bulk CdTe [79]. While some early studies were complicated by surface morphology, that by Luria *et al.* [79] made a tomographic study by progressive erosion of the surface material using a diamond AFM tip to reveal a 3D conductive boundary network. This appears to be good evidence that CdTe grain boundaries in PV device material are conductive.

(iv) *Spatially resolved measurements of carrier collection.* EBIC measurements in the SEM are capable of forming maps of current collection from solar cell devices. The development of methods to make clean cross sections has added power to the method. For CdTe, EBIC of argon ion-polished cross sections shows bright grain boundary contrast [80–84]. This implies that the grain boundaries collect current more strongly than the grains. However, if the cross sections are prepared by focused ion beam milling using gallium, then the results are inverted: the contrast is dark, implying that there is reduced collection from the grain boundaries [85]. It appears that sample preparation can dominate the results from EBIC measurements, and that they are not conclusive.

(v) *Modelling of grain boundary collection effects.* Modelling of 2D collection effects including grain boundaries has been reported that uses parameters appropriate for CIGS [86] and CdTe [87,88]. Both indicate that if there is downward band bending, then it can act to increase the local collected current, but that there is an inevitable decrease in V_{oc} also. The V_{oc} loss outweighs any J_{sc} gain, making grain boundaries deleterious overall. Importantly, Jin and Dunham [87,88] noted that bright EBIC contrast does *not* correlate to enhanced PV performance in devices.

(vi) *Cell efficiency as a function of grain size.* There are two reports of devices fabricated as a function of CdTe grain size and they concurred that the FF, J_{sc}, V_{oc} and efficiency all increased as the grain size increased. For CSS-grown devices, there was a plateau in performance when the grains became large enough to occupy the full thickness of the film. For sputtered material, when the benign $\Sigma = 3$ coherent twins were excluded, there was a smooth increase in efficiency with grain size [89], as shown in Figure 2.12.

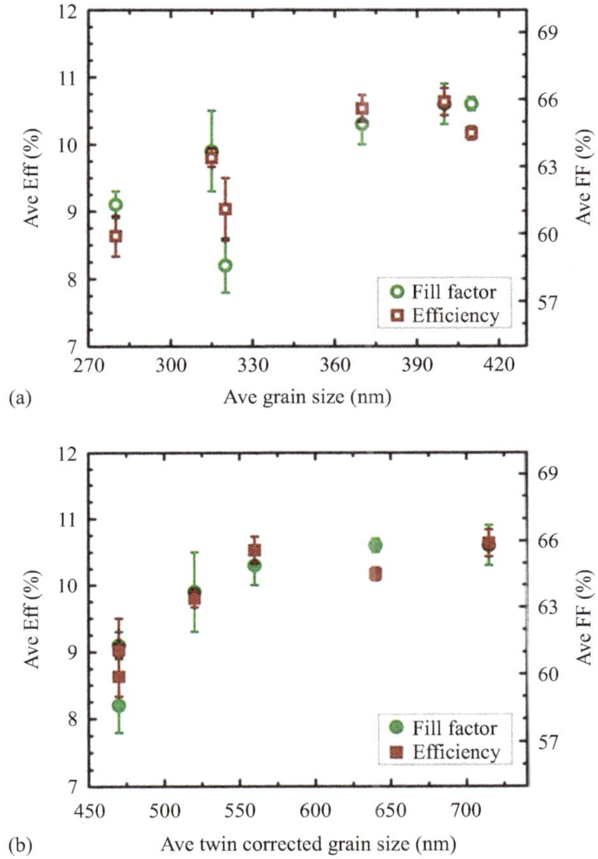

Figure 2.12 Variation of solar cell efficiency and fill factor with grain size as reported by Nowell et al. [89] for sputtered CdTe solar cells. The efficiency increases with increasing grain size: (a) includes grains having all relationships whereas (b) has the harmless coherent $\Sigma = 3$ *boundaries removed. Reproduced from Nowell et al. [89] with permission from Cambridge University Press*

These results give a clear demonstration that larger grains, not smaller, lead to more efficient solar cells, and therefore grain boundaries must be deleterious.

(vii) *Single crystal and epitaxial CdTe devices.* The ultimate test of the influence of grain boundaries should be to make a single crystal device. There are two experimental approaches:

The first is to dope single crystals of CdTe and to form solar cells on them. This was attempted in the 1990s by Alnajjar *et al.* who phosphorous-doped vapour-grown CdTe crystals [45] and formed solar cells on them by evaporation of CdS achieving $V_{oc} = 0.740$ V and an efficiency of 6.8% [90]. More recently, Burst *et al.* [47] used Bridgman-grown phosphorous-doped CdTe with an n-type CdS:In overlayer (followed by ZnO and ZnO:Al) to achieve $V_{oc} = 1.017$ V and 15.2% efficiency.

The second is to grow epitaxial structures by MBE. Burst *et al.* [47] grew a double heterostructure device on an InSb substrate using $Mg_xCd_{1-x}Te$ partner layers. V_{oc} reached 1.096 V and the efficiency was 17.0%. The carrier lifetime was (400 ns) was also exceptionally high.

Brinkman *et al.* reviewed epitaxial CdTe solar cell devices in 1992 [91].

Since single crystals devices have yielded the highest voltages yet recorded (approximately 150 mV greater than for thin-film polycrystalline materials), this is clear evidence that grain boundaries are harmful to V_{oc}.

Overall, it may be concluded that grain boundaries decrease and *do not increase solar cell performance*. Despite the fact that chloride treatment confers some degree of electrical passivation to the grain boundaries, they still promote recombination. If the potentials surrounding the grain boundaries do indeed assist in current collection, and assuming the grain boundaries to be conductive, then nevertheless the modelling shows that the benefits to current will be outweighed by voltage losses.

2.4.6 Alternative chloride treatments

Chloride treatment of CdTe is an essential part of the processing of all practical CdTe solar cell devices. The first and most widely used treatment is with $CdCl_2$ which has been successfully applied as vapour from solution and as a solid source – there are many reports. There is also a long history of alternative processes. For example, HCl and even elemental Cl_2 have been trialled, notwithstanding the difficulties in handling them. Chlorine-containing freons are effective [92]. Of these, difluorochloromethane (R-22) has been the most successful [93,94], but it has not been widely adopted. Many salts of chlorine have been trialled, with early work being done on NaCl [95] and $ZnCl_2$ (Romeo, private communication, 1998). Alternative salts became more popular after the report of a successful process using $MgCl_2$ [96] which has zero toxicity and is used in bath salts. Equal results in terms of device performance were achieved with both $MgCl_2$ and $CdCl_2$, with $MgCl_2$ having the advantage of being non-toxic. This work was part of a wider survey of chlorides, including NH_4Cl [97]. It appears that so long as the cations do not introduce harmful electronic levels into the CdTe, then their chlorides may be suitable for solar cell processing.

2.5 Contacts to CdTe and rear surface field engineering

2.5.1 Contacts to p-CdTe

CdTe has a significant fundamental disadvantage in that its electron affinity is unusually high at $\chi = 4.5$ eV. Hence for p-CdTe the ionisation potential may be as large as $\chi + E_g = 5.95$ eV. To avoid an electron barrier and hence form an Ohmic contact, a metal having a work function higher than this value is necessary. There is no such metal, and so simple electrical contact to CdTe solar cells is always compromised and presents a barrier to current flow in forward bias – there is always a Schottky junction. The effect of this on the J–V curve is to limit the forward-bias current, giving a characteristic 'rollover' effect as shown in Figure 2.13(a). Stollwerck and Sites [98] described this using a two-diode model (Figure 2.13(b)) in which a diode representing the contact barrier is in opposition to that representing the main junction. Further analysis of rollover (and of light/dark J–V crossover) is given by Niemegeers and Burgelman [99].

Experimentally the band bending at the contact (and other interfaces) has been measured by photoemission profiling [100] as shown in Figure 2.14. Clearly photoemission, although powerful, is not routine, and a practical method to measure the barrier height was developed by Batzner *et al.* [101]. If temperature dependent J–V curves are measured it is possible to extract the series resistance temperature profile from them, which may be fitted to (2.1)

$$R_s(T) = R_{\Omega 0} + \frac{\partial R_{s\Omega}}{\partial T}.T + \frac{C_{SK}}{T^2}e^{\frac{-\varphi_b}{kT}} \quad (2.1)$$

where the last term describes thermionic emission over a barrier having height ϕ_b, which along with $R_{\Omega 0}$, $R_{\Omega 0}$ and C_{SK} is one of the adjustable parameters in the fit. Experimentally, this yields barrier heights of 0.2–0.5 eV. Practically speaking, the consequences for solar cell operation are not serious if the barrier height is less than about 0.5 eV.

The most popular strategy to reduce the impact of contact barriers to CdTe is to dope the back surface heavily p-type to reduce the depletion width of the Schottky barrier and hence encourage tunnelling. The effectiveness of this is limited by the difficulties in doping CdTe heavily p-type (see Section 2.4.2), and also the

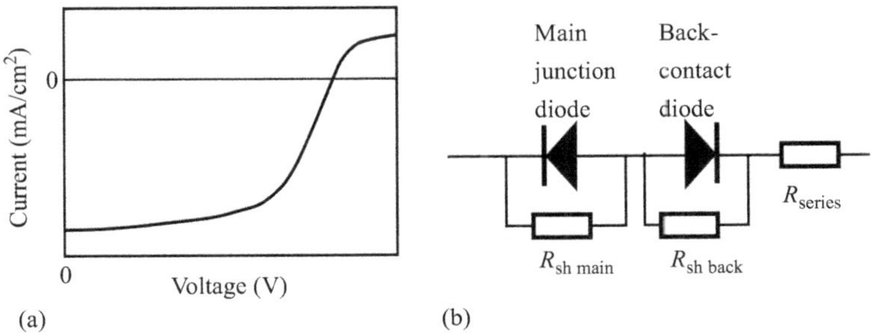

Figure 2.13 *(a) Forward current limitation effects or 'rollover' occur in CdTe solar cells when a back-contact barrier is present. (a) shows the effect on J–V curves, (b) shows the two-diode model proposed by Stollwerck and Sites [98], and for which a second diode in opposition to that of the photoactive p-n junction is used to describe the current limitation. Redrawn from Stollwerck and Sites [98]*

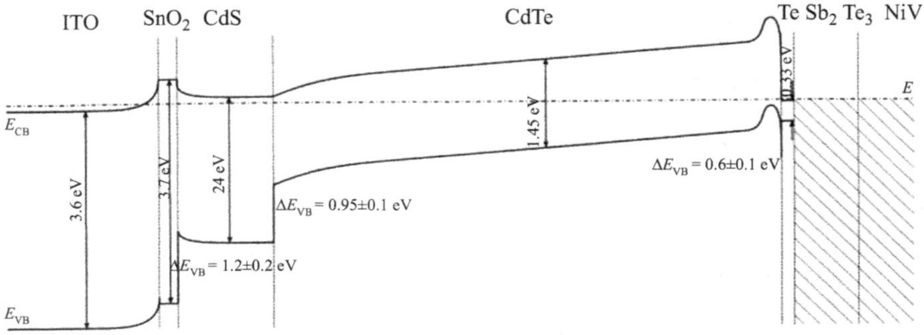

Figure 2.14 Experimental band diagram obtained from photoemission experiments for a CdS/CdTe of the type manufactured by ANTEC GmbH in the 1990s–2000s. Reproduced from Fritsche et al. [100] with permission from Elsevier

tendency of p-dopants such as Cu to diffuse to unwanted locations. Hence there have been a plurality of approaches and there is no universal recipe in use:

(i) *Gold.* Gold has a high work function and is often used in research labs but not in production. The principal impurities in gold are copper and silver, and hence its use is accompanied by unintentional doping.
(ii) *Te-rich surfaces.* Etching to make a Te-rich surface is considered to impart V_{Cd} p-conductivity and is most often reported using a mixture of nitric and phosphoric acids [102], the so-called 'N-P' etch. Its effectiveness, or not, is variously reported [103,104], and although it is used in combination with other approaches in this list, its uptake is not universal. Use of Te in forming electron-reflecting layers is mentioned in Section 2.5.2.
(iii) *Cu doping.* Intentional doping with copper coupled with etching to make a Te-rich surface can form Cu_xTe ($1 < x < 2$) phases, those for which $x > 1.4$ being stable and giving acceptable performance [105,106]. Stability issues have led to the adoption of some novel processes for the formation of the required phase [106].
(iv) *ZnTe:Cu.* Zinc telluride doped with copper provides a method of forming a low-barrier contact while stabilising the copper [107–112].
(v) *Oxides.* Some oxides have high electron affinities and allow for transparency giving bifacial functionality in the solar panels [113,114]. However, MoO_x [115,116] while effective can be difficult to control experimentally.
(vi) *Carbon and graphite.* Carbon paste with dopants is practical for laboratory use (see, e.g. [117]).
(vii) *Organic materials.* More recently, organic contact materials have been used, and they offer a range of band line up options that differ from the inorganic partner layer choices. For example, poly(3-hexylthiophene) (P3HT) [118] and copper thiocyanate [119], poly(3,4-ethylenedioxythiophene) (PEDOT-PSS)

[120]. Spin coating of organics has the advantage of making a conformal coating that blocks pinholes [118].

2.5.2 Rear surface field engineering – electron reflectors and interface passivation layers

An alternative strategy to simply forming a contact to the back surface of the CdTe is to take the opportunity to enhance the efficiency by introducing an electron reflector so as to prevent back surface recombination [121]. Reports are relatively few, but so far the approach has been to use (Cd,Mg)Te [122] alone or in combination with a Te layer [123].

It has also been demonstrated recently that Al_2O_3 acts as an excellent passivation layer for CdTe and CdSeTe. Double heterostructures have been used for tests and surface recombination velocities below 100 $cm{\cdot}s^{-1}$ have been reported [124].

Both approaches have the potential to increase device performance.

2.6 n-CdS partner layer and alternatives for it

A heterojunction partner for CdTe is essential in a thin-film solar cell for two reasons – the first being the need to create a p-n junction and the second being to prevent recombination at the absorber's surface. It is the high absorption coefficient of CdTe (i.e. high levels of near-surface photo-injection of charge) that make homojunctions unfeasible. It is therefore the function of the n-type partner layer to passivate the CdTe surface and to act as a 'window layer'.

2.6.1 CdS window layers

As-deposited CdS is naturally n-type making it compatible with the p-CdTe absorber. Its band alignments with p-CdTe contain no deleterious barriers or spikes. However, CdS has a significant disadvantage: photoabsorption in the CdS does not make a contribution to the photocurrent from the PV devices. Instead there is parasitic absorption [125], which is revealed in the spectral response curves shown in Figure 2.3. It is thought to take place either by (i) the CdS being highly conductive such that the junction is forced into the CdTe (indeed the CdTe/CdS junction is widely thought to operate by means of a shallow CdTe n-p homojunction), or (ii) by the hole lifetime in the CdS being sufficiently low as to kill any photocurrent. Whatever the cause, for the highest efficiency devices, it is essential to use the thinnest CdS films possible, certainly not thicker than 50–100 nm. Greater thicknesses increase the level of parasitic absorption and photocurrent loss that is seen in the external quantum efficiency spectrum series as shown in Figure 2.3. Fundamental studies of the solid solution CdS_xTe_{1-x} show that (i) CdS is soluble in CdTe and *vice versa*, and although there is a miscibility gap, films with all compositions have been prepared and are metastable, (ii) the bandgap–composition curve is bowed, and the bandgap of material having low S compositions dips below that of CdTe, (iii) CdS and CdTe interdiffuse in solar cell device material, and by excessive treatment it is even possible to dissolve the entirety of a thin CdS film into the CdTe.

Interdiffusion in device material has been studied widely, with the first studies being done by x-ray diffraction (XRD) and modelling by McCandless *et al.* [126]. Such mixing is seen as being beneficial to solar cell device operation by reducing interfacial stress and forming a low-gap phase near the front wall of the device. Sulphur diffusion is expected to be faster along the grain boundaries and there is some indication that is it has an electrical impact [127].

Use of CdS has been superseded by CdS:O or $Mg_xZn_{1-x}O$.

2.6.2 Oxide bilayers with CdS or 'high resistance transparent' layers

Inclusion of a transparent but resistive layer (high resistance transparent [HRT] layer) on top of the transparent conductor and beneath the n-CdS window layer has the paradoxical effect of ensuring better V_{oc} outcomes from devices. There are a wide range of hypotheses: (i) it may block harmful diffusion, e.g. out-diffusion of In from ITO TCOs in early devices, (ii) it may provide more favourable band line ups, (iii) it may promote more favourable crystalline texture in the overlying crystalline films and (iv) it may act to reduce the deleterious impact of pinholes or weak diodes in the CdS/CdTe junction. More than one factor may act in any given case. There have been many studies and systematic surveys. Major and Durose [128] found that selection of the combined CdS and HRT thickness was important. Kephart *et al.* [129] gave a brief review and presented an experimental survey, including $Mg_xZn_{1-x}O$ (see later). Replacement of CdS/HRT layer combinations with either CdS:O or $Mg_xZn_{1-x}O$ is becoming more widespread. Use of SnO_2 directly is also an option.

2.6.3 MZO interfacial layers

MZO has a wider bandgap and a lower electron affinity than CdS, both of which are desirable properties in device design [130]. Kephart and Sampath reported studies to optimise the composition and found $Mg_xZn_{1-x}O$ ($x = 0.11$) to be the most effective. MZO has further been used with CdTe directly, with CdS [131] in an HRT configuration (see 2.6.2 above) and with CdSe/CdTe films (see Section 2.7). Composition-dependent band gap and crystallinity are reported for $Mg_xZn_{1-x}O$ ($0 < x < 0.25$) [132]. MZO is usually made by sputtering, but at the time of writing not all labs are able to successfully make high efficiency devices with it. Perhaps there is a hidden variable in film formation. Post-growth high temperature annealing of MZO gives improved outcomes.

2.6.4 Oxygenated CdS–CdS:O

Inclusion of oxygen into CdS increases the band gap dramatically from 2.45 eV to beyond the visible, creating highly transparent films. This possibility to overcome the parasitic absorption of CdS itself by using CdS:O was exploited by Wu *et al.* [133]

and this generated a world record device that stood for a decade. However, the use of CdS:O was not taken up more widely for some years.

The increase in optical band gap from oxygenation is considered by using isolation of small CdS grains, leading to a quantum confinement effect [133]. x-ray photoelectron spectroscopy (XPS) showed that $CdSO_4$ is present and this is what probably breaks up the grains [134].

CdS:O is most often made by sputtering using a CdS target, and a mixed O_2/Ar plasma [135,136]. Post-growth thermal annealing of single films of CdS:O can lead to its reversion to CdS [137]. However, it behaves differently in-situ as a device layer, where it assumes a band gap of 2.15 eV. Nevertheless, an optical advantage is retained from CdS:O since processing causes thinning of the film [137].

To conclude on n-type partner layers in CdTe-based devices, higher transparency alternatives to CdS give higher performance by allowing higher photocurrent generation. Both MZO and CdS:O have been shown to be effective, although both have their challenges and complexities in growth and/or processing.

2.7 Devices with graded Cd(Se,Te) junctions

The use of graded gap absorbers comprising $CdTe_xSe_{1-x}$ films has been developed commercially and patented, and at the time of writing the academic community is following up with further investigations. Bulk crystals of $CdTe_xSe_{1-x}$ are the subject of a separate strand of literature reporting their use in radiation detector applications [138,139]. The solid solution has a bowed bandgap-composition diagram, with the bandgap of the Se-rich compositions dipping below that of CdTe [140] while retaining the zinc blende crystal lattice (CdSe has the wurtzite lattice). Moreover, the solubility of CdSe in CdTe is greater than that of CdS, and so it is possible to form graded junctions by first depositing CdSe and then CdTe, and allowing them to diffuse. Initial works used combined CdS/CdSe layers [141], then the CdS was omitted and the CdSe grew direct onto SnO_2:F/SnO_2. Baines *et al.* noted that growth onto CdS would retain the parasitic optical loss, and trialled a range of oxides: growth of CdSe directly onto TiO_2, ZnO, SnO_2:F produced devices which were all inferior to growth onto SnO_2 [142]. Further to this, Mia *et al.* reported a comparison of the electrical behaviours of CdS and CdSe containing devices [143] and concluded that those with CdSe are defect-rich. Further university-based lab work on devices having $CdTe_xSe_{1-x}$ films is anticipated but it is widely considered that the state of knowledge on this topic presently lags behind what has been developed commercially.

2.8 Cadmium and tellurium issues

2.8.1 Cadmium

CdTe has received significant scrutiny with regard to its cadmium content, and the scarcity of tellurium. It is undeniable that Cd is a toxic heavy metal. However, the bond strength of CdTe is high and is stable to both aqueous dissolution and to light.

Manufacturing requires processing with $CdCl_2$ and this is potentially a greater hazard if not controlled well during production, since it is a water-soluble powder, although with a low vapour pressure. Alternatives, most notably $MgCl_2$, have recently come to light, although they have yet to be used in production. Fthenakis and Kim [144] have made a thorough environmental evaluation of the Cd issue. Comparisons of the Cd emissions from PV power generation with CdTe PV and a wide range of other technologies (fossil fuels; from energy consumption for Si PV production and others) gives a very favourable account of CdTe, with there being a total of 67 mg of Cd emission per GW-h of power generation [145]. Since Cd is a by-product from the distillation of Zn, it has also been argued that locking Cd into the production of stable CdTe PV modules is an environmental asset.

2.8.2 Tellurium

Tellurium is relatively rare in the Earth's crust and since it is not used in high volumes outside of the PV and radiation detector industries, its annual production is limited to just several hundred tonnes (estimates vary). It has been suggested that scale up of CdTe solar cell production would ultimately be limited by the availability of tellurium. Nevertheless, Andersson [146] estimated that production of CdTe modules would only be limited if it exceeded 20 GWp per annum, this being far in excess of annual production of several GWp at the time of writing. (Further analysis of the element resource issues concerning PV power generation by all technologies was given in the study by Feltrin and Freundlich [147]). Notwithstanding this, some commentators have observed that much of the evaluation of the tellurium supply has been based on annual production rather than the actual limits of the global mineral resources. Presently, there are few large-scale uses for tellurium: if the demand for it increased, then there would be a strong motivation for surveying for new mineral resources. Given that little exploratory work done to date for tellurium, it is likely that the 'supply issue' and its limit to the PV industry have been severely overstated.

2.9 Conclusions and outlook

Recently the field of CdTe PVs has been re-invigorated by changes made to Bonnet's 'superstrate' CdTe/CdS/TCO/glass solar cell design (Figure 2.2). Although it was the main focus of industry and research for nearly 40 years, it suffered from parasitic light loss in the n-CdS window layer. The redesign has both replaced the window with a higher gap alternative and included a graded gap absorber. A surge in efficiency gains resulted (Figure 2.1), driven by higher currents from increased photon harvesting at both short and long wavelengths (Figure 2.3). However, the voltage has not increased and remains an issue.

A combination of both high p-type carrier concentrations and long carrier lifetimes is needed to achieve high voltages (Figure 2.8). Throughout the history of CdTe PV, there has been an apparent 'glass ceiling' for V_{oc} of ~850 mV. Current devices use chlorine to passivate the grain boundaries and copper to achieve p-type doping. Grain boundaries do not increase the device efficiency overall, despite there being a literature to the contrary (see Section 2.4.5). Group V doping of CdTe has recently been more widely re-evaluated as a line of investigation to provide an alternative to the state of the art. Impressive results have been achieved with phosphorous and arsenic-doped single crystal- and epitaxial-based devices: voltages greater than 1 V have been demonstrated which indicate that significant increases in V_{oc} are possible beyond the present-day limit. It is clear that the group V doping concept will be translated more widely into thin-film devices in the near future. Nevertheless, the doping density of p-CdTe will always be limited by self-compensation. While comparatively high doping (e.g. 10^{18} cm^{-3}) might become achievable through appropriate thermal scheduling, it is not expected to be stable. Hence some questions about the viability of group V doped CdTe solar cells remain at the time of writing.

Despite such physical limitations, CdTe solar PVs are full expected to retain a significant position in the global PV market. The 'learning curve' (manufacturing cost vs cumulative sales) for thin-film PV lies beneath that for crystalline silicon. Therefore, if CdTe continues to increase its volume production methodology, it may expect to retain a cost advantage. This will of course require continual technological innovation to follow on from the developments outlined in this review.

Acknowledgements

The author would like to thank the many authors and colleagues whose work has informed this review. He also thanks those who gave advice on the manuscript, including Jim Sites, Walajabad Sampath, Tim Gessert and Jon Major.

References

[1] Triboulet R, Siffert P. *CdTe and Related Compounds; Physics, Defects, Hetero- and Nano-structures, Crystal Growth, Surfaces and Applications*. Amsterdam, Elsevier; 2009.

[2] Bonnet D, Rabenhorst H, editors. New results on the development of thin film p-CdTe/n-CdS heterojunction solar cell. *Proc 9th IEEE Photovoltaic Specialists Conf.* Silver Springs, MD: IEEE; 1972.

[3] Britt J, Ferekides C. Thin-film CdS/CdTe solar-cell with 15.8% efficiency. *Appl Phys Lett*. 1993;62(22):2851–2.

[4] Wu XZ. High-efficiency polycrystalline CdTe thin-film solar cells. *Sol Energy*. 2004;77(6):803–14.

[5] Basol BM. High-efficiency electroplated heterojunction solar-cell. *J Appl Phys*. 1984;55(2):601–3.

[6] Basol BM, Ou SS, Stafsudd OM. Type conversion, contacts, and surface effects in electroplated CdTe films. *J Appl Phys*. 1985;58(10):3803–13.

[7] Kremheller A, Faria S, Goldberg P, Bracco DJ. Retention of chloride in zinc sulfide during phosphor preparation. *J Electrochem Soc*. 1960;107(9):749–53.

[8] Kremheller A, Faria S, Levine AK. Study of ball milling and the determination of lattice chloride in zinc sulfide. *J Electrochem Soc*. 1960;107(9):753–8.
[9] Aven M, Prener JS. *Physics and Chemistry of II–VI Compounds*. Amsterdam: North Holland; 1967.
[10] Best research-cell efficiency chart 2019. Available from: https://www.nrel.gov/pv/cell-efficiency.html.
[11] First Solar builds the highest efficiency thin film PV cell on record [press release]. 05 August 2014.
[12] First Solar achieves world Record 18.6% thin film module conversion efficiency [press release]. 15 June 2015.
[13] Green MA, Hishikawa Y, Dunlop ED, *et al.* Solar cell efficiency tables (Version 53). *Prog Photovoltaics*. 2019;27(1):3–12.
[14] Green MA, Emery K, Hishikawa Y, Warta W, Dunlop ED. Solar cell efficiency tables (Version 45). *Prog Photovoltaics*. 2015;23(1):1–9.
[15] Aramoto T, Kumazawa S, Higuchi H, *et al.* 16.0% efficient thin-film CdS/CdTe solar cells. *Jpn J Appl Phys Part 1 – Regul Pap Short Notes Rev Pap*. 1997;36(10):6304–5.
[16] Luschitz J, Siepchen B, Schaffner J, *et al.* CdTe thin film solar cells: interrelation of nucleation, structure, and performance. *Thin Solid Films*. 2009;517(7):2125–31.
[17] Barna PB, Adamik M. Fundamental structure forming phenomena of polycrystalline films and the structure zone models. *Thin Solid Films*. 1998;317(1–2):27–33.
[18] Cousins MA, Durose K. Grain structure of CdTe in CSS-deposited CdTe/CdS solar cells. *Thin Solid Films*. 2000;361:253–7.
[19] Durose K, Cousins MA, Boyle DS, Beier J, Bonnet D. Grain boundaries and impurities in CdTe/CdS solar cells. *Thin Solid Films*. 2002;403–404:396–404.
[20] Saraie J, Kitagawa M, Ishida M, Tanaka T. Liquid-phase epitaxial-growth of CdTe in CdTe-$CdCl_2$ system. *J Crystal Growth*. 1978;43:13.
[21] Andronik IK, Kuleva P, Suschkevich K. Feasibility of obtaining CdTe crystals by zone refining from a solution in fused $CdCl_2$ *Inorg Mater*. 1976;12:759.
[22] Oleinik GS, Mizetskii PA, Nuzhnaya TP. Liquidus surface of the ternary system $CdCl_2$-CdS-CdTe. *Inorg Mater*. 1986;22(1):164.
[23] AlAllak HM, Brinkman AW, Richter H, Bonnet D. Dependence of CdS/CdTe thin film solar cell characteristics on the processing conditions. *J Crystal Growth*. 1996;159(1–4):910–5.
[24] Major JD, Bowen L, Treharne R, Durose K. Assessment of photovoltaic junction position using combined focused ion beam and electron beam-induced current analysis of close space sublimation deposited CdTe solar cells. *Prog Photovoltaics*. 2014;22(10):1096–104.
[25] Ma J, Kuciauskas D, Albin D, *et al.* Dependence of the minority-carrier lifetime on the stoichiometry of CdTe using time-resolved photoluminescence and first-principles calculations. *Phys Rev Lett*. 2013;111(6):067402.

[26] Sites J, Pan J. Strategies to increase CdTe solar-cell voltage. *Thin Solid Films*. 2007;515(15):6099–102.

[27] Zhang LX, Da Silva JLF, Li JB, Yan YF, Gessert TA, Wei SH. Effect of copassivation of Cl and Cu on CdTe grain boundaries. *Phys Rev Lett*. 2008;101(15).

[28] *Properties of Narrow Gap Cadmium-Based Compounds*, editor P Capper. EMIS Datareview Series. IEE/INSPEC, London; 1994.

[29] Meyer BK, Stadler W, Hofmann DM, Omling P, Sinerius D, Benz KW. On the nature of the deep 1.4 eV emission bands in CdTe – a study with photo-luminescence and ODMR spectroscopy. *J Crystal Growth*. 1992;117(1–4):656–9.

[30] Emziane M, Durose K, Bosio A, Romeo N, Halliday DP. Effect of the purity of CdTe starting material on the impurity profile in CdTe/CdS solar cell structures. *J Mater Sci*. 2005;40(6):1327–31.

[31] Emziane M, Durose K, Halliday DP, Bosio A, Romeo N. Role of substrate and transparent conducting oxide in impurity evolvement in polycrystalline thin-film devices. *Appl Phys Lett*. 2005;87(25):1–3.

[32] Emziane M, Durose K, Halliday DP, Bosio A, Romeo N. Efficiency improvement in thin-film solar cell devices with oxygen-containing absorber layer. *Appl Phys Lett*. 2005;87(26):1–2.

[33] Emziane M, Durose K, Halliday DP, Bosio A, Romeo N, Does CdTe deposition affect the impurity profile in sputtered CdS window layers? Materials Research Society Symposium Proceedings Volume 865, 2005, Article number F14.11, Pages 411–416 Materials Research Society Spring Meeting; San Francisco, CA; United States; 29 March–1 April 2005.

[34] Emziane M, Durose K, Halliday DP, Romeo N, Bosio A. The distribution of impurities in the interfaces and window layers of thin-film solar cells. *J Appl Phys*. 2005;97(11):114910.

[35] Emziane M, Durose K, Romeo N, Bosio A, Halliday DP. Effect of $CdCl_2$ activation on the impurity distribution in CdTe/CdS solar cell structures. *Thin Solid Films*. 2005;480–1:377–81.

[36] Emziane M, Durose K, Romeo N, Bosio A, Halliday DP. A combined SIMS and ICPMS investigation of the origin and distribution of potentially electrically active impurities in CdTe/CdS solar cell structures. *Semicond Sci Technol*. 2005;20(5):434–42.

[37] Emziane M, Halliday DP, Durose K, Romeo N, Bosio A, editors. On the origins of impurities in CdTe-based thin film solar cells. Materials Research Society Symposium Proceedings Volume 1012, 2007, Pages 195–200 Materials Research Society Spring Meeting; San Francisco, CA; United States; 9 April–13 April 2007.

[38] Emziane M, Ottley CJ, Durose K, Halliday DP. Impurity analysis of $CdCl_2$ used for thermal activation of CdTe-based solar cells. *J Phys D Appl Phys*. 2004;37(21):2962–5.

[39] Galloway SA, Edwards PR, Brinkman AW, Durose K. Characterisation of thin film CdTe/CdS solar cells using electron beam induced current. *Solar Energy Mater Solar Cells*. 1997;57(1):61–74.

[40] Woods LM, Levi DH, Kaydanov V, Robinson GY, Ahrenkiel RK, editors. Electrical characterisation of CdTe grain boundary properties from as processed CdTe/CdS solar cells. *2nd World Conference on Photovoltaic Solar Energy Conversion*. Vienna: WIP; 1998.

[41] Durose K, Edwards PR, Halliday DP. Materials aspects of CdTe/CdS solar cells. *J Cryst Growth*. 1999;197:733–42.

[42] Gleiter H. Segregation of copper at high angle grain boundaries in lead. *Acta Metallurgica*. 1970;18(1):117.

[43] Li C, Poplawsky J, Yan YF, Pennycook SJ. Understanding individual defects in CdTe thin-film solar cells via STEM: from atomic structure to electrical activity. *Mater Sci Semicond Process*. 2017;65:64–76.

[44] Mao D, Wickersham CE, Gloeckler M. Measurement of chlorine concentrations at CdTe grain boundaries. *IEEE J Photovolt*. 2014;4(6):1655–8.

[45] Alnajjar AA, Watson CCR, Brinkman AW, Durose K. Post-growth doping of bulk CdTe crystals with phosphorus. *J Crystal Growth*. 1992;117 (1–4):385–90.

[46] Burst JM, Farrell SB, Albin DS, *et al.* Carrier density and lifetime for different dopants in single-crystal and polycrystalline CdTe. *APL Mater*. 2016;4(11):6.

[47] Burst JM, Duenow JN, Albin DS, *et al.* CdTe solar cells with open-circuit voltage breaking the 1 V barrier. *Nat Energy*. 2016;1:7.

[48] Berrigan RA, Maung N, Irvine SJC, Cole-Hamilton DJ, Ellis D. Thin films of CdTe/CdS grown by MOCVD for photovoltaics. *J Crystal Growth*. 1998;195(1–4):718–24.

[49] Barrioz V, Rowlands RL, Jones EW, Irvine SJC, Zoppi G, Durose K. A comparison of in situ As doping with ex situ $CdCl_2$, treatment of CdTe solar cells. In: Shafarman W, Gessert T, Niki S, Siebentritt S, editors. *Thin-Film Compound Semiconductor Photovoltaics. Materials Research Society Symposium Proceedings*. Volume 865. Warrendale: Materials Research Society; 2005. pp. 67–72.

[50] Proskuryakov YY, Major JD, Durose K, *et al.* Comparative study of trap densities of states in CdTe/CdS solar cells. *Appl Phys Lett*. 2007;91(15):3.

[51] Kartopu G, Oklobia O, Turkay D, *et al.* Study of thin film poly-crystalline CdTe solar cells presenting high acceptor concentrations achieved by in-situ arsenic doping. *Solar Energy Mater Solar Cells*. 2019;194:259–67.

[52] Farrell S, Barnes T, Metzger WK, Park JH, Kodama R, Sivananthan S. In situ arsenic doping of CdTe/Si by molecular beam epitaxy. *J Electron Mater*. 2015;44(9):3202–6.

[53] Marfaing Y. Self-compensation in II–VI compounds. *Prog Cryst Growth Charact Mater*. 1981;4(4):317–43.

[54] Park CH, Chadi DJ. Bulk lattice instability in II–VI semiconductors and its effect on impurity compensation. *Phys Rev Lett*. 1995;75(6):1134.

[55] Yang JH, Yin WJ, Park JS, *et al.* Enhanced p-type dopability of P and As in CdTe using non-equilibrium thermal processing. *J Appl Phys*. 2015; 118(2):6.

[56] Yang JH, Yin WJ, Park JS, Ma J, Wei SH. Review on first-principles study of defect properties of CdTe as a solar cell absorber. *Semicond Sci Technol.* 2016;31(8):22.

[57] Holt DB, Yacobi BG. *Extended Defects in Semiconductors: Electronic Properties, Device Effects and Structures.* Cambridge, Cambridge University Press; 2007.

[58] Durose K. Extended defects in CdTe. In: Triboulet R, Siffert P, editors. *CdTe and Related Compounds; Physics, Defects, Hetero-and Nano-Structures, Crystal Growth, Surfaces and Applications.* Amsterdam, Elsevier; 2009.

[59] Brown PD, Durose K, Russell GJ, Woods J. The absolute determination of CdTe crystal polarity. *J Crystal Growth.* 1990;101(1–4):211–5.

[60] Moseley J, Metzger WK, Moutinho HR, *et al.* Recombination by grain-boundary type in CdTe. *J Appl Phys.* 2015;118(2):9.

[61] Durose K, Russell GJ. Lateral twins in the sphalerite structure. *Inst Phys Conf Ser.* 1987(87):327–32.

[62] Durose K, Russell GJ. Twinning in CdTe. *J Crystal Growth.* 1990;101 (1–4):246–50.

[63] Schreiber J, Horing L, Uniewski H, Hildebrandt S, Leipner HS. Recognition and distribution of A(g) and B(g) dislocations in indentation deformation zones on {111} and {110} surfaces of CdTe. *Phys Status Solidi A-Appl Res.* 1999;171(1):89–97.

[64] Humphreys FJ, Hatherley M. *Recrystallization and Related Annealing Phenomena.* Amsterdam, Pergamon (copyright Elsevier); 2012.

[65] Hiie J. CdTe:$CdCl_2$:O_2 annealing process. *Thin Solid Films.* 2003;431:90–3.

[66] Hiie J, Valdna V, Taklaja A. Modeling of flux composition for thermal $CdCl_2$: O_2 annealing of polycrystalline CdTe. In: Noufi R, Shafarman WN, Cahen D, Stolt L, editors. *Compound Semiconductor Photovoltaics. Materials Research Society Symposium Proceedings.* Volume 763. Warrendale: Materials Research Soc; 2003. pp. 397–402.

[67] Moutinho HR, Al-Jassim MM, Levi DH, Dippo PC, Kazmerski LL. Effects of $CdCl_2$ treatment on the recrystallization and electro-optical properties of CdTe thin films. *J Vac Sci Technol A-Vac Surf Films.* 1998; 16(3):1251–7.

[68] Qi B, Kim DW, Williamson DL, Trefny JU. Effects of postdeposition heat-treatment on morphology and microstructure of CdTe grown by electro-deposition. *J Electrochem Soc.* 1996;143(2):517–23.

[69] Quinones SA, Ammu SM, Escobedo A, *et al.* SEM characterization of CdTe growth on CdTe(111) by close-spaced sublimation. *J Mater Sci-Mater Electron.* 2007;18(11):1085–91.

[70] Moutinho HR, Dhere RG, Romero MJ, Jiang CS, To B, Al-Jassim MM. Electron backscatter diffraction of CdTe thin films: effects of $CdCl_2$ treatment. *J Vac Sci Technol A.* 2008;26(4):1068–73.

[71] Stechmann G, Zaefferer S, Konijnenberg P, *et al.* 3-Dimensional micro-structural characterization of CdTe absorber layers from CdTe/CdS thin film solar cells. *Solar Energy Mater Solar Cells.* 2016;151:66–80.

[72] Stechmann G, Zaefferer S, Schwarz T, *et al.* A correlative investigation of grain boundary crystallography and electronic properties in CdTe thin film solar cells. *Solar Energy Mater Solar Cells*. 2017;166:108–20.

[73] Durose K, Watson CCR. Cathodoluminescence of CdTe, (CdZn)Te and (CdSe)Te. In: Capper P, editor. *Properties of Narrow Gap Cadmium-Based Compounds. EMIS Data review Series*. London, IEE/INSPEC; 1994. p. 446.

[74] Mendis BG, Bowen L, Jiang QZ. A contactless method for measuring the recombination velocity of an individual grain boundary in thin-film photovoltaics. *Appl Phys Lett*. 2010;97(9):3.

[75] Mendis BG, Gachet D, Major JD, Durose K. Long lifetime hole traps at grain boundaries in CdTe thin-film photovoltaics. *Phys Rev Lett*. 2015;115 (21):5.

[76] Al-Jassim MM, Hasoon FS, Jones KM, Keyes BM, Matson RJ, Moutinho HR, editors. The morphology, microstructure and luminescent properties of CdS/CdTe thin film solar cells. *Twenty Third IEEE Photovoltaic Specialists Conference*, Louisville, Kentucky, USA, 10–14 May 1993.

[77] Yan YF, Al-Jassim MM, Jones KM. Characterization of extended defects in polycrystalline CdTe thin films grown by close-spaced sublimation. *Thin Solid Films*. 2001;389(1–2):75–7.

[78] Visoly-Fisher I, Cohen SR, Ruzin A, Cahen D. How polycrystalline devices can outperform single-crystal ones: thin film CdTe/CdS solar cells. *Adv Mater*. 2004;16(11):879–83.

[79] Luria J, Kutes Y, Moore A, Zhang LH, Stach EA, Huey BD. Charge transport in CdTe solar cells revealed by conductive tomographic atomic force microscopy. *Nat Energy*. 2016;1:6.

[80] Li C, Wu YL, Poplawsky J, *et al.* Grain-boundary-enhanced carrier collection in CdTe solar cells. *Phys Rev Lett*. 2014;112(15):5.

[81] Poplawsky JD, Guo W, Paudel N, *et al.* Corregium: structural and compositional dependence of the CdTexSe1−x alloy layer photoactivity in CdTe-based solar cells. *Nat Commun*. 2016;7:1.

[82] Poplawsky JD, Guo W, Paudel N, *et al.* Structural and compositional dependence of the $CdTe_xSe_{1-x}$ alloy layer photoactivity in CdTe-based solar cells. *Nat Commun*. 2016;7:9.

[83] Poplawsky JD, Paudel NR, Li C, *et al.* Direct imaging of Cl- and Cu-induced short-circuit efficiency changes in CdTe solar cells. *Adv Energy Mater*. 2014;4(15):8.

[84] Zywitzki O, Modes T, Morgner H, *et al.* Effect of chlorine activation treatment on electron beam induced current signal distribution of cadmium telluride thin film solar cells. *J Appl Phys*. 2013;114(16):5.

[85] Major JD, Bowen L, Mumtaz A, Durose K. Unpublished results. 2019.

[86] Rau U, Taretto K, Siebentritt S. Grain boundaries in $Cu(In,Ga)(Se,S)_2$ thin-film solar cells. *Appl Phys A-Mater Sci Process*. 2009;96(1):221–34.

[87] Jin Y, Dunham ST. The impact of charged grain boundaries on CdTe solar cell: EBIC measurements not predictive of device performance. *IEEE J Photovolt*. 2017;7(1):329–34.

[88] Jin Y, Dunham ST. Enhanced EBIC response but degraded solar cell performance for CdTe grain boundaries. *2016 IEEE 43rd Photovoltaic Specialists Conference. IEEE Photovoltaic Specialists Conference*. New York: IEEE; 2016. pp. 2188–90.

[89] Nowell MM, Scarpulla MA, Paudel NR, Wieland KA, Compaan AD, Liu XX. Characterization of sputtered CdTe thin films with electron backscatter diffraction and correlation with device performance. *Microsc microanal*. 2015;21(4):927–35.

[90] Alnajjar AA. Single crystal CdS/CdTe:P solar cells, PhD Thesis. Durham; 1992.

[91] Brinkman AW, Al Allak HM, Awan GR, *et al.* Epitaxial CdTe-based solar cells. *Int J Solar Energy*. 1992;12(1–4):233–45.

[92] Bosio A, Romeo N, Mazzamuto S, Canevari V. Polycrystalline CdTe thin films for photovoltaic applications. *Prog Cryst Growth Charact Mater*. 2006;52(4):247–79.

[93] Mazzamuto S, Vaillant L, Bosio A, Romeo N, Armani N, Salviati G. A study of the CdTe treatment with a Freon gas such as CHF_2Cl. *Thin Solid Films*. 2008;516(20):7079–83.

[94] Leoncini M, Artegiani E, Lozzi L, *et al.* Difluorochloromethane treated thin CdS buffer layers for improved CdTe solar cells. *Thin Solid Films*. 2019;672:7–13.

[95] Durose K, Albin D, Ribelin R, *et al.* The influence of NaCl on the microstructure of CdS films and CdTe solar cells. In: Cullis AG, Beanland R, editors. *Microscopy of Semiconducting Materials 1999, Proceedings*. Institute of Physics Conference Series. Bristol: IOP Publishing Ltd; 1999. pp. 207–10.

[96] Major JD, Treharne RE, Phillips LJ, Durose K. A low-cost non-toxic post-growth activation step for CdTe solar cells. *Nature*. 2014;511(7509):334.

[97] Major JD, Al Turkestani M, Bowen L, *et al.* In-depth analysis of chloride treatments for thin-film CdTe solar cells. *Nat Commun*. 2016;7:13231.

[98] Stollwerck G, Sites JR, editors. Analysis of CdTe back-contact barriers. *13th European Photovoltaic Solar Energy Conference*. Nice, France, 1995.

[99] Niemegeers A, Burgelman M. Effects of the Au/CdTe back contact on IV and CV characteristics of Au/CdTe/CdS/TCO solar cells. *J Appl Phys*. 1997;81(6):2881–6.

[100] Fritsche J, Kraft D, Thissen A, Mayer T, Klein A, Jaegermann W. Band energy diagram of CdTe thin film solar cells. *Thin Solid Films*. 2002;403:252–7.

[101] Batzner DL, Oszan ME, Bonnet D, Bucher K. Device analysis methods for physical cell parameters of CdTe/CdS solar cells. *Thin Solid Films*. 2000;361:288–92.

[102] Sarlund J, Ritala M, Leskela M, Siponmaa E, Zilliacus R. Characterization of etching procedure in preparation of CdTe solar cells. *Solar Energy Mater Solar Cells*. 1996;44(2):177–90.

[103] Proskuryakov YY, Durose K, Taele BM, Welch GP, Oelting S. Admittance spectroscopy of CdTe/CdS solar cells subjected to varied nitric–phosphoric etching conditions. *J Appl Phys*. 2007;101(1):8.

[104] Song T, Sites JR, Role of tellurium buffer layer on CdTe solar cells' absorber/back-contact interface. *2017 IEEE 44th Photovoltaic Specialist Conference*. New York: IEEE; 2017. pp. 1308–11.

[105] Wu X, Zhou J, Duda A, *et al.* Phase control of Cu_xTe film and its effects on CdS/CdTe solar cell. *Thin Solid Films*. 2007;515(15):5798–803.

[106] Romeo N, Bosio A, Romeo A. An innovative process suitable to produce high-efficiency CdTe/CdS thin-film modules. *Solar Energy Mater Solar Cells*. 2010;94(1):2–7.

[107] Mondal A, Birkmire RW, McCandless BE. *A New Method for Forming ZnTe Contacts for CdTe Solar Cells*. New York: IEEE; 1991. pp. 1126–30.

[108] Niles DW, Hochst H, Rioux D. Valence band discontinuity at the ZnTe/CdTe interface – making ohmic contact to p-type CdTe. In: Noufi R, editor. *Photovoltaic Advanced Research & Development Project. AIP Conference Proceedings*. 268. Woodbury: AIP Press; 1992. pp. 279–84.

[109] Rioux D, Niles DW, Hochst H. ZnTe – a potential interlayer to form low-resistance back contacts in CdS/CdTe solar-cells. *J Appl Phys*. 1993;73(12):8381–5.

[110] Gessert TA, Li X, Coutts TJ, Mason AR, Matson RJ. Dependence of the material properties of radiofrequency magnetron-sputtered, Cu-doped, ZnTe thin-films on deposition conditions. *J Vac Sci Technol A-Vac Surf Films*. 1994;12(4):1501–6.

[111] Tang J, Feng L, Mao D, Song W, Zhu Y, Trefny JU. Study of ZnTe:Cu back contacts on CdTe/CdS thin film solar cells. In: Ginley D, Catalano A, Schock HW, Eberspacher C, Petersen TM, Wada T, editors. *Thin Films for Photovoltaic and Related Device Applications. Materials Research Society Symposium Proceedings*. Volume 426. Warrendale: Materials Research Society; 1996. pp. 355–60.

[112] Gessert TA, Metzger WK, Dippo P, Asher SE, Dhere RG, Young MR. Dependence of carrier lifetime on Cu-contacting temperature and ZnTe:Cu thickness in CdS/CdTe thin film solar cells. *Thin Solid Films*. 2009;517(7):2370–3.

[113] Romeo A, Khrypunov G, Galassini S, Zogg H, Tiwari AN. Bifacial configurations for CdTe solar cells. *Solar Energy Mater Solar Cells*. 2007;91(15–16):1388–91.

[114] Calnan S, Uphadhyaya HM, Buecheler S, *et al.* Application of high mobility transparent conductors to enhance long wavelength transparency of the intermediate solar cell in multi-junction solar cells. *Thin Solid Films*. 2009;517(7):2340–3.

[115] Paudel NR, Compaan AD, Yan Y. Sputtered CdS/CdTe solar cells with MoO_{3-x}/Au back contacts. *Solar Energy Mater Solar Cells*. 2013;113:26–30.

[116] Gretener C, Perrenoud J, Kranz L, *et al.* Development of MoOx thin films as back contact buffer for CdTe solar cells in substrate configuration. *Thin Solid Films*. 2013;535:193–7.

[117] Rose DH, Hasoon FS, Dhere RG, *et al.* Fabrication procedures and process sensitivities for CdS/CdTe solar cells. *Prog Photovoltaics*. 1999;7(5): 331–40.

[118] Major JD, Phillips LJ, Al Turkestani M, *et al.* P3HT as a pinhole blocking back contact for CdTe thin film solar cells. *Solar Energy Mater Solar Cells*. 2017;172:1–10.

[119] Paudel NR, Yan YF. Application of copper thiocyanate for high open-circuit voltages of CdTe solar cells. *Prog Photovoltaics*. 2016;24(1): 94–101.

[120] Wang WN, Paudel N, Yan YF, Duarte F, Mount M. PEDOT:PSS as back contact for CdTe solar cells and the effect of PEDOT:PSS conductivity on device performance. *J Mater Sci-Mater Electron*. 2016;27(2):1057–61.

[121] Hsiao KJ, Sites JR. Electron reflector to enhance photovoltaic efficiency: application to thin-film CdTe solar cells. *Prog Photovoltaics*. 2012;20 (4):486–9.

[122] Song T, Kanevce A, Sites JR. Design of epitaxial CdTe solar cells on InSb substrates. *IEEE J Photovolt*. 2015;5(6):1762–8.

[123] Song T, Moore A, Sites JR. Te layer to reduce the CdTe back-contact barrier. *IEEE J Photovolt*. 2018;8(1):293–8.

[124] Kuciauskas D, Kephart JM, Moseley J, Metzger WK, Sampath WS, Dippo P. Recombination velocity less than 100 cm/s at polycrystalline Al2O3/ CdSeTe interfaces. *Appl Phys Lett*. 2018;112(26):5.

[125] Dhere R, Rose D, Albin D, *et al.*, editors. Influence of CdS/CdTe interface properties on the device properties. *EU Photovolt Spec Conf 1997*. 1997.

[126] McCandless BE, Moulton LV, Birkmire RW. Recrystallization and sulfur diffusion in $CdCl_2$-treated CdTe/CdS thin films. *Prog Photovoltaics*. 1997;5(4):249–60.

[127] Edwards PR, Galloway SA, Durose K. *Carrier Density Microscopy of CdS/CdTe Solar Cells*. Alford NM, editor. London: Inst Materials; 1998. pp. 50–7.

[128] Major JD, Durose K. Co-sputtered $Zn_{1-X}Sn_XO$ buffer layers for CdTe thin film solar cells. *2013 IEEE 39th Photovoltaic Specialists Conference*. New York: IEEE; 2013. pp. 1146–9.

[129] Kephart JM, McCamy JW, Ma Z, Ganjoo A, Alamgir FM, Sampath WS. Band alignment of front contact layers for high-efficiency CdTe solar cells. *Solar Energy Mater Solar Cells*. 2016;157:266–75.

[130] Kephart JM, Sampath WS. Gallium-doped magnesium zinc oxide (GMZO) transparent conducting oxide layers for CdTe thin-film photovoltaics. *2015 IEEE 42nd Photovoltaic Specialist Conference*. New York: IEEE; 2015.

[131] Yang S, Fang XH, Ma X, Wu LL, Zhang JQ, Feng LH. Apparent quantum efficiency effects under forward bias in MZO/CdS/CdTe solar cells. *Mater Res Express*. 2018;5(11):7.

[132] Wang AL, Tang TY, Ren SQ, *et al.* Characterization of co-sputtered $Mg_xZn_{1-x}O$ thin films and their application in CdTe solar cells. *Mater Sci Semicond Process*. 2019;94:28–34.

[133] Wu X, Dhere RG, Yan Y, *et al. High-Efficiency Polycrystalline CdTe Thin-Film Solar Cells with an Oxygenated Amorphous CdS (a-CdS:O) Window Layer*. New York: IEEE; 2002. pp. 531–4.

[134] Duncan DA, Kephart JM, Horsley K, *et al.* Characterization of sulfur bonding in CdS:O buffer layers for CdTe-based thin-film solar cells. *ACS Appl Mater Interfaces*. 2015;7(30):16382–6.

[135] Kephart JM, Geisthardt R, Sampath WS, *Sputtered, Oxygenated CdS Window Layers for Higher Current in CdS/CdTe Thin Film Solar Cells*. New York: IEEE; 2012. pp. 854–8.

[136] Kephart JM, Geisthardt RM, Sampath WS. Optimization of CdTe thin-film solar cell efficiency using a sputtered, oxygenated CdS window layer. *Prog Photovoltaics*. 2015;23(11):1484–92.

[137] Meysing DM, Reese MO, Warren CW, *et al.* Evolution of oxygenated cadmium sulfide (CdS:O) during high-temperature CdTe solar cell fabrication. *Solar Energy Mater Solar Cells*. 2016;157:276–85.

[138] Roy UN, Bolotnikov AE, Camarda GS, *et al.* High compositional homogeneity of $CdTe_xSe_{1-x}$ crystals grown by the Bridgman method. *APL Mater*. 2015;3(2):7.

[139] Gul R, Roy UN, Egarievwe SU, *et al.* Point defects: their influence on electron trapping, resistivity, and electron mobility-lifetime product in $CdTe_xSe_{1-x}$ detectors. *J Appl Phys*. 2016;119(2):6.

[140] Tit N, Obaidat IM, Alawadhi H. Origins of bandgap bowing in compound-semiconductor common-cation ternary alloys. *J Phys-Condes Matter*. 2009;21(7):6.

[141] Paudel NR, Yan YF. Enhancing the photo-currents of CdTe thin-film solar cells in both short and long wavelength regions. *Appl Phys Lett*. 2014;105 (18):5.

[142] Baines T, Zoppi G, Bowen L, *et al.* Incorporation of CdSe layers into CdTe thin film solar cells. *Solar Energy Mater Solar Cells*. 2018;180:196–204.

[143] Mia MD, Swartz CH, Paul S, *et al.* Electrical and optical characterization of CdTe solar cells with CdS and CdSe butters – a comparative study. *J Vac Sci Technol B*. 2018;36(5):9.

[144] Fthenakis VM, Kim HC. Photovoltaics: life-cycle analyses. *Sol Energy*. 2011;85(8):1609–28.

[145] Fthenakis VM. Life cycle impact analysis of cadmium in CdTePV production. *Renew Sust Energ Rev*. 2004;8(4):303–34.

[146] Andersson BA. Materials availability for large-scale thin-film photovoltaics. *Prog Photovoltaics*. 2000;8(1):61–76.

[147] Feltrin A, Freundlich A. Material considerations for terawatt level deployment of photovoltaics. *Renew Energy*. 2008;33(2):180–5.

Chapter 3
$Cu(In,Ga)Se_2$ and related materials

Angus Rockett[1]

3.1 Overview of CIGS semiconductors

"CIGS" is the common term used to describe a broad class of chalcopyrite materials consisting of a group I element cation, a group III cation, and a group V anion. The materials are alloys with compositions such as $(Ag,Cu)(In,Ga)(Se,S)_2$—in other words, they are mixtures of two elements from groups I, III, and V and are a subset of a larger set of I, III, VI_2 compounds. We will use "CIGS" as short form for them. The materials are attractive for the semiconductor optical absorber layer in photovoltaics because they have very high optical absorption coefficients and span a wide range of energy gaps.

3.1.1 Optoelectronic properties of CIGS

The prototypical material in this class of alloys is $CuInSe_2$ (CIS) [1]. The related alloys behave very similarly, so we will summarize some properties of CIS to give a general sense of the material. CIS is a direct-gap semiconductor with a band structure recently calculated as in Figure 3.1. Many earlier calculations exist as well [3]. A particularly good example of earlier work is by Jaffe and Zunger [4], where a detailed discussion is provided about methods to calculate band structures of these materials. That work also gives a review of many experimental results for properties of chalcopyrite semiconductors and the interaction of various atomic orbitals. The upper valence band is dominated by the Se 4p orbitals (above about −1.5 eV), and the lower valence band (below about −2 eV) is primarily due to the Cu 3d states. The p–d repulsion reduces the energy gap relative to related materials such as ZnSe [4]. The single branch of the conduction band around the minimum is primarily associated with the In 4s orbital. The electron affinity is about 4.6 eV, resulting in a band offset in the conduction band of ~0.3 eV with respect to CdS, which is the most common material used in heterojunctions as discussed later.

The band offsets are particularly important for devices. The heterojunction with CdS produces a "spike"-type band offset in the conduction band. Since CdS is moderately n-type as deposited for these devices, the spike band offset promotes

[1]Department of Metallurgical and Materials Engineering, Colorado School of Mines, Golden, CO, USA

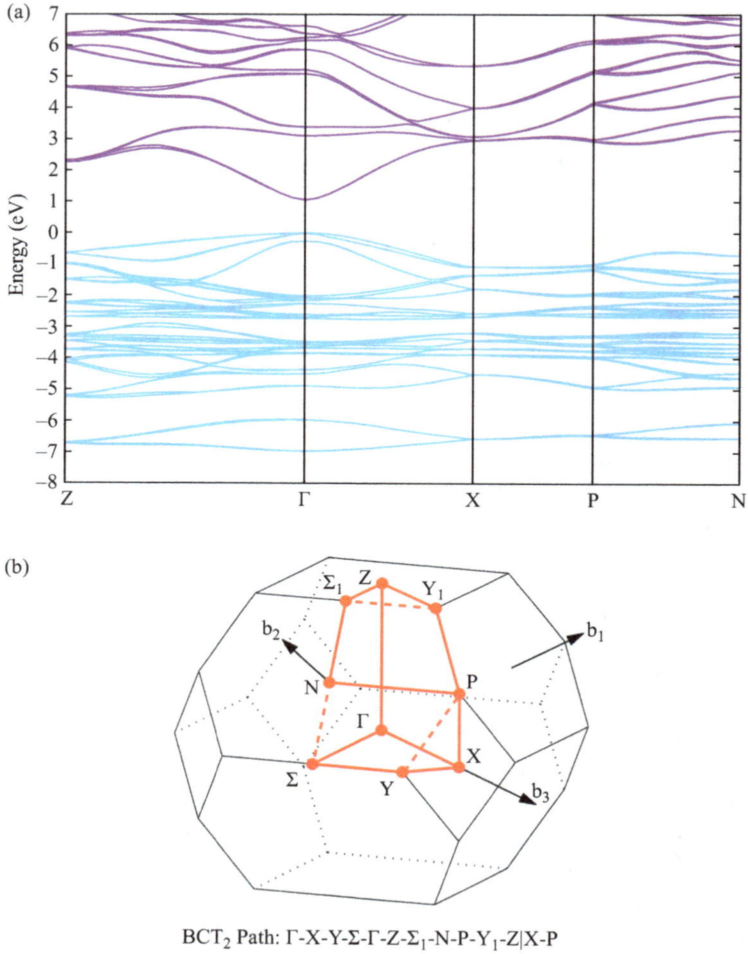

Figure 3.1 *(a) The band structure for $CuInSe_2$ showing a direct energy gap at the Γ point. This was calculated with the VASP computer code using the HSE06 hybrid functional with 32% exact-exchange and spin orbit effects included. PAW pseudopotentials, not including In d orbitals were used [J. Varley and V. Lordi, Electronic structure of $CuInSe_2$, private communication]. (b) The first Brillouin zone for $CuInSe_2$ [2]*

electron transfer into the CIGS; hence, it promotes type inversion of the CIGS near the surface, creating the p–n junction. Too large, a band offset has two effects. First, electrons collected by the field in the CIGS to the heterojunction need to cross the spike to reach the transparent contact in the device. If the spike significantly exceeds 0.3 eV, then thermal energy will not suffice to permit charge transfer. At the same time, as the device enters forward bias during operation, the depletion region shrinks and the field in the CIGS is reduced. At some point, the top of the spike rises above the conduction band edge in the bulk of the CIGS, which typically limits the open-circuit voltage (V_{oc}). For these reasons, it is important not to increase the electron affinity of the CIGS through alloying because it would increase the height of the spike in the band edges. Fortunately, the most effective alloys increase the energy gap of CIGS and lower the electron affinity, decreasing the size of the band offset with CdS. Ultimately, the limit to performance in high-Ga alloys may be the reversal of the sign of this band offset, which may limit inversion of the carrier type in the CIGS.

Some limited intermixing of the heterojunction could potentially spread out the band offset. This would still result in an equivalent amount of type inversion of the CIGS surface, but without the detrimental aspects of the spike. However, deliberate intermixing of the heterojunction does not clearly improve the device. In some good devices, Cu is found in CdS when used as a heterojunction partner material while Cd is found in the CIGS. This indicates that at least a modest amount of intermixing may occur without hurting the device and may be beneficial [5].

The band offset in the valence band is typically large and produces a "cliff"-type discontinuity. This generally has no effect on devices other than to keep holes out of the heterojunction partner material. However, there is potential for benefit if holes can be extracted from the heterojunction partner material and injected into the CIGS. For most device configurations, the energy difference between the valence bands of CdS and CIGS—combined with the built-in voltage of the junction—exceeds the bandgap of the CIGS. Thus, one might expect carrier multiplication and greater-than-unity quantum efficiency to be possible. However, CdS typically produces no photocurrent at all. An alternate heterojunction partner material could achieve carrier multiplication and significant gains in photocurrent.

The relative effective mass of electrons in the conduction band of CIS is ~0.09; and for holes in the valence band, values of 0.71 (heavy) and 0.09 (light) have been reported. At 300 K, these values give band-edge effective densities of states of $\sim 7 \times 10^{17}$ cm^{-3} for the conduction band and $\sim 1.5 \times 10^{19}$ cm^{-3} for the valence band. The bandgap exhibits a temperature dependence of -2×10^{-4} eV/K and a pressure dependence of 3×10^{-11} eV/Pa [6].

Carrier mobilities in CIGS are high. Both bulk and epitaxial thin-film single crystals of Cu(In,Ga)Se$_2$ have been produced and studied by temperature-dependent Hall effect. Lyahovitskaya *et al.* studied slightly Se-rich and Cu-deficient bulk single crystals grown by the traveling-heater method and found room-temperature electron mobilities in n-type material up to 780 cm^2/V-s [7]. Schroeder *et al.* studied epitaxial thin films by temperature-dependent Hall effect and observed room-temperature hole mobilities typically near 250 cm^2/V-s, with peak mobilities exceeding 1,000 cm^2/V-s at 77 K [8]. They performed a detailed analysis of the temperature dependence and determined that the primary scattering mechanism was phonon scattering in spite of the significant deviation from stoichiometry that would have favored impurity scattering. The phonon-scattering behavior was found even in materials with hole concentrations exceeding the band-edge effective density of states. Hole mobilities in thin-film epitaxial single crystals of $CuGaSe_2$ were studied by Arushanov *et al.* and found to be between 50 and 150 cm^2/V-s at 300 K, increasing to a maximum exceeding 250 cm^2/V-s in some samples at 180 K [9]. The temperature dependence is generally consistent with phonon scattering at higher temperatures, as was observed by Schroeder for $CuGaSe_2$. Both Arushanov and Schroeder observed steep decrease in hole mobility below a peak temperature, which was interpreted as a transition to defect band conduction. Karunagaran and Ramasamy studied hole transport in $AgInSe_2$ bulk single crystals grown by the Bridgeman method and found hole mobilities of 383 cm^2/V-s at 300 K [10]. Typically, hole mobilities in polycrystals are much lower than in single crystals—in the order of 10 cm^2/V-s—as would be expected when carriers would have to cross grain boundaries. Thus, it seems appropriate to use the higher values in device simulations where carriers are not required to cross grain boundaries to be collected. All of the above results for a variety of CIGS-type materials suggest, at ~300 K, a hole mobility in the range of 200 ± 100 cm^2/V-s and electron mobilities of 500 ± 300 cm^2/V-s and that at least hole scattering would be limited by phonon scattering near room temperature.

The low-frequency dielectric constant for CIGS is 13.6. Taking this value and the effective masses of electrons and holes, one can estimate a hydrogenic state energy for substitutional impurities as 6 meV for electrons and 52 meV for holes. These values represent the theoretical ionization energy of a donor or acceptor state resulting from a single carrier charge bound to the opposite nuclear charge. Although it is attractive to apply these values to interpretation of defect states in the material, it is noted that the primary defects are intrinsic point defects that do not lend themselves to a hydrogenic model. However, the lower value is close to the exciton binding energies observed for $CuInSe_2$ and $CuGaSe_2$ of 5.1 and 9.5 meV, respectively [11]. Furthermore, these values should be taken in context of the band-edge Urbach energies, estimated for the wider band edge of 10–20 meV at 300 K [12–14]. The corresponding width of the band tails makes it difficult to assess the nature of observed defect states with ionization energies of this order of magnitude or less. Band-tail widths and corresponding band-edge fluctuations have been observed directly by scanning tunneling spectroscopy [15] and have been linked to device performances [16].

A very attractive aspect of CIGS is its high optical absorption coefficient [4,17,18]. Early studies of single crystals showed optical absorption coefficients exceeding 10^3 cm^{-1} only 50 meV above the band edge [19,20]. The value rises rapidly to 10^5 cm^{-1} by 500 meV above the band edge [4,17]. These are some of the highest absorption coefficients for any common semiconductor. For comparison of the CIGS absorption coefficient with those of several other semiconductors used in photovoltaics, see Figure 3.2 [21]. The absorption coefficient is high enough that photovoltaic devices based on CIGS have been produced with high photocurrent densities with thicknesses well under 1 μm [22–25].

Device-quality CIGS material typically has an exceptionally consistent carrier concentration in p-type material, within an order of magnitude of 5×10^{15} cm^{-3} based on capacitance/voltage measurements and simulations of device behaviors.

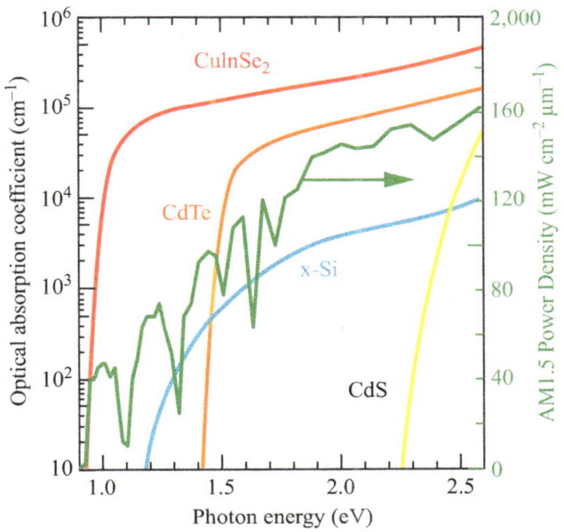

Figure 3.2 A comparison of optical absorption coefficients for several semiconductors used in photovoltaics

This is in spite of wide variations in stoichiometry. It appears that the interaction of Fermi level with defect formation energy in the presence of many point defects and defect clusters results in a very stable doping level. However, in a recent study, Nikolaeva *et al.* showed that there are significant fluctuations in doping from one grain to another [26]. In addition, in cathodoluminescence measurements, it is observed that individual grains in CIGS exhibit peak luminescence at different wavelengths relative to grain boundaries [27]. Such grain-to-grain variations would be expected to limit device performances. Narrow-energy-gap grains would be expected to produce cells with high current and low voltage, whereas wide-gap grains would produce cells with high voltage and low current. Adjacent grains act in parallel, so the low voltage grains and low current grains would dominate performance. Improving the uniformity should improve the devices if this can be achieved.

3.1.2 Phases and compounds

The alloys spanning the complete phase space appear to be broadly soluble, so phase separation in the deposited materials is not generally observed. The compounds have a I–III–VI$_2$ nominal stoichiometry and chalcopyrite structure, but will dissolve excess group III and group VI elements. The result is that at equilibrium they follow a tie line between I–III–VI$_2$ and III$_2$–VI$_3$ compounds (Figure 3.3). For example, excess group I element is generally not soluble and results in precipitation of a I$_2$–VI second phase on the surface of a thin film growing from the vapor phase [28]. The ternary equilibrium phase diagram for CuInSe$_2$ has been reported in detail (Figure 3.4) [29]. However, most thin films of these compounds are deposited well

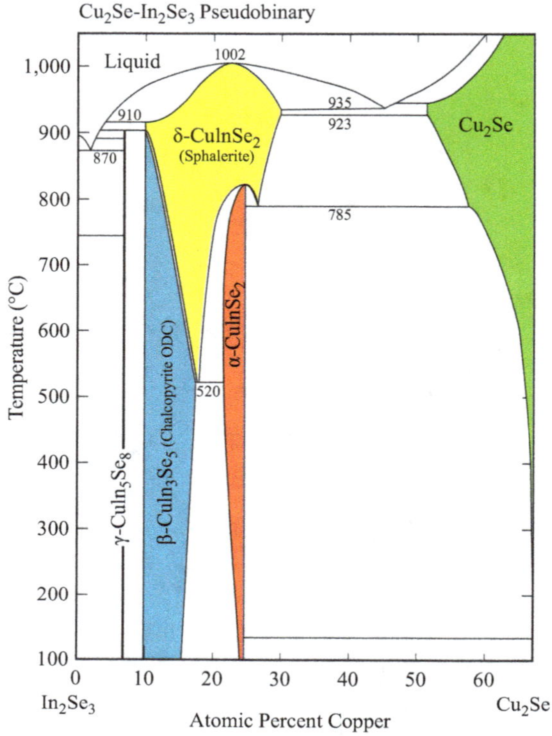

Figure 3.3 *The pseudobinary phase diagram connecting In_2Se_3 and Cu_2Se along the diagonal of the ternary phase diagram and encompassing $CuInSe_2$, $CuIn_3Se_5$, and $CuIn_5Se_8$*

away from equilibrium and can exhibit metastable single-phase compositions [33]. Therefore, one should expect, as is typically observed, that the properties of CIGS films may vary significantly from one lab to another based on the details of the growth process.

The implication of the phase behavior represented by the ternary phase diagram is that the compounds at equilibrium favor valence compensation in intrinsic point defects. It has been concluded from both theoretical calculations and experimental measurements that the defects consisting of group III elements on group I sites (III_I) and group I vacancies (V_I) have relatively low formation energies in typical p-type materials. These are expected to form charge-neutral defect clusters $[(III_I)^{+2} + 2(V_I)^{-1}]^0$ [34,35]. Thus, the solubility of excess group III element accompanied by increasing group V atom fraction is simply a reflection of the addition of more of these neutral defect complexes.

As their concentration increases, ordered defect compounds (ODCs) are observed, and in some cases altered structures are found [36–38]. A particularly detailed discussion of phase relationships and ordered defect structures in CIGS

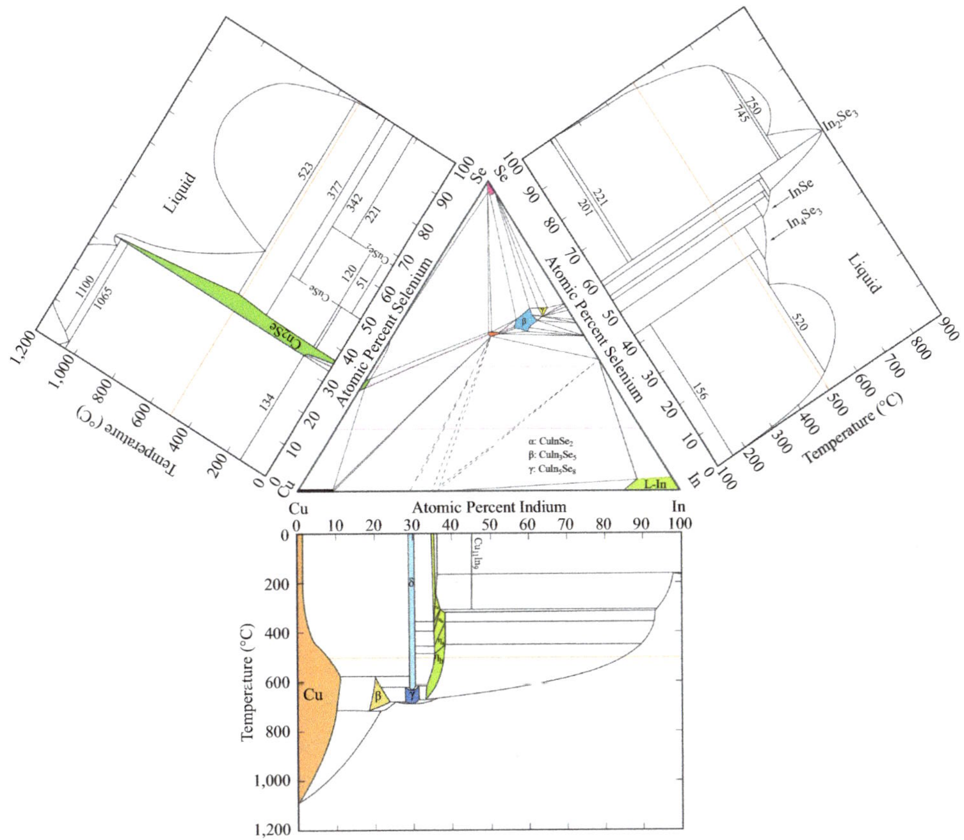

Figure 3.4 *Binary and ternary phase diagrams for the Cu-In-Se ternary system. Data replotted from Godecke and others [29–31] and binary phase diagrams in the ASM Handbook v.3 [32]*

may be found in Stanbery [39]. Altered structures may include CuAu ordering in which the cations disproportionate onto alternating close-packed (100) planes [36]. This is particularly common in $CuInS_2$ and may contribute to lower performance in devices based on that material [40–43]. CuPt ordering has also been observed in some cases [37,44]. The normal ODC with compositions near $CuIn_3Se_5$ has an increased energy gap relative to the perfect chalcopyrite material [38]. Some researchers have concluded that a thin surface layer of ODC is present on group III-rich CIGS and is responsible for the behavior of the device [45]. This remains controversial and limited evidence for the ODC on the surface of the films is available [45,46]. However, as noted earlier, discrepancies in observations may be due to differences in the deposition method. It remains unclear how important the ODC is to the performance of the best CIGS devices.

One of the unique features of CIGS (and the II–VI materials in general) is that devices generally work better when made from polycrystalline material. Recently, however, Nishinaga *et al.* reported a device fabricated from a single-crystal CIGS layer that achieved 20.0% conversion efficiency [47]. The V_{oc} was 815 meV, well above the 744 meV of the current champion cell [48], but short of the highest reported value of 960 meV [49]. The device also exhibited a short-circuit current density (J_{sc}) of 32.6 mA/cm^2 and a fill factor (FF) of 75.3%, the latter also exceeding the champion cell performance. This is an extraordinary result and demonstrates that the relatively low performance of single-crystal epitaxial CIGS layers in devices is primarily a difference in the processing, rather than a fundamental limitation to single crystals or epitaxial materials. It seems likely that further optimization will lead to a champion cell from an epitaxial layer.

One of the remarkable features of CIGS is its preference for the two close-packed polar (112) crystal planes. This has been demonstrated by growth of epitaxial films on both orientations (Ga- or As-terminated) of the polar GaAs (111) surface [50]. On the nonpolar GaAs (110) surface, epitaxial CIGS spontaneously faceted into (112) planes [51]. By analysis of internal voids trapped inside CIGS grains, a Wulff construction for the relative surface energies of Cu(In,Ga)Se_2 was produced, as shown in Figure 3.5 [52]. This result is remarkable and important in several respects. First, the polar surfaces should have relatively high energies because they should have an intrinsic electric charge. Typically, semiconductors either prefer nonpolar surfaces (as in GaAs) or reconstruct the surface to achieve charge neutrality. Several reconstructions have been proposed for CIGS [53], but to date none has been observed despite several scanning tunneling microscopy studies. Second, the preference for (112) surfaces also relates to an observation that in some cases a (110) surface orientation produces improved devices. The implication of the surface science is that (110) CIGS surfaces essentially do not exist, although the average orientation of the surface can be (110). The actual heterojunction is almost entirely constructed from inclined (112) facet planes.

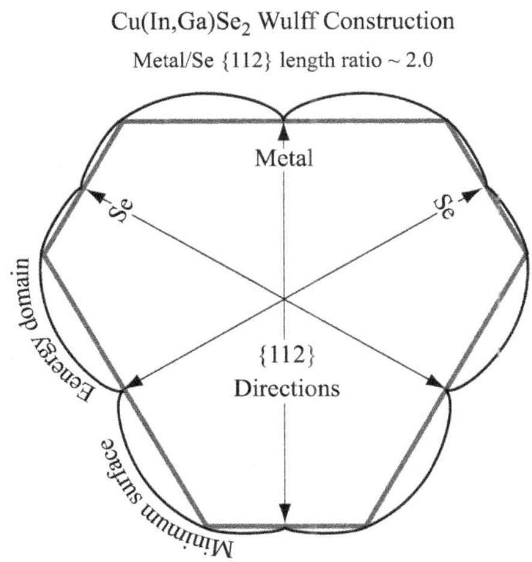

Figure 3.5 The Wulff construction for $Cu(In,Ga)Se_2$. Note that the surface energies are inversely proportional to the length of the arrows shown [52]

3.1.3 CIGS alloys

As noted earlier, "CIGS" refers to a whole spectrum of miscible semiconductor alloys covering the compositions $(Ag,Cu)(In,Ga)(Se,S)_2$—but even this does not cover the full range of compositions. Al has been studied as a substitution on the group III site and Te has been considered as a substitution on the VI site. Al tends to be difficult to work with because of its strong tendency to oxidize, but it has shown some significant successes [54,55]. Most CIGS synthesis processes are not conducted in ultrahigh vacuum, so oxidation is a consistent issue. Te tends to reduce the energy gap too much to be technologically valuable. Furthermore, the greater the size and electronic structure differences among the atoms, the lower the miscibility. Therefore, the CIGS community normally restricts itself to $(Ag,Cu)(In,Ga)(Se,S)_2$.

There are additional practical restrictions on this alloy space. Alloys with high Ga or S contents appear to produce lower-quality devices. In the case of Ga alloys, the cause is not clear. However, it seems most likely that intrinsic defects that produce states near the band edges in $CuInSe_2$ result in much deeper states in $CuGaSe_2$ that are therefore more detrimental to devices [56]. In the case of S alloys, various defect structures are more common in high S compositions than in the selenides [40,43]. Nonetheless, some high-performance results have been obtained with $CuInS_2$ [49]. Due to the difficulty achieving high performances, high Ga and S alloys are generally avoided. This is unfortunate because these alloys have energy gaps in the range that would be ideal for the high-gap portion of a multijunction device. Research continues to find a way to make good devices from materials with higher energy gap. In general, more complex alloys with additional alloying components tend to produce better devices for a given energy gap. This may be due to reduced ordering of defects, analogous to the concepts behind high-entropy metal alloys. (Note that "high entropy" is a misnomer, underrepresenting the reasons for improved stability and solubility in these alloys.)

A primary advantage of CIGS alloys is that bandgap engineering for devices is relatively straightforward. CIGS alloys have the interesting property that the bottom of the conduction band edge is almost entirely composed of group III element valence s orbitals [4,57]. Thus, alloys that replace In with Ga primarily affect the conduction band edge or electron affinity of the CIGS [58]. For the $Cu(In,Ga)Se_2$ alloy, the energy gap ranges from ~1.0 to ~1.7 eV with a small bowing coefficient of ~0.2 eV. The valence band-edge energy is primarily composed of the anion (Se, S) valence p orbitals. The $CuIn(S,Se)_2$ alloys exhibit almost no bowing between the energy gaps of $CuInSe_2$ (1.0 eV) and $CuInS_2$ (1.45 eV) and nearly all of the band-edge shift is in the valence band. However, the valence band energies are significantly affected by the shallow valence d orbitals in the group I elements. Replacing Cu with Ag increases the energy gap, despite the increased lattice constant (1% and 5% for the a and c axes, respectively). A lattice constant increase would suggest that the energy gap should decrease [59]. However, the different energies and interactions of the shallow d orbitals with the anion p orbitals result in energy gaps for $CuInSe_2$ of ~1.0 eV and $AgInSe_2$ of ~1.23 eV with a bowing coefficient of ~0.2 eV. This value decreases with addition of Ga [60]. A comprehensive review and excellent discussion of bandgap bowing in $Cu(In,Ga)(S,Se)_2$ alloys and several other important alloy systems can be found in Schnohr [61]. A detailed study of bandgaps and bowing coefficients in $(Ag,Cu)(In,Ga)Se_2$ alloys is also available in Shafarman [60]. The small bowing coefficients of these alloys explain the high miscibility because these are linked through the electronic structure of the solids.

The result of the earlier is that the CIGS photovoltaic device community typically assumes that the (In, Ga) alloy can be used to shift the conduction band edge with little effect on the valence band whereas the (S, Se) alloy is assumed to be useful to shift the valence band edge almost exclusively. To achieve the highest device performances, a common approach is to increase the Ga content of CIGS absorbers toward the back of the device. This produces a potential increase in the conduction band and reduces the chance that minority-carrier electrons reach and recombine at the back contact. The front of the device can be treated with S to replace some of the Se, which increases the energy gap near the junction and increases the driving force for holes to move away from the junction where they could recombine. It is also possible to increase the energy gap near the front of the device by adding Ga, although this results in a less favorable band offset with the heterojunction partner material, so S alloying is preferred. In any case, it is important that the front alloyed layer does not extend beyond the depletion region because this would result (for similarly p-type alloys) in a barrier to minority electron collection at the front of the device. An example simulation is shown in Figure 3.6, illustrating the behavior of this barrier under bias. The depletion region

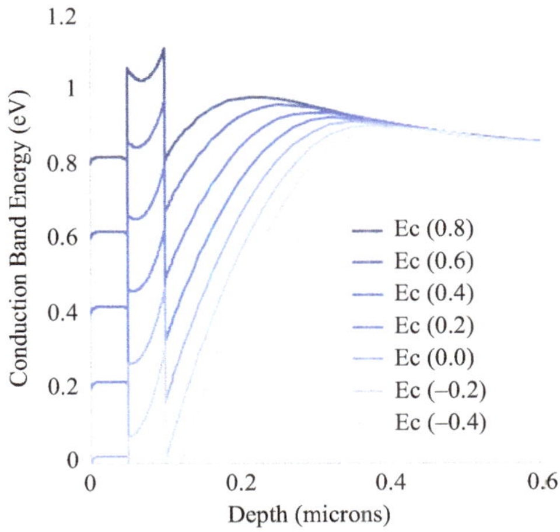

Figure 3.6 SCAPS simulations of the conduction band edge in a CIGS device where the wide-gap material penetrates too far below the surface. Even for −0.4 V bias, there is a barrier to electron collection. (Simulation used default parameters for CIGS junctions available with SCAPS.)

is bias-dependent, so this can cause problems for the device under operating conditions.

3.2 CIGS devices

3.2.1 Overview

The CIGS photovoltaic device typically consists of an opaque and preferably reflective metal back contact deposited on glass, stainless steel, polyimide or other material, the CIGS absorber layer, a heterojunction partner material, and a transparent conductor. Each of these layers except the substrate will be discussed briefly as follows.

CIGS devices have been far more successful in "substrate" configurations, as described earlier, unlike the "superstrate" geometry used in CdTe devices. The CdTe/CdS heterojunction and similar heterojunctions are stable and any mixing during deposition does not affect the device performance. The CIGS/CdS heterojunction and junctions with other partner materials are not stable under conditions of high deposition temperature. When the junction is annealed significantly above ~200 °C, CIGS/CdS devices degrade rapidly. Surprisingly, annealing the device after CdS deposition in air for 2 min at 200 °C has been shown in some cases to improve the device [62]. When junctions based on epitaxial $CuInSe_2$ and polycrystalline CdS were annealed to 500 °C for 120 min, it was found that there was no detectable intermixing of the anions (S, Se). However, Cu exchanged with Cd across the heterojunction to form an identifiable compound, $Cu_3Cd_2In_3Se_8$ [63]. This is consistent with conclusions from other studies that Cd atoms at or near the surface of the CIGS play a key role in the performance of the junction. This also presumably contributes to the lower performance of other heterojunction partner materials not containing Cd [64]. The temperature limitation of the junction effectively eliminates the option for deposition of the CIGS on CdS or other heterojunction partner materials. In spite of this behavior, long-term operation does not cause degradation as does higher temperature annealing.

CIGS currently produces devices with champion efficiencies comparable with the major competing technologies—polycrystalline Si, CdTe, and the hybrid perovskites—all of which now lag somewhat behind the champion single-crystal Si devices [65]. The devices have achieved an efficiency of 22.9% in single-junction, one-sun configurations [65]. Devices exceeding 20% efficiency have been deposited at relatively low temperatures (350 °C) on flexible polyimide substrates using a post-deposition alkali halide annealing process [66]. The devices have been manufactured as complete modules with efficiencies in full-scale modules of 15.7% and smaller modules up to 19.2% [65]. The modules are typically found to be quite stable in operation [67], although shadowing and other issues can lead to degradation [68,69].

Devices with optimized antireflection coatings have virtually 100% external quantum efficiency (EQE) across a wide spectral range [65]. Typically, the EQE shows a lower value below wavelengths of ~580 nm due to absorption when a sufficiently thick CdS heterojunction partner layer is used. Thinner CdS or wider-gap heterojunction partner materials (e.g., by partial or complete replacement of the Cd with Zn or the S with O) can eliminate this absorption [66]. The result is that the short-circuit current in champion CIGS devices is typically near the theoretical limit. FFs exceeding 81% have been achieved [62], but the V_{oc} typically lags behind theoretical limits significantly. Presently the champion device is reported to exhibit a V_{oc} of 744 meV [48]. An example of V_{oc} deficits relative to a range of energy gaps is illustrated in Figure 3.7 for a series of test conditions at Solar Frontier [48]. The values are typical for high-performance CIGS devices. However, in a recent work reviewing alkali-metal treatment effects on CIGS devices, voltage deficits in the range of 170 meV have been achieved when the CIGS is produced close to the stoichiometric composition (Figure 3.8) [70]. These low voltage deficits are accompanied by other problems such that the achieved efficiencies are not at champion levels. For example, the current champion device exhibits a FF of 79.5% whereas some devices have achieved more than 81%. The best-reported values of J_{sc}, V_{oc}, and FF for CIGS devices suggest that there is significant room for improvement in device performances.

CIGS devices exhibit a number of metastable changes in performance which, although they are not a major concern, are potential limiters to device efficiency. Ultimately, metastability will affect the competitiveness of the technology unless resolved. It has been observed that under some temperature and bias conditions, a long-lived metastable state can be induced. This is easily removed by heating the device to only 330 K, a temperature that would be common in an operating device under full-sun conditions. However, there are also metastabilities that occur under

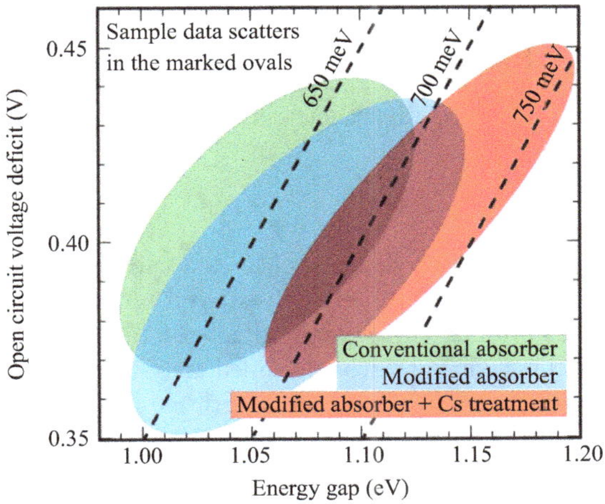

Figure 3.7 Voltage deficits in Solar Frontier CIGS photovoltaic devices as a function of minimum energy gap (the absorber energy gap is graded) comparing three processing approaches. The results show the importance of considering the absorber process when evaluating defects and recombination mechanisms [48]

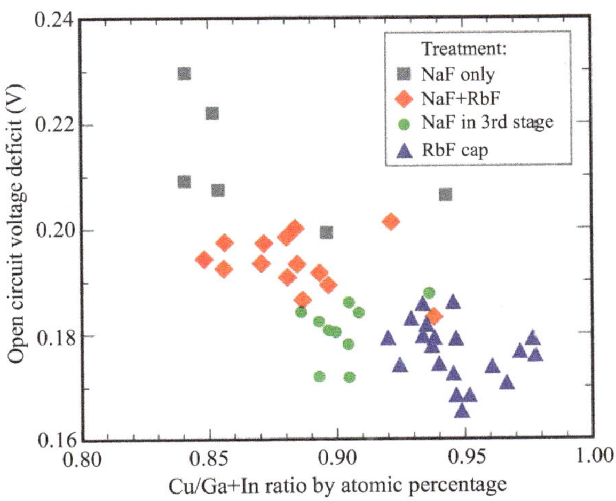

Figure 3.8 Results showing the effect of various alkali halide post-deposition processes on voltage deficit for different Cu/(III) ratios [70]

normal operating conditions that can affect the current, voltage, and FF in either positive or negative ways and which depend highly on process. These have been connected with midgap defects in the case of voltage metastability and have been specifically attributed to Cu-Se divacancies in the case of current metastability, as discussed later.

3.2.2 The back contact

CIGS devices are universally produced on Mo as a back contact. This material is not ideal because it has relatively high resistivity and is relatively expensive. However, the choices for alternatives are limited as follows. Alkali metals are too reactive to underlie the device and would immediately selenize during processing. The group Ib and IIIa elements are isovalent with the elements in the device and would dissolve in the CIGS during deposition. The group IIb and IVa elements could produce uncontrolled doping and might react with CIGS to form other compounds, as is the case with Cd. Furthermore, the group IVa elements are not sufficiently conductive. The remainder of the group "a" elements are similarly a problem. This leaves compounds and transition elements. Most transition elements are too reactive or soluble with one or more components of CIGS. Mo has several advantages. It has almost no solubility for group Ib or IIIa elements. It reacts with Se to produce $MoSe_2$ oriented with the atomic planes perpendicular to the surface of the Mo [71,72]. This leaves dangling bonds at the sheet ends that bond well to CIGS and form an effectively ohmic contact. Thus, some reaction of Mo with Se is desirable. MoO_2, which is commonly formed if some oxygen is present during deposition of Mo, is an intercalation host for NaO, forming $NaMoO_3$ [73]. Since NaO is a component of the soda-lime glass used as a substrate and because Na turns out to improve the CIGS performance, this is a benefit to the devices. Many other materials have been tried as back contacts including some compounds such as transition metal nitrides, but none performs as well as Mo.

3.2.3 The front contact

The heterojunction with CIGS is formed from CdS in most of the highest performance devices. This is not ideal because of the toxicity of the Cd. A very wide spectrum of heterojunction partner materials has been explored. An excellent review was published by Hariskos *et al.* [64] who compared the many materials considered. The most successful alternative to date is $Zn(S,O,OH)_x$ [74]. Various deposition methods have been considered for these, with chemical bath deposition and physical vapor deposition being the most widely adopted. It appears that ion exchange between a group II element and Cu on the CIGS surface is important and can result in doping of the surface layer(s), converting the surface of the CIGS to n-type [5,75,76]. This inversion helps to explain why the details of the heterojunction structure are not critical. Typically, the CdS or equivalent material deposited by chemical bath deposition is nanocrystalline and may contain a significant number of secondary phases and impurities [77]. In spite of this, it makes an excellent heterojunction. In some cases, a thin or relatively thick epitaxial layer of CdS may form on the CIGS and makes an excellent, but not distinguishably better, heterojunction than the poor-quality nanocrystalline materials [75,78]. It appears that the basic requirements for a successful front contact to a CIGS device are that it dope the surface of the CIGS heavily n-type and not permit significant exciton recombination there. Otherwise, a wide spectrum of deposition methods and resulting materials microstructures seems to produce at least good heterojunctions.

3.3 Intrinsic defects in CIGS

CIGS may be several percent off stoichiometry in good devices and may contain significant variations in composition and hence variations in average local point-defect density [79]. As noted earlier, the compositions observed suggest high concentrations (1% or more) of indium on copper sites, In_{Cu} and copper vacancies, V_{Cu}. These form defect clusters in the material which, when dissolved, move the composition along the tie line between In_2Se_3 and $CuInSe_2$ (or similar phases, where alloying replaces the Cu, In, or Se), so In_{Cu} and V_{Cu}, particularly as clusters, are considered ubiquitous in the material (see Figure 3.4).

The pseudobinary phase diagram between Cu_2Se and In_2Se_3 demonstrates that the energy favoring ordering of the cation sublattice (chalcopyrite vs. sphalerite) is low (Figure 3.3). The disordering phase transformation is close to the higher deposition temperatures (e.g., 625 °C) used in some cases to make good devices, especially when group III-rich. Therefore, it is reasonable to expect Cu_{In} antisite defects, as well. Se vacancies, V_{Se}, are thought to be common in clusters with V_{Cu}. These divacancy clusters are suspected to cause some of the metastable response in the devices, as discussed later [80]. Other defect clusters have also been described, but they are not expected to be common [81].

CIGS is known to be p-type, although in the late 1980s and early 1990s there was a consistent understanding that Cu-rich CIGS was p-type whereas In-rich CIGS was n-type (Figure 3.9) [82]. In the mid-1990s, the understanding of the material changed—from that time onward, CIGS has been considered p-type in all compositions. It is unclear what changed, but the transition occurred around the time that Ga-containing alloys became increasingly popular. It is possible that a change in deposition conditions was implemented that increased the Se flux. However, extensive work to optimize thin-film growth of CIGS had been performed prior to the change from predominantly n-type to predominantly p-type In-rich $CuInSe_2$ and Se flux was part of those studies [83]. It is also possible that the change was due to increasing addition of Na to the deposited material.

In any case, there was a time when group III-rich CIGS was uniformly considered to be n-type and highly resistive [38]. This is significant because it implies that in group III-rich material there may be a defect state above the middle of the bandgap that could have been dominating carrier density, and thus, the Fermi level. Such states are still observed, as discussed later. It seems likely that rather than eliminating a donor defect, the process conditions now favor formation of more acceptors, thus making p-type material. CIGS is often found to be heavily compensated, suggesting that both

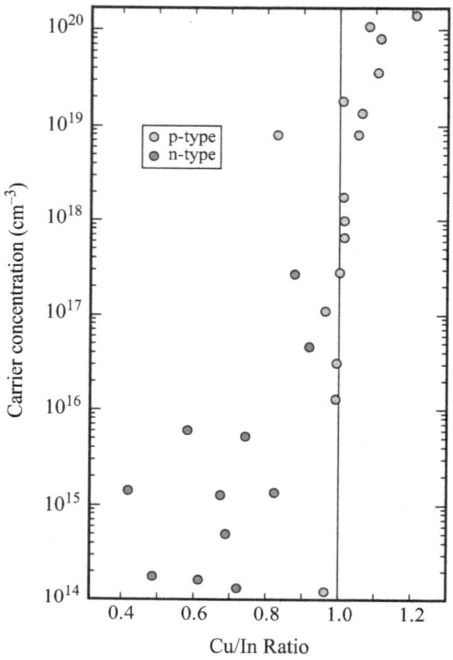

Figure 3.9 Carrier concentration vs. Cu/In ratio in $CuInSe_2$ showing the common understanding at the time (before the mid-1990s) that In-rich $CuInSe_2$ is n-type. Data after Noufi et al. [82]

acceptor and donor defects are common. The presence of a deep donor associated with the compensation could indicate the existence of a defect that would promote minority-carrier trapping and recombination.

The history of n-type behavior in group III-rich material is further significant because the heterojunction is thought to be strongly inverted, with the surface of the CIGS being converted to n-type [46]. This could be facilitated by active doping by a donor such as Cd on a Cu site (Cd_{Cu}) or by changes in the intrinsic acceptor or donor concentrations, resulting in an n-type material [84,85]. Most current understanding is that it is primarily the result of Cd_{Cu} donors, which explains why non-Cd-containing heterojunction partners tend to produce inferior devices unless very carefully optimized to produce a similar effect. However, the deep donor may also play a role.

If there is a deep donor that depends on Fermi energy, this could also lead to Fermi-level pinning, which is rarely accounted for in most current device models or in interpretation of capacitance-based measurements of devices [86]. Therefore, it is important to determine if such pinning exists in the devices to build a complete model as the basis on which to optimize the device architecture and materials synthesis.

A number of states are observed in the energy gap of CIGS [87]. General agreement is that there are at least two relatively shallow acceptor states and at least one compensating donor [6,8,57,87,88]. Three acceptor states have been identified with ionization energies of 40 $\pm$ 20 meV, 70 $\pm$ 20 meV, and 160 $\pm$ 20 meV, although in some materials the ionization energies are outside these ranges. The concentrations of the defects also vary significantly by deposition process. In some epitaxial single crystals, the 160-meV defect dominates, and at least one of the shallower defects is also observed [8]. In all CIGS materials studied to date by admittance spectroscopy, a capacitance step is found that can be interpreted as an acceptor state with an activation energy of 250 $\pm$ 40 meV [89,90].

The most likely sources of acceptor states near the valence band edge are $V_{Cu}{}^{-1}$ and Cu-on-In antisite defects ($Cu_{In}{}^{-1}$ and $Cu_{In}{}^{-2}$) [6,91]. This would account for three states, two of which should track each other in concentration. Normally, the shallowest acceptor is interpreted as due to $V_{Cu}{}^{-1}$. Generally, both the 40- and 70-meV states are not observed together in Hall effect measurements and device results, although both may be observed in photoluminescence (PL). These may not be distinct states, but rather two chemistries associated with the $V_{Cu}{}^{-1}$ defect with different ionization energies due to surrounding defects. It is noteworthy that the 40-meV ionization energy is close to the predicted hydrogenic state energy for these materials, although none of the intrinsic defects would be expected to behave as a simple hydrogenic state. It could perhaps represent an exciton binding energy. The 180- and 250-meV defects may be the two states due to Cu_{In}, but this remains to be established. Certainly, any state associated with a $Cu_{In}{}^{-1}$ defect would be expected to be accompanied by an additional deeper state for $Cu_{In}{}^{-2}$ because this is a divalent antisite defect.

In all materials studied by PL or temperature-dependent Hall effect, there is a donor defect that compensates some or most of the acceptors [8,92,93]. Various energies have been proposed for one or more donor states in the energy gap. Siebentritt and collaborators have described a shallow donor about 7 meV from the conduction band edge [94]. This state has been observed for many years based on similar PL measurements along with several other proposed donors [87]. However, given the Urbach energy of the bands, it is unclear whether a state that is shallow would be distinct from the band edge. This may simply represent a portion of the band tail that is sufficiently common to produce luminescence and sufficiently rare to be below the mobility edge for the band tail [15]. The PL measurements are somewhat indirect because the defect energies that result are based on the energy of the emitted photon, which may include contributions due to phonons and may include phonon replicas [95].

To some extent, capacitance measurements can be more reliable than PL because, if properly interpreted and modeled, they can measure direct trap escape energies. Examples of such measurements include deep-level optical spectroscopy (DLOS), deep-level transient spectroscopy (DLTS), transient photocapacitance (TPC), and more. TPC, DLTS, and DLOS have been applied to measurement of defect states in CIGS [13,96–99]. DLTS can be effective to measure states in the lower half of the energy gap whereas DLOS and TPC are best for measuring states in the upper portion of the energy gap. Many other more obscure techniques such as photothermal deflection spectroscopy have also been applied and provide useful complementary results [100–102].

The various capacitance methods yield evidence consistent with some of the donor states identified by PL. Using TPC, Heath *et al.* found evidence of two donor states—one ~800 meV above the valence band edge and another ~1,000 meV above the edge—with both remaining roughly at the same energy regardless of Ga/In+Ga content in device-quality Cu(In,Ga)Se_2 polycrystals [90,103]. Similar states were observed in epitaxial layers [103]. Arehart and collaborators found only one state using the closely related DLOS at ~970 meV above the valence band [98]. The observations of these defects suggest a state that does not vary with Ga content. Therefore, it seems unlikely to be due to a group III atom on a group I site (e.g., In_{Cu}), even though those defects are common. A more likely explanation would be V_{Se}. As with In_{Cu}, this is a divalent intrinsic point defect and should produce two donor states in the energy gap. The main argument against this assignment is that the state is not clearly changed by varying the growth temperature or Se activity. However, the recent arguments in favor of a V_{Cu}–V_{Se} divacancy would suggest that V_{Se} may be common [104,105].

It is observed that the V_{oc} of high-quality CIGS devices as a function of temperature exhibits a nearly linear behavior that extrapolates to the energy gap of the CIGS at 0 K [106]. This has been interpreted as indicating that interface recombination is not the dominant recombination mechanism in these devices [107]. However, it is also common to find that between 150 and 250 K, the voltage progressively falls below the linear dependence and transitions to a new trend extrapolating to a lower voltage, as shown in Figure 3.10. The V_{oc} gives a measure of the splitting of the quasi-Fermi levels for electrons and holes. If one assumes that the hole quasi-Fermi level is approximately equal to the equilibrium Fermi level, then the new extrapolated limit to V_{oc} primarily reflects the electron quasi-Fermi level. When this value is far from any pinning energy, the voltage should rise with decreasing temperature. However, if a pinning state exists, then that would limit further increases in quasi-Fermi level. Therefore, it seems reasonable to expect that the lower extrapolated V_{oc} would represent some structure pinning the electron quasi-Fermi level. This could be a donor or trap state in the material.

Recently Paul *et al.* have observed a dependence of the 970-meV donor defect concentration on Se overpressure in CIGS films deposited by physical vapor deposition [108]. It was suggested that the defect was due to the V_{Cu}–V_{Se} divacancy, in agreement with assignment by Igalson and collaborators [109–112]. The concentration of that defect was observed to correlate with metastable changes in J_{SC} under light soaking, as expected for reorganization of the divacancy defect under light [113].

In some cases, device improvement has been observed after air annealing of CIGS devices that have been completed to the point of deposition of the heterojunction partner material. For some time in the 1990s, it was found that such annealing helped devices produced in North America, but did not help devices produced in Europe. The reasons for this difference were never clear. Oxygen does

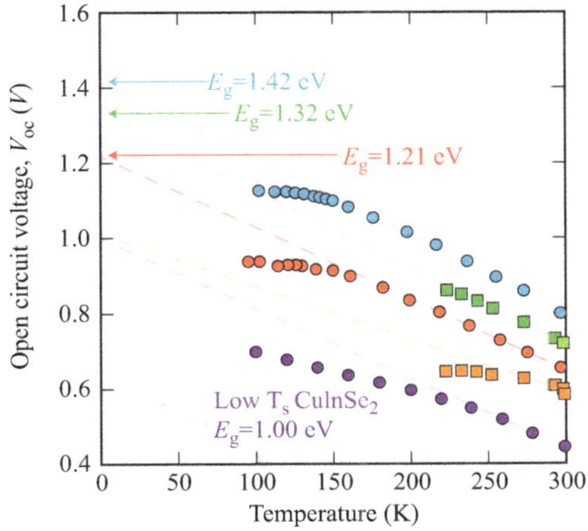

Figure 3.10 Temperature-dependent open-circuit voltage of several photovoltaic devices under air-mass 1.5 solar intensity (100 mW/cm^2) for absorbers with different energy gaps as marked. The square orange data points are the same absorber as the red circular points, but for a film deposited without Na addition [W. Shafarman, Temperature dependence of open circuit voltage in Cu(In,Ga)Se_2, private communication]

have a significant effect on elemental redistribution at CdS/CIGS heterojunctions when CdS is used as a heterojunction partner material [5]. The effect of air annealing on the absorber itself has been studied, as well [104,114]. It was found that such an anneal reduces the Cu concentration at the film surface and results in type inversion. This would be consistent with improved device performance, especially when accounting for changes to the Cd movement across the junction.

Extended defects, common in CIGS, include twins and stacking faults, along with grain boundaries [115]. As noted earlier, twins and stacking faults do not seem to have obvious detrimental effects on the devices and are ubiquitous in CIGS single-crystal and polycrystalline materials. Dislocations were proposed to be a problem but are rare in the device-quality material. By contrast, they are common in single-crystal epitaxial layers, which may explain the typically poorer performance of those layers. In one study, a twin termination in a polycrystalline commercial cell material showed a series of partial dislocations that had expanded dislocation cores that formed nanopipes through the material with internal diameters of about one atom scale [52]. This indicates that free surfaces of CIGS are more energetically favorable than at least some dislocation cores.

Twins and stacking faults are common in the materials and probably are not highly detrimental [37,52]. The true impact of twins and stacking faults remains to be established. For example, they do not show up strongly in electron-beam-induced current images, which would not be the case if they were affecting carrier properties strongly. Twins and stacking faults change the ordering of the cation sublattice, as do the many point-defect clusters in the typical device material. The number of those clusters is far higher than the density of twin boundaries. Thus, perhaps it is not surprising that twins and stacking faults would not degrade the devices significantly.

A number of studies have been performed on the polycrystals in an effort to determine the behavior of grain boundaries in the materials and how they affect devices [27,33,116,117]. The results are inconclusive, with some investigators finding a modest change in potential at grain boundaries [116,118–120] and some finding little or none [121]. At the same time, there are some observations that grain boundaries exhibit changes in composition and others that find none. Some impurities, notably the alkali metals, are found to segregate to the grain boundaries [122]. The surface of CIGS has been found to be Cu-deficient in some cases and it is expected that the grain boundaries would be as well [123]. This would produce an expanded energy gap in the boundary region and would repel minority carriers. The mechanism was proposed as an explanation for why the grain boundaries would not be detrimental [124]. The behavior of grain boundaries is discussed further later.

Certainly, one observation is that single-crystal CIGS contains far more dislocations than polycrystals, where the dislocations have sufficient mobility to move into grain boundaries, leaving nearly dislocation-free grains behind [79,125,126]. In many cases, dislocations cannot be observed in polycrystals at all. At the same time, dislocations are very common in epitaxial and bulk single crystals. Note that extreme care was taken to grow the epitaxial layer that produced the 20%-efficient device described earlier, and the resulting material probably contained relatively few dislocations. In the end, the basis for improved performance in polycrystalline devices remains unclear. It seems likely that dislocations may be a problem for CIGS.

Although it seems that devices operate better in many respects as they approach stoichiometry, this is not a panacea. For example, it is known that Cu-rich CIGS, which consists of two phases—nearly stoichiometric CIGS and Cu_2Se—produces very sharp PL lines, suggestive of highly perfect material [127]. Nonetheless, etching the Cu_2Se off the surface does not result in high-performance devices. This may or may not be the result of residual Cu_2Se; nevertheless, this method for producing CIGS very near stoichiometry does not appear to help.

A careful series of studies was performed by Siebentritt and collaborators to evaluate the potential of producing devices from initially Cu-rich material; see reviews of this series in [128] and [129]. Briefly, the group found that they could produce devices with efficiencies equivalent to what could be produced from Cu-poor material by etching away the residual Cu_2Se with KCN. However, there appeared to be a problem with interface recombination. The overall conclusions from the project indicated that (1) KCN etching produces a Se-related surface defect that promotes recombination there and (2) the defect variously identified in the 800–1,000-meV range (see later) above the valence band edge was apparently more common in the Cu-rich material [130–133]. It is not intuitively obvious if that defect is related to the Cu-Se divacancy, as some groups have suggested; however, the difference in processing conditions could easily have resulted in a different behavior, as is common in CIGS.

3.4 Extrinsic impurities in CIGS

In addition to intrinsic point defects, a number of studies were concerned with extrinsic impurity-related defects and behaviors in CIGS. Alkali metals have long been known to improve the performance of CIGS photovoltaics [134], primarily by improving the V_{oc} [135–137]. This has contributed to the choice of soda-lime glass as a substrate material, which provides NaO as a natural impurity to the film through the Mo back contact of the device [73]. There is an ideal amount of Na for improving the films [138,139]. Therefore, many groups have examined the possibility of applying a diffusion barrier to the substrate or using a Na-free substrate material such as stainless steel and supplying Na from an evaporation source. Diffusion barriers are difficult to implement to provide sufficient protection from Na from the soda-lime glass. The barrier layer typically has pinholes that allow Na to enter the film. Furthermore, Na diffuses rapidly, so even a modest density of pinholes results in strong Na contamination of the film. Na-free substrates have been used, most notably stainless steel, which is used at this time by MiaSolé and Global Solar Energy, among others. In this case, Na has successfully been added before or during deposition. Controlled studies have been performed by evaporation of Na compounds before and during deposition.

The incorporation and diffusion of Na in CIGS has been studied by a variety of groups. Implantation of Na into CIGS polycrystals and subsequent annealing showed that Na diffuses to the surface of CIGS films [73,140]. Studies of CIGS on soda-lime glass showed that rinsing the films with deionized water effectively removed Na from the surface, but gentle heating in vacuum caused Na to return to the surface from the bulk of the film. Diffusion of Na in bulk single crystals of $CuInSe_2$ was studied by Forest *et al.* and the diffusion coefficient was determined [141]. Several groups have shown that Na tends to diffuse to grain boundaries in CIGS polycrystals. One of the clearest studies of this type was conducted by Stokes *et al.* using the atom probe to show high concentrations in the grain boundaries [122]. A similar study showed a correlation between Na and O concentration in grain boundaries, and it was suggested that one role of Na is to bring O into the boundary and compensate for Se vacancies in the boundaries [142]. Although this is an attractive concept, further studies have not verified a correlation between grain-boundary oxygen and improved performance. Similar improvements may be obtained if Na is supplied to the film in the form of Na_2Se when films are grown on stainless steel. One of the interesting observations about Na in grain boundaries is that its presence is not correlated with a decrease in concentration of other matrix species from the CIGS; so, an obvious pathway to substitution for a matrix element has not been found. However, it has been proposed that a surface phase, $NaInSe_2$, exists and that Na on film surfaces is important in passivating Cu vacancies [143]

Although it may exist on surfaces, no bulk phase of this composition has been observed and the evidence for the surface phase is weak. For example, when Na was shown to diffuse onto CIGS surfaces by vacuum annealing, no corresponding reduction in other species was found.

The electronic effects of Na in CIGS have been studied in single crystals by diffusion of Na into their bulk [144]. Temperature-dependent Hall effect showed that the main effect was to reduce the density of donor states in the material. No apparent doping was found, and no acceptor or donor state was correlated with the amount of Na introduced. Since Na apparently acts through the grain boundaries, it seems likely that it acts either by production of a grain-boundary potential there or by changing the chemical potential of a donor defect in the bulk, thus causing a reduction in that defect and an improved device performance as a result.

In recent years, much interest has focused on the effect of other alkali metals on CIGS devices, sparked primarily by demonstrations of low-temperature-grown, high-efficiency CIGS devices by Tiwari and collaborators [70,145–148]. In particular, Carron *et al.* provides an excellent review of alkali metal treatments for fabricating high-efficiency photovoltaics [70]. Notably, the use of post-deposition treatment of CIGS films with RbF has been shown to reduce the V_{oc} deficit and consequently to improve device efficiency.

The recent result producing a champion CIGS device at Solar Frontier included the use of a Cs post-deposition treatment (see Figure 3.7) [48]. The authors noted that heavier alkali metals produce improved performances, in agreement with Tiwari and collaborators. Segregation of impurities to surfaces and grain boundaries is driven by electronic and size differences, with oversized atoms segregating more strongly to grain boundaries. Larger alkali-metal atoms have a greater size mismatch and are also more distinct electronically; thus, they would be less likely to enter the grains or even surface lattice sites themselves. The report also noted that Cs post-deposition treatment allowed for reduced chemical-bath-deposited CdS heterojunction partner thickness—from 20 to 10 nm—which reduced absorption in the 400–500-nm photon wavelength range. The reduced thickness may be the result of the segregated layer of alkali metal promoting a smoother CdS layer, presumably through enhanced nucleation. Certainly, depositing a thickness of only 10 nm would require very rapid nucleation and a very conformal layer of CdS. Time-resolved photoluminescence (TRPL) on the Cs-treated material showed a carrier lifetime of 124 ns. Such a long lifetime requires that recombination at surfaces and grain boundaries to be very slow. The report did not describe whether the "bare absorber" tested by TRPL had been treated with Cs or coated with CdS. In any case, the conditions used resulted in very recombination-inactive grain boundaries [48].

Currently, considerable research is following up on these studies to establish the details of the effects and how to optimize the resulting devices. None of the alkali metals appears to act within the CIGS grains. As with Na, the behavior appears to be as a surfactant to improve crystal quality or point-defect organization such that recombination is reduced.

The high point-defect density in CIGS makes normal doping of the material difficult. In general, a dopant atom that would shift the Fermi energy would cause a corresponding change in the electrical activity of the existing point defects—for example, by decomposition of defect clusters. A typical reaction might be $(\mathrm{In_{Cu}}^{+2} + 2\mathrm{V_{Cu}}^{-1})^0$ decomposing to $(\mathrm{In_{Cu}}^{+1} + \mathrm{V_{Cu}}^{-1})^0 + \mathrm{V_{Cu}}^{-1}$. This would be expected in n-type material and could result in Fermi-level pinning. For sufficiently high doping levels, it is possible that the ability of the material to respond in this way could be overwhelmed. This is the expectation for surface doping of CIGS by Cd to produce an n-type material near the heterojunction. Similar results are expected with Zn, although this level of doping appears to be more difficult to produce. Nonetheless, Cd-free CIGS photovoltaics with high performance have been produced using a combination of ZnO/Zn(O,S,OH) as the heterojunction partner material [149,150]. Since oxygen is isovalent with selenium, it seems likely that the success of the device is primarily based on Zn doping of the CIGS surface. The mechanism for the operation of these devices remains to be established in detail. Other Cd-free heterojunction partner materials have been explored and performances have typically been lower than with CdS [64]. That the devices work to some extent suggests that at least some surface n-type doping should occur or that electrons are supplied by the heterojunction partner in sufficient quantity to produce a p–n junction.

A number of other impurities have been studied to some extent. O, Cr, Zn, Al, and Se were ion implanted into Cu(In,Ga)Se_2 epitaxial single crystals and the resulting films were studied by temperature-dependent Hall effect [8]. The samples were annealed to remove implant damage and then reanalyzed. The data showed that all samples exhibited dramatic degradation of mobility and increase in carrier concentration after implantation. However, annealing the samples at 525 °C for 25 min resulted in a complete restoration of the pre-implant properties for the samples implanted with Se and Cr. The samples implanted with O, Zn, and Al did not recover their properties significantly on annealing. DC photoconductivity showed no loss of carrier lifetime in the annealed Cr-implanted sample, indicating that Cr is not a recombination center in CIGS. Nonetheless, a study by Pianezzi *et al.* on complete devices by admittance spectroscopy showed that Cr can introduce deep levels in CIGS and reduce device performances [151]. The authors also compared the effects of Ni and Cr on CIGS devices and found that although both degrade the performance of the devices, Ni has a significantly greater effect. This is consistent with the results of other groups.

Secondary-ion mass spectrometry studies of CIGS devices deposited on stainless-steel substrates have shown some evidence of Fe, Cr, and Ni in the resulting films. Comparison of these impurities with the performance of the resulting devices clearly shows evidence that Fe has a detrimental effect on the devices [152,153]. Wuerz *et al.* also showed that SiO_x could be an effective diffusion barrier that prevented Fe ingress into the films.

It is clear that stainless steel can be used successfully as a substrate for CIGS devices based on the commercial products available. For example, MiaSolé produces products regularly on these substrates and held the record for CIGS module performance at one point. It seems important to use a Ni-free stainless steel and to avoid high levels of Fe (above 0.08 ppm) and Cr in the device.

Few detailed studies of impurities in CIGS have been completed. Certainly, it appears that Mo is not a concern. Halides may affect the device but are more likely to have an impact through altering the surface compositions of the grains. It is unlikely that halides have a significant negative effect because many of the alkali post-deposition treatments are based on an alkali halide source material. Other transition metals probably have effects similar to Fe, Cr, and Ni to the extent that they affect the material at all. The alkali metals appear beneficial up to a point. CIGS can be successfully produced in process tools from a variety of materials and in high vacuum or less. Thus, it seems unlikely that the material can be highly sensitive to any impurity, which is expected in a material that contains numerous point defects.

References

[1] A. Rockett, F. Abou-Elfotouh, D. Albin, *et al.*, Structure and chemistry of $CuInSe_2$ for solar cell technology: current understanding and recommendations, *Thin Solid Films*, 237 (1994) 1–11.

[2] W. Setyawan, S. Curtarolo, High-throughput electronic band structure calculations: challenges and tools, *Computational Materials Science*, 49 (2010) 299–312.

[3] R.Z. Chen, C. Persson, Band structure and optical properties of $CuInSe_2$, in: *2014 4th International Conference on Advanced Materials Research, ICAMR 2014*, January 22–23, 2014, Trans Tech Publications, Macau, China, 2014, pp. 254–258.

[4] J.E. Jaffe, A. Zunger, Electronic structure of the ternary chalcopyrite semiconductors $CuAlS_2$, $CuGaS_2$, $CuInS_2$, $CuAlSe_2$, $CuGaSe_2$, and $CuInSe_2$, *Physical Review B*, 28 (1983) 5822–5847.

[5] X.Q. He, T. Paulauskas, P. Ercius, *et al.*, Cd doping at PVD-CdS/$CuInGaSe_2$ heterojunctions, *Solar Energy Materials and Solar Cells*, 164 (2017) 128–134.

[6] A. Rockett, R.W. Birkmire, $CuInSe_2$ for photovoltaic applications, *Journal of Applied Physics*, 70 (1991) 81–97.

[7] V. Lyahovitskaya, S. Richter, F. Frolow, *et al.*, Growth of single $CuInSe_2$ crystals by the traveling heater method and their characterization, *Journal of Crystal Growth*, 197 (1999) 177–185.

[8] D.J. Schroeder, J.L. Hernandez, G.D. Berry, A.A. Rockett, Hole transport and doping states in epitaxial $CuIn_{1-x}Ga_xSe_2$, *Journal of Applied Physics*, 83 (1998) 1519–1526.

[9] E. Arushanov, S. Siebentritt, T. Schedel-Niedrig, M.C. Lux-Steiner, Defect band transport in p-type $CuGaSe_2$, *Journal of Physics: Condensed Matter*, 17 (2005) 2699–2704.

[10] N. Karunagaran, P. Ramasamy, Synthesis, growth and characterization of $AgInSe_2$ single crystals, *Materials Science in Semiconductor Processing*, 40 (2015) 591–595.

[11] A.V. Mudryi, I.V. Bodnar, V.F. Gremenok, I.A. Victorov, A.I. Patuk, I.A. Shakin, Free and bound exciton emission in $CuInSe_2$ and $CuGaSe_2$ single crystals, *Solar Energy Materials and Solar Cells*, 53 (1998) 247–253.

[12] S.M. Wasim, C. Rincon, G. Marin, *et al.*, Effect of structural disorder on the Urbach energy in Cu ternaries, *Physical Review B*, 64 (2001) 195101–195101.

[13] J.T. Heath, J.D. Cohen, W.N. Shafarman, Correlation between deep defect states and device parameters in $CuIn_{1-x}Ga_xSe_2$ photovoltaic devices, in: Conference Record of the Twenty-Ninth IEEE Photovoltaic Specialists Conference 2002 (Cat. No.02CH37361), IEEE, Piscataway, NJ, USA, 2002, pp. 596–599.

[14] S. Siebentritt, Why are kesterite solar cells not 20% efficient? *Thin Solid Films*, 535 (2013) 1–4.

[15] M.A. Mayer, L.B. Ruppalt, D. Hebert, J. Lyding, A.A. Rockett, Scanning tunneling microscopic analysis of Cu(In,Ga)Se_2 epitaxial layers, *Journal of Applied Physics*, 107, 034906 (2010); https://doi.org/10.1063/1.3304919

[16] I. Repins, L. Mansfield, A. Kanevce, *et al.*, Wild band edges: the role of bandgap grading and band-edge fluctuations in high-efficiency chalcogenide devices, in: *2017 IEEE 44th Photovoltaic Specialists Conference (PVSC)*, June 25–30, 2017, IEEE, Piscataway, NJ, USA, 2017, pp. 6.

[17] W. Horig, H. Neumann, H. Sobotta, B. Schumann, G. Keuhn, Optical properties of $CuInSe_2$ thin films, *Thin Solid Films*, 48 (1978) 67–72.

[18] T. Ikari, K. Yoshino, T. Shimizu, *et al.*, Dependence of Cu/In ratio on the optical properties of $CuInSe_2$ epitaxial layers examined by piezoelectric photoacoustic spectroscopy, in: *Ternary and Multinary Compounds. ICTMC-11*, September 8–12, 1997, Institute of Physics Publishing, Salford, UK, 1998, pp. 511–514.

[19] H. Neumann, Optical properties and electronic band structure of $CuInSe_2$, *Solar Cells*, 16 (1986) 317–333.

[20] J. Gonzalez, C. Rincon, Optical absorption and phase transitions in $CuInSe_2$ and $CuInS_2$ single crystals at high pressure, *Journal of Applied Physics*, 65 (1989) 2031–2034.

[21] J.E. Jaffe, A. Zunger, Anion displacements and the band-gap anomaly in ternary ABC_2 chalcopyrite semiconductors, *Physical Review B*, 27 (1983) 5176–5179.

[22] L.M. Mansfield, A. Kanevce, S.P. Harvey, *et al.*, Efficiency increased to 15.2% for ultra-thin Cu(In,Ga)Se_2 solar cells, *Progress in Photovoltaics: Research and Applications*, 26 (2018) 949–954.

[23] M. Schmid, Review on light management by nanostructures in chalcopyrite solar cells, *Semiconductor Science and Technology*, 32 (2017) 043003 (043017 pp.).

[24] V. Gusak, O. Lundberg, E. Wallin, S.O. Katterwe, U. Malm, L. Stolt, Optimization of alkali supply and Ga/(Ga+In) evaporation profile for thin

(0.5 m) CIGS solar cells, in: *2018 IEEE 7th World Conference on Photovoltaic Energy Conversion (WCPEC)*, June 10–15, 2018, IEEE, Piscataway, NJ, USA, 2018, pp. 1641–1644.

[25] S. Marsillac, V. Ranjan, K. Aryal, *et al.*, Toward ultra thin CIGS solar cells, in: *38th IEEE Photovoltaic Specialists Conference, PVSC 2012*, June 3–8, 2012, IEEE, Austin, TX, USA, 2012, pp. 1492–1494.

[26] A. Nikolaeva, M. Krause, J. Marquez, *et al.*, Fluctuations in net doping and lifetime in Cu(In,Ga)Se_2 solar cells, in: *2018 IEEE 7th World Conference on Photovoltaic Energy Conversion (WCPEC)*, June 10–15, 2018, IEEE, Piscataway, NJ, USA, 2018, pp. 2512–2514.

[27] M.A. Contreras, I. Repins, W.K. Metzger, M. Romero, D. Abou-Ras, Se activity and its effect on Cu(In,Ga)Se_2 photovoltaic thin films, *Physica Status Solidi A*, 206 (2009) 1042–1048.

[28] C.H. Lei, A.A. Rockett, I.M. Robertson, N. Papathanasiou, S. Siebentritt, Interface reactions and Kirkendall voids in metal organic vapor-phase epitaxy grown Cu(In,Ga)Se_2 thin films on GaAs, *Journal of Applied Physics*, 100, 114915 (2006); https://doi.org/10.1063/1.2397282.

[29] T. Godecke, T. Haalboom, F. Ernst, Phase equilibria of Cu–In–Se. I. Stable states and nonequilibrium states of the In_2Se_3–Cu_2Se subsystem, *Zeitschrift fur Metallkunde*, 91 (2000) 622–634.

[30] T. Godecke, T. Haalboom, F. Ernst, Phase equilibria of Cu-In-Se. II. The In-In_2Se_3-Cu_2Se-Cu subsystem, *Zeitschrift fur Metallkunde*, 91 (2000) 635–650.

[31] T. Godecke, T. Haalboom, F. Ernst, Phase equilibria of Cu-In-Se. III. The In_2Se_3-Se-Cu_2Se subsystem, *Zeitschrift fur Metallkunde*, 91 (2000) 651–662.

[32] ASM International, *ASM Handbook*, ASM International, Materials Park, OH, USA, 2009.

[33] C. Lei, C.M. Li, A. Rockett, I.M. Robertson, Grain boundary compositions in Cu(InGa)Se_2, *Journal of Applied Physics*, 101, 024909 (2007); https://doi.org/10.1063/1.2426962.

[34] A. Zunger, S.B. Zhang, W. Su-Huai, Revisiting the defect physics in $CuInSe_2$ and $CuGaSe_2$, in: *26th IEEE Photovoltaic Specialists Conference—1997*, September 29–October 3, 1997, IEEE, New York, NY, USA, 1997, pp. 313–318.

[35] S.B. Zhang, S.H. Wei, A. Zunger, Stabilization of ternary compounds via ordered arrays of defect pairs, *Physical Review Letters*, 78 (1997) 4059–4062.

[36] B.J. Stanbery, C.H. Chang, S. Kim, *et al.*, Epitaxial growth of CuAu-ordered $CuInSe_2$ structural polytypes by migration enhanced epitaxy, in: *Self-Organized Processes in Semiconductor Alloys*. Symposium, November 29–December 2, 1999, Materials Research Society, Boston, MA, USA, 2000, pp. 195–200.

[37] C.J. Kiely, R.C. Pond, G. Kenshole, A. Rockett, A TEM study of the crystallography and defect structures of single-crystal and polycrystalline copper indium diselenide, *Philosophical Magazine A*, 63 (1991) 1249–1273.

[38] H.Z. Xiao, L.-C. Yang, A. Rockett, Structural, optical, and electrical properties of epitaxial chalcopyrite $CuIn_3Se_5$ films, *Journal of Applied Physics*, 76 (1994) 1503–1510.
[39] B. Stanbery, Heteroepitaxy and nucleation control for the growth of metal chalcogenides using activated reactant sources, PhD thesis, Chemical Engineering, University of Florida, Gainesville, FL, 2001, pp. 400.
[40] D.S. Su, W. Neumann, R. Hunger, *et al.*, CuAu-type ordering in epitaxial $CuInS_2$ films, *Applied Physics Letters*, 73 (1998) 785–787.
[41] R. Hunger, D. Su, A. Krost, D. Ellmer, H.J. Lewerenz, R. Scheer, Structure of extended defects in epitaxial $CuInS_2$/Si(111), in: *1999 E-MRS Conference, Symposium O: Chalcogenide Semiconductors for Photovoltaics*, June 1–4, 1999, Elsevier, Switzerland, 2000, pp. 437–442.
[42] R. Scheer, R. Klenk, J. Klaer, I. Luck, $CuInS_2$ based thin film photovoltaics, *Solar Energy*, 77 (2004) 777–784.
[43] R. Scheer, Open questions after 20 years of $CuInS_2$ research, *Progress in Photovoltaics: Research and Applications*, 20 (2012) 507–511.
[44] M.H. Bode, M.M. Al-Jassim, J. Tuttle, D. Albin, Microcharacterization of $CuInSe_2$ grown by coevaporation and selenization, in: *Microcrystalline Semiconductors: Materials Science and Devices Symposium*, November 30–December 4, 1992, Materials Research Society, Boston, MA, USA, 1993, pp. 927–932.
[45] D. Schmid, M. Ruckh, F. Grunwald, H.W. Schock, Chalcopyrite/defect chalcopyrite heterojunctions on the basis of $CuInSe_2$, *Journal of Applied Physics*, 73 (1993) 2902–2909.
[46] A.J. Nelson, P.E. Sobol, A.M. Gabor, M.A. Contreras, P. Asoka-Kumar, Correlation of polycrystalline Cu(In,Ga)Se$_2$ device efficiency with homojunction depth and interfacial structure: X-ray photoemission and positron annihilation spectroscopic characterization, Published by the U.S. Department of Commerce NTIS, 1994.
[47] J. Nishinaga, T. Nagai, T. Sugaya, H. Shibata, S. Niki, Single-crystal Cu(In, Ga)Se$_2$ solar cells grown on GaAs substrates, *Applied Physics Express*, 11, 082302 (2018).
[48] T. Kato, W. Jyh-Lih, Y. Hirai, H. Sugimoto, V. Bermudez, Record efficiency for thin-film polycrystalline solar cells up to 22.9% achieved by Cs-treated Cu(In,Ga)(Se,S)$_2$, *IEEE Journal of Photovoltaics*, 9 (2019) 325–330.
[49] H. Hiroi, Y. Iwata, K. Horiguchi, H. Sugimoto, 960-mV open-circuit voltage chalcopyrite solar cell, *IEEE Journal of Photovoltaics*, 6 (2016) 309–312.
[50] D. Liao, A. Rockett, The structure and morphology of (112)-oriented Cu(In, Ga)Se$_2$ epitaxial films, *Journal of Applied Physics*, 104, 094908 (2008); https://doi.org/10.1063/1.3009961.
[51] D.X. Liao, A. Rockett, Effect of surface orientation on the growth and properties of Cu(In,Ga)Se$_2$, in: *Conference Record of the Twenty-Ninth IEEE Photovoltaic Specialists Conference, 2002*, New Orleans, LA, USA, 2002, pp. 515–518, doi: 10.1109/PVSC.2002.1190615.

[52] C. Lei, A. Rockett, I.M. Robertson, W.N. Shafarman, M. Beck, Void formation and surface energies in Cu(InGa)Se_2, *Journal of Applied Physics*, 100, 073518 (2006); https://doi.org/10.1063/1.2357422.

[53] J.E. Jaffe, A. Zunger, Defect-induced nonpolar-to-polar transition at the surface of Cu$InSe_2$, *Journal of Physics and Chemistry of Solids*, 64 (2003) 1547–1552.

[54] S. Marsillac, P.D. Paulson, M.W. Haimbodi, R.W. Birkmire, W.N. Shafarman, High-efficiency solar cells based on Cu(InAl)Se_2 thin films, *Applied Physics Letters*, 81 (2002) 1350–1352.

[55] W.N. Shafarman, S. Marsillac, T. Minemoto, P.D. Paulson, R.W. Birkmire, Critical issues for Cu(InAl)Se_2 thin film solar cells, in: *Proceedings of 3rd World Conference on Photovoltaic Energy Conversion*, May 12–16, 2003, Arisumi Printing Inc, Osaka, Japan, 2003, pp. 2869–2873.

[56] S. Siebentritt, Wide gap chalcopyrites: material properties and solar cells, *Thin Solid Films*, 403–404 (2002) 1–8.

[57] D. Haneman, Properties and applications of copper indium diselenide, *CRC Critical Reviews in Solid State and Materials Sciences*, 14 (1988) 377–413.

[58] S.H. Wei, S.B. Zhang, A. Zunger, Effects of Ga addition to Cu$InSe_2$ on its electronic, structural, and defect properties, *Applied Physics Letters*, 72 (1998) 3199–3201.

[59] W.A. Harrison, *Electronic Structure and the Properties of Solids: The Physics of the Chemical Bond*, Freeman, San Francisco, 1980.

[60] W.N. Shafarman, Novel approaches to wide bandgap Cu$InSe_2$ based absorbers, 2011, 45 p, Contract FG36-08GO18019 Sponsored by Department of Energy, Washington, DC.

[61] C.S. Schnohr, Compound semiconductor alloys: from atomic-scale structure to bandgap bowing, *Applied Physics Reviews*, 2 (2015) 031304 (031343 pp.).

[62] I. Repins, M.A. Contreras, B. Egaas, *et al.*, 19.9%-efficient ZnO/CdS/ CuInGaSe_2 solar cell with 81.2% fill factor, *Progress in Photovoltaics: Research and Applications*, 16 (2008) 235–239.

[63] A. Aquino, A. Rockett, Chemical reactions at CdS heterojunctions with Cu$InSe_2$, *Journal of Vacuum Science & Technology A*, 31, 021202 (2013); https://doi.org/10.1116/1.4775341.

[64] D. Hariskos, S. Spiering, M. Powalla, Buffer layers in Cu(In,Ga)Se_2 solar cells and modules, *Thin Solid Films*, 480–481 (2005) 99–109.

[65] M.A. Green, Y. Hishikawa, E.D. Dunlop, *et al.*, Solar cell efficiency tables (Version 53), *Progress in Photovoltaics*, 27 (2019) 3–12.

[66] A. Chirila, P. Reinhard, F. Pianezzi, *et al.*, Potassium-induced surface modification of Cu(In,Ga)Se_2 thin films for high-efficiency solar cells, *Nature Materials*, 12 (2013) 1107–1111.

[67] J.A. Del Cueto, Stability of CIS/CIGS modules at the outdoor test facility over two decades. Proceedings of the 2008 33rd IEEE Photovoltaic Specialists Conference, IEEE, Piscataway, NJ, USA, 2008, pp. 18.

[68] T.J. Silverman, L. Mansfield, I. Repins, S. Kurtz, Damage in monolithic thin-film photovoltaic modules due to partial shade, in: *2017 IEEE 44th Photovoltaic Specialists Conference (PVSC)*, June 25–30, 2017, IEEE, Piscataway, NJ, USA, 2017, pp. 6.

[69] E. Palmiotti, S. Johnston, A. Gerber, *et al.*, Identification and analysis of partial shading breakdown sites in $CuIn_xGa_{1-x}Se_2$ modules, *Solar Energy*, 161 (2018) 1–5.
[70] R. Carron, S. Nishiwaki, T. Feurer, *et al.*, Advanced alkali treatments for high-efficiency $Cu(In,Ga)Se_2$ solar cells on flexible substrates, *Advanced Energy Materials*, 9 (2019) 1900408.
[71] F.S. Hasoon, H.A. Al-Thani, The formation of the $MoSe_2$ layer at Mo/CIGS interface and its effect on the CIGS device performance, in: *44th IEEE Photovoltaic Specialist Conference, PVSC 2017*, June 25–30, 2017, IEEE, Washington, DC, USA, 2017, pp. 1–5.
[72] S. Nishiwaki, N. Kohara, T. Negami, T. Wada, $MoSe_2$ layer formation at Cu (In,Ga)Se_2/Mo interfaces in high efficiency $Cu(In_{1-x}Ga_x)Se_2$ solar cells, *Japanese Journal of Applied Physics, Part 2 (Letters)*, 37 (1998) 71–73.
[73] M. Bodegard, K. Granath, L. Stolt, A. Rockett, The behaviour of Na implanted into Mo thin films during annealing, *Solar Energy Materials and Solar Cells*, 58 (1999) 199–208.
[74] H. Sugimoto, T. Yagioka, M. Nagahashi, *et al.*, Achievement of over 17% efficiency with 3030 cm^2-sized $Cu(InGa)(SeS)_2$ submodules, in: *37th IEEE Photovoltaic Specialists Conference, PVSC 2011*, June 19–24, 2011, IEEE, Seattle, WA, USA, 2011, pp. 003420–003423.
[75] T. Nakada, Nano-structural investigations on Cd-doping into $Cu(In,Ga)Se_2$ thin films by chemical bath deposition process. Thin Solid Films, v. 361–362, pp 346–352 (2000).
[76] D. Liao, A. Rockett, Cu depletion at the $CuInSe_2$ surface, *Applied Physics Letters*, 82 (2003) 2829–2831.
[77] A. Kylner, J. Lindgren, L. Stolt, Impurities in chemical bath deposited CdS films for $Cu(In,Ga)Se_2$ solar cells and their stability, *Journal of the Electrochemical Society*, 143 (1996) 2662–2669.
[78] X.Q. He, P. Ercius, J. Bailey, *et al.*, Cu rich domains and secondary phases in PVD-CdS/PVD-$CuIn_{1-x}Ga_xSe_2$ heterojunctions, in: *2015 IEEE 42nd Photovoltaic Specialist Conference (PVSC)*, 2015.
[79] C.-H. Lei, C.-M. Li, A. Rockett, I.M. Robertson, W. Shafarman, Application of advanced microstructural and microchemical microscopy techniques to chalcopyrite solar cells, in: *Thin-Film Compound Semiconductor Photovoltaics*, Materials Research Society, 2005, pp. 93–103.
[80] A. Urbaniak, M. Igalson, S. Siebentritt, Kinetics of charge trapping and emission in CIGS solar cells, in: Volume 1012, Symposium Y – Thin-Film Compound Semiconductor Photovoltaics, 1012-Y12-16. Materials Research Society, Warrendale, PA, USA, 2007, pp. 385–392.
[81] J. Krustok, J. Raudoja, J.H. Schon, M. Yakushev, H. Collan, The role of deep donor–deep acceptor complexes in CIS-related compounds, in: *1999 E-MRS Conference, Symposium O: Chalcogenide Semiconductors for Photovoltaics*, June 1–4, 1999, Elsevier, Switzerland, 2000, pp. 406–410.
[82] R. Noufi, R. Axton, C. Herrington, S.K. Deb, Electronic properties versus composition of thin films of $CuInSe_2$, *Applied Physics Letters*, 45 (1984) 668–670.

[83] H.R. Moutinho, D.J. Dunlavy, L.L. Kazmerski, R.K. Ahrenkiel, F.A. Abou-Elfotouh, Effects of processing on the electronic defect levels in $CuInSe_2$, in: *Proceedings of 23rd IEEE Photovoltaic Specialists Conference*, May 10–14, 1993, IEEE, New York, NY, USA, 1993, pp. 572–576.

[84] D. Liao, A. Rockett, Cd doping at the $CuInSe_2$/CdS heterojunction, *Journal of Applied Physics*, 93 (2003) 9380–9382.

[85] T. Maeda, T. Wada, First-principles studies on Cd doping in $CuInSe_2$ and related compounds during chemical bath deposition of CdS buffer layer, *Japanese Journal of Applied Physics*, 52 (2013) 061201.

[86] M. Igalson, C. Platzer-Bjorkman, The influence of buffer layer on the transient behavior of thin film chalcopyrite devices, *Solar Energy Materials and Solar Cells*, 84 (2004) 93–103.

[87] F. Abou-Elfotouh, D.J. Dunlavy, L.L. Kazmerski, D. Albin, K.J. Bachman, R. Menner, Photoluminescence spectra of some ternary and quaternary chalcopyrite semiconductors. *Journal of Vacuum Science and Technology* v. 6(3), 1515 (1988). https://doi.org/10.1116/1.575352

[88] S. Siebentritt, M. Igalson, C. Persson, S. Lany, The electronic structure of chalcopyrites—bands, point defects and grain boundaries, *Progress in Photovoltaics: Research and Applications*, 18 (2010) 390–410.

[89] V. Alberts, S. Zweigart, J.H. Schon, H.W. Schock, E. Bucher, Characterization of polycrystalline Cu(In,Ga)Se_2 thin films produced by rapid thermal processing, *Japanese Journal of Applied Physics, Part 1 (Regular Papers, Short Notes & Review Papers)*, 36 (1997) 5033–5039.

[90] J.T. Heath, J.D. Cohen, W.N. Shafarman, Bulk and metastable defects in $CuIn_{1-x}Ga_xSe_2$ thin films using drive-level capacitance profiling, *Journal of Applied Physics*, 95 (2004) 1000–1010.

[91] S.M. Wasim, Transport properties of $CuInSe_2$, *Solar Cells*, 16 (1986) 289–316.

[92] I. Dirnstorfer, D.M. Hofmann, D. Meister, B.K. Meyer, W. Riedl, F. Karg, Postgrowth thermal treatment of Cu(In,Ga)Se_2: characterization of doping levels in In-rich thin films, *Journal of Applied Physics*, 85 (1999) 1423–1428.

[93] S. Siebentritt, N. Papathanasiou, M.C. Lux-Steiner, Potential fluctuations in compensated chalcopyrites, *Physica B: Condensed Matter*, v. 376–377, 2006, pp. 831–833.

[94] N. Rega, S. Siebentritt, J. Albert, M.L. Steiner, Photoluminescence of Cu $(In_{1-x},Ga_x)Se_2$ epitaxial thin films grown by MOVPE, in: *Compound Semiconductor Photovoltaics. Symposium*, April 22–25, 2003, Materials Research Society, San Francisco, CA, USA, 2003, pp. 183–188.

[95] S. Niki, Y. Makita, A. Yamada, *et al.*, Photoluminescence properties of $CuInSe_2$ grown by molecular beam epitaxy, in: *7th International Photovoltaic Science and Engineering Conference*, November 22–26, 1994, The Netherlands, 1994, pp. 141–147.

[96] M. Igalson, P. Zabierowski, Electron traps in Cu(In,Ga)Se_2 absorbers of thin film solar cells studied by junction capacitance techniques, *Opto-Electronics Review*, 11 (2003) 261–267.

[97] J.D. Cohen, J.T. Heath, W.N. Shafarman, New junction capacitance methods for the study of defect distributions and carrier properties in the copper indium diselenide alloys, in: *Proceedings of the Materials Research Society Symposum B, Compound Semiconductor Photovoltaics*, ed. by R. Noufi, D. Cahen, W. Shafarman, and L. Stolt, V. 763. Materials Research Society, Warrendale, PA, USA, 2003, pp. 429–440.

[98] P.K. Paul, K. Aryal, S. Marsillac, T.J. Grassman, S.A. Ringel, A.R. Arehart, Identifying the source of reduced performance in 1-stage-grown Cu(In, Ga)Se_2 solar cells, in: *2016 IEEE 43rd Photovoltaic Specialists Conference (PVSC)*, June 5–10, 2016, IEEE, Piscataway, NJ, USA, 2016, pp. 3641–3644.

[99] P.K. Paul, D.W. Cardwell, C.M. Jackson, *et al.*, Direct nm-scale spatial mapping of traps in CIGS, in: *2015 IEEE 42nd Photovoltaic Specialists Conference (PVSC)*, June 14–19, 2015, IEEE, Piscataway, NJ, USA, 2015, pp. 3.

[100] M.Y. Dong, X.F. Chen, H. Deng, Optical properties of $CuInSe_2$ film semiconductor studied with photothermal deflection spectroscopy, in: *Solar Optical Materials XVI*, July 22, 1999, SPIE-International Society for Optical Engineering., USA, 1999, pp. 131–137.

[101] H.W. Schock, Polycrystalline compound semiconductor thin films in solar cells, in: *Polycrystalline Semiconductors, Grain Boundaries and Interfaces. Proceedings of the International Symposium*, August 29–September 2, 1988, Springer-Verlag, Malente, West Germany, 1989, pp. 246–256.

[102] M. Stutzmann, Laser characterization of semiconductors, in: *1st International Symposium on Semiconductor Processing and Characterization with Lasers*, April 18–20, 1994, Switzerland, 1995, pp. 141–152.

[103] J.T. Heath, J.D. Cohen, W.N. Shafarman, D.X. Liao, A.A. Rockett, Effect of Ga content on defect states in $CuIn_{1-x}Ga_xSe_2$ photovoltaic devices, *Applied Physics Letters*, 80 (2002) 4540.

[104] S. Niki, R. Suzuki, S. Ishibashi, *et al.*, Anion vacancies in $CuInSe_2$, in: *European Materials Research Society Symposium N, Thin Film Chalcogenide Materials, Thin Solid Films*, v. 387(1–2), pp. 129–134 (2001).

[105] R. Suzuki, T. Ohdaira, S. Ishibashi, *et al.*, Characterization of intrinsic defects in $CuInSe_2$ films by monoenergetic positron beams, in: *Ternary and Multinary Compounds. Proceedings of the 11th International Conference, ICTMC-11*, September 8–12, 1997, Institute of Physics Publishing, Salford, UK, 1998, pp. 757–760.

[106] C.P. Thompson, S. Hegedus, W. Shafarman, D. Desai, Temperature dependence of V_{OC} in CdTe and Cu(InGa)$(SeS)_2$-based solar cells, in: *2008 33rd IEEE Photovoltaic Specialists Conference*, San Diego, CA, USA, 2008, pp. 1–6, doi: 10.1109/PVSC.2008.4922548.

[107] R. Scheer, Activation energy of heterojunction diode currents in the limit of interface recombination, *Journal of Applied Physics*, 105, 104505 (2009); https://doi.org/10.1063/1.3126523.

[108] P.K. Paul, K. Aryal, S. Marsillac, T.J. Grassman, S.A. Ringel, A.R. Arehart, Identifying the source of reduced performance in 1-stage-grown Cu(In,Ga)Se_2 solar cells, in: *44th IEEE Photovoltaic Specialist Conference, PVSC 2017*, June 25–30, 2017, IEEE, Washington, DC, USA, 2017, pp. 1–4.

[109] M. Igalson, M. Cwil, M. Edoff, Metastabilities in the electrical characteristics of CIGS devices: experimental results vs theoretical predictions, *Thin Solid Films*, 515 (2007) 6142–6146.

[110] D. Przado, M. Igalson, R. Bacewicz, M. Edoff, The influence of metastabilities on the luminescence in the Cu(In,Ga)Se_2 solar cells, *Acta Physica Polonica A*, 112 (2007) 183–189.

[111] M. Cwil, M. Igalson, P. Zabierowski, S. Siebentritt, Charge and doping distributions by capacitance profiling in Cu(In,Ga)Se_2 solar cells, *Journal of Applied Physics*, 103 (2008) 063701–063701.

[112] M. Igalson, P. Zabierowski, D. Przado, A. Urbaniak, M. Edoff, W.N. Shafarman, Understanding defect-related issues limiting efficiency of CIGS solar cells, *Solar Energy Materials and Solar Cells*, 93 (2009) 1290–1295.

[113] S. Lany, A. Zunger, Light- and bias-induced metastabilities in Cu(In,Ga)Se_2 based solar cells caused by the (V_{Se}–V_{Cu}) vacancy complex, *Journal of Applied Physics*, 100 (2006) 113725–113721.

[114] U. Rau, D. Braunger, H.-W. Schock, Air-annealing effects on polycrystalline Cu(In,Ga)Se_2 heterojunctions [solar cells], in: *Diffusion and Defect Data Part B Proceedings of the Fifth International Conference on Polycrystalline Semiconductors (POLYSE '98)*, September 13–18, 1998, *Solid State Phenomena* v. 67–68, pp. 409–414 (1999).

[115] F.S. Hasoon, Y. Yan, H. Althani, *et al.*, Microstructural properties of Cu(In,Ga)Se_2 thin films used in high-efficiency devices, *Thin Solid Films*, 387 (2001) 1–5.

[116] S. Siebentritt, S. Sadewasser, M. Wimmer, C. Leendertz, T. Eisenbarth, M.C. Lux-Steiner, Evidence for a neutral grain-boundary barrier in chalcopyrites, *Physical Review Letters*, 97 (2006) 146601–146601.

[117] C. Lei, M. Duch, I.M. Robertson, A. Rockett, Effects of solution-grown CdS on Cu(InGa)Se_2 grain boundaries, *Journal of Applied Physics*, 108, 114908 (2010); https://doi.org/10.1063/1.3512966.

[118] C.-S. Jiang, R. Noufi, K. Ramanathan, *et al.*, Does the local built-in potential on grain boundaries of Cu(In,Ga)Se_2 thin films benefit photovoltaic performance of the device? *Applied Physics Letters*, 85 (2004) 2625–2627.

[119] C.S. Jiang, I.L. Repins, L.M. Mansfield, *et al.*, Electrical conduction channel along the grain boundaries of Cu(In,Ga)Se_2 thin films, *Applied Physics Letters*, 102, 253905 (2013); https://doi.org/10.1063/1.4812827.

[120] C.S. Jiang, B. To, S. Glynn, H. Mahabaduge, T. Barnes, M.M. Al-Jassim, Recent progress in nanoelectrical characterizations of CdTe and Cu(In,Ga)Se_2, in: *44th IEEE Photovoltaic Specialist Conference, PVSC 2017*, June 25–30, 2017, IEEE, Washington, DC, USA, 2017, pp. 1–6.

[121] A.J. Hall, D. Hebert, C. Lei, A. Rockett, S. Siebentritt, Epitaxial growth of very large grain bicrystalline Cu(In,Ga)Se_2 thin films by a hybrid sputtering method, *Journal of Applied Physics*, 103, 083540 (2008); https://doi.org/10.1063/1.2907446.

[122] A. Stokes, M. Al-Jassim, D. Diercks, B. Gorman, Alkali segregation and matrix concentrations in thin film Cu(In,Ga)Se_2 at targeted interfaces characterized in 3-D at the nanoscale, in: *2015 IEEE 42nd Photovoltaic Specialists Conference (PVSC)*, June 14–19, 2015, IEEE, Piscataway, NJ, USA, 2015, pp. 1–4.

[123] D.X. Liao, A. Rockett, Cu depletion at the $CuInSe_2$ surface, *Applied Physics Letters*, 82 (2003) 2829–2831.

[124] C. Persson, A. Zunger, Anomalous grain boundary physics in polycrystalline $CuInSe_2$: the existence of a hole barrier, *Physical Review Letters*, 26(1), p. 266401 (2003).

[125] F.S. Hasoon, Y. Yan, K.M. Jones, *et al.*, Microstructural properties of the surface of Cu(In,Ga)Se_2 thin films, in: *Proceedings of 28th IEEE Photovoltaic Specialists Conference*, September 15–22, 2000, IEEE, Piscataway, NJ, USA, 2000, pp. 513–516.

[126] C.-M. Li, C.-H. Lei, I.M. Robertson, A. Rockett, Microstructural and microchemical analysis of chalcopyrite Cu(In,Ga)Se_2 films, in: *MRS Proceedings: Compound Semiconductor Photovoltaics*, April 22–25, 2003, Materials Research Society, San Francisco, CA, USA, 2003, pp. 169–174.

[127] P. Fons, S. Niki, A. Yamada, M. Nishitani, T. Wada, T. Kurafuji, The effects of KCN etching on Cu-rich epitaxial $CuInSe_2$ thin films, in: *Thin Films for Photovoltaic and Related Device Applications*. Symposium, April 8–11, 1996, Materials Research Society, San Francisco, CA, USA, 1996, pp. 213–218.

[128] S. Siebentritt, L. Gutay, D. Regesch, Y. Aida, V. Depredurand, Why do we make Cu(In,Ga)Se_2 solar cells non-stoichiometric? *Solar Energy Materials and Solar Cells*, 119 (2013) 18–25.

[129] H. Elanzeery, F. Babbe, M. Melchiorre, A. Zelenina, S. Siebentritt, Potassium fluoride ex situ treatment on both Cu-rich and Cu-poor $CuInSe_2$ thin film solar cells, *IEEE Journal of Photovoltaics*, 7 (2017) 684–689.

[130] V. Depredurand, Y. Aida, J. Larsen, T. Eisenbarth, A. Majerus, S. Siebentritt, Surface treatment of CIS solar cells grown under Cu-excess, in: *2011 37th IEEE Photovoltaic Specialists Conference (PVSC 2011)*, June 19–24, 2011, IEEE, Piscataway, NJ, USA, 2011, pp. 000337–000342.

[131] T. Bertram, V. Depredurand, S. Siebentritt, In-Se surface treatment of Cu-rich grown $CuInSe_2$, in: *2014 IEEE 40th Photovoltaic Specialists Conference (PVSC)*, June 8–13, 2014, IEEE, Piscataway, NJ, USA, 2014, pp. 3633–3636.

[132] Y. Aida, V. Depredurand, J.K. Larsen, *et al.*, Cu-rich $CuInSe_2$ solar cells with a Cu-poor surface, *Progress in Photovoltaics: Research and Applications*, 23 (2015) 754–764.

[133] F. Babbe, L. Choubrac, S. Siebentritt, Quasi Fermi level splitting of Cu-rich and Cu-poor Cu(In,Ga)Se_2 absorber layers, *Applied Physics Letters*, 109 (2016) 082105.

[134] L. Stolt, J. Hedstrom, J. Kessler, M. Ruckh, K.-O. Velthaus, H.-W. Schock, ZnO/CdS/$CuInSe_2$ thin-film solar cells with improved performance, *Applied Physics Letters*, 62 (1993) 597–599.

[135] B.J. Stanbery, S. Kincal, S. Kim, *et al.*, Role of sodium in the control of defect structures in CIS, in: *28th IEEE Photovoltaic Specialists Conference, PVSC 2000*, September 15–22, 2000, IEEE, Anchorage, AK, USA, 2000, pp. 440–445.

[136] A. Rockett, The effect of Na in polycrystalline and epitaxial single-crystal $CuIn_{1-x}Ga_xSe_2$, *Thin Solid Films*, 480 (2005) 2–7.

[137] A. Rockett, K. Granath, S. Asher, *et al.*, Na incorporation in Mo and $CuInSe_2$ from production processes, *Solar Energy Materials and Solar Cells*, 59 (1999) 255–264.

[138] A. Rockett, J.S. Britt, T. Gillespie, *et al.*, Na in selenized Cu(In,Ga)Se_2 on Na-containing and Na-free glasses: distribution, grain structure, and device performances, *Thin Solid Films*, 372 (2000) 212–217.

[139] W.K. Batchelor, M.E. Beck, R. Huntington, *et al.*, Substrate and back contact effects in CIGS devices on steel foil, in: *Conference Record of the Twenty-Ninth IEEE Photovoltaic Specialists Conference 2002*, 2002, pp. 716–719.

[140] A. Rockett, M. Bodegard, K. Granath, L. Stolt, Na incorporation and diffusion in $CuIn_{1-x}Ga_xSe_2$, in: *Conference Record of the Twenty Fifth IEEE Photovoltaic Specialists Conference—1996*, 1996, pp. 985–987.

[141] R.V. Forest, B.E. McCandless, X.Q. He, *et al.*, Diffusion of sodium in single crystal $CuInSe_2$, *Journal of Applied Physics*, 121, 245102 (2017); https://doi.org/10.1063/1.4986635.

[142] V. Lyahovitskaya, Y. Feldman, K. Gartsman, H. Cohen, C. Cytermann, D. Cahen, Na effects on $CuInSe_2$: distinguishing bulk from surface phenomena, *Journal of Applied Physics*, 91 (2002) 4205–4212.

[143] W. Su-Huai, S.B. Zhang, A. Zunger, Effects of Na on the electrical and structural properties of $CuInSe_2$, *Journal of Applied Physics*, 85 (1999) 7214–7218.

[144] D.J. Schroeder, A.A. Rockett, Electronic effects of sodium in epitaxial $CuIn_{1-x}Ga_xSe_2$, *Journal of Applied Physics*, 82 (1997) 4982–4985.

[145] P. Reinhard, A. Chirila, P. Blosch, *et al.*, Review of progress toward 20% efficiency flexible CIGS solar cells and manufacturing issues of solar modules, *IEEE Journal of Photovoltaics*, 3 (2013) 572–580.

[146] D. Rudmann, F.-J. Hang, M. Kaelin, H. Zogg, A.N. Tiwari, G. Bilger, Low temperature growth of CIGS thin films for flexible solar cells, in: *II–VI Compound Semiconductor Photovoltaic Materials*, April 16–20, 2001, Materials Research Society, San Francisco, CA, USA, 2001, pp. 381–386.

[147] E. Avancini, R. Carron, T.P. Weiss, *et al.*, Effects of rubidium fluoride and potassium fluoride postdeposition treatments on Cu(In,Ga)Se_2 thin films and solar cell performance, *Chemistry of Materials*, 29 (2017) 9695–9704.

[148] P. Reinhard, B. Bissig, F. Pianezzi, *et al.*, Features of KF and NaF post-deposition treatments of Cu(In,Ga)Se_2 absorbers for high efficiency thin film solar cells, *Chemistry of Materials*, 27 (2015) 5755–5764.

[149] K.F. Tai, R. Kamada, T. Yagioka, T. Kato, H. Sugimoto, From 20.9 to 22.3% $Cu(In,Ga)(S,Se)_2$ solar cell: reduced recombination rate at the heterojunction and the depletion region due to K-treatment, *Japanese Journal of Applied Physics*, 56 (2017) 08MC03.
[150] H. Sugimoto, High efficiency and large volume production of CIS-based modules, in: *40th IEEE Photovoltaic Specialist Conference, PVSC 2014*, June 8–13, 2014, IEEE, Denver, CO, USA, 2014, pp. 2767–2770.
[151] F. Pianezzi, S. Nishiwaki, L. Kranz, *et al.*, Influence of Ni and Cr impurities on the electronic properties of $Cu(In,Ga)Se_2$ thin film solar cells, *Progress in Photovoltaics: Research and Applications*, 23 (2015) 892–900.
[152] R. Wuerz, A. Eicke, F. Kessler, F. Pianezzi, Influence of iron on the performance of CIGS thin-film solar cells, *Solar Energy Materials and Solar Cells*, 130 (2014) 107–117.
[153] R. Wuerz, A. Eicke, M. Frankenfeld, *et al.*, CIGS thin-film solar cells on steel substrates, *Thin Solid Films*, 517 (2009) 2415–2418.

Chapter 4
Perovskite solar cells

Fei Zhang[1] and Kai Zhu[1]

4.1 Overview of perovskite solar cells

Perovskite solar cells (PSCs) represent an emergent photovoltaic (PV) technology, and recently they have been considered a potential economically and environmentally feasible renewable technology option to compete with traditional solar cell technologies in addressing global challenges within the areas of energy generation and climate change [1]. The synthesis of organometal halide perovskites, reported in the 1970s [2,3], represents a major milestone in the discovery of organic–inorganic hybrid materials. The structural understanding, solution processing, and optoelectronic properties of this family of hybrid materials were further developed in the 1990s [4], but these materials did not find widespread interest until the first few reports of using methylammonium lead triiodide ($MAPbI_3$) in solar cells in 2009–2012 [5–7]. Miyasaka *et al.* pioneered the first $MAPbI_3$ and $MAPbBr_3$ PSCs, with a resulting power conversion efficiency (PCE) of only 3.81% in an iodide-based liquid-electrolyte configuration. The disadvantage of liquid-electrolyte-based PSCs is that the perovskite nanocrystals will be dissolved or decomposed in the electrolyte, resulting in rapid cell degradation within a few minutes [8]. A promising strategy to solve this problem is to use a solid-state hole-transport layer (HTL) to avoid electrolyte leakage and corrosion from hole-transporting materials. Park and colleagues first employed a solid-state HTL—2,2′,7,7′-tetrakis(*N*,*N*-di-*p*-methoxyphenylamine)-9,9′-spirobifluorene (spiro-OMeTAD)—as an alternative to liquid electrolyte to give a PCE of 9.7% with hundreds of hours of storage stability [6]. In the same period, Snaith and others also reported spiro-OMeTAD-based solid-state PSCs with PCEs >10% [7]. This pioneer work triggered an explosive research effort worldwide to improve the device architecture and optimize fabrication [9–22]. At present, the certified record PCE for single-junction PSCs has reached to 24.2% [23]. Due to the rapid increase in PCEs within a short period of time, as shown in Figure 4.1, PSCs have been considered a contender in opening up a new and promising avenue for next-generation PV development.

[1]Chemistry and Nanoscience Center, National Renewable Energy Laboratory, Golden, Colorado, USA

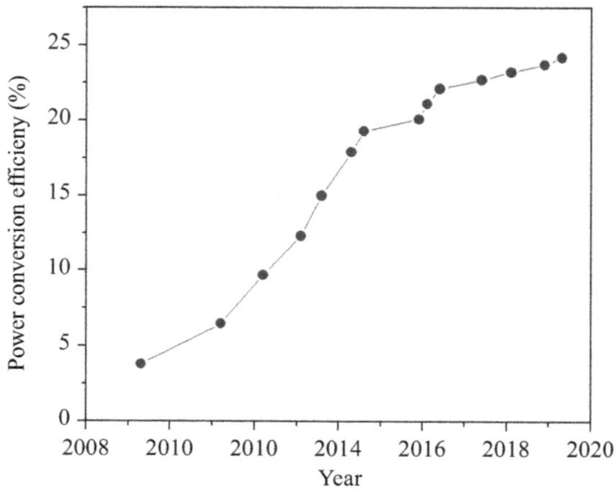

Figure 4.1. Progress of PCE for single-junction PSCs

In the following sections of this chapter, we first discuss the structural and optoelectronic properties as well as the defect tolerance of halide perovskites for solar cell applications. We then compare the various common device architectures for PSCs and discuss several perovskite fabrication strategies, including solution deposition and vacuum processes, that are key to preparing high-quality perovskite thin films for high-performance PSCs. Many research and development (R&D) challenges exist that must be addressed to ready PSCs for practical applications. We review key issues on stability (moisture, thermal, light, and chemical compatibility), material toxicity (Pb and Pb-free PSCs), and scaling up (material, device architecture, and coating approach selection). A key focus area for further PSC development is characterization. Since perovskite behaves differently than conventional semiconductors, specific characterization protocols must be established to reliably evaluate the progress among different research groups. Stability characterization needs special attention given the complexity of perovskite in response to various external stress factors. Finally, we provide an outlook on the research trends in PSC development toward commercialization.

4.2 Structural and optoelectronic properties of perovskites

4.2.1 Crystal structure

The standard organic–inorganic hybrid halide perovskites have a general structure of ABX_3, which is similar to the crystal structure of $CaTiO_3$. For the halide perovskites, a divalent metal cation (B site) is coordinated to six halide anions (X site) to form a BX_6 octahedral framework. Twelve large monovalent cations (A site) occupy the centers of four BX_6 octahedra, as shown in Figure 4.2(a). In general, the A-site

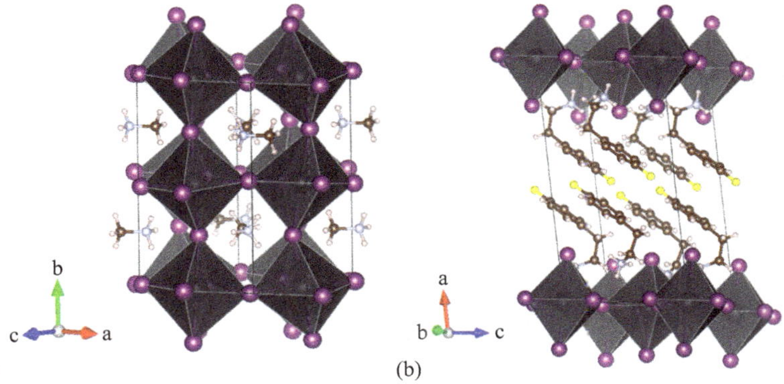

(a) (b)

Figure 4.2. Crystal structure of normal (a) three-dimensional perovskite $CH_3NH_3PbI_3$ [24] and (b) two-dimensional perovskite $(F\text{-}PEA)_2PbI_4$ [25]

monovalent cations could be either organic cations [e.g., $CH_3NH_3^+$ (methylammonium, MA^+) and $CH(NH_2)_2^+$ (formamidinium, FA^+)] or inorganic cations (e.g., Cs^+ and Rb^+); the B-site divalent metal cations are normally Pb^{2+} and/or Sn^{2+}; and the X-site halide anions are normally one or more from I^-, Br^-, and Cl^-. The standard halide perovskite has a three-dimensional (3D) crystal structure, which is shown to have good light absorption and charge-transport properties. In recent years, two-dimensional (2D) halide perovskites have also quickly become a promising perovskite structure for solar cell applications [26]. The 2D perovskites consist of alternating sheets of the corner-sharing metal halide octahedra interspersed with monolayers or bilayers of bulky organic cations. One advantage of 2D perovskite is the improved stability against moisture due to the steric hindrance for moisture adsorption or intrusion from the large spacer layer. Although the lateral extent of 2D perovskite is confined by the size of the 2D metal halide lattice, the large organic A-site cations can be arbitrarily long, enabling the placement of large, high-aspect ratio cations such as those based on aliphatic or aromatic groups. Furthermore, the geometry of the 2D octahedral arrangement typically involves a BX_4^{2-} inorganic repeat unit (as opposed to a BX_3^- unit in the 3D structure); therefore, the negative charge associated with the extra anion must be balanced by an additional positive charge. For the most commonly encountered layered lead halide perovskites such as those based on 4-fluorophenethylammonium $[F - C_6H_5(CH_2)_2NH_3^+$ or $F\text{-}PEA^+]$ [25], a bilayer of monovalent cations forms between two adjacent lead halide sheets, creating a van der Waals gap between them (Figure 4.2(b)); such compositions possess A_2BX_4 stoichiometry [27]. Alternatively, each pair of cations in the above may be replaced by a single divalent cation with tethering groups at either end to connect to adjacent halide sheets, leading to ABX_4 stoichiometry. In this chapter, we mainly focus on the standard 3D perovskites.

For 3D perovskites, each of the A, B, and X sites can consist of multiple elements, leading to flexibility in tuning the structural and optoelectronic properties of perovskites. A simple guide, based on the sizes of the elements involved in forming perovskite, can be used to evaluate whether certain compositions can form a stable perovskite structure. The restrictions based on consideration of ionic sizes are generally expressed in terms of the Goldschmidt tolerance factor, t, given in terms of the ionic radii r_A, r_B, and r_X [28]:

$$t = \frac{r_A + r_X}{\sqrt{2}(r_B + r_X)}$$

Empirically, the 3D perovskite structure is favored for values of t between 0.8 and 1.0 [29]. Also, of importance is the octahedral factor, $\mu = \frac{r_B}{r_X}$, which assesses whether the B-site atoms will prefer an octahedral coordination of X-site atoms (as opposed to favoring larger or smaller coordination numbers); this condition is satisfied for values of μ between 0.4 and 0.9 [30].

4.2.2 Optoelectronic properties

Organic–inorganic halide perovskites (e.g., $MAPbX_3$-based materials) generally exhibit attractive optical and electrical properties. The p-p transition and direct bandgap result in a higher absorption coefficient (α) than GaAs, as shown in Figure 4.3(a) [31]. The high α value ($>10^5$ cm^{-1}) provides great potential to fully utilize photon energy higher than the bandgap and delivers high short-circuit photocurrent density (J_{sc}) from thin-film (e.g., 300 nm) devices. Compared with CIGS-, GaAs-, and Si-based solar materials, $MAPbX_3$-based materials could achieve the highest PCE for a given thickness, as shown in Figure 4.3(b). Moreover, the high open-circuit voltage (V_{oc}) value was achieved on $MAPbI_3$ devices relative to its optical bandgap (~1.55 eV). The small V_{oc} deficit of the $MAPbI_3$ device can be attributed to fewer deep defect states and efficient interface contacts. From the optical measurement, the sharp absorption edge measured from photothermal deflection spectroscopy and Fourier-transform photocurrent spectroscopy on the $MAPbI_3$ material shows the well-ordered microstructure and negligible deep states. These features demonstrated with $MAPbI_3$ have been found as general characteristics for a variety of perovskite compositions.

With regard to the electrical property, perovskite materials have shown efficient ambipolar carrier transport behavior and long carrier lifetime. Through electron-beam-induced current (EBIC) measurement on the perovskite device, the carriers can be collected on both electrodes when the electron beam is focused on either side of the perovskite layer, which is consistent with the good ambipolar transport behavior observed in $MAPbX_3$-based materials [34]. Also, diffusion length of more than 1 μm and carrier lifetime of more than hundreds of nanoseconds were reported during the early stage of perovskite research [35]. These parameters were further improved greatly with advances in developing perovskite fabrication and defect passivation strategies. These results point to a special general characteristic of perovskites: the non-radiative recombination process is limited in polycrystalline perovskite thin films. This, together with the efficient ambipolar

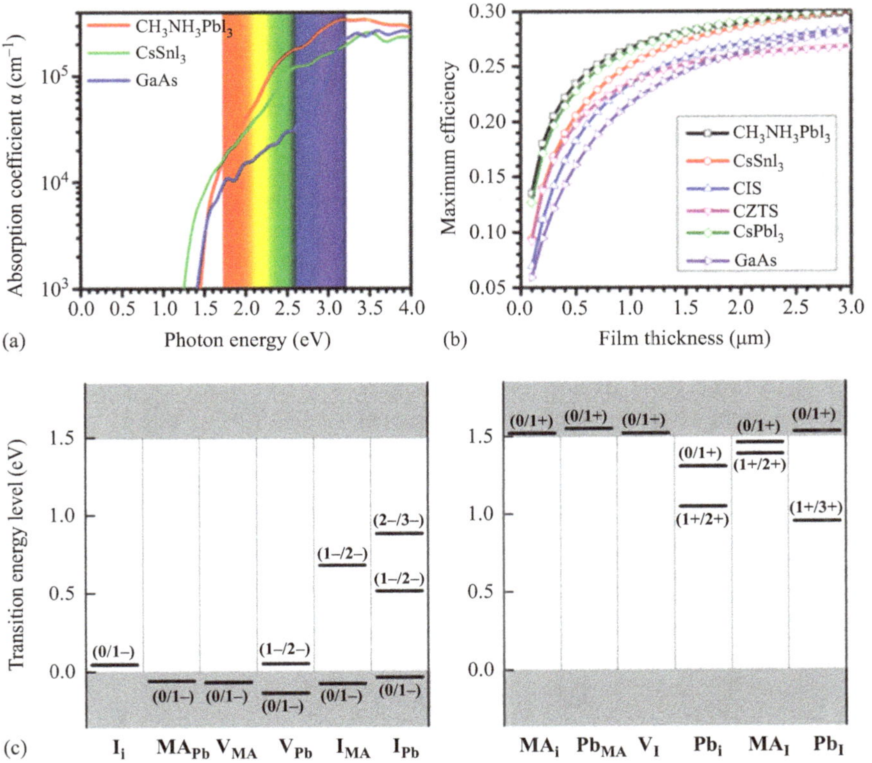

Figure 4.3. *(a) The optical absorption spectra of $CH_3NH_3PbI_3$, $CsSnI_3$, and GaAs. (b) Calculated maximum efficiencies of halide perovskites, CIS, CZTS, and GaAs as a function of film thickness. Reproduced with permission from [32]. (c) Defect diagram of $MAPbI_3$ perovskite from density functional theory calculation: (left) the intrinsic acceptors and (right) intrinsic donors. Reproduced with permission from [33]*

transport behavior, have enabled perovskites to be used in a variety of device architectures with high device performance.

4.2.3 Defect characteristics

For solution deposition to form polycrystalline semiconductors, in general, it is inevitable to have the wide range of defects that are formed. In principle, this is also true for perovskite formation, especially considering the rapid crystal growth (e.g., seconds to minutes) at relatively high growth temperature (e.g., 100–150 °C). Theoretical calculations are a powerful tool to investigate the PV properties of perovskites as influenced by defects, as well as to provide valuable guidance to defect engineering in material design and device fabrication.

As an example, the ideal $MAPbI_3$ crystal structure has 12 possible intrinsic point defects: three vacancies (V_{MA}, V_{Pb}, V_I), three interstitials (MA_i, Pb_i, I_i), and six substitutions (MA_{Pb}, Pb_{MA}, MA_I, Pb_I, I_{MA}, I_{Pb}) [33,36]. All these point defects create a transition level, which is a Fermi-level position located near the conduction band minimum (CBM) or valence band maximum (VBM) or within the bandgap. The transition level introduces potential charge recombination centers (or traps) to localize free electrons or holes. A defect-induced transition level near the band edges is often referred to as a shallow-level trap state, which normally does not cause significant non-radiative recombination because the trapped charges have a high probability of being thermally delocalized. According to the energy-level location of these point defects in the bandgap of $MAPbI_3$, all the point defects could be further divided into electron acceptors (I_i, MA_{Pb}, V_{MA}, V_{Pb}, I_{MA}, and I_{Pb}) and electron donors (MA_i, Pb_{MA}, V_I, Pb_i, MA_I, and Pb_I), as shown in Figure 4.4(c), respectively. An interesting conclusion drawn from early calculation work is that the defects with low formation energy (i.e., easy to form) only create shallow traps whereas the mid-gap traps (strong recombination centers) have high formation energies.

The theoretical work provides some insight to the unique feature of defect tolerance for hybrid halide perovskites. This feature has become a signature feature for perovskites. The defect tolerance aspect was further verified with experimental

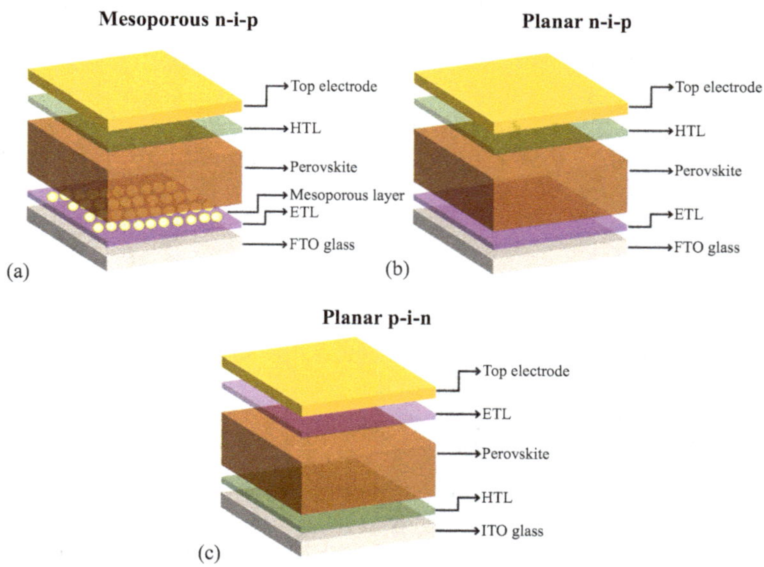

Figure 4.4. Schematics of three typical PSC architectures: (a) mesoporous n-i-p, (b) planar n-i-p, and (c) planar p-i-n

evidence [37]. In this study, X-ray photoemission spectroscopy was used to induce and track dynamic chemical and electronic transformations in the $MAPbI_3$ perovskite. Compositional changes were found to begin immediately with exposure to X-ray irradiation; however, the predominant electronic structure of perovskite appears tolerant to the formation of compensating defect pairs of V_I and V_{MA} and for a large range of I/Pb ratios (from 3:1 to about 2.5:1).

4.3 PSC device architectures and fabrication approaches

4.3.1 Typical device architectures of PSCs

The most common device architectures for PSCs can be categorized into three types: mesoporous n-i-p, planar n-i-p, and planar p-i-n, as schematically illustrated in Figure 4.4 [38]. In general, PSCs consist of the perovskite active layer that is sandwiched between an electron-transporting layer (ETL) and a hole-transporting layer (HTL). Following the illumination direction, if the light goes through ETL before entering the perovskite layer, it is described as n-i-p structure. The opposite structure should be p-i-n, which is generally referred to as an inverted structure because the carrier extraction layers are inverted with respect to the n-i-p structure, as shown in Figure 4.4(c). Typically, mesoscopic device structure also follows the general n-i-p configuration with the distinction of the ETL consisting of a mesoporous structure. An example is FTO glass/compact ETL (e.g., TiO_2, SnO_2)/mesoporous ETL (e.g., mesoporous TiO_2, SnO_2, Al_2O_3, ZnO)/perovskite/HTL/top electrode (Figure 4.4(a)). The mesoporous represents an early version of the device structure that developed from the mesoporous photoelectrode structure used in dye-sensitized solar cells (DSSCs). The planar-type PSCs are further divided into two configurations: n-i-p planar [e.g., FTO glass/compact ETL (e.g., TiO_2, SnO_2, PCBM)/perovskite/HTL/ electrode] and p-i-n planar (e.g., ITO glass/HTL [e.g., poly TPD, PEDOT:PSS, NiO)/ perovskite/compact ETL (e.g., C_{60}, BCP, PCBM)/top electrode].

Note that the determination of either n-i-p or p-i-n device architectures is still under debate, mainly because it has not been concluded whether the perovskite layer should be viewed as an intrinsic material or as a doped semiconductor. Some efforts have tried to elucidate the device operation mechanism by using Kelvin probe force microscopy (KPFM) or other related characterization techniques (e.g., impedance spectroscopy) to determine the electric field distribution across the device stack [39,40]. This will likely become more important as research efforts collectively move into the engineering/optimization of the contact/interface layer. This progress, along with work on doping control in perovskites, will highlight the need for better understanding the basic device operation principle. Characterizations that can provide nanometer-resolution distribution of electrical field, composition, density of state, and/or Fermi level along with surface-sensitive characterization of the work function of contact materials will become increasingly important to establish a reliable energy-level diagram that can be used to guide further development of PSCs. Device modeling based on simple diffusion and continuity equations will also find more use for future device development.

4.3.2 Common device fabrication approaches

As shown in Figure 4.4, PSCs consist of multiple functional layers including charge transport/contact layers and the perovskite absorber layer as well as other current collection layers (e.g., transparent conducting oxide and the metal contact). Thus, fabricating a PSC requires a systematic and balanced approach to prepare each layer under the conditions that will not undermine other layers. However, the rapid progress in performance demonstrated over the past several years is mainly the result of advances in deposition strategies for the perovskite absorber layer with much-improved quality control (e.g., coverage, thickness, crystallinity, grain morphology, composition). Many of the other functional layers have adopted deposition protocols previously established from other PV fields (e.g., DSSC and organic PV).

The two major general approaches for preparing perovskite absorber layers are solution processing and vacuum deposition. Solution processing methods represent the most common approaches used in the PSC field because of their intrinsic advantage of being low cost and compatible with printing technology for roll-to-roll device production. For both solution process and vacuum deposition, researchers have developed various steps or procedures for preparing perovskite thin films with higher film quality and device efficiency and stability. The most common procedures for vacuum deposition include one-step precursor deposition, sequential vapor deposition, and dual-source vacuum deposition processes [41–43]. For the solution-processing approach, some of the most common methods are one-step deposition, two-step deposition, and vapor-assisted solution processing (VASP) deposition (see Figure 4.5) [44,45]. A brief account of some typical deposition methods is discussed later.

4.3.2.1 One-step solution deposition

This solution-based deposition method involves the precipitation from perovskite precursor solution, where the metal halide and organic halide precursors are dissolved in an organic solvent to produce perovskite crystals. The precursor solution is often prepared by mixing MAI or MACl with PbI_2 in a polar aprotic solvent such as *N,N*-dimethylformamide (DMF), dimethylsulfoxide (DMSO), or gamma-butyrolactone (GBL). The precursor solution is usually spin-coated on the substrate to form a precursor film, which is subsequently annealed at an elevated temperature (e.g., 100–150 °C) to form the perovskite crystals. However, it is generally difficult to control film properties such as coating uniformity, film coverage, and grain morphology by using this method. To address these issues, many groups have developed various modifications to this general one-step solution deposition approach by either introducing additional coating steps or modifying precursor ink compositions; successful examples include antisolvent extraction [49], gas quenching [50], hot casting [46], additive engineering [51], drop casting [21], self-seeding growth [47], and vacuum flash-assisted process [48]. These methods (or the combination of them) along with perovskite composition tuning have greatly improved the perovskite film quality that has contributed to the rapid progress of PSC performance.

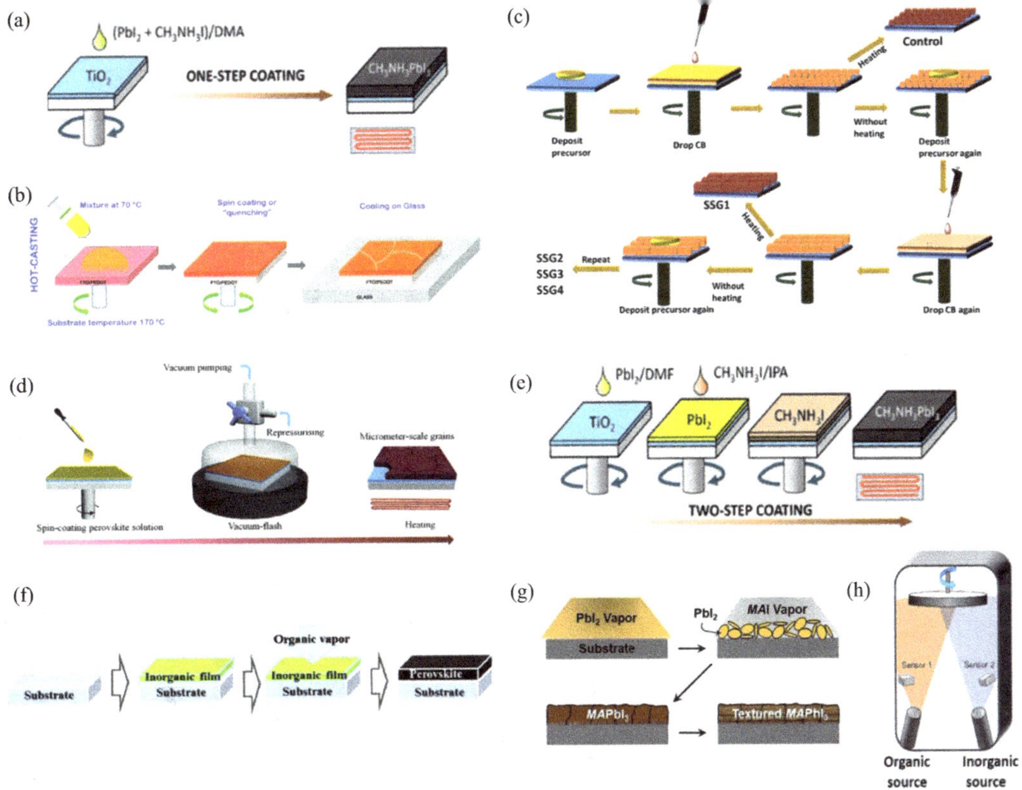

Figure 4.5. *Typical methods for preparing perovskite active layers. (a) One-step solution deposition, reproduced with permission from [45]. (b) Hot-casting deposition, reproduced with permission from [46]. (c) Self-seeding growth deposition, reproduced with permission from [47]. (d) Vacuum flash-assisted solution processing, reproduced with permission from [48]. (e) Two-step solution deposition, reproduced with permission from [45]. (f) VASP, reproduced with permission from [44]. (g) Sequential-vapor deposition, reproduced with permission from [41]. (h) Dual-source vacuum deposition, reproduced with permission from [43]*

4.3.2.2 Two-step solution deposition

This method was introduced primarily to address the challenge of poor film coverage associated with one-step solution deposition during the early stage of PSC development. In typical two-step solution deposition, a metal halide (e.g., PbI_2) layer is first spin coated on the substrate, which is subsequently converted to perovskite (e.g., $MAPbI_3$) by either coating with an additional layer of organic halide salt (e.g., MAI) or soaking in an organic halide salt solution (normally dissolved in IPA). The color of the film turns from yellowish to dark brown during the conversion step (or the second solution process step). Then, the resulting film is normally annealed (e.g., 10 min at 90 °C) to improve the film morphology and crystallinity. The composition of the perovskite films can be adjusted by changing the initial metal halide composition (e.g., PbI_2, $PbBr_2$, and/or $PbCl_2$) during the first step of the process as well as the organic halide salts (e.g., MAI, MABr, and/or MACl) during the second conversion step. The second conversion step involves diffusion of organic cations (e.g., MA^+) into the metal halide matrix to undergo substantial volume change to form perovskite film. Thus, controlling the morphology (e.g., mesoporous) of the metal halide matrix as well as the volume change from metal halide matrix to the final perovskite film have been shown to be critical to the speed and completeness of the conversion process—which, in turn, affects the quality of the resulting perovskite film.

4.3.2.3 Vapor-assisted solution process

This procedure involves the deposition of PbI_2 film from a solution first (e.g., by spin coating), followed by annealing the film in a chamber filled with MAI vapor at 150 °C for 2 h. Devices based on VASP have outstanding V_{oc} and show no sign of current leakage through the absorber film. It has exhibited perovskite films with larger grain sizes, which help to reduce the presence of grain boundaries, trap density, and charge-carrier scattering, leading to lower charge recombination and enhanced device performance. VASP also significantly improved the reproducibility of making high-quality perovskite films with high purity and it is compatible with large-area perovskite fabrication with fine control over film thickness and morphology for achieving high PSC performance.

4.3.2.4 Vacuum deposition methods

These methods allow the deposition of perovskite precursors in a vacuum atmosphere. It is possible to use this method to deposit one precursor at a time (single step) or to deposit two or more perovskite precursors simultaneously (co-deposition). The advantage of this method is that the deposition occurs in a clean environment and avoids potential contamination from solvents. It allows precise control over the film thickness and offers better film uniformity. In addition, the low fabrication temperature makes vacuum deposition compatible with a wide range of substrates, including flexible substrates and textiles. One of the advantages of dual-source vapor deposition is the possibility of preparing highly crystalline perovskite films of arbitrary thickness and the ability to monitor material growth in real time. Generally, in addition to the better control of the uniformity of the perovskite thin films by vapor-phase deposition, it is also solvent-free deposition; hence, it contains fewer impurities than film produced by solution processing. However, the role played by solvents in solution processing is still unclear. Thus, a comparison of vapor-phase deposition and solution processing to grow perovskites with the same composition can help elucidate the specific roles (either positive or negative) to the performance and stability of the perovskite absorber layer in PSCs.

4.3.2.5 Perovskite film characterization

The rapid progress in performance for PSCs is largely ascribed to the improved control in preparing perovskite thin films with higher qualities. The improved optoelectronic properties (e.g., mobility, carrier lifetime, defect density) are often correlated to the better film morphologies and grain structures. The influence of grain size and grain boundaries on material properties and devices have been widely studied in the field of materials science. These features can be studied using characterization techniques such as scanning electron microscopy (SEM), atomic force microscopy (AFM), transmission electron microscopy (TEM), and electron backscatter diffraction (EBSD) [52]. However, the fragile (or soft) nature of the hybrid organic–inorganic perovskite structure presents a challenge for microstructural characterizations with the previously mentioned techniques because these techniques themselves may have direct impact on the microstructures and properties of perovskites. For example, high-energy electron beams can cause significant degradation of perovskites during electron-probe-based experiments (e.g., EBSD and cathodoluminescence) [53]. Note that strong electric fields can also change material properties (e.g., ion migration) to alter the composition distribution [54]. In general, new measurement conditions need to be established to allow certain conventional techniques to be applied to perovskites. For example, the electron dose needs to be minimized while ensuring sufficient signal-to-noise ratio. Samples should also be compared before and after certain measurements to ensure that the experimental conclusions are not based on artifacts.

4.4 R&D challenges of PSCs

4.4.1 Stability

At present, the most challenging issue in PSCs is long-term stability, which must be ensured before the practical application of PSCs. The stability of PSCs when exposed to severe environment (e.g., moisture, thermal, light illumination) appears to be the bottleneck that impedes their further commercialization [55]. The decrease of stability of PSCs could be attributed to two primary reasons: the degradation of perovskite material itself or interfacial degradation.

4.4.1.1 Moisture stability

The organic cations used in PSCs are generally hygroscopic. It has been suggested that water molecules form weak hydrogen bonds with the cations and that this compromises the structural stability of the crystal. This can lead to the formation of a hydrated perovskite phase, causing reversible structural change. However, when there is too much moisture penetration into the perovskite film, the perovskite crystal often decomposes, which is often irreversible. For example, a nonreversible degradation in the standard $MAPbI_3$ perovskite leads to the formation of an aqueous form of MAI and PbI_2. During the decomposition process, hydroiodic acid (HI) and methylamine are produced. When oxygen is present, water can also be generated. The formation of additional water will accelerate perovskite decomposition in a cyclic process, leading to further moisture-induced degradation. The presence of excessive PbI_2 can negatively impact the perovskite active layer. One way to reduce this effect is to increase the strength of the bonds between the organic component and the metal halides, such as by using additives engineering or defect passivation [56,57].

4.4.1.2 Thermal stability

Thermal stability of PSCs refers to the degradation of PSCs when subjected to elevated temperature, which can often occur in PV devices operating in outdoor conditions. Any accelerated degradation at a high temperature would raise serious concern for the practical application of PSCs in the field. It is well known that changing the environmental temperature can alter perovskite's crystal structure and phase. It has been reported that perovskite undergoes a phase change from tetragonal to cubic at about 54–56 °C [58]. Solar modules will be exposed to elevated temperature during operation under an international standard (IEC 61646 climatic chamber tests). The solar cell must demonstrate thermal stability up to 85 °C [59]. Conings *et al.* found that perovskite could decompose into PbI_2 when heated in nitrogen at 85 °C for 24 h [60]. Note that slightly overheating perovskite could lead to the formation of a small amount of PbI_2, which could enhance the device performance due to the passivation of defects at perovskite surfaces or grain boundaries [61]. Heating too much could lead to serious decomposition and poor device performance of perovskite.

4.4.1.3 Light stability

The effect of light illumination on PSC stability is most significant when combined with other factors (e.g., moisture or oxygen exposure). Some studies showed that for certain perovskite compositions, in oxygen-rich conditions, large PbI_2 structures can be observed in the perovskite film, whereas the same is not true for perovskites prepared in a nitrogen-rich atmosphere [62]. However, it has been shown that $MAPbI_3$ can degrade to PbI_2 under ultraviolet light in the absence of moisture or oxygen. In addition, the stability of selective contacts should be carefully considered when designing the device structure with good long-term stability. Unlike traditional PV materials such as silicon, the organic–inorganic hybrid perovskites show significant ionic characteristics, which could affect the long-term stability of perovskite materials due to the relatively low activation energy for ion migration within the perovskite layer [63]. Ion migration has been observed to become severe when the device is subjected to thermal stress, external electric bias, or illumination.

4.4.1.4 Chemical compatibility

Chemical reaction between perovskite and the electrode will also result in some interface degradation. Recently, some traditional metal electrode materials (e.g., Al and Ag) were found to react with perovskite materials [64]. Insulating perovskites from the electrode (Ag/Al) could enhance the stability. However, diffusion of halide ions through the transport layer could undermine the stability of PSCs. The most commonly used ETL in PSCs cells is TiO_2 in a n-i-p device structure. The instability of TiO_2 under ultraviolet exposure was ascribed to the surface oxygen vacancy generated by the reaction of photo-induced hole and oxygen radicals which further induce deep surface traps, inhibiting photogenerated charge collection [65]. One way is mitigate this is to modify TiO_2 or replace it with alternatives, such as SnO_2 and ZnO. HTL also has an impact on PSC's stability. Spiro-OMeTAD is the most widely used HTL in PSCs, but it requires some additives or dopants to achieve suitable conductivity. However, the dopant (e.g., lithium salt) is often hygroscopic and can migrate throughout the device stack; thus, it is not compatible with long-term operation of PSCs under environmental factors (e.g., air and moisture) [66]. Dopant-free HTLs and inorganic HTLs with high mobility will be necessary [67].

4.4.1.5 Stability characterization

One thing presently lacking in the scientific community is the ability to compare results of stability across different labs/groups. There are a whole host of parameters, including the spectrum of light illuminating the cells, temperature of device operation, device aging under load or at open circuit, atmosphere, humidity, cell encapsulation, materials/methods for encapsulation, and many other parameters. Before establishing a standard stability test protocol, it is critical to clearly describe the conditions used for stress testing when evaluating stability. This will ensure proper comparison among different labs/groups to reach a more realistic determination of the device stability.

In addition to the challenge associated with using proper stress conditions for stability tests, it is also challenging to have proper measurements (e.g., current density–voltage [J–V] and external quantum efficiency [EQE]) to determine the actual device performance of a PSC. PSCs operate differently than conventional PV cells. For example, PSC can exhibit strong hysteresis behavior with cell efficiency derived from J–V measurements depending strongly on J–V measurement conditions (e.g., voltage sweep direction, sweep rate, precondition before J–V measurement). Mobile ions in perovskites are one key factor contributing to this complexity. There are efforts to standardize the measurement protocols. For efficiency determination, the common practice is to either measure a stabilized steady-state power output (measure at a fixed voltage over time and take the current until it stabilizes without changing) or maximum power point (MPP) tracking (measuring the maximum power during continuous cell operation), which is an essential component for quantifying efficiency [68].

Finally, since PSC is a young technology, outdoor stability testing is insufficient to track the actual lifetime of PSCs operating in the field. Thus, it is important to determine the acceleration factors under various stress conditions. These acceleration factors, if understood and designed properly, can be used to predict the lifetime of perovskite PV under outdoor operation. This is likely a very challenging task because perovskite composition and device architecture are still undergoing frequent changes with progress in the field.

4.4.2 Material toxicity

Presently, perovskite composition is primarily based on a Pb-containing component. However, the potential toxicity of Pb-based perovskite materials could form a barrier for PSCs to enter the PV market. Although the amount of Pb used in PSCs is not significant, the Pb in PSC is soluble in water. This feature, along with the perception of using a toxic element, could make it difficult to commercialize PSCs. Thus, developing low-toxicity Pb-free materials is of interest to the future PSC application, assuming that device performance will not be compromised. Ideal Pb-free candidates for a solar cell absorber should have low toxicity, narrow direct bandgaps, high optical absorption coefficients, high mobilities, low exciton-binding energies, long charge-carrier lifetimes, good stability, and be scalable. Some low-toxicity constituents with a perovskite structure and with suitable optoelectronic character have attracted research attention, such as Sn/Ge-based halides, some double perovskites, and some Bi/Sb-based halides with perovskite-like structure. However, the performance of devices based on these Pb-free absorbers is generally low compared with the high efficiency demonstrated by Pb-based PSCs.

So far, only the Sn-based perovskites appear promising for achieving high performance in the near future [69]. However, Sn-based perovskites exhibit significantly worse material stability compared with Pb-based PSCs, even with encapsulation. It is still too early to judge whether there will be viable Pb-free stable alternative perovskite absorbers. Nevertheless, PCEs of 10% or higher for pure-Sn-based PSCs are likely on the horizon. Once the Sn^{4+} oxidation issue is fully addressed and photocarrier recombination rates are suppressed to the levels of the pure-Pb-based perovskites, V_{oc} values of about 0.8 to 1.00 V should be achievable which, in turn, will open the path to PCEs beyond 15% and will dramatically improve their future prospect as viable Pb-free contenders. Other studies have shown that Ge-based perovskites are not ideal candidate for Pb-free PSCs. However, the mixed Ge/Sn-based perovskites appear to be promising if future work can boost efficiencies above 10% to excite more interest in pursuing this direction. Overall, the current development of Pb-free perovskite materials is facing a poor dichotomy: (1) a path to high efficiency but poor stability (Sn^{2+}-based) or (2) good stability but low performance (Sn^{4+}/Sb/Bi-based). More research is needed to close this gap.

One viable option for reducing the hazards of Pb exposure is to develop effective encapsulation strategies. If commercial solar panels are to last for more than 25 years, then the modules have to be extremely well encapsulated against moisture and oxygen ingress. For the most robust encapsulation protocols, this is typically achieved by laminating the modules between two sheets of glass and polymer foil, followed by careful edge sealing. Furthermore, at the end of the module lifetime, it is important to ensure that all materials, especially toxic and expensive components, can be recycled; this scenario has already been realized for commercial thin-film PV technologies such as CdTe.

4.4.3 Scaling up

To make a PSC in the laboratory, researchers deposit perovskite precursor chemicals onto a substrate. Spin coating, the most commonly used deposition method in the laboratory, produces devices with the highest efficiency. However, the spin-coating process wastes more than 90% of the chemicals used. In addition, spin coating works best on cells smaller than 4 in [2]. To fully realize the potential of PSCs, it is critical to develop coating strategies over a large area. Scalable solution deposition methods for perovskite growth include, but are not limited to, blade coating, slot-die coating, meniscus coating, spray coating, inkjet printing, screen printing, and electrodeposition [70]. Here, we briefly compare these deposition methods as well as their use for PSCs and module development.

Figure 4.6 shows several common scalable deposition approaches. (1) *Blade coating* uses a blade to spread the chemical solution on substrates to form wet thin films. The process can be adapted for roll-to-roll manufacturing, with flexible substrates moving on a roller beneath a stationary blade similar to how newspapers are printed. Blade coating wastes less of the ink than spin coating. (2) *Slot-die coating* relies on a reservoir to supply the precursor ink to apply ink over the substrate. The process has not been as well explored as other methods, and so far it has demonstrated lower efficiency than blade coating. But the reproducibility of slot-die coating is better than blade coating when the ink is well developed. So this is more applicable for roll-to-roll manufacturing. (3) *Inkjet printing* uses a small nozzle to disperse the precursor ink. The process has been used to make small-scale solar cells, but whether it is suitable for the high-volume, large-area production will depend on the printing speed and device structure. (4) *Screen printing* uses a patterned mesh screen to hold and transfer ink to the substrate. The unwanted area of the mesh screen is blocked by the exposed photosensitive polymer emulsion, and the open holes of the mesh hold the viscous ink as a squeegee spreads ink across the screen. The ink is then transferred to the substrate to form the desired pattern. The thickness of the resulting film is determined by the mesh size and thickness of the emulsion layer. (5) *Spray coating* is another approach that is compatible with large-scale deposition of the perovskite layer. Common spray coating includes ultrasonic spraying and electro-spraying. During spray coating, the deposited perovskite area can be repeatedly coated with the perovskite precursor. Thus, it is important to specifically control the balance of solvent drying, perovskite redissolution, perovskite regrowth, coverage, and thickness.

One challenge for the scaling-up effort is to determine the perovskite composition and the general coating method and strategy. Perovskite composition has undergone continuous optimization for higher efficiency and stability. At present,

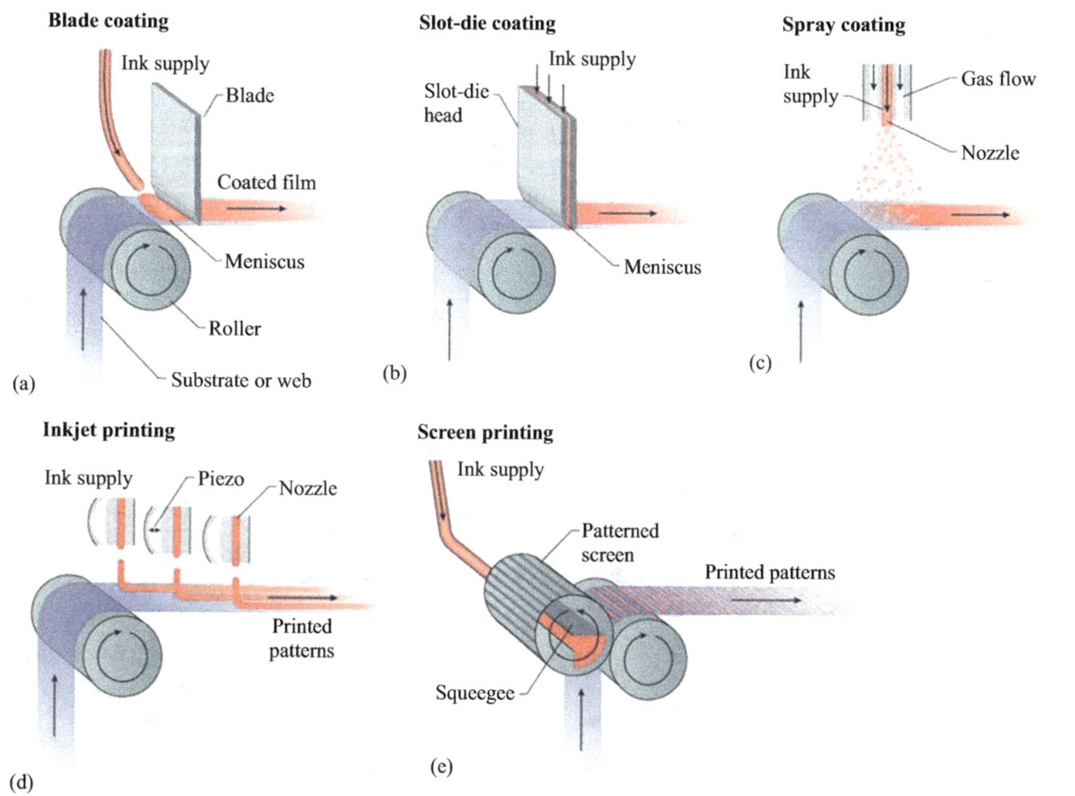

Figure 4.6. Common scalable solution deposition methods for the roll-to-roll fabrication of PSCs. Reproduced with permission from [70]

the composition has become increasingly complicated in comparison with earlier $MAPbI_3$. A popular example is the triple-cation-based mixed-halide perovskite such as $(FA_{1-x-y}Cs_xMA_y)Pb(I_{1-z}Br_z)_3$ [71]. The ratio of individual perovskite elements varies among groups. In addition, all three common device architectures (Figure 4.4) can produce high-efficiency PSCs. For solution processing, the most popular antisolvent extraction step developed for spin coating is difficult to be implemented at a large scale. These factors add to the complexity for deciding on the perovskite composition and solvent system. Moreover, the different scalable coating methods will not have the same processing environment, which can strongly affect the outcome of the solution processing. Due to these diverse factors involved in scalable deposition, it is difficult to determine the best approaches for scaling up; those chosen deposition approaches and perovskite compositions will likely change with further advances of PSCs at the lab scale.

4.5 Future research trend of PSCs

During the past 10 years, PSCs have witnessed unprecedented rapid progress, surpassing the performance of most conventional PV technologies, and perovskite is now in a very strong position to be a contender for utility-scale PV applications. Despite the certified record efficiency of 24.2%, single-junction PSCs have the potential to further improve near 30%, similar to III–V solar cells. In addition, perovskite-based tandem or multijunction solar cells are expected to break the single-junction Shockley–Queisser efficiency limit. The following areas will receive more attentions moving forward to push perovskite PV technology to the market.

Development of efficient and cost-effective HTL will continue to be an attractive topic in an effort to reduce the cost of some expensive and commonly used materials (e.g., spiro-OMeTAD) while seeking to improve the reliability of PSCs, especially at high operating temperature and humidity. Another area of active research is to find Pb-free perovskite compositions without sacrificing device performance. Current efforts using less-toxic Pb-alternative metals (e.g., Se, Cu, Sn) have suffered from poor performance. It is also attractive to develop strategies to recycle Pb or prevent the entry of Pb into the environment. Toward scaling up PSCs or modules, more efforts will focus on developing fully scalable fabrication of all device layers (not just the perovskite absorber layer) as well as reliable module interconnection schemes. These efforts will be closely coupled to the continuous development of absorber compositions and device architectures.

Further improvement of PSC efficiency will likely shift focus more toward all-perovskite tandem structures. With recent material innovations in low-bandgap (1.1–1.3 eV) and wide-bandgap (1.7–1.9 eV) perovskite absorbers and transparent conductive interconnection layers, the use of polycrystalline halide perovskites for both the top and bottom cells in a tandem structure represents a promising combination for developing ultrahigh-efficiency dual-junction thin-film tandem solar cells. In principle, such a device structure can be solution processed and promises to achieve efficiencies of >35% at 1-sun. For low-bandgap perovskites, research will be focused on increasing carrier lifetime and reducing defect densities for Sn-Pb-based perovskites. For wide-bandgap perovskites, efforts will be focused on the bandgap range of ~1.8–1.9 eV to complement the absorption from the current low-bandgap perovskites (~1.2–1.3 eV). A special target is to reduce voltage loss with respect to bandgap. Developments on the interconnection layer will focus on optical transmittance, electric conductance, and robustness to protect the existing perovskite device stack from damage during deposition of the top perovskite cell stack.

Stability is always a major component need to be improved for PSCs for both single-junction and tandem or multijunction device architectures. Materials development (e.g., perovskite absorber, contact layer, encapsulation) will continue. However, what has become more critical to the field is the lack of standardized characterization protocols to properly evaluate the progress from different groups. We do not have sufficient information to know if various stability characterizations under different stress conditions (e.g., temperature, humidity, light) are the proper tests to provide PSC lifetimes under actual outdoor operation conditions. It is important to determine the degradation acceleration factors of each stress condition corresponding to PSCs operating in the field. Such work is truly at the core of needed research to gain a good understanding of device stability and reliability. At present, discrepancies exist within the research community related to stability characterization, such as the correct methodology for stressing PSCs, how to report stability to ensure proper comparison among different groups, and what characterization standards should be followed to determine whether or not a PSC is stable. Thus, it is important for the community to develop a set of standardized characterization protocols that will be followed by the entire research community. To enhance the stability of perovskites and PSCs, several factors must be considered for more systematic development, including structure design, charge-transport materials, electrode material preparation, and encapsulation methods. Encapsulation plays a vital role in improving the stability of PSCs and will help to accelerate the technology to the goal of commercialization. It is not enough to just modify the current perovskite materials or interface to achieve the performance goal concerning both efficiency and stability. It is important to develop new materials and designs for fully encapsulated modules with high stability under severe conditions.

In summary, organic–inorganic halide perovskites offer an unprecedented opportunity to produce a highly efficient, low-cost, and scalable PV technology that has the potential for deployment at the terawatt scale. Although maximum efficiency for single-junction solar cells has climbed above 24% in early 2019, efforts are needed to address issues on large-scale reproducibility, current–voltage hysteresis, operation stability, and characterization protocols to properly evaluate progress and underlying mechanisms. Nonetheless, with recent progress being made in chemical synthesis, device engineering, film deposition, and theoretical modeling, the field of PSCs should have a bright, rapidly evolving, and practical future.

Acknowledgments

The work was supported by the U.S. Department of Energy (DOE) under contract no. DE-AC36-08GO28308 with Alliance for Sustainable Energy, Limited Liability Company (LLC), the Manager and Operator of the National Renewable Energy Laboratory. We acknowledge the support from the De-risking Halide Perovskite Solar Cells program of the National Center for Photovoltaics, funded by the U.S. Department of Energy, Office of Energy Efficiency and Renewable Energy, Solar Energy Technologies Office. The views expressed in this chapter do not necessarily represent the views of the DOE or the U.S. Government.

References

[1] F. Zhang, S. Wang, X. Li and Y. Xiao, *Current Nanoscience*, 2016, **12**, 137–156.
[2] D. Weber, *Z. Naturforsch. B.*, 1978, **33**, 1443.
[3] D. Weber, *Z. Naturforsch. B.*, 1979, **34**, 939.
[4] D. B. Mitzi, C. A. Feild, W. T. A. Harrison and A. M. Guloy, *Nature*, 1994, **369**, 467–469.
[5] A. Kojima, K. Teshima, Y. Shirai and T. Miyasaka, *Journal of the American Chemical Society*, 2009, **131**, 6050–6051.
[6] H.-S. Kim, C.-R. Lee, J.-H. Im, *et al.*, *Scientific Reports*, 2012, **2**, 591.
[7] M. M. Lee, J. Teuscher, T. Miyasaka, T. N. Murakami and H. J. Snaith, *Science*, 2012, **338**, 643–647.
[8] J.-H. Im, C.-R. Lee, J.-W. Lee, S.-W. Park and N.-G. Park, *Nanoscale*, 2011, **3**, 4088–4093.
[9] J. Tong, Z. Song, D. H. Kim, *et al.*, *Science*, 2019, **364**, 475–479.
[10] D. Bi, C. Yi, J. Luo, *et al.*, *Nature Energy*, 2016, **1**, 16142.
[11] F. Zhang, D. Bi, N. Pellet, *et al.*, *Energy & Environmental Science*, 2018, **11**, 3480–3490.
[12] Q. Jiang, Y. Zhao, X. Zhang, *et al.*, *Nature Photonics*, 2019, **13**, 460–466.
[13] N. J. Jeon, H. Na, E. H. Jung, *et al.*, *Nature Energy*, 2018, **3**, 682–689.
[14] W. S. Yang, B.-W. Park, E. H. Jung, *et al.*, *Science*, 2017, **356**, 1376–1379.
[15] S. S. Shin, E. J. Yeom, W. S. Yang, *et al.*, *Science*, 2017, **356**, 167–171.
[16] W. S. Yang, J. H. Noh, N. J. Jeon, *et al.*, *Science*, 2015, **348**, 1234–1237.
[17] N. J. Jeon, J. H. Noh, W. S. Yang, *et al.*, *Nature*, 2015, **517**, 476.
[18] M. Abdi-Jalebi, Z. Andaji-Garmaroudi, S. Cacovich, *et al.*, *Nature*, 2018, **555**, 497.
[19] N. Ahn, D.-Y. Son, I.-H. Jang, S. M. Kang, M. Choi and N.-G. Park, *Journal of the American Chemical Society*, 2015, **137**, 8696–8699.
[20] J. Burschka, N. Pellet, S.-J. Moon, *et al.*, *Nature*, 2013, **499**, 316.
[21] A. Mei, X. Li, L. Liu, *et al.*, *Science*, 2014, **345**, 295–298.
[22] E. H. Jung, N. J. Jeon, E. Y. Park, *et al.*, *Nature*, 2019, **567**, 511–515.

[23] *Best Research-Cell Efficiency Chart*, https://www.nrel.gov/pv/cell-efficiency.html, 2019.
[24] Y. Yamada, T. Yamada, L. Q. Phuong, *et al.*, *Journal of the American Chemical Society*, 2015, **137**, 10456–10459.
[25] F. Zhang, D. H. Kim, H. Lu, *et al.*, *Journal of the American Chemical Society*, 2019, **141**, 5972–5979.
[26] W. A. Dunlap-Shohl, Y. Zhou, N. P. Padture and D. B. Mitzi, *Chemical Reviews*, 2019, **119**, 3193–3295.
[27] L. Mao, W. Ke, L. Pedesseau, *et al.*, *Journal of the American Chemical Society*, 2018, **140**, 3775–3783.
[28] V. M. Goldschmidt, *Naturwissenschaften*, 1926, **14**, 477–485.
[29] B. Saparov and D. B. Mitzi, *Chemical Reviews*, 2016, **116**, 4558–4596.
[30] C. Li, X. Lu, W. Ding, L. Feng, Y. Gao and Z. Guo, *Acta Crystallographica Section B*, 2008, **64**, 702-707.
[31] T.-B. Song, Q. Chen, H. Zhou, *et al.*, *Journal of Materials Chemistry A*, 2015, **3**, 9032–9050.
[32] W.-J. Yin, T. Shi and Y. Yan, *Advanced Materials*, 2014, **26**, 4653–4658.
[33] W.-J. Yin, T. Shi and Y. Yan, *Applied Physics Letters*, 2014, **104**, 063903.
[34] E. Edri, S. Kirmayer, S. Mukhopadhyay, K. Gartsman, G. Hodes and D. Cahen, *Nature Communications*, 2014, **5**, 3461.
[35] S. D. Stranks, G. E. Eperon, G. Grancini, *et al.*, *Science*, 2013, **342**, 341–344.
[36] C. Ran, J. Xu, W. Gao, C. Huang and S. Dou, *Chemical Society Reviews*, 2018, **47**, 4581–4610.
[37] K. X. Steirer, P. Schulz, G. Teeter, V. Stevanovic, M. Yang, K. Zhu and J. J. Berry, *ACS Energy Letters*, 2016, **1**, 360–366.
[38] R. Wang, M. Mujahid, Y. Duan, Z.-K. Wang, J. Xue and Y. Yang, *Advanced Functional Materials*, 2019, **29**, 1808843.
[39] C.-S. Jiang, M. Yang, Y. Zhou, *et al.*, *Nature Communications*, 2015, **6**, 8397.
[40] P. Cui, D. Wei, J. Ji, *et al.*, *Nature Energy*, 2019, **4**, 150–159.
[41] H. Hu, D. Wang, Y. Zhou, *et al.*, *RSC Advances*, 2014, **4**, 28964–28967.
[42] C.-W. Chen, H.-W. Kang, S.-Y. Hsiao, P.-F. Yang, K.-M. Chiang and H.-W. Lin, *Advanced Materials*, 2014, **26**, 6647v6652.
[43] M. Liu, M. B. Johnston and H. J. Snaith, *Nature*, 2013, **501**, 395.
[44] Q. Chen, H. Zhou, Z. Hong, *et al.*, *Journal of the American Chemical Society*, 2014, **136**, 622–625.
[45] J.-H. Im, H.-S. Kim and N.-G. Park, *APL Materials*, 2014, **2**, 081510.
[46] W. Nie, H. Tsai, R. Asadpour, *et al.*, *Science*, 2015, **347**, 522–525.
[47] F. Zhang, C. Xiao, X. Chen, *et al.*, *Joule*, 2019, **3**, 1452–1463.
[48] X. Li, D. Bi, C. Yi, *et al.*, *Science*, 2016, **353**, 58–62.
[49] N. J. Jeon, J. H. Noh, Y. C. Kim, W. S. Yang, S. Ryu and S. I. Seok, *Nature Materials*, 2014, **13**, 897.
[50] M. Zhang, J. S. Yun, Q. Ma, *et al.*, *ACS Energy Letters*, 2017, **2**, 438–444.
[51] C. Zuo and L. Ding, *Nanoscale*, 2014, **6**, 9935–9938.
[52] Y. Zhou, H. Sternlicht and N. P. Padture, *Joule*, 2019, **3**, 641–661.

[53] C. Xiao, Z. Li, H. Guthrey, *et al.*, *The Journal of Physical Chemistry C*, 2015, **119**, 26904–26911.
[54] M. U. Rothmann, W. Li, J. Etheridge and Y.-B. Cheng, *Advanced Energy Materials*, 2017, **7**, 1700912.
[55] X. Qin, Z. Zhao, Y. Wang, J. Wu, Q. Jiang and J. You, *Journal of Semiconductors*, 2017, **38**, 011002.
[56] F. Zhang, W. Shi, J. Luo, *et al.*, *Advanced Materials*, 2017, **29**, 1606806.
[57] H. Zhu, F. Zhang, Y. Xiao, S. Wang and X. Li, *Journal of Materials Chemistry A*, 2018, **6**, 4971–4980.
[58] T. Baikie, Y. Fang, J. M. Kadro, *et al.*, *Journal of Materials Chemistry A*, 2013, **1**, 5628–5641.
[59] H.-S. Kim, J.-Y. Seo and N.-G. Park, *ChemSusChem*, 2016, **9**, 2528–2540.
[60] B. Conings, J. Drijkoningen, N. Gauquelin, *et al.*, *Advanced Energy Materials*, 2015, **5**, 1500477.
[61] Q. Chen, H. Zhou, T.-B. Song, *et al.*, *Nano Letters*, 2014, **14**, 4158–4163.
[62] A. J. Pearson, G. E. Eperon, P. E. Hopkinson, *et al.*, *Advanced Energy Materials*, 2016, **6**, 1600014.
[63] E. Bi, H. Chen, F. Xie, *et al.*, *Nature Communications*, 2017, **8**, 15330.
[64] J. You, L. Meng, T.-B. Song, *et al.*, *Nature Nanotechnology*, 2015, **11**, 75.
[65] T. Leijtens, G. E. Eperon, S. Pathak, A. Abate, M. M. Lee and H. J. Snaith, *Nature Communications*, 2013, **4**, 2885.
[66] F. Zhang, C. Yi, P. Wei, *et al.*, *Advanced Energy Materials*, 2016, **6**, 1600401.
[67] F. Zhang, X. Liu, C. Yi, *et al.*, *ChemSusChem*, 2016, **9**, 2578–2585.
[68] K. Domanski, E. A. Alharbi, A. Hagfeldt, M. Grätzel and W. Tress, *Nature Energy*, 2018, **3**, 61–67.
[69] R. Wang, J. Wang, S. Tan, Y. Duan, Z.-K. Wang and Y. Yang, *Trends in Chemistry*, 2019, **1**, 368–379
[70] Z. Li, T. R. Klein, D. H. Kim, *et al.*, *Nature Reviews Materials*, 2018, **3**, 18017.
[71] M. Saliba, T. Matsui, J.-Y. Seo, *et al.*, *Energy & Environmental Science*, 2016, **9**, 1989–1997.

Chapter 5

Photovoltaic device modeling: a multi-scale, multi-physics approach

Marco Nardone[1]

5.1 Introduction

This chapter provides the essential features of advanced photovoltaic (PV) device modeling and specific examples for cadmium telluride (CdTe), copper indium gallium diselenide (CIGS), and silicon (Si)-based technologies. It is instructive to compare polycrystalline thin-film and crystalline devices.

Modeling can serve for several purposes, including:

- Prediction of device performance under standard and field operating conditions,
- Calculation of performance variations over time under various stress conditions,
- Simulation of characterization techniques to assist with data analysis, and
- Quantitative hypothesis testing in comparison with experimental data.

In all cases, modeling includes determining and solving the pertinent physical equations under a set of known assumptions. Assumptions and hypotheses are checked by comparison of the model calculations with empirical data. This provides a quantitative means to evaluate mechanisms that plausible based on all of the available data.

A versatile approach allows for multi-physics, multidimensional, and multi-scale functionality. Multi-physics enables customizable equations related to electrical, optical, thermal, time-dependent, and defect reaction phenomena and their coupling. For example, temperature dependence is critical when modeling characterization experiments that use temperature as an independent variable and in reliability tests that stress devices over a range of temperatures. Time-dependent capabilities enable the simulation of degradation rates and characterization techniques such as time-resolved photoluminescence (TRPL). Studying lateral non-uniformities such as grain boundaries (GBs), shunts, and light gradients requires two-dimensional (2D) and three-dimensional (3D) capabilities. Cell and module

[1]Department of Physics and Astronomy, Bowling Green State University, Bowling Green, Ohio, USA

models can be developed using a multi-scale approach where the results of small-spatial-scale analysis can be fed into larger-scale simulations. At the cell and module scales, simulation of electroluminescence (EL) and thermographic techniques is possible.

Numerical solutions of model equations can be obtained through in-house coding or by using available software packages, which have widely varying cost, benefits, and capabilities. Regardless of the approach, a thorough understanding of the underlying physics and judicious selection of parameters is most important. The art of this work is inclusion of the most important physics while neglecting that which makes the model overly complicated without significant benefit. A naïve approach would be attempting to model the real system exactly, which is usually impossible.

In this chapter, Section 5.2 provides the fundamental physical equations that govern device behavior at the micro-diode, cell, and module scales. At the level of micrometers, the optical and electronic properties of the various device layers and interfaces are defined, and the electric potential and current densities are determined in response to voltage, light, and temperature; this is the micro-diode scale. Those calculations provide the current density–voltage (JV), quantum-efficiency (QE), and capacitance–voltage (CV) characteristics, as well as insight on the spatial distribution of recombination rates, charge-carrier densities, and so on. Such detail in the equations cannot be considered at the cell scale due to the very large aspect ratio ($\geq 10^4$) of thin, large-area devices. Therefore, the micro-diode responses are used as inputs to the cell-scale models, which determine the JV behavior under various conditions. A similar approach is used at the module scale except that attention must be paid to how cells are connected (e.g., monolithic, soldered). Sections 5.3, 5.4, and 5.5 provide case studies for degradation-related defect kinetics in CdTe, electrothermal runaway in shaded CIGS modules, and electron-beam-induced current (EBIC) analysis of GBs in multicrystalline (mc) Si, respectfully. Finally, Section 5.6 discusses the challenges and opportunities associated with PV device modeling.

5.2 Physics for multi-scale PV device simulation

This chapter reviews the physics underlying numerical simulation of semiconductor devices in general, with emphasis on PV devices at the micro-diode, cell, and module scales. Charge-transport, optical, thermal, and mass-transport physics, as well as their coupling, are briefly reviewed with an emphasis on PV devices. The reader is directed to thorough reviews of fundamental semiconductor device physics [1], PV device physics [2], and specialization in chalcogenide thin-film PVs [3]. Other references to optical, thermal, and diffusion physics will be provided later. Numerical modeling of semiconductor devices is expounded by Selberherr [4] and Baets *et al.* [5]. In this chapter, the modeling methodology is based on the finite-element method (FEM).

5.2.1 Micro-diode scale modeling

Numerical modeling at the micro-diode scale can be conducted in a one-dimensional (1D) space or in 2D/3D space for devices with aspect ratio of less than about 100 (going much beyond that is limited by meshing the large aspect ratio). Detailed semiconductor physics models are employed at this scale, as described later. Subsequent sections will explain how these results are extended to the cell and module scales.

5.2.1.1 Charge transport

Predicting PV device performance requires solving the transport equations, including the Poisson equation and those for electron and hole current continuity. The basic purpose is to calculate the electric potential, $\phi(x, y, z, t)$, electron concentration, $n(x, y, z, t)$, and hole concentration, $p(x, y, z, t)$, as functions of space and time (time dependence can be neglected in stationary studies). With those three variables, all pertinent outputs can be calculated, including current–voltage characteristics, quantum efficiency, and capacitance. The coupled transport equations are:

$$\nabla \cdot (\varepsilon_s \nabla \phi) = -\rho \tag{5.1}$$

$$\frac{\partial n}{\partial t} = \frac{1}{q} \nabla \cdot \boldsymbol{J}_n - R_n + G_n \tag{5.2}$$

$$\frac{\partial p}{\partial t} = \frac{1}{q} \nabla \cdot \boldsymbol{J}_p - R_p + G_p \tag{5.3}$$

where q is the elementary charge, ε_s is the semiconductor permittivity, R_n and R_p are recombination rates, G_n and G_p are generation rates for electrons and holes, respectively. The charge density is given by:

$$\rho = q\left(p - n - N_A^- + N_D^+ \pm N_t^{+/-}\right). \tag{5.4}$$

Spatially fixed charges are charged acceptors, N_A, donors, N_D, and trap states, N_t, depending on their occupation probability. The electron and hole current densities, $\boldsymbol{J_n}$ and $\boldsymbol{J_p}$, have both drift and diffusion components given by:

$$\boldsymbol{J}_n = -q\mu_n n \nabla \phi + qD_n \nabla n \tag{5.5}$$

$$\boldsymbol{J}_p = -q\mu_p n \nabla \phi - qD_p \nabla p \tag{5.6}$$

where μ_n and μ_p are the electron and hole mobilities, respectively. $D = \mu kT/q$ is the diffusion coefficient with Boltzmann constant k and temperature T. The total current density is $\boldsymbol{J} = \boldsymbol{J_n} + \boldsymbol{J_p}$.

The boundary conditions (BCs) on the currents depend on the type of metal contacts being considered or possibly insulating/charged surfaces. For the case of Schottky contacts, the BCs depend on the carrier concentrations and recombination velocities, S_n or S_p, at the metal/semiconductor interface according to:

$$\boldsymbol{J}_n \cdot \widehat{\boldsymbol{n}} = -qS_n(n - n_0) \tag{5.7}$$

$$\boldsymbol{J}_p \cdot \widehat{\boldsymbol{n}} = qS_p(p - p_0), \tag{5.8}$$

where $\widehat{\boldsymbol{n}}$ is the unit normal vector at the boundary and n_0 and p_0 are the equilibrium electron and hole concentrations, respectively. For the electric potential, the BC is $\phi = -\boldsymbol{\Phi}_m + V_a$, where $\boldsymbol{\Phi}_m$ is the metal work function and V_a is the applied voltage. Ideal ohmic BCs can also be considered when a good ohmic contact to a device can be assumed.

All the above assumes a nondegenerate semiconductor such that $E_c - E_f \gg kT$ (n-type) or $E_f - E_v \gg kT$ (p-type), where E_c, E_v, and E_f are the conduction band, valence band, and Fermi energies, respectively. Degenerate semiconductors can also be accounted for as necessary and are especially important in highly doped regions when the doping concentration is close to or greater than the effective density of states. In such cases, Fermi–Dirac statistics are employed rather than Maxwell–Boltzmann statistics [1].

Equations (5.1)–(5.8) constitute the transport equations along with specific BCs; but expressions for the recombination rate, R, and the generation rate, G, are still required.

5.2.1.2 Recombination

Modeling can provide insight to the spatial distribution and relative importance of various recombination processes. Several mechanisms can be at play, such as trap-assisted, direct, and Auger recombination. In general, trap-assisted, non-radiative recombination depends on the carrier concentration and properties of the trapping defects, such as the capture cross sections for electrons and holes, σ_n and σ_p, thermal velocity, v_{th}, and trap density, N_t, according to:

$$R_n = \sigma_n v_{th} N_t [n(1 - f_t) - n_1 f_t] \tag{5.9}$$

$$R_p = \sigma_p v_{th} N_t [p f_t - p_1(1 - f_t)]. \tag{5.10}$$

The factors $n_1 = n_i \exp[(E_t - E_i)/kT]$ and $p_1 = n_i \exp[-(E_t - E_i)/kT]$ are related to the intrinsic density, n_i, and provide information on the energy level of the trap, E_t, relative to the intrinsic level, E_i. Trap energy levels can be discrete or continuous as defined by the density of states. Dependence on the bandgap, E_g, and the effective densities of states in the conduction and valence bands, N_c and N_v, are given by the intrinsic concentration, $n_i = \sqrt{N_c N_v} \exp(-E_g/2kT)$. In time-dependent scenarios (e.g., when calculating capacitance features with AC bias), the trap occupation probability, f_t, is determined by:

$$N_t \frac{\partial f_t}{\partial t} = R_n - R_p. \tag{5.11}$$

The charge density of the traps must also be computed to determine its effect on the electric potential and capacitance. For example, a given concentration of donor-type, $N_t^{(+/0)}$, and acceptor-type, $N_t^{(-/0)}$, defect species would result in a charge density:

$$Q_t = q(1 - f_t)N_t^{(+/0)} - qf_tN_t^{(-/0)}. \tag{5.12}$$

In steady-state conditions, $R_n = R_p$ and (5.9)–(5.11) reduce to the Shockley–Read–Hall (SRH) expression for recombination:

$$R_{\mathrm{SRH}} = \frac{np - n_i^2}{\tau_p(n + n_1) + \tau_n(p + p_1)}, \tag{5.13}$$

where $\tau_n = 1/\sigma_n v_{th} N_t$ and $\tau_p = 1/\sigma_p v_{th} N_t$ are average electron and hole lifetimes, respectively, for this mechanism.

Direct band-to-band recombination is a radiative process. The net rate of direct recombination is given by:

$$R_d = C\left(np - n_i^2\right), \tag{5.14}$$

where the rate constant, $C \approx 10^{-10}$ cm^3 s^{-1} is typical for some thin films (direct-absorption bandgap), but for crystalline silicon (c-Si), $C = 4.73 \times 10^{-15}$ cm^3 s^{-1} (indirect bandgap) [6,7]. The rate constant is a material parameter that can depend on temperature, dopant concentrations, and injection density (due to Coulomb forces between carriers) [8].

Auger recombination is a three-body mechanism typical at higher injection levels defined by:

$$R_A = \left(g_{eeh} C_n n + g_{ehh} C_p p\right)\left(np - n_i^2\right). \tag{5.15}$$

For c-Si, $C_n = 2.8 \times 10^{-31}$ cm^6 s^{-1} and $C_p = 9.9 \times 10^{-32}$ cm^6 s^{-1} [9]. The enhancement factors, g_{eeh} and g_{ehh}, account for Coulomb attractions. Empirically derived expressions for the carrier concentration and temperature dependencies of g_{eeh}, g_{ehh}, C_n, and C_p for c-Si are provided in Altermatt *et al.* [10].

The earlier described mechanisms constitute the most important bulk recombination mechanisms, but surface recombination can also be critical, especially near the illuminated or highly defective surfaces. Surface mechanisms are similar to SRH but with the lifetimes replaced by surface recombination velocities, S_p and S_n, and carrier concentrations replaced by surface densities, n_s and p_s, in units of cm^{-2}. The resulting steady-state expression is:

$$R_{\mathrm{SRH}} = \frac{n_s p_s - n_i^2}{1/S_p(n_s + n_1) + 1/S_n(p_s + p_1)}. \tag{5.16}$$

For example, given a c-Si surface passivated by 100 nm of SiO_2, typical values are $S_n = S_p = 2{,}000$ cm/s [11]. Surface recombination velocities can also be related to specific defects by $S_n = N_{st}\sigma_n v_{th}$, where N_{st} is the density of surface states. As with bulk charge given by (5.12), surface charge can play an important role.

5.2.1.3 Optical physics: photogeneration

A critical component in PV device modeling is the contribution of optical absorption to the generation rate, G, in (5.1)–(5.3). Electron–hole pairs may also be generated by electron or laser beams. This section focuses on uniform light absorption, but similar physics applies to electron beams used in EBIC or laser beams used in TRPL. Calculation of G can be accomplished by several means depending on the complexity of the model and the level of detail required. In any case, it is typical to assume an equal generation rate for electrons and holes, $G = G_n = G_p$. For a simple 1D case with photo-absorption along the z direction, one can employ the exponential decay expression:

$$G(\lambda, z) = \alpha(\lambda)\Gamma_0(\lambda)\exp(-\alpha z), \tag{5.17}$$

where λ is the wavelength and Γ_0 is the incident photon flux. The absorption coefficient can be obtained from measurement or approximated by $\alpha = A\sqrt{h\nu - E_g}$ for direct optical absorption, where h is Planck's constant and ν is light frequency (indirect absorption and the case of degenerate semiconductors requires some simple modifications [1]).

G from (5.17) can be added to the numerical model as an algebraic expression and summed over all wavelengths. When reflective surfaces and multiple internal reflections are considered, the expression includes several additional terms with the thickness of each layer and the reflectance of each surface. Programs such as SCAPS-1D [12] and AMPS [13] employ that approach to calculate G for a given device. Those calculations can also be conducted in generic math programs such as MATLAB® or Mathematica®, or the PV-specialized optical simulation software e-ARC [14].

An alternate means of determining G avoids assembly of the algebraic expressions instead by solving the differential equation:

$$\frac{\partial \Gamma}{\partial x}\hat{x} + \frac{\partial \Gamma}{\partial y}\hat{y} + \frac{\partial \Gamma}{\partial z}\hat{z} = -\alpha\Gamma(\lambda, x, y, z), \tag{5.18}$$

which provides the spatial variation of the photon flux, Γ. Equation (5.18) essentially leads to the exponential attenuation of light, and the photo-generation rate for each wavelength is obtained by $G = \alpha\sqrt{\Gamma_x^2 + \Gamma_y^2 + \Gamma_z^2}$. The advantages of using (5.18) are that it can be employed in 2D or 3D and is quickly solved by the FEM in the same model as the semiconductor problem. The disadvantage is that it does not include multiple reflections.

The most rigorous way to compute G is by starting from Maxwell's equations. This approach can include plasmonic effects, diffraction, multiple internal reflections, various angles of incidence, and other fine details of optical physics. The spatial variation of the optical electric field is determined by:

$$\nabla \times \left(\nabla \times \boldsymbol{E}_{opt}\right) - k_0^2\varepsilon_r\boldsymbol{E}_{opt} = 0, \tag{5.19}$$

which is solved in the frequency domain with the wave vector $\boldsymbol{k}_0$. The relative permittivity is related to the complex refractive index $\varepsilon_r = (n_r - ik_e)^2$, which comprises a real part, n_r, and an imaginary part, k_e, also known as the extinction coefficient. The absorption coefficient is given by $\alpha = 4\pi k_e/\lambda$. Light intensity (proportional to $\boldsymbol{E}^2_{opt}$) decreases as it is absorbed, resulting in a power dissipation density, Q. From there, the generation rate can be computed, $G = Q\lambda/hc$, where c is the speed of light.

Only a single wavelength is considered in (5.19), but typically the incident light must be input as a spectrum (such as AM1.5) with the intensity specified for each wavelength interval. Total G is the integral of individual G values over wavelength of the full spectrum. When using (5.19), port BCs are employed to specify input power and angle of incidence, and periodic BCs allow for efficient modeling of large geometries. There are several software packages capable of handling this level of optical simulation, including COMSOL Mutliphysics and Lumerical DEVICE®.

Example solutions of (5.19) given by the resulting generation rates are shown in Figure 5.1 for 500- and 800-nm light incident on a CdTe/MgZnO/SnO_2 PV device. Most of the optical-field absorption is within the first micron of the CdTe layer.

5.2.1.4 Temperature dependence

Exponential temperature dependence is inherent in the solution of the carrier concentrations (n and p), but explicit temperature dependencies of other parameters

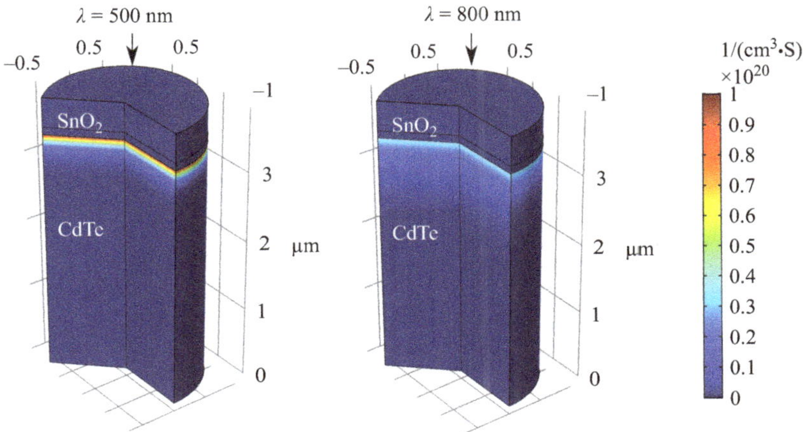

Figure 5.1 Generation rates as functions of depth in CdTe PV device at the two wavelengths shown. Rates were calculated using (5.19). The cylindrical geometry represents a columnar grain. The 500-nm wavelength is absorbed more strongly at the CdTe surface and the 800-nm light penetrates deeper. There is a 100-nm-thick MgZnO layer between the tin oxide (SnO_2) and the CdTe

must be added manually and are material-specific. Approximate temperature-dependent expressions for the effective density of states in the conduction band, $N_c(T)$, valence band, $N_v(T)$, and thermal velocity, $v_{th}(T)$ are as follows:

$$N_c = N_{c0}\left(\frac{T}{T_0}\right)^{3/2} \text{ where } N_{c0} = 2\left(\frac{m_e^* k T_0}{2\pi\hbar}\right)^{3/2} \tag{5.20}$$

$$N_v = N_{v0}\left(\frac{T}{T_0}\right)^{3/2} \text{ where } N_{v0} = 2\left(\frac{m_h^* k T_0}{2\pi\hbar}\right)^{3/2} \tag{5.21}$$

$$v_{th} = v_{th0}\sqrt{T/T_0}. \tag{5.22}$$

In (5.20)–(5.22), m_e^* and m_h^* are the effective electron and hole masses, k is Boltzmann's constant, and $\hbar$ is the reduced Planck constant. For example, baseline values N_{c0}, N_{v0}, and v_{th0} can be determined from the literature at a reference temperature of $T_0 =$ 293.15 K. The baseline thermal velocity is $v_{th,0} = \sqrt{3kT_0/m^*} \approx 10^7$ cm/s. Temperature dependence of the bandgap can also be included. For example, in c-Si, the bandgap depends on temperature according to [15]:

$$E_{g0} = 1.206 - 2.73 \times 10^{-4}T(\text{eV}) \text{ for } T > 250 \text{ K}. \tag{5.23}$$

At $T = 300$ K, $E_g = 1.124$ eV. Other temperature-dependent expressions specific to c-Si are discussed in Section 5.5.

5.2.1.5 Bandgap grading and narrowing

Control of the bandgap as a function of position (grading) is an advantageous feature in thin films such as CIGS, where the gap is controlled by gallium (Ga) content relative to indium (In). Secondary-ion mass spectroscopy (SIMS) or glow-discharge optical emission spectroscopy (GDOES) can provide the depth-dependent Ga/(Ga+In) (GGI) profile. In graded bandgap models, it is important to know how the conduction and valence bands shift with changes in the gap, which can be understood in terms of variations of the electron affinity, χ. For CIGS, the bandgap and affinity can be determined from Stokes *et al.* [16]:

$$E_g = E_g^{CIS} + x\left(E_g^{CGS} - E_g^{CIS}\right) - bx(1-x) \tag{5.24}$$

$$\chi = \chi_g^{CIS} - \beta x\left[E_g^{CGS} - E_g^{CIS} - b(1-x)\right], \tag{5.25}$$

where $x =$ GGI, ${E_g}^{CIS} = 1.04$ eV, ${E_g}^{CGS} = 1.68$ eV, $b = 0.15 - 0.24$, $\chi^{\text{CIS}} = 4.5$ eV, and $\beta = 0.94$. GGI measurements for three CIGS cell and the resulting profiles for $b = 0.15$ are shown in Figure 5.2 [17]. Caution is required when using bandgap grading because the energy difference between the Fermi level and the valence band (for p-type) or the conduction band (for n-type) can be determined by the doping concentration, which will overrule the expected affinity grading from (5.25).

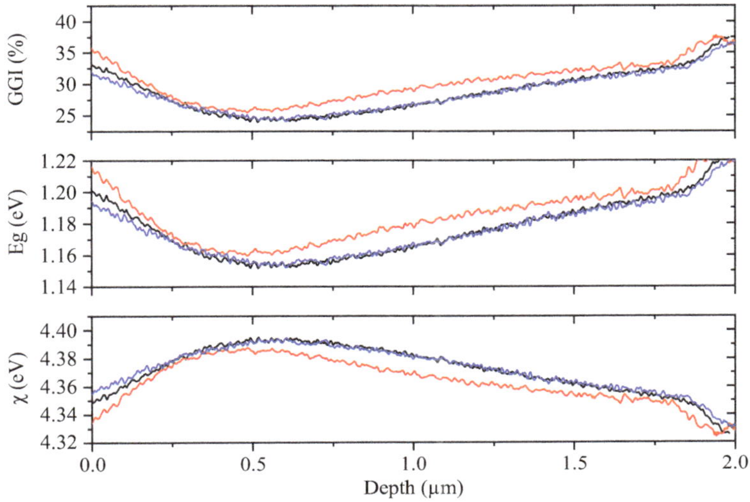

Figure 5.2 GGI as a function of depth in the CIGS layer from GDOES data along with corresponding bandgap and electron affinity profiles for three CIGS cells. Zero depth is at the buffer/CIGS interface [17]

Bandgap narrowing (BGN), ΔE_g, can also occur when the doping concentration, N, is sufficiently high. For Si, the empirical model of Yan and Cuevas [18,19] appropriate for Fermi–Dirac statistics implies:

$$\Delta E_g = \Delta E_{slope}\left[ln\frac{N}{N_{onset}}\right]^b \text{ for } N > N_{onset}, \tag{5.26}$$

with values of $\Delta E_{slope} = 4.20 \times 10^{-5}$ eV in phosphorous-doped Si, $\Delta E_{slope} = 4.72 \times 10^{-5}$ eV in boron-doped Si, $N_{onset} = 10^{14}$ cm^{-3}, and $b = 3$. At 10^{19} cm^{-3} of p-doping, $\Delta E_g \approx 0.05$ eV. A comprehensive BGN model was provided by Schenk [20] and further described in McIntosh and Altermatt [21]. However, that model is quite complex, and the Yan and Cuevas model provides a good fit for the data over a wide range of doping concentrations at $T = 300$ K.

Variations in the bandgap leads to an "effective" intrinsic concentration given by:

$$n_{i,eff} = n_i \exp\frac{\Delta E_g}{2kT}, \tag{5.27}$$

where $n_i = 9.65 \times 10^9$ cm^{-3} for c-Si [22]. When BGN is considered, the substitution $n_i \rightarrow n_{i,eff}$ should be made in all the earlier equations.

5.2.1.6 Ionic drift-diffusion reaction

Migration of ions can strongly affect the performance and degradation of PV devices. The motion of ions of concentration, c, is influenced by the concentration gradient, ∇c, diffusion coefficient, D, electric field, $\boldsymbol{E} = -\nabla\phi$, and the reaction rate, R, as dictated by:

$$\frac{\partial c}{\partial t} = \nabla \cdot (D\nabla c) - \nabla \cdot (c\mu \boldsymbol{E}) + R. \tag{5.28}$$

The ion mobility, μ, is related to the diffusivity by $D = \mu kT/q$, which is thermally activated: $D = D_0 exp(-E_a/kT)$, where E_a is thermal energy. Ion transport therefore has strong temperature dependence.

Coupling of (5.28) to the transport equations (5.1)–(5.3) occurs through the electric field, which both depends on and affects the ion distribution; basically, c from (5.28) becomes part of the charge density, ρ, in (5.4). Ion migration can occur through bulk material, through GBs, and along surfaces. Reactions between ions and lattice defects can also be specified as differential reaction rate equations embodied by the term R in (5.28). Examples of that approach related to CdTe devices can be found in Guo *et al.* [23]. A challenge in employing (5.21) is the determination of reasonable values for D and R. Other subtle complications include the segregation and precipitation of impurities at GBs and other extended defects. A thorough description of diffusion physics in solids can be found in the study by Mehrer [24].

An example of positive ion (e.g., Na^+) transport in a 2D CIGS model with a GB is shown in Figure 5.3, in which the conditions are 1-sun light intensity, zero bias (short circuit), and $T = 293.15$ K. Diffusion coefficients for sodium in the CIGS bulk and along the GB are, respectively, $D = 9.7 \times 10^{-9}$ exp(-0.36 eV/kT) cm^2/s and $D_{gb} = 6.5 \times 10^{-9}$ exp(-0.21 eV/kT) cm^2/s [25]. The center graph shows how ion migration along the GB is strongly affected by the built-in field; it is suppressed by the field at about 2.8 μm from the back contact. Correspondingly, the band diagram clearly shows the location of the electric-field boundary at the edge of the depletion region. The same physics described by (5.28) accounts for any chemical species.

5.2.1.7 GB effects

In addition to acting as preferential pathways for ion migration, GBs can have significant electronic effects related to localized charge and recombination properties that impact performance, degradation, and the interpretation of experimental measurements. As shown in Figure 5.4, a positively charged GB in a CIGS device can reduce the p-n junction field resulting in lower V_{oc}. The result implies that a localized decrease in the p-n junction barrier height leads to weak diode behavior (i.e., lower V_{oc} without shunting) [45]. The book by Möller [26] provides detailed information on GBs in PV materials.

These types of calculations are important because GB properties can make it difficult to experimentally distinguish bulk from GB features. For example, Figure 5.5 shows a 2D axisymmetric model of a CdTe micro-diode with a vertical GB as the outer surface of the cylinder. The GB recombination velocity, S_{gb}, and acceptor-type charge density in the GB, $N_{a,gb}$, are specified independently of the bulk mid-gap defect density, N_t, and acceptor doping, N_a. Various combinations of

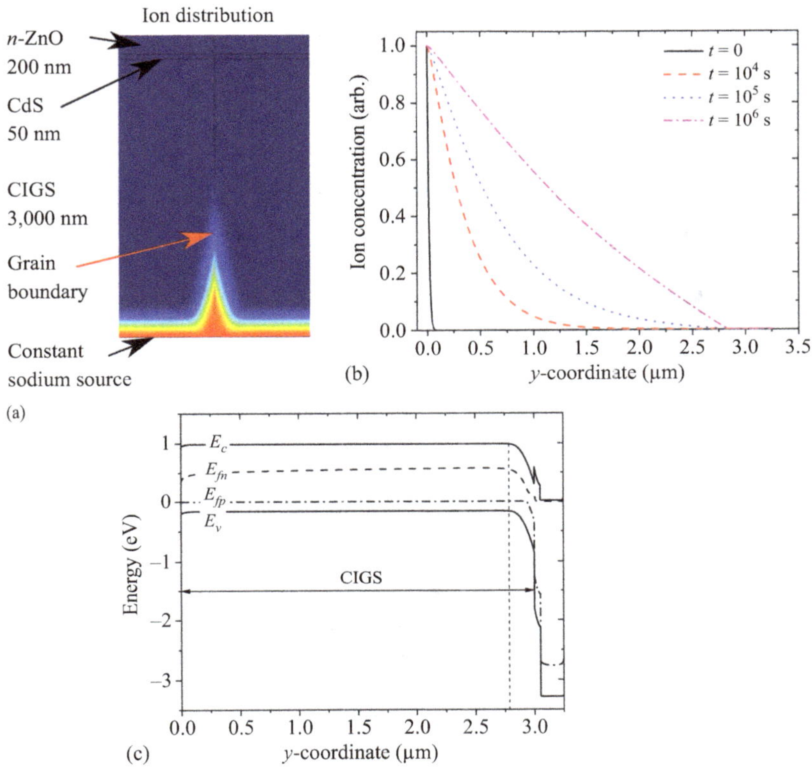

Figure 5.3 2D simulation of a CIGS device with a GB. Ion transport is modeled with a constant source of sodium at the bottom. (a) Model geometry and ion concentration color map at $t = 10^5$ s. (b) Ion concentration along GB at different times; ion transport stopped by built-in electric field. (c) Calculated band diagram for CIGS at 1-sun light intensity illumination and short-circuit conditions; dashed line indicates the field boundary at the edge of the depletion region

those four parameters (plus grain size, d) affect the JV, QE, and CV characteristics in different ways. The properties for three hypothetical cases are listed in Table 5.1, all with $d = 1$ μm. Cases A and B have the same bulk properties, but case B has higher GB recombination. Cases A and C have the same recombination rates, whereas acceptor doping is dominant in the bulk for case A and in the GB for case C. The JV and QE curves in Figure 5.6 show that GB recombination has a significant impact on performance; yet, there is equally good performance for doping in either the bulk or GB. In Figure 5.7, the CV results, shown as acceptor doping vs depletion width (W_d), indicate that it is not possible using this technique to determine if the doping is in the bulk (case A) or GB (case B). Grain size is also critical

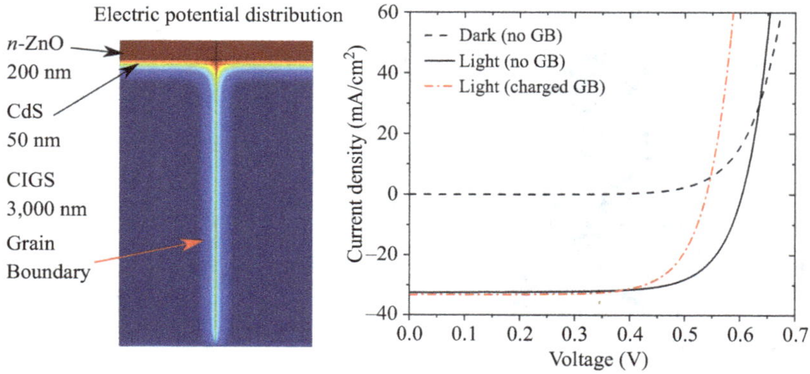

Figure 5.4 Left: Electric potential in a CIGS device with a positively charged GB (red to blue is 0 to 0.95 V). Right: the impact of the positively charged GB on the JV curve—local weak diode behavior and loss of V_{oc} with a very slight increase in J_{sc}

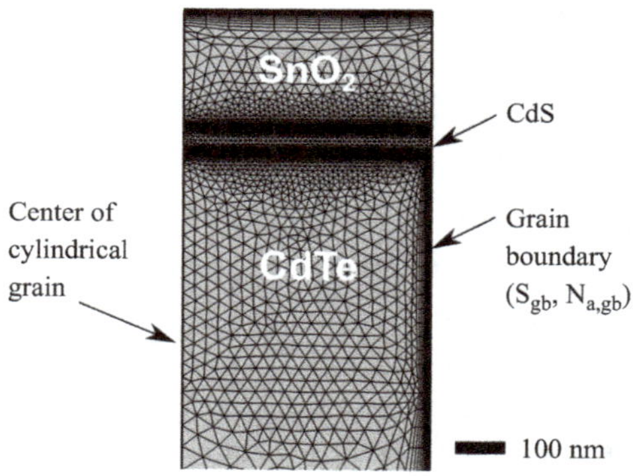

Figure 5.5 Cylindrical grain model of CdTe PV device showing the mesh in a zoomed region. GB properties defined on the surface of the cylinder, S_{gb}, surface recombination velocity; $N_{s,gb}$, acceptor-type doping in GB. Note the very fine mesh at all surfaces and interfaces

Table 5.1 Bulk and GB parameters for a CdTe micro-diode device model

	Bulk		**Grain boundary**	
Case	N_a **(cm^{-3})**	N_t **(cm^{-3})**	$N_{a,gb}$ **(cm^{-3})**	S_{gb} **(cm/s)**
A	10^{15}	10^{13}	0	10^5
B	10^{15}	10^{13}	0	10^7
C	10^{13}	10^{13}	3×10^{15}	10^5

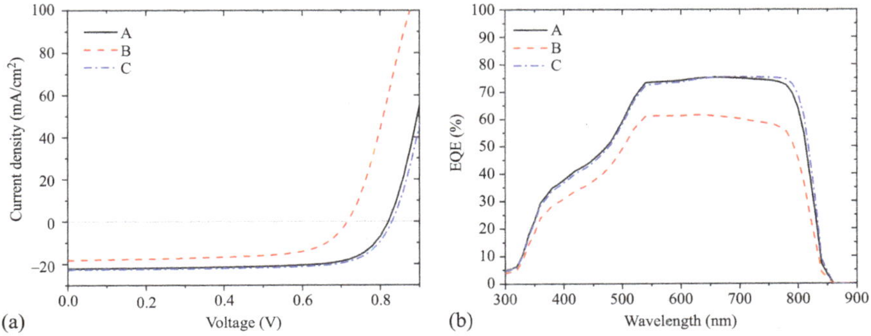

Figure 5.6 JV and EQE calculations for the CdTe micro-diode in Figure 5.5 with the parameters listed in Table 5.1. Higher GB recombination (case B) reduces performance significantly. Cases A and C indicate that equal performance is obtained with doping at GBs or in the bulk

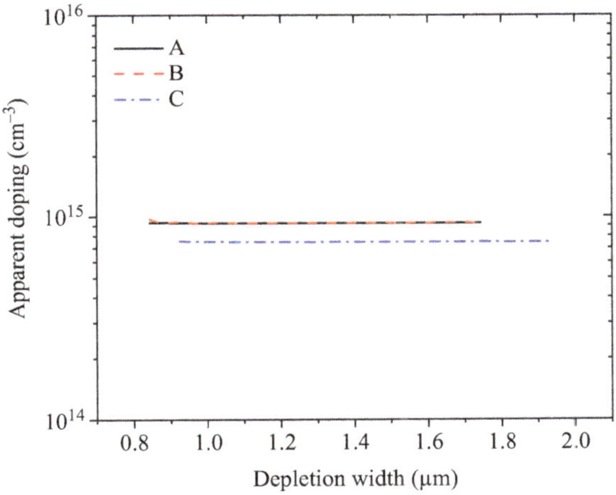

Figure 5.7 CV (shown as apparent doping vs depletion width) calculations for the CdTe micro-diode in Figure 5.5 with the parameters listed in Table 5.1. Doping in the bulk (case C) cannot be distinguished from doping in the GBs

for determining performance and degradation because it defines the overall density of GBs in the material.

5.2.2 Cell-scale modeling

5.2.2.1 Thin-film cells

By solving the equations described in Section 5.2.1, one can simulate all measurable outputs typical of PV cells. However, given the impracticality of solving the semiconductor equations at the cell scale, the fundamental physics is calculated at the micron scale and those outputs are used as inputs to the cell scale. Specifically, the current density, j, as a function of voltage, V, temperature, T, and illumination intensity, Γ are calculated at the micron scale, and those values are used to build an interpolating function, $j(V,T,\Gamma)$. This interpolating function is then used as an input to the cell-scale model, as described later. Alternatively, any analytical expression for $j(V,T,\Gamma)$ can be used (e.g., based on the diode equation).

This approach exploits the fact that at the cell scale, we are mainly concerned with lateral currents through the electrically conductive layers (e.g., transparent conductive oxide (TCO) layer for the case of thin films). Therefore, we can employ the equivalent-circuit model of a thin-film solar cell shown in Figure 5.8. The micro-diodes, which are parallel to each other, feed a certain amount of current density per unit length (or area in 2D), $j(V,T,\Gamma)$, into the top TCO layer, which has some finite sheet resistance, R_{TCO}. The metal contact on the bottom is assumed to be a perfect conductor on this scale. In the limit of infinitesimal lateral diode spacing, we obtain the current conservation equation $\nabla \cdot \boldsymbol{J} = j/d$, which provides the current density in the TCO layer and where j serves as a source term in units of A/m^2 (in 2D). Using Ohms law, $\boldsymbol{J} = -\sigma\nabla\phi$, we obtain:

$$-\nabla \cdot (\sigma\nabla\phi) = j/d, \tag{5.29}$$

which is solved for the electric potential, ϕ, throughout the resistive electrode with the TCO conductivity given by $\sigma = 1/(d\ R_{TCO})$. The TCO thickness, d, is usually less than 1 μm. Equation (5.22) can be solved efficiently by using the FEM, which has several advantages over circuit network solvers, such as SPICE, including easier coupling to thermal effects and modeling arbitrary geometries. Using FEM, BCs on ϕ can be set to study any metal-contact shape or location, such as an outer-ring contact on a circular cell. This approach can also be used to study the effects of nonuniformities in the TCO conductivity and the micro-diode currents, j, as shown in Figure 5.9.

Figure 5.9 shows a 1-cm diameter CdTe cell with 40 randomly placed "weak diodes," identified by the very fine mesh regions. The weak micro-diodes have $V_{oc} = 0.62$ V, whereas the rest of the cell has $V_{oc} = 0.86$ V. From Figure 5.9, it can be observed that as the number of weak diodes in the cell increases, the V_{oc} of the cell decreases. Also shown is the JV effect of a full shunt (shunt resistance $R_{shunt} = 0\ \Omega$) in the center of the cell. Using this approach, we can calculate the

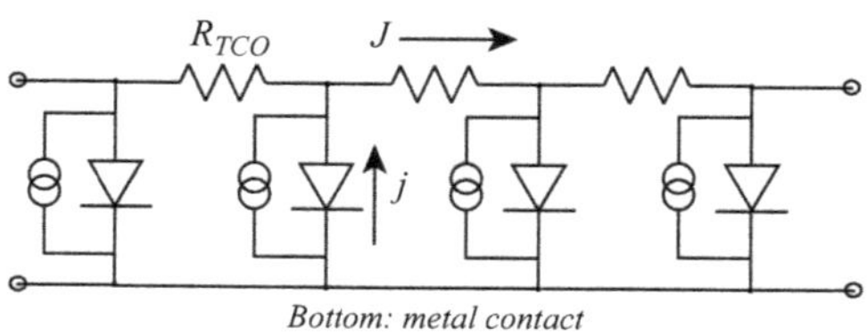

Figure 5.8 Equivalent-circuit model of a solar cell used to derive (5.29)

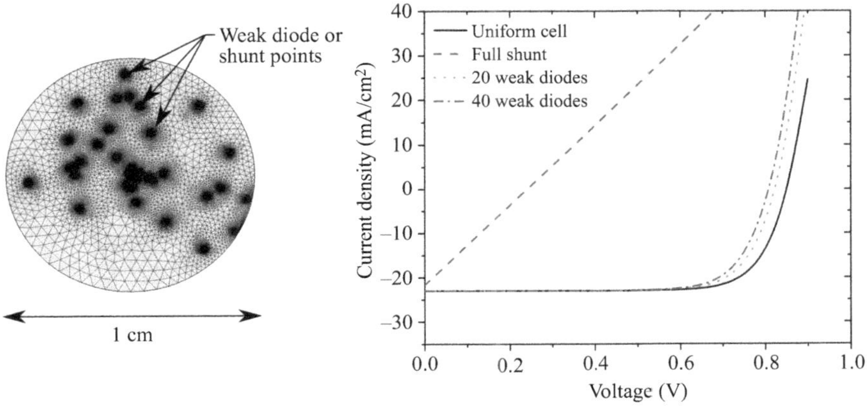

Figure 5.9 *Right: mesh of a 1-cm-diameter CdTe cell with 40 randomly placed weak diodes (weak diodes have $V_{oc} = 0.62$ V while $V_{oc} = 0.86$ V for the rest of the cell). Left: JV curves for a cell with 0, 20, and 40 randomly placed weak diodes. The effect of a full shunt ($R_{shunt} = 0\ \Omega$) through the center of the cell is also shown*

effects of the severity and locations of weak diodes and shunts in any type of cell. A similar approach can be used for modules, as described in Section 5.2.3.

5.2.2.2 Si cells

As with thin-film PV at the cell scale, we are mainly concerned with lateral currents that flow through the electrically conductive front-surface materials. The main difference here is that we are dealing with a metal-grid pattern and conductive, highly doped Si emitter layer as the front surface (typically), rather than a TCO. Yet, the lateral current is still determined by (5.29), where the conductivity, σ, and thickness, d, depend on the specific material (metal or semiconductor) carrying the current. For c-Si devices, $\sigma = 1/(d\ R_{sq})$, where d is the thickness of the front doped emitter layer and $R_{sq} \approx 10 - 100\ \Omega$/sq is its sheet resistance.

Geometric inputs to the c-Si cell model include the cell dimensions, finger width, spacing, and thickness, and bus bar width, spacing, and thickness. Material parameters include Si sheet resistance and resistivity of the metal. A specific cell layout is shown in Figure 5.10, along with the electric potential distribution at 1-sun light intensity, short-circuit condition. The light and dark JV curves are also shown in Figure 5.10. Dark diode behavior is assumed below the metal-grid pattern. Any metal grid or cell shape can be studied in this way, along with nonuniformities such as shunts and weak diodes. Parameter values and cell properties can be found in Altermatt and colleagues [6,11].

5.2.3 Module-scale modeling

There are basically two types of modules, depending on how the individual cells are electrically connected, either soldered or monolithically integrated. Our

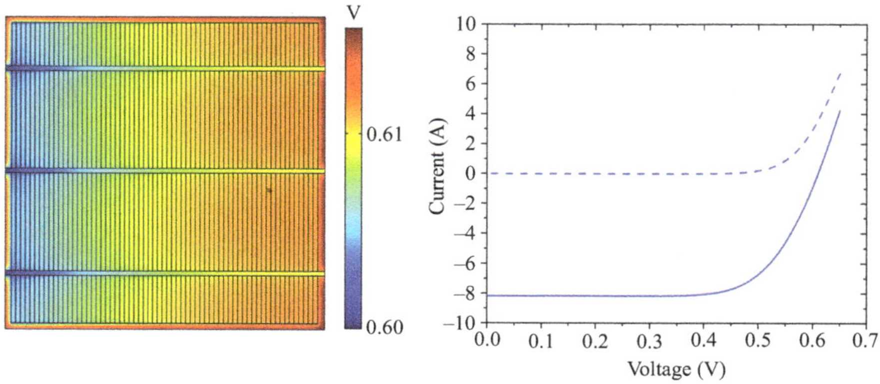

Figure 5.10 Left: Electric potential distribution (color legend in volts) in a 15.6 × 15.6 cm² Si cell under 1-sun light intensity and applied bias of 0.6 V. Bias is applied to the left ends of the three bus bars. Right: Calculated dark and light IV curves for the Si cell shown on the left at 1-sun light intensity and T = 293 K

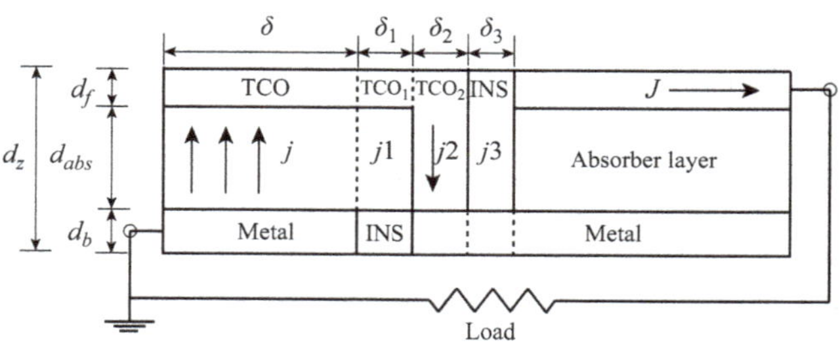

Figure 5.11 Monolithic module schematic with essential features for simulation. Dashed lines indicate current domains, not material breaks. INS is insulating material. Light is incident from the top. This schematic is a simplified representation of a typical substrate-configuration device. For a superstrate device (such as commercial CdTe), the TCO and metal regions would be interchanged [40]

current discussion focuses on the latter, which is typically how thin-film PV modules are constructed. These modules are made by depositing the active materials over the entire surface of the substrate and sequential scribing of the contacts using laser ablation or mechanical means. A schematic of a monolithic module section used as a model for the calculations is shown in Figure 5.11. The dashed lines indicate different current domains, not breaks in the material. Typical scribe widths (δ_1, δ_2, δ_3) are on the order of 100 μm, and the active region is $\delta < 1$ cm. Many of the pertinent parameter values for monolithic modules are provided by Silverman *et al.* [27].

Simulation of a monolithic module is similar to that of a cell in which we ultimately solve (5.29). But now, we must deal with two layers—the front contact and the back contact—to account for the interconnection geometry. Therefore, (5.22) is required for both layers, and the two equations must be solved simultaneously for the electric potentials of the front contact (*fc*) and back contact (*bc*): $\phi_{fc}(x,y)$ and $\phi_{bc}(x,y)$. Separate material properties must be specified for both layers. The two contact layers are connected by currents that act as a source current for one layer and a corresponding sink for the other layer (e.g., j and j_2 in Figure 5.11). In all, there are four different current source domains, as indicated by the lowercase j's in Figure 5.11. j is the same micro-diode current $j(V,T,\Gamma)$ discussed in Section 5.2.2 (cell model). In the other current domains, $j_1 = j_3 = 0$, and j_2 is an ohmic current, $j_2 = (\phi_{fc} - \phi_{bc})\sigma_2/d_{abs}$, where σ_2 is the electrical conductivity of the connective material of width δ_2 in Figure 5.11 and d_{abs} is the absorber layer thickness.

An example of the electric potential distribution and the dark and light JV curves for a 10-cm × 10-cm CIGS mini-module are shown in Figure 5.12. The dimensions are $\delta = 0.5$ cm, $\delta_1 = 100$ μm, $\delta_2 = 100$ μm, and $\delta_3 = 100$ μm (refer to Figure 5.11); these are commercially relevant dimensions for thin-film PV modules.

A useful feature of this finite-element modeling approach for both cells and modules is that it enables coupling with thermal physics models (described in Section 5.2.3.1). The latter point is particularly crucial for studying how modules are affected by hot spots, shunts, and partial shading. An example calculation of thermal runaway in CIGS modules is provided in Section 5.4.

5.2.3.1 Electrothermal physics for cell and module simulations

The lateral current flow model of (5.29) can be employed at both the cell and module scales to simulate electrical performance. However, electric current and

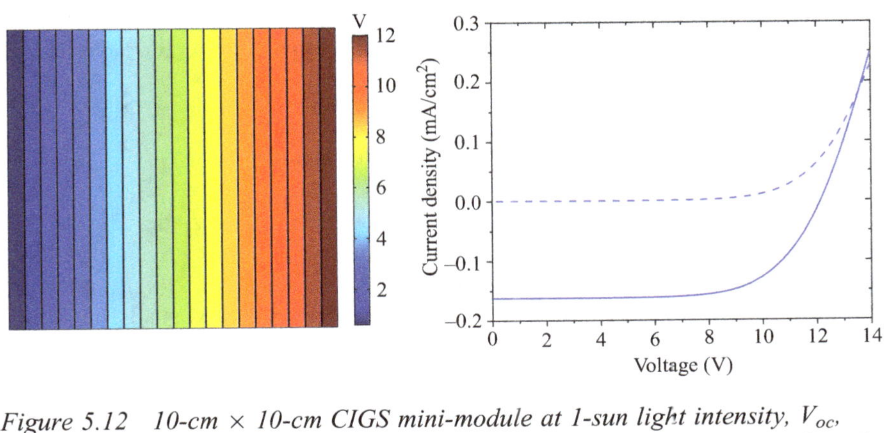

Figure 5.12 10-cm × 10-cm CIGS mini-module at 1-sun light intensity, V_{oc}, $T = 293$ K. Left: color plot of electric potential distribution (in volts) in the TCO layer relative to ground. Right: dark and light JV

temperature are directly coupled, and thermal effects can have major implications for device performance and degradation. The temperature distribution in a cell or module, which depends on environmental conditions, material properties, and device architecture, is governed by the heat equation:

$$\rho C \frac{\partial T}{\partial t} - \nabla \cdot (\kappa \nabla \mathrm{T}) = \mathrm{Q}, \tag{5.30}$$

where ρ is the material density, C is the specific heat capacity, and κ is the thermal conductivity. The sum of all heat sources, Q in units of W/m^2, can include several terms:

$$Q_d = Q_L + j(\phi_{fc} - \phi_{bc})/d_{abs} \text{ Radiant heat}(Q_L) \text{ and diode joule heat} \tag{5.31}$$

$$Q_J = \boldsymbol{J} \cdot \boldsymbol{E} \text{ Joule heat through contacts (e.g., TCO, metal)} \tag{5.32}$$

$$Q_h = h_{conv}(T - T_\infty) \text{ Convective heat transfer} \tag{5.33}$$

$$Q_\varepsilon = \varepsilon \sigma_{SB}(T^4 - T_\infty^4) \text{ Radiative heat transfer} \tag{5.34}$$

$\boldsymbol{J}$ and $\boldsymbol{E}$ are the lateral current densities and electric fields, respectively, in each of the contact layers, h_{conv} is the convective heat-transfer coefficient in Newton's Law of cooling, T_∞ is ambient temperature, ε is the emissivity, and $\sigma_{SB} = 5.67 \times 10^{-8}$ W/m^{2}K^4 is the Stefan–Boltzmann constant. The radiant heat at 1 sun is $Q_L = 1{,}000$ W/m^2, j is the diode current, and $\phi_{fc} - \phi_{bc}$ is the difference in potential between the front and back contacts, as shown in Figure 5.11 for the module model. It should be noted that photocurrent is negative, resulting in a cooling affect via (5.24) when the PV device dissipates power externally. Thermal effects of module packaging can be included in the convective and emissive factors in (5.33) and (5.34). Equations (5.30)–(5.34) can also be used at the cell scale, where only one contact and corresponding potential, ϕ, would be considered. Coupling between the electrical and thermal models is accomplished through the voltage and temperature dependence of the diode current j (and all other temperature-dependent parameters).

5.3 Modeling defect kinetics: CdTe device example

There are several mechanisms by which PV modules can degrade. But in this section, we are primarily concerned with the gradual, transient effects that have been observed in both CdTe cells and modules upon exposure to light, elevated temperatures, and/or voltage bias [28–30]. These phenomena are usually thermally activated in nature. In the following, two transient mechanisms are described with numerical simulations results compared with data: (1) defect kinetics driven by nonequilibrium charge density and (2) back-contact degradation.

Hypothesis testing requires that a physically reasonable model must predict how key performance metrics, such as efficiency (η), open-circuit voltage (V_{oc}), short-circuit current (J_{sc}), and fill factor (*FF*), change with time. As a more stringent test, postulated degradation mechanisms must account for the observed variations of QE and CV, as well as for other measurements. Here, it is also demonstrated how modeling results can be compared with Kelvin probe force microscopy (KPFM) data. We begin by describing the parameters and geometries specific to baseline CdTe thin-film PV models.

5.3.1 Baseline CdTe model

Baseline thin-film CdTe devices follow the description by Rose *et al.* [31] with some modifications: the intrinsic SnO_2 layer is neglected and a thin (100-nm) layer of interdiffused $CdTe_{1-x}S_x$ material is included between the CdTe and CdS layers. The superstrate configuration of a CdTe cell, along with the calculated band diagram, is shown in Figure 5.13. Baseline model parameter values are provided by Gloeckler *et al.* [32], but they will vary depending on the device under study.

The model includes a $CdTe_{1-x}S_x$ layer as an interdiffused region because it may play an important role in degradation. The presence of a thin (~100 nm) interdiffused region has been identified [33], along with the presence of a strong electric field confined to that narrow layer; that feature is evident in the model band diagram in Figure 5.13 for $7.9 < x < 8.0$ μm. The material parameters of the interdiffused layer were kept the same as CdTe but with an increase in both the acceptor (N_a) and mid-gap defect (N_t) concentrations from 2×10^{14} cm^{-3} to 2×10^{16} cm^{-3} (to account for the relatively high defect concentration and field

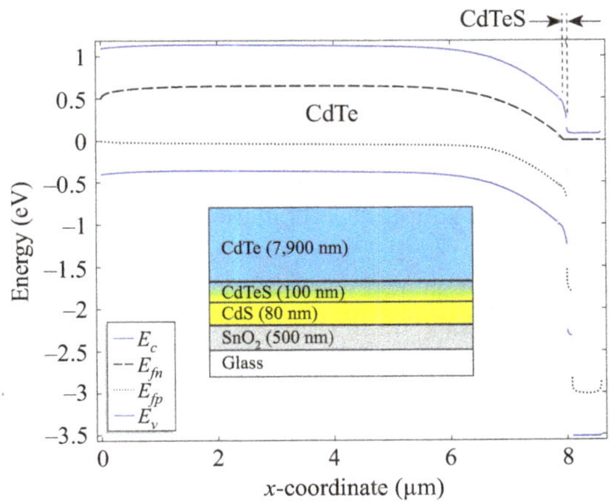

Figure 5.13 Diagram of baseline superstrate CdTe cell structure and calculated energy band diagram under 1-sun light intensity and short-circuit condition. $CdTe_{1-x}S_x$ region within $7.9 < x < 8.0$ μm. E_c, E_v, E_{fn}, and E_{fp} are, respectively, the conduction band, valence band, electron quasi-Fermi, and hole quasi-Fermi levels. The reference energy is the equilibrium Fermi level at $E_f = 0$ eV [30]

there). N_a was assumed to be fully ionized, whereas the energy level and electron/hole capture cross sections for N_t were 0.75 eV above the valence band (at mid-gap), $\sigma_n = 10^{-12}$ cm^2, and $\sigma_p = 10^{-15}$ cm^2, respectively.

5.3.2 Defect kinetics: charge-induced degradation

In this model, excess charge carriers generated by light and/or voltage bias act as the drivers of degradation. Basically, these nonequilibrium charge carriers (electrons and/or holes) cause new defects to form (or defect transformations) via large lattice relaxations, which can increase recombination and alter the built-in field. In turn, those changes affect charge-carrier concentrations and a feedback loop is established. This type of degradation mechanism has been studied in many types of semiconductors [34] and has been evaluated with respect to CdTe [30,35].

If we assume that the defect generation rate is linear in the nonequilibrium charge-carrier concentration, n, then the rate equation is given by Harju *et al.* [36]:

$$\frac{\partial N_t}{\partial t} = \alpha n(N_t) - \beta N_t, \tag{5.35}$$

where α and β are, respectively, defect creation and annihilation rates with thermally activated nature:

$$\alpha = \alpha_0 \exp(-E_a/kT) \text{ and } \beta = \beta_0 \exp(-E_b/kT). \tag{5.36}$$

The activation energies, E_a and E_b, depend on the nature of atomic transitions in the material, and the prefactors, α_0 and β_0, can be estimated as the characteristic atomic frequency, $v_0 \approx 10^{13}$ s^{-1}. The term with β accounts for annealing effects that lead to saturation or reversal of defect formation. When defect reactions are involved, the rate coefficients, α and β, may also depend on the concentrations of impurities or other factors that affect lattice energetics.

Due to recombination processes, carrier concentration is a function defect density, $n = n(N_t)$, which makes (5.35) nonlinear. An analytical calculation is possible given the approximations of a simple recombination rate, $R = C_n N_t n = G$, where C_n is the average recombination coefficient ($G = R$ implies a quasi-stationary carrier concentration). For the case of short time, $\beta t \ll 1$, such that the defect reaction does not saturate, then the defect concentration evolves according to:

$$N = N_0\left(1 + \frac{t}{\tau}\right)^{1/2} \text{ where } \tau = \frac{N_0^2 C_n}{2\alpha G}, \tag{5.37}$$

where N_0 is the initial defect concentration and τ is the characteristic degradation time. Equation (5.37) and similar expressions provide a means for fitting data to interpret defect kinetics.

In the numerical approach, (5.35) is coupled to the time-dependent transport in (5.1)–(5.3) through the defect concentration, N_t, which appears in both the charge density and the recombination rate. An important point is that the numerical approach considers the spatial dependence of all parameters, whereas the analytical estimates assume spatially uniform values.

5.3.3 Back-contact degradation

Back-contact degradation is considered for comparison with defect reaction mechanisms. Traditionally, a nearly ohmic back contact is created in a CdTe cell first by chemical etching of the CdTe layer and application of a Cu-doped, conductive paste followed by annealing at 260 °C. This process may lead to a region with a high concentration of Cu substitutions for Cd vacancies (Cu_{Cd}), which acts as a thin p^+-layer [12]. Such a layer can reduce the back barrier to holes by an amount given by Sze and Ng [1]:

$$\Delta\Phi = \frac{q}{\varepsilon_s}\sqrt{\frac{ca}{4\pi}}, \tag{5.38}$$

where c is the doping concentration (related to Cu here), a is the thickness of the p^+-layer (in the order of 10 nm), q is the elementary charge, and ε_s is the permittivity. A reaction of the type $Cu_{Cd} \rightarrow Cu_i^+$ could release positive interstitial Cu ions over time due to the nonequilibrium concentration of Cu_{Cd} near the back contact after the annealing process. Cu ions migrate away from the contact as the reaction proceeds (most rapidly via GBs) and the barrier-lowering effect of (5.38) is weakened, thereby increasing the back barrier, which is detrimental to device performance. A reduction in FF and an increase in the saturation of the forward current, also known as "JV curve rollover," are often associated with back barrier increase [46,47].

The loss of Cu at the back contact is governed by the reaction/drift/diffusion process described in Section 5.2.1.6. A simple approximation is considered here with a linearly increasing barrier height over time given by:

$$\Phi = \Phi_{bp0}(1 + \gamma t), \tag{5.39}$$

where Φ_{bp0} is the initial barrier height to holes and γ is a fitting parameter that embodies the pertinent transport and reaction rates. Equation (5.39) can be included in the time-dependent numerical model.

Nardone and Albin [30] compared the charge-induced defect kinetics and back-barrier degradation mechanisms to experimental data for a set of six CdTe cells that were stressed at 1-sun light intensity, $V = V_{oc}$, $T = 65$ °C. The model predictions for performance metrics over time are compared with the experimental data in Figure 5.14(a). The results indicate that typical degradation modes observed for cells stressed under light-soak and open-circuit voltage can be reasonably predicted by these mechanisms. In Figure 5.14(a), the points are from the data, the solid lines are the defect kinetics mechanism (referred to as junction model), with a defect activation energy of $E_a = 1$ eV (see (5.35) and (5.36)), and the dashed line also includes the effect of back-barrier height increase with $\gamma = 10^{-9}$ s^{-1} (see (5.38) and (5.39)). The latter resulted in a back-barrier increase from 0.40 to 0.42 eV after 700 h of light-soak stress. The back-barrier effect also resulted in slight FF reduction and an increase in JV curve rollover.

To account for the data, the defect reaction model in this case produced both deep recombination centers and shallow acceptors in response to charge injection.

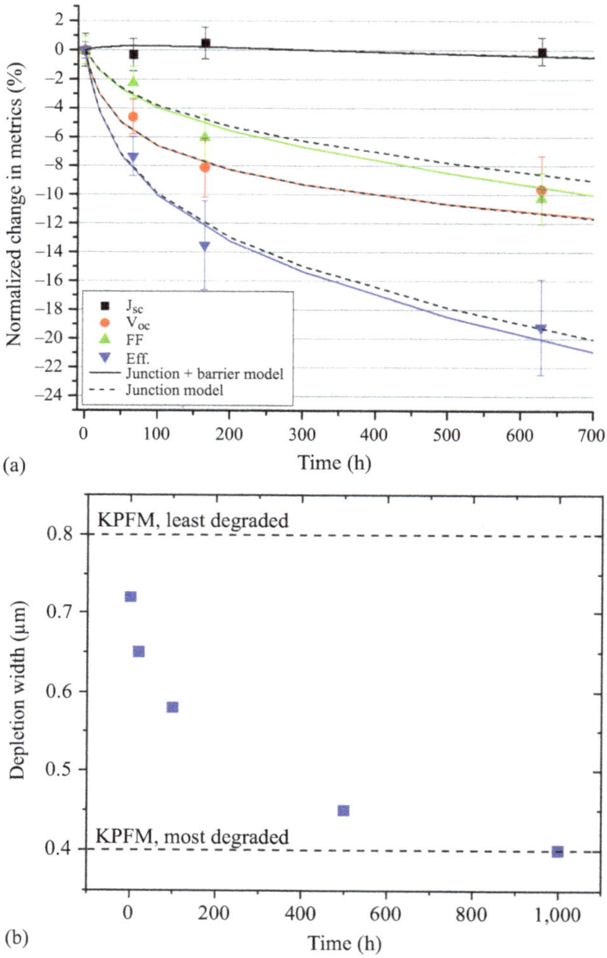

Figure 5.14 Degradation predictions for CdTe device. (a) Performance metrics vs stress time at 1-sun light intensity, $V = V_{oc}$, and $T = 65\ ^\circ C$. Points are measurements for six cells with error bars for one standard deviation. Dashed lines are the charge-induced (junction) model, and solid lines include back-barrier degradation model [14]. (b) Simulated depletion width (points) as a function of time under light/heat stress (1-sun light intensity, 100 °C, open-circuit bias) compared with KPFM measurements (dashed lines) for the least- and most-degraded conditions [35]

Although the deep defects explain the V_{oc} loss, the shallow acceptors are required for the stable J_{sc} (sometimes increasing) and narrowing of the depletion width. Verification of the depletion-width shrinkage measured by KPFM is compared with the model results shown in Figure 5.14(b) [35].

5.4 Electrothermal runaway: shading of CIGS modules

If an illuminated module is partially shaded, the shaded portions experience reverse bias. CIGS modules subjected to reverse-bias stress can develop white, worm-like features caused by hot-spot ignition events [27,37,38]. Such ignition events were correlated with voltage and thermography pulses under dark reverse-bias stress and by EL within seconds during partial-shading stress tests. The resulting damage caused increased shunt conductance and module power loss. The nature of shading-induced degradation in thin-film modules is a critical problem [39], especially because bypass diodes are difficult to integrate in a monolithic module.

This section describes electrothermal numerical simulation of shading-induced failure in thin-film modules using CIGS as a case study [40], but the approach is valid for other technologies. The results demonstrate an example of thermal instability as a positive-feedback mechanism in a nonuniform thin film, the physics of which was described by Karpov [41]. An important aspect is the presence of nonuniformities, such as weak diodes and shunts, that can cause thermal runaway.

The monolithic module model and its coupling to thermal physics is described in Section 5.2.3. As shown in Figure 5.15, the test case is a 10×10 cm^2 monolithic CIGS mini-module with 20 series-connected cells that are each 5 mm wide with scribe dimensions of 100 μm (see Section 5.2.3). Domains were discretized by triangular mesh elements in the cell areas and quadrilateral elements in the scribe areas to better handle the large differences in widths. About 20,000 mesh elements were used and the mesh was highly refined around shunt points. Details on the electrical and thermal parameter values, and the diode characteristics, are provided in Nardone *et al.* [40]. The thermal effects of module encapsulation can also be controlled via the convective and emissive parameters of the model (cf. (5.33) and (5.34)). In this case study, both encapsulated and unencapsulated modules are considered. Regarding the voltage experienced by each cell, shading 4 of the 20 cells (see Figure 5.15) results in 0.6 V across the illuminated cells (1-sun light intensity) and -2.4 V for the shaded cells (completely dark).

Low reverse-breakdown voltages in the -2 to -4 V range have been observed in small CIGS cells. In a module, spots with lower reverse-breakdown voltage can accumulate current from the surrounding cell area causing a temperature increase, which further increases current (exponentially) until thermal runaway occurs. Such weak points will be referred to as nonohmic shunts and 20 of them were added to the module model at random locations (see Figure 5.15).

A simulated temperature distribution including the nonohmic shunts is shown in Figure 5.15 for the 20% shaded, unencapsulated module at 450 s after exposure to 1-sun light intensity at an ambient $T = 293$ K and $V = 0$ V (short-circuit condition). Normal operating temperatures of 315 K are attained in most of the module, but the shaded weak points reach about 335 K. Thermal runaway occurs soon later. Temperature evolution of the hottest point in the module is shown for the encapsulated and unencapsulated cases shown in Figure 5.16. Rapid thermal runaway is observed at 100 and 490 s for the encapsulated and unencapsulated cases,

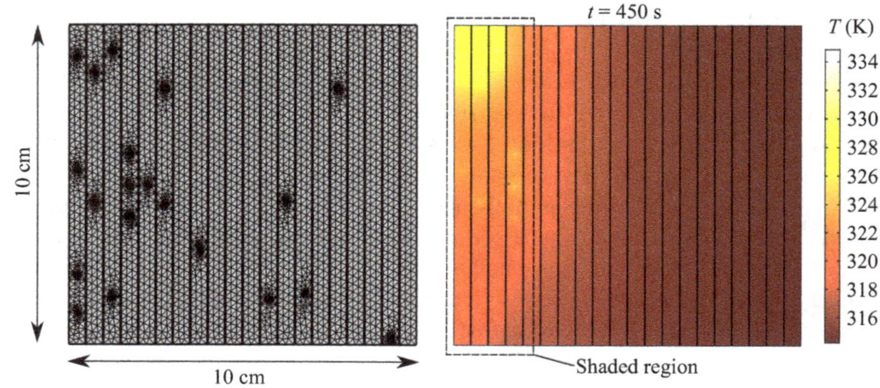

Figure 5.15 CIGS mini-module with weak reverse diodes subjected to partial shade. Left: meshed geometry showing finer mesh at 20 random weak spots. Right: temperature distribution after 450 s of exposure to 1-sun light intensity at V = 0 V. Cells in the shaded region (dashed box) are under reverse bias [40]

respectively. In both cases, runaway is triggered close to 350 K. Quantitative comparison of the heating terms in (5.31)–(5.34) indicates that joule heating in the TCO layer is the largest heat source at runaway, with current density in the order of 10^8 A/m^2 in the 1-μm-thick electrode (see inset in Figure 5.16).

In summary, electrothermal modeling of cells and modules can assist with understanding the impact of thermal effects and nonuniformities. Simulated temperature distributions can be compared with thermal-imaging analysis such as lock-in thermography. EL can also be simulated by using the relation $\phi_{EL} \propto \exp(qV/kT)$, where ϕ_{EL} is the EL intensity and V is the calculated junction potential.

5.5 EBIC simulation: mc-Si device with GBs

Numerical simulation can be employed to test various hypotheses in the analysis of EBIC data. The example shown here is based on the study of temperature-dependent EBIC to elucidate recombination mechanisms at extended defects [42]. In this section, we describe the methodology and concepts required for simulation of EBIC measurements, including the electronic properties and geometry of the device under study, temperature dependencies, and GB features. Although this case study features multicrystalline (mc) Si technology, the same approach can be used for polycrystalline thin films.

5.5.1 Device model and GB recombination

In the mc-Si device model, the current is calculated using the same transport equations given in (5.1)–(5.8), except that the electron–hole pairs are generated by a

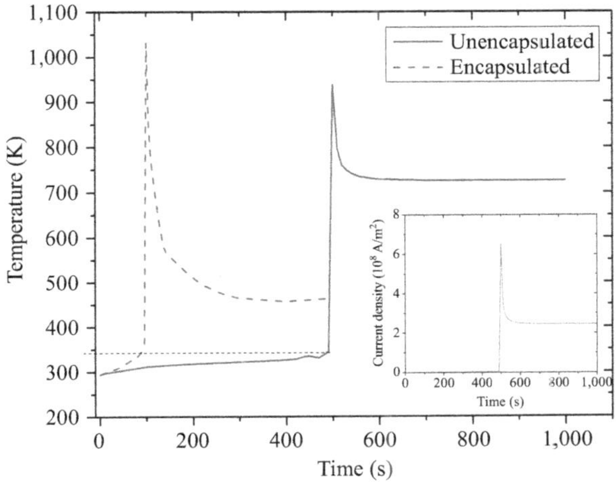

Figure 5.16 Maximum temperature of the module (at the hottest weak spot) as a function of time assuming unencapsulated and encapsulated conditions. Exposure to 1-sun light intensity begins at t = 0 s. Onset of thermal runaway shown by dashed line at T ~ 350 K. Inset: maximum current density of the TCO layer as a function of time [40]

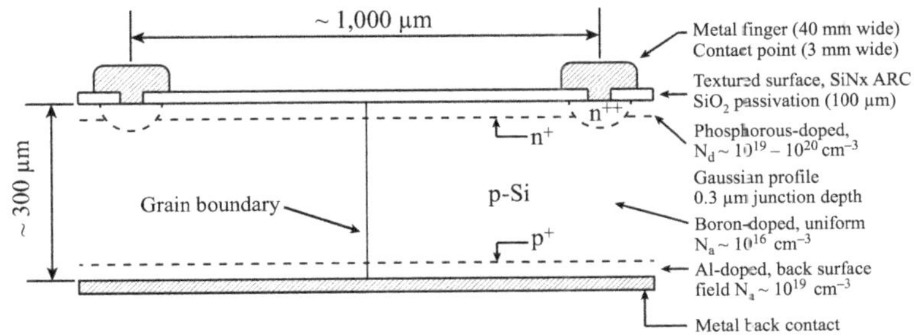

Figure 5.17 Schematic of baseline Si PV micro-diode with a GB down the center. Electronic properties of the bulk crystal and GBs can be set independently. All dimensions and values are approximate

focused electron beam instead of light. For this example, the semiconductor equations are solved assuming Fermi–Dirac statistics (rather than Maxwell–Boltzmann) due to the relatively high doping concentrations near the front surface of a conventional Si-based PV device where doping concentrations are 10^{19}–10^{20} cm^{-3}.

A schematic for a baseline Si device is shown in Figure 5.17. It includes the typical uniform, flat back-contact, but other architectures, such as passivated emitter and rear locally diffused (PERL), and passivated emitter rear contacted

(PERC) cells, can also be modeled. Material properties are based on the review by Altermatt [6] and the references therein. For simulating EBIC at extended defects, the model can include a GB, dislocation, and/or stacking fault at any location with individually specified electronic properties.

The geometry used for EBIC simulation should allow for efficient calculations while capturing the most important phenomena. In the following example, a 100-μm-thick × 100-μm-wide domain was used, and it was assumed that symmetry allows for a 2D model. The 2D assumption enables more rapid calculation than a 3D model due to the fine finite-element mesh required to accurately calculate e-beam-induced generation rates that vary by several orders of magnitude over short distances confined within the top 10 μm of the device. The model domain also included one edge that served as a GB where recombination mechanisms were specified.

Standard bulk recombination mechanisms of direct band-to-band (radiative), Auger, and SRH (see Section 5.2.1.2) were employed. However, the primary hypothesis testing was related to the recombination mechanism at the GB and how it affects the EBIC contrast as a function of temperature and e-beam intensity, $C(T, I_b)$. Previous work has demonstrated that coupled defect states within GBs allow charge carriers not only to move between the continuous bands and the defect levels, but also among coupled discrete defect levels in the gap [43]. The model described later follows that approach, with coupled defects at shallow band-tail levels and deep energy levels in the gap. A diagram of the process is provided in Figure 5.18.

Simulation of EBIC over a wide temperature range requires that the temperature dependence of pertinent material parameters is included in the device model. A general description of how the effective densities of states, bandgap, and thermal

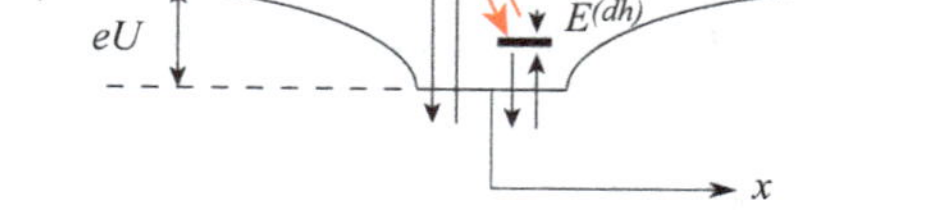

Figure 5.18 Recombination pathways for localized gap states in a GB. Solid arrows are non-radiative processes, red/bold arrows indicate transitions between coupled states, and the dashed arrow is radiative. Charged defect states can create a potential barrier, eU. The valence and conduction bands are E_v and E_c. Other variables are described in the running text [42]

velocities depend on temperature is provided in Section 5.2.1.4. Specific expressions, including the temperature dependence of mobility, for the case of mc-Si are also required [42].

5.5.2 *EBIC generation and contrast calculations*

This model assumes a spherical electron-beam generation volume. Following the approach described by Schroder [44], an electron beam of energy E_b generates N_{eh} electron–hole pairs per incident electron according to:

$$N_{eh} = \frac{E_b}{E_{eh}}\left(1 - \gamma\frac{E_{bs}}{E_b}\right), \tag{5.40}$$

where $E_{eh} = 3.6$ eV is the ionization energy for an electron–hole pair in Si, γ is the back-scattering coefficient, and E_{bs} is the mean energy of the back-scattered electrons. The second term in parenthesis is about 0.1 for Si. The total rate of e–h pair generation, $I_b N_{eh}/e$, depends on the given e-beam current I_b, where e is the elementary charge.

The electron-beam penetration depth (in cm) is related to the beam energy (in eV) and material density ρ_{Si} (in g/cm^3) by:

$$R_e = \frac{2.41 \times 10^{-11}}{\rho_{Si}} E_b^{1.75}. \tag{5.41}$$

Assuming a spherical generation volume, V_{sph}, of diameter R_e, the generation rate per unit volume can be obtained from:

$$G_e = \frac{I_b N_{eh}}{V_{sph} e}. \tag{5.42}$$

With experimentally relevant $E_b = 30$ keV and $I_b = 3.77$ nA, we obtain $N_{eh} = 7{,}500$, $R_e = 7$ μm, and $G_e \approx 10^{24}$ cm^{-3} s^{-1}. This generation rate is used in the transport equations, (5.1)–(5.3), but it is specified only within the volume V_{sph}. The generation volume can also be allowed to move with time to simulate EBIC scanning.

Measured EBIC values depend on the collection efficiency of the device, η_e, by $I = I_b N_{eh} \eta_e$. The presence of defects can decrease the collection efficiency and cause differences in the measured currents. For a given defect-free reference current, I_0, such differences are quantified by the contrast:

$$C = \frac{I_0 - I}{I_0}. \tag{5.43}$$

5.5.3 *Temperature-dependent EBIC: simulation and measurement*

EBIC contrast data at three GBs over a temperature range of 75–300 K are shown in Figure 5.19. Decreasing contrast with temperature is observed, but two GBs

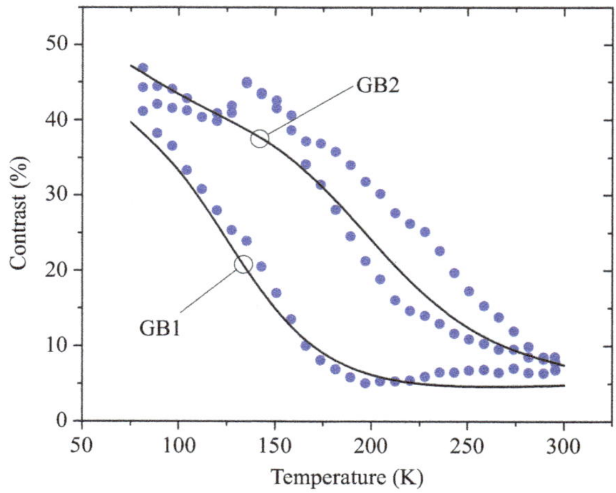

Figure 5.19 EBIC contrast data (points) vs temperature at three GBs in a mc-Si device. Model results (lines) with two different gap-state distributions, GB1 and GB2 (see Table 5.2) [42]

exhibit very similar changes whereas the other exhibits a more rapid decrease. Therefore, data were compared with two models, labeled GB1 and GB2, by adjusting two parameters: mid-gap defect density, $N^{(m)}$ and shallow energy levels, $E^{(de)} = E^{(dh)}$. The densities of shallow states, $N^{(de)}$ and $N^{(dh)}$, were determined as a function of their energy levels and so were not independent parameters [43]. Parameter values are provided in Table 5.2. Both GB1 and GB2 had close-to-uniform energetic defect distributions, with concentrations in the order of 10^{10}–10^{11} cm^{-2}, except that GB2 had a higher concentration of shallow states. That range of defect density represents relatively "clean" GBs because the contrast was only about 5% at room temperature; contaminated GBs and dislocations can show opposite temperature dependence and higher contrast values. However, such clean GBs are not entirely benign, and device models incorporating the properties of GB1 indicated a 5.4% relative loss in efficiency compared with a GB-free model.

5.6 Challenges and opportunities

Numerical techniques to simulate PV devices are continually advancing. Capabilities to model 2D and 3D geometries, time dependence, and the coupling of electrical, optical, and thermal physics provide opportunities to simulate device operation under realistic conditions and analyze a wide range of characterization techniques. Postulates of physical mechanisms, such as defect kinetics, GB effects, and degradation, can be quantitatively tested, providing useful feedback between experiment and theory. From a design perspective, it also enables development and optimization of new device architectures.

Table 5.2 GB defect parameters

Parameter	Unit	GB1	GB2
$E^{(de)}, E^{(dh)}$	eV	0.085	0.140
$N^{(de)}$	cm^{-2}	5.4×10^{10}	9.0×10^{10}
$N^{(dh)}$	cm^{-2}	9.0×10^{10}	1.5×10^{11}
$N^{(m)}$	cm^{-2}	1.0×10^{10}	1.0×10^{10}

In addition to the mechanisms described in this chapter, there are many other physical processes that can be important for a particular study. For example, the effects of impact ionization, electron tunneling at interfaces, and reverse-breakdown mechanisms can be included in a model if the computational method is flexible enough. In that case, it is useful to have a software tool that allows for manipulation of the fundamental equations and addition of new ones.

Along with the advancements comes a greater degree of caution required to understand the system under study and to implement the most appropriate physical model. Model complexity is not desirable because each component introduces parameters that may be difficult to measure or coupled in unknown ways. Fortunately, parallel advancements in characterization techniques and *ab initio* calculations provide more insight to material properties. Judicious parameter selection is best achieved by designing models to study specific, controlled experiments. Ultimately, the results of calculations are most valuable when they can be directly tested against experimental data.

References

[1] S. M. Sze and K. K. Ng, *Physics of Semiconductor Devices*. John Wiley & Sons, 2006.

[2] A. Fahrenbruch and R. Bube, *Fundamentals of Solar Cells: Photovoltaic Solar Energy Conversion*. Elsevier, 2012.

[3] R. Scheer and H.-W. Schock, *Chalcogenide Photovoltaics: Physics, Technologies, and Thin Film Devices*. John Wiley & Sons, 2011.

[4] S. Selberherr, *Analysis and Simulation of Semiconductor Devices*. Springer Science & Business Media, 2012.

[5] R. Baets, J. Barker, J. A. Barnard *et al.*, *Semiconductor Device Modelling*. Springer Science & Business Media, 2012.

[6] P. P. Altermatt, "Models for numerical device simulations of crystalline silicon solar cells—A review," *J. Comput. Electron.*, vol. 10, no. 3, pp. 314–330, 2011.

[7] T. Trupke, M. A. Green, P Würfel *et al.*, "Temperature dependence of the radiative recombination coefficient of intrinsic crystalline silicon," *J. Appl. Phys.*, vol. 94, no. 8, pp. 4930–4937, 2003.

[8] P. P. Altermatt, F. Geelhaar, T. Trupke, X. Dai, A. Neisser, and E. Daub, "Injection dependence of spontaneous radiative recombination in c-Si: Experiment, theoretical analysis, and simulation," in *Proceedings of the 5th International Conference on Numerical Simulation of Optoelectronic Devices*, 2005, vol. 47.

[9] J. Dziewior and W. Schmid, "Auger coefficients for highly doped and highly excited silicon," *Appl. Phys. Lett.*, vol. 31, no. 5, pp. 346–348, 1977.

[10] P. P. Altermatt, J. Schmidt, G. Heiser, and A. G. Aberle, "Assessment and parameterisation of Coulomb-enhanced Auger recombination coefficients in lowly injected crystalline silicon," *J. Appl. Phys.*, vol. 82, no. 10, pp. 4938–4944, 1997.

[11] P. P. Altermatt, G. Heiser, A. G. Aberle *et al.*, "Spatially resolved analysis and minimization of resistive losses in high-efficiency Si solar cells," *Prog. Photovolt. Res. Appl.*, vol. 4, no. 6, pp. 399–414, 1996.

[12] M. Burgelman, P. Nollet, and S. Degrave, "Modelling polycrystalline semiconductor solar cells," *Thin Solid Films*, vol. 361, pp. 527–532, 2000.

[13] S. J. Fonash, *Amps-1d for Windows' 95/Nt*. The Electronic Materials and Processing Research Laboratory, Penn State University, 1997.

[14] "AIST: Research Center for Photovoltaics—Provided Services." [Online]. Available: https://unit.aist.go.jp/rcpv/cie/service/index.html [Accessed: March 18, 2019].

[15] M. A. Green, "Intrinsic concentration, effective densities of states, and effective mass in silicon," *J. Appl. Phys.*, vol. 67, no. 6, pp. 2944–2954, 1990.

[16] A. Stokes, M. Al-Jassim, A. Norman, D. Diercks, and B. Gorman, "Nanoscale insight into the p-n junction of alkali-incorporated Cu(In,Ga)Se_2 solar cells," *Prog. Photovolt. Res. Appl.*, vol. 25, no. 9, pp. 764–772, 2017.

[17] M. Nardone, Y. Patikirige, C. Walkons *et al.*, "Baseline models for three types of CIGS cells: Effects of buffer layer and Na content," presented at the *World Conference on Photovoltaic Energy Conversion*, Waikoloa, HI, 2018.

[18] D. Yan and A. Cuevas, "Empirical determination of the energy band gap narrowing in highly doped n+ silicon," *J. Appl. Phys.*, vol. 114, no. 4, p. 044508, 2013.

[19] D. Yan and A. Cuevas, "Empirical determination of the energy band gap narrowing in p+ silicon heavily doped with boron," *J. Appl. Phys.*, vol. 116, no. 19, p. 194505, 2014.

[20] A. Schenk, "Finite-temperature full random-phase approximation model of band gap narrowing for silicon device simulation," *J. Appl. Phys.*, vol. 84, no. 7, pp. 3684–3695, 1998.

[21] K. R. McIntosh and P. P. Altermatt, "A freeware 1D emitter model for silicon solar cells," in *Photovoltaic Specialists Conference (PVSC), 2010 35th IEEE*, 2010, pp. 002188–002193.

[22] P. P. Altermatt, A. Schenk, F. Geelhaar, and G. Heiser, "Reassessment of the intrinsic carrier density in crystalline silicon in view of band-gap narrowing," *J. Appl. Phys.*, vol. 93, no. 3, pp. 1598–1604, 2003.

[23] D. Guo, T. Fang, A. Moore *et al.*, "Numerical simulation of copper migration in single crystal CdTe," *IEEE J. Photovolt.*, vol. 6, no. 5, pp. 1286–1291, 2016.

[24] H. Mehrer, *Diffusion in Solids: Fundamentals, Methods, Materials, Diffusion-Controlled Processes*, vol. 155. Springer Science & Business Media, 2007.

[25] A. Laemmle, R. Wuerz, T. Schwarz, O. Cojocaru-Mirédin, P.-P. Choi, and M. Powalla, "Investigation of the diffusion behavior of sodium in Cu(In,Ga) Se_2 layers," *J. Appl. Phys.*, vol. 115, no. 15, p. 154501, 2014.

[26] H. J. Möller, *Semiconductors for Solar Cells*. Artech House Publishers, 1993.

[27] T. J. Silverman, M. G. Deceglie, X. Sun *et al.*, "Thermal and electrical effects of partial shade in monolithic thin-film photovoltaic modules," *IEEE J. Photovolt.*, vol. 5, no. 6, pp. 1742–1747, 2015.

[28] J. F. Hiltner and J. R. Sites, "Stability of CdTe solar cells at elevated temperatures: Bias, temperature, and Cu dependence," in *AIP Conference Proceedings*, 1999, vol. 462, pp. 170–175.

[29] D. S. Albin, "Accelerated stress testing and diagnostic analysis of degradation in CdTe solar cells," in *Reliability of Photovoltaic Cells, Modules, Components, and Systems*, International Society for Optics and Photonics. 2008, vol. 7048, p. 70480N.

[30] M. Nardone and D. S. Albin, "Degradation of CdTe solar cells: Simulation and experiment," *IEEE J. Photovolt.*, vol. 5, no. 3, pp. 962–967, 2015.

[31] D. H. Rose, F. S. Hasoon, R. G. Dhere *et al.*, "Fabrication procedures and process sensitivities for CdS/CdTe solar cells," *Prog. Photovolt. Res. Appl.*, vol. 7, no. 5, pp. 331–340, 1999.

[32] M. Gloeckler, A. L. Fahrenbruch, and J. R. Sites, "Numerical modeling of CIGS and CdTe solar cells: Setting the baseline," in *Photovoltaic Energy Conversion, 2003. Proceedings of 3rd World Conference on*, 2003, vol. 1, pp. 491–494.

[33] R. G. Dhere, Y. Zhang, M. J. Romero *et al.*, "Investigation of junction properties of CdS/CdTe solar cells and their correlation to device properties," in *2008 33rd IEEE Photovoltaic Specialists Conference*, 2008, pp. 1–5.

[34] D. Redfield and R. H. Bube, *Photo-Induced Defects in Semiconductors*, vol. 4. Cambridge University Press, 2006.

[35] S. Johnston, D. S. Albin, P. Hacke *et al.*, "Imaging, microscopic analysis, and modeling of a CdTe module degraded by heat and light," *Sol. Energy Mater. Sol. Cells*, vol. 178, pp. 46–51, 2018.

[36] R. Harju, V. G. Karpov, D. Grecu, and G. Dorer, "Electron-beam induced degradation in CdTe photovoltaics," *J. Appl. Phys.*, vol. 88, no. 4, pp. 1794–1801, 2000.

[37] J. E. Lee, S. Bae, W. Oh *et al.*, "Investigation of damage caused by partial shading of $CuIn_xGa_{1-x}Se_2$ photovoltaic modules with bypass diodes," *Prog. Photovolt. Res. Appl.*, vol. 24, no. 8, pp. 1035–1043, 2016.

[38] P.-O. Westin, U. Zimmermann, L. Stolt, and M. Edoff, "Reverse bias damage in CIGS modules," in *24th European Photovoltaic Solar Energy*

Conference, September 21–25, 2009, Hamburg, Germany, 2009, pp. 2967–2970.

[39] I. Kozinsky, B. Bob, and R. Jones-Albertus, "Challenges and opportunities for improving thin-film photovoltaics," *MRS Adv.*, vol. FirstView, pp. 1–6, June 2016.

[40] M. Nardone, S. Dahal, and J. M. Waddle, "Shading-induced failure in thin-film photovoltaic modules: Electrothermal simulation with non-uniformities," *Sol. Energy*, vol. 139, pp. 381–388, 2016.

[41] V. G. Karpov, "Coupled electron–heat transport in nonuniform thin film semiconductor structures," *Phys. Rev. B*, vol. 86, no. 16, p. 165317, 2012.

[42] M. Nardone, J. Moseley, S. Dahal, A. V. Parikh, and J. M. Waddle, "Numerical simulation of EBIC for analysis of extended defects," in *IEEE Photovoltaics Specialists*. IEEE, Washington, DC, 2017.

[43] V. Kveder, M. Kittler, and W. Schröter, "Recombination activity of contaminated dislocations in silicon: A model describing electron-beam-induced current contrast behavior," *Phys. Rev. B*, vol. 63, no. 11, p. 115208, 2001.

[44] D. K. Schroder, *Semiconductor Material and Device Characterization*. John Wiley & Sons, 2006.

[45] V. G. Karpov, A. D. Compaan, and D. Shvydka, "Random diode arrays and mesoscale physics of large-area semiconductor devices," *Phys. Rev. B*, vol. 69, no. 4, pp. 045325, 2004.

[46] S. H. Demtsu, and J. R. Sites, "Effect of back-contact barrier on thin-film CdTe solar cells," *Thin Solid Films*, vol. 510, no. 1–2, pp. 320–324, 2006.

[47] Y. Roussillon, V. G. Karpov, D. Shvydka, J. Drayton, and A. D. Compaan, "Back contact and reach-through diode effects in thin-film photovoltaics," *J. Appl. Phys.*, vol. 96, no. 12, pp. 7283–7288, 2004.

Chapter 6

Luminescence and thermal imaging of thin-film photovoltaic materials, devices, and modules

Dana B. Sulas-Kern[1] *and Steve Johnston*[1]

6.1 Advantages of thermal and luminescence imaging for thin-film technologies

The quality of photovoltaic (PV) cells and modules, along with their degradation over time, is most often monitored using the device power output and/or current–voltage characteristics. However, it is important to note that monitoring the electrical output of the device is a bulk measurement that does not provide information about spatial heterogeneity or about which device components are most responsible for power loss. For more targeted approaches to optimizing and maintaining devices, imaging techniques have been gaining popularity, including photoluminescence (PL), electroluminescence (EL), and lock-in thermography (LIT). These spatially resolved measurement techniques offer rapid, large-area analysis of PV materials and devices, allowing identification of defects such as shunts [1], variations in series resistance [2], recombination centers [3], weak diode areas [4,5], composition/bandgap inhomogeneities [6–8], partial-shading breakdown sites [9], and unpassivated grain boundaries [10]. In commercial-scale applications, imaging techniques are attractive for quality management or troubleshooting field failure mechanisms. At the research and development level, imaging is often used for process improvement by identifying the points of failure and then tuning recipes and fabrication methods accordingly.

Luminescence imaging was initially applied to silicon PV cells in 1963 [11], and the technique was further adapted for widespread use in 2005 when Fuyuki *et al.* demonstrated the ability to use a silicon charge-coupled device (CCD) camera for imaging silicon luminescence [12,13]. Camera-based LIT imaging was also developed and applied to silicon PV cells around 2000 [14,15]. Since then, the benefits of combining luminescence and thermal imaging have been recognized increasingly [16] and both techniques have been applied to thin-film technologies.

Indeed, thin-film PV particularly benefits from spatially resolved imaging methods such as PL, EL, and LIT for several reasons. First, the rates of degradation

[1]Materials Science Center, National Renewable Energy Laboratory, Golden, CO, USA

for commercialized thin-film products are greater compared with silicon [17], despite marked improvements over recent years. Therefore, it is especially important to accurately identify and address the points of failure in thin-film devices to improve their long-term stability and help them compete in the silicon-dominated market. Additionally, the performance of thin-film PV devices depends heavily on their microstructure, and the amorphous or polycrystalline nature of thin-film active materials leads to significantly greater inherent device inhomogeneity. Uncontrolled inhomogeneity or defect formation is known to negatively impact device performance and stability [18–21]. Inhomogeneities and defects can further evolve upon degradation caused by weathering in the field or lab-based accelerated stress testing, and it is essential to address the spatial evolution in material parameters under such environmental conditions.

In this chapter, we describe luminescence and thermal imaging in the context of thin-film PV devices. First, we discuss both traditional and developing experimental designs, including dark LIT (DLIT), illuminated LIT, PL using different illumination configurations, traditional EL performed under forward-bias current injection, and contactless EL induced with sub-cell illumination patterns. Next, we provide some brief theoretical background to highlight relationships between fundamental materials properties and the resulting luminescence and thermal images. Finally, we give examples of thin-film PV PL, EL, and LIT images to demonstrate the utility of these imaging methods ranging from understanding the microstructure of active layers to managing the quality and reliability of full PV modules.

6.2 Design of thermal and luminescence imaging systems

6.2.1 PL and EL imaging

Figure 6.1 shows two different configurations for camera-based luminescence imaging of thin-film PV devices. Both configurations contain the basic components

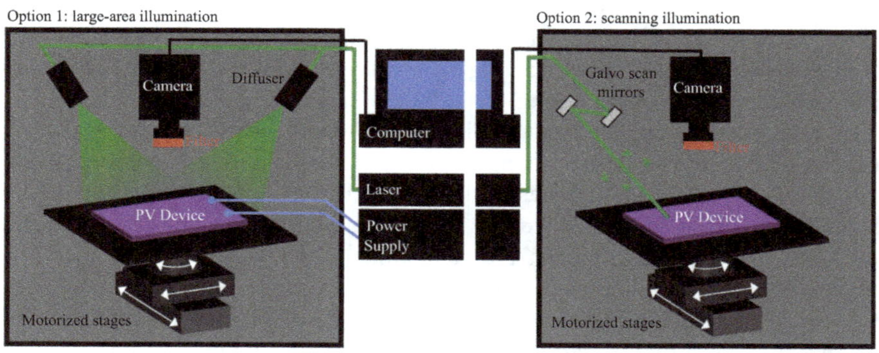

Figure 6.1. Schematic diagram for setup of luminescence imaging, highlighting two different options for PL imaging including large-area illumination and scanning-laser illumination

of an excitation source paired with a camera equipped with an appropriate long-pass filter to block any excitation light. In addition to optical excitation to generate PL, EL can also be imaged in the same configuration using forward-bias current injection. To block ambient light, the camera and PV device are enclosed in a dark box where all objects inside the enclosure are typically matte black to minimize reflections.

The cameras used for luminescence imaging usually employ a silicon or InGaAs sensor, depending on the bandgap of the PV material to be imaged. Silicon cameras are used for higher-energy emission (<1,000 nm), whereas InGaAs cameras are used for lower-energy emission (1,000–1,600 nm). Although the PL spectra of many thin-film PV materials fall in the Si detection range, near-infrared (NIR) detection with InGaAs cameras is increasingly important to characterize low-energy defect states or for emerging PV materials such as narrow-gap perovskites [22].

The lower cost of silicon sensors enables the use of cameras with a greater pixel number at comparable price point, increasing obtainable image resolution. In many cases, the responsivity at the low-energy edge of modern Si detectors is often sufficient to image NIR-emitting samples using the high-energy edge of their luminescence onset. In fact, extremely low-cost imaging systems in the order of US$100 have been designed using Si detectors commonly employed in cell-phone cameras. For example, Figure 6.2 shows an inexpensive programmable system available from Raspberry Pi, which employs a US$35 computer unit and US$30 Sony IMX219 complementary metal-oxide-semiconductor (CMOS) sensor with the infrared-blocking filter removed (marketed as the NoIR camera).

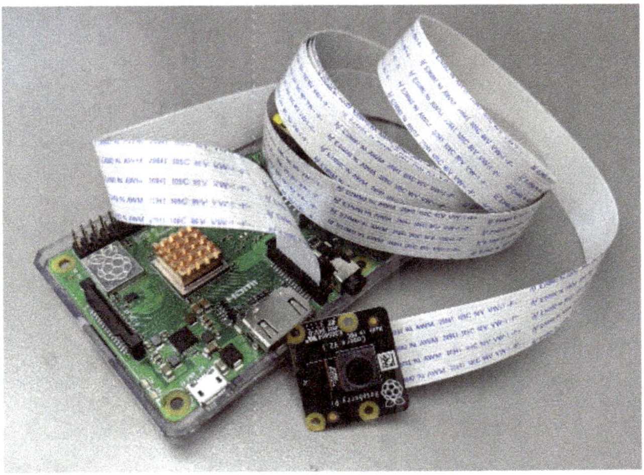

Figure 6.2. Low-cost imaging solution using a cell-phone camera (Sony IMX219 sensor) with no IR-blocking filter run by a Raspberry Pi computer module

Consistent images can be collected by directly controlling acquisition and output parameters with available Python packages. Although such an inexpensive imaging system is not often considered for scientific applications, the Raspberry Pi camera has been used for quantitative analysis in atmospheric science applications and in a nanometer-resolution spectrometer [23–25]. Recently, a similar system has been applied to PVs for *in-situ* imaging during PV mini-module accelerated stress testing [26].

Figure 6.3 shows a comparison between EL images collected using the low-cost Raspberry Pi imaging system versus images collected using a scientific-grade Si CCD camera. The images correspond to the central area of a commercial thin-film PV module, where the series-connected cells appear as horizontal stripes. The Raspberry Pi camera has lower sensitivity, with about 500 times lower pixel counts compared with the scientific-grade camera when the images in Figure 6.3 are normalized for their respective integration times. It is also important to note that the

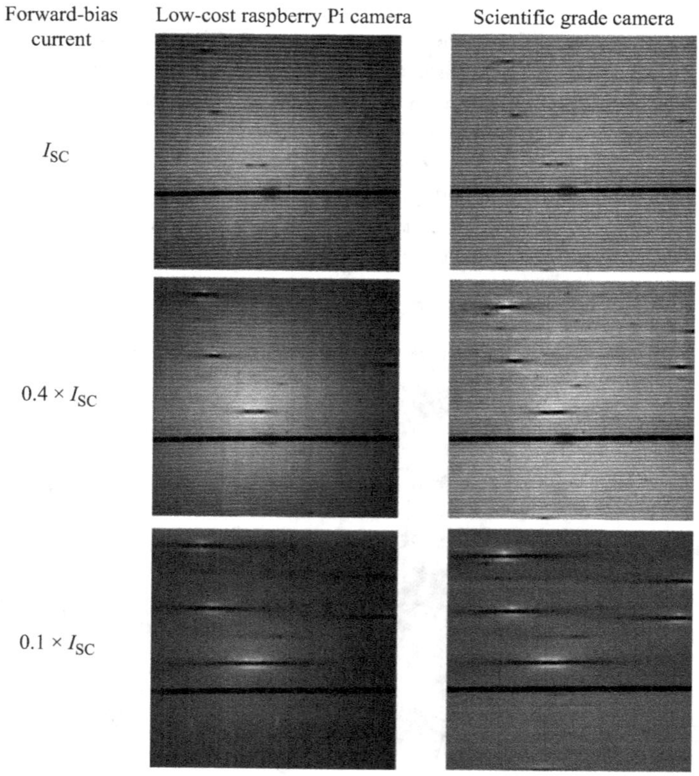

Figure 6.3. Comparison of thin-film module EL images at various forward-bias injection conditions relative to the short-circuit current (I_{SC}) collected using either the low-cost Raspberry Pi camera system or a scientific-grade camera

Raspberry Pi camera is hardware-limited to a 10-s integration time, so additional processing such as image stacking may be necessary to detect weaker signals.

Despite the lower sensitivity of the inexpensive cell-phone camera, it is remarkable that the low-cost system is sufficient to record EL images of a thin-film PV module even at forward-bias current injection equivalent to a tenth of the short-circuit current. Furthermore, as shown in Figure 6.3, the low-cost system is able to resolve similar trends in the injection dependence of the EL images, such as the increasing impact of series resistance at higher current densities. At low injection, where the effects of series resistance are minimal, several cells show large dark areas as the luminescence is quenched by carrier motion to the shunted regions, and there are bright regions next to the shunts due to current crowding. At high injection, the impact of series resistance increases, and the dark areas around each shunt become smaller in size while the images become dominated by the bright cell edges that have been attributed to the sheet resistance of the front contact (resulting in a distinct horizontal striped pattern) [27,28]. These results demonstrate the promise of low-cost imaging systems for applications where cost is a major factor or where camera degradation is expected to occur (such as *in-situ* imaging in environmental stress chambers) [26].

In addition to the example of EL imaging in Figure 6.3, PL imaging is useful to characterize material quality without dependence on electrical contacts. In this case, an illumination source is required. The illumination source is most often a laser excitation or an array of light-emitting diodes (LEDs). Laser excitation sources are often preferred for their greater intensity and narrowly defined spectral range. However, band-pass filters are still usually necessary to ensure the appropriate excitation wavelengths, regardless of whether LEDs or laser sources are used.

In Figure 6.1, we highlight two possible configurations for PL imaging. The configuration labeled as Option 1 shows a PL imaging setup in which the excitation source is spread evenly over the entire cell using homogenization optics such as diffusers. This imaging configuration is desirable because the homogeneous illumination mimics outdoor operating conditions, it limits lateral current flow that can be induced by inhomogeneously generated photovoltage, and it enables bias-dependent PL measurements.

Option 2 in Figure 6.1 shows a PL imaging configuration that uses a scanning laser beam as an excitation source. In this example, the laser beam is rastered across the sample by reflecting it from a dual-axis scanning galvanometer mirror assembly. The camera will generally be set to a long exposure time to capture a homogeneous luminescence image after the laser beam has scanned over the entire sample area. The advantage of this configuration is the lower laser cost for a given excitation power density, as the total laser power is spread over a smaller area. However, it is also important to note that a sub-cell excitation spot size can induce spatial variations in photovoltage, which can have important implications for lateral carrier motion and the resulting images. These effects have been previously discussed in both confocal and wide-field imaging designs, where lateral carrier motion away from the illumination spot can affect the luminescence intensity to varying extents depending on the material quality [29,30].

While the effects of sub-cell illumination patterns have largely been discussed on micrometer size scales, it is important to note that lateral carrier motion can occur on much larger size scales, such as through conductive layers in commercial thin-film PV modules. In silicon PV modules, using patterned illumination such as that shown in Option 2 of Figure 6.1 has been recognized as a method to measure "contactless EL" in the non-illuminated areas [31–34], though this technique has not yet been widely implemented for thin-film modules. However, Figure 6.4 shows contactless EL emission from a commercial thin-film module, demonstrating that patterned illumination can provide novel module inspection capabilities for thin-film PV.

Figure 6.4 illustrates the concept of measuring contactless EL on a full-sized commercial thin-film PV module. A light source is incident on the module outside of the detection area, which induces a voltage on the cell. The spatial difference in photogenerated voltage causes photoexcited carriers to move into the non-illuminated cell regions, mimicking forward-bias current injection. As shown in the three images taken with an LED light source at different positions along the edge of the module, the voltage spreads laterally across the partially illuminated cells, whereas the resistance is too large for carrier injection into neighboring cells.

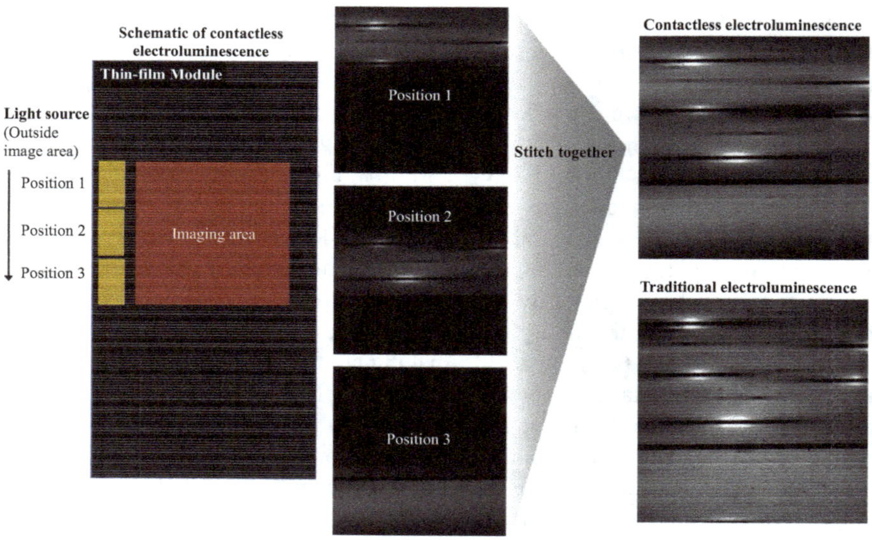

Figure 6.4. Schematic of contactless EL imaging on a thin-film module demonstrating the concept of inducing luminescence outside of the illuminated area. Images collected while the illumination source scans down the module are continuously stitched together into complete module images that are reminiscent of traditional (contacted) EL images

In practice, the coordinated movement of the light source with the camera's field of view can be automated, with the coordinates of the step motor providing the necessary spatial information to continuously stitch together full-module contactless EL images. As shown in the right side of Figure 6.4, the contactless EL images resemble traditional (contacted) EL images under low injection conditions (in this case, forward-bias current equivalent to 0.1 times the short-circuit current). The equivalent current density for contactless EL can be approximated by scaling the absorbed fluence of the light source by the ratio of non-illuminated to illuminated cell areas, assuming that most carriers redistribute across the cell area before they recombine [33]. While both traditional and contactless EL imaging can provide similar information about the spatial distribution of defects that is essential for module performance/failure analysis, contactless EL is an appealing alternative because it can be performed without accessing the module's electrical contacts and it opens a possibility for portable and remote (drone-based) screening.

6.2.2 LIT imaging

Thermal imaging can be carried out in a similar experimental setup to PL and EL imaging, and it is often desirable to use the same type of stimulus (current or light) so that the power dissipation maps from thermal images can be directly compared with the voltage distributions implied by luminescence images. However, unlike PL and EL, a dark enclosure is not required because thermal cameras do not respond to visible light.

Cameras used for thermal imaging are either sensitive to mid-wave infrared (MWIR, 3–5 μm) or long-wave infrared (LWIR, 8–14 μm). Common MWIR sensors are made of indium antimonide, lead sulfide, and lead selenide, and LWIR sensors are most often made of vanadium oxide or amorphous silicon. Mercury cadmium telluride (HgCdTe) can detect across both MWIR and LWIR windows. Figure 6.5 shows a comparison of the MWIR and LWIR regions for detector sensitivity in relation to the Planck blackbody curve at two different temperatures, as well as the transmission spectra for calcium fluoride (CaF_2) and fused silica windows, which are discussed later. Sensors employing MWIR detection typically provide greater signal contrast due to larger differences in the blackbody curve with temperature in this spectral range. However, LWIR detectors can detect a larger dynamic range of temperatures because lower temperatures are detectable in the LWIR but not in the MWIR, and the peak of the blackbody curve shifts to shorter wavelengths at increasing temperature. This shift effectively increases the temperature at which the LWIR sensor will be saturated because at high temperatures only the tail of the spectrum will be in the LWIR region.

MWIR and HgCdTe sensors are generally cooled, whereas LWIR detectors are most often uncooled microbolometer sensors. This results in lower cost, weight, and size for LWIR detectors, making them the preferred option for portable steady-state thermal imaging in the field. MWIR detectors are a good option for lab-based, research-grade LIT measurements. Cooled MWIR cameras may be an order of magnitude more expensive than some LWIR cameras. Interestingly, an inexpensive

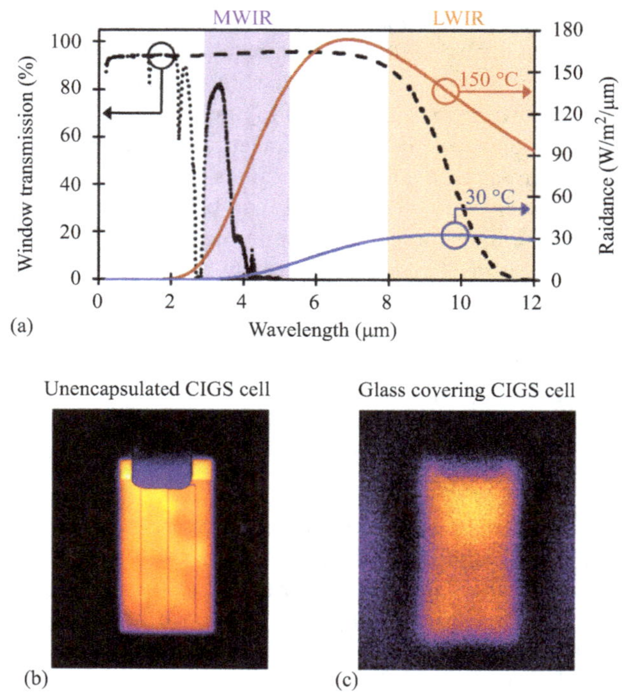

Figure 6.5. Transmission spectra of CaF_2 (dashed trace) and fused silica (dotted trace) windows overlaid with the relative detector sensitivity of an InSb camera (gray trace) and the spectra for blackbody radiance at 150 °C (red trace) and 30 °C (blue trace). The bottom row shows a comparison between thermal images for an unencapsulated copper indium gallium diselenide (CIGS) cell and a CIGS cell that is encapsulated with a glass cover slide

option for thermal imaging pricing less than US$1,000 has been demonstrated as a proof of concept with images of mechanical defects and dental caries using a cell-phone attachment and custom lock-in acquisition code [35].

Another important concern for thermal imaging of PV modules is the opacity of glass to IR wavelengths. Figure 6.5(a) shows the transmission spectra of fused silica glass as well as calcium fluoride, demonstrating that standard glass encapsulants do not transmit light in the LWIR region and minimally transmit light in the MWIR region. Materials that are transparent to IR wavelengths, such as calcium fluoride, barium fluoride, or zinc selenide, are possible window layers that could be used for research studies requiring thermal imaging. This has been demonstrated, for example, in thermal imaging of battery materials [36].

However, most studies, especially at the module level, will suffer from blurring effects when imaging through the glass encapsulation. Figure 6.5(b) and (c) shows the substantial blurring in a thermal image when a copper indium gallium diselenide (CIGS) cell is encapsulated in glass. The heat profile becomes blurred because the images represent blackbody emission from the glass layer that has been heated by the cell, rather than imaging the blackbody emission from the cell itself, so heat diffusion and the thermal mass of glass play an important role. In silicon PV, this issue is circumvented to some extent by imaging through the module backsheet, which is much thinner and has a lower thermal mass than glass. However, thin-film modules tend to be encapsulated with glass on both sides. For this reason, more in-depth studies of hot spots in thin-film modules may rely on extracting areas of interest using mechanical coring [37] or laser micromachining [38] procedures. Thermoreflectance imaging has also been proposed as a method to obtain high spatial resolution of heating even through glass, where heating-induced changes in the reflectivity of shorter-wavelength visible light can produce 200–400-nm spatial resolution of hot spots [39–41].

Thermal imaging can be carried out in either steady-state or lock-in acquisition modes. Steady-state thermal imaging is more facile and less expensive, and it can be a useful method for identifying module failures in the field during module operation in daylight [42]. LIT is an appealing alternative over steady-state methods in the lab to obtain greater spatial resolution of the defective points where heat is initially generated. Lock-in methods measure heating caused by an oscillating stimulus such as a pulsed forward bias (for DLIT) or a pulsed illumination (for illuminated LIT), where any steady-state heat background is effectively subtracted out of the image. We further discuss the lock-in technique in the theoretical background section.

6.3 Theoretical background

6.3.1 PL and EL

In this section, we provide some brief theoretical background with a purpose of developing a basic understanding of the relationship between fundamental materials properties and luminescence images. For a more thorough overview of the physics behind solar cell operation and the processes leading to luminescence and recombination losses in PV devices, the reader is directed to several in-depth texts developed for that purpose [43–48].

The utility of luminescence imaging stems from the necessary equality between the rate of photon absorption and emission in a semiconducting material, which places a thermodynamic limit on the chemical potential of the system. In 1917, this detailed balance relationship was shown by Einstein as a way to derive Planck's law, which has been the basis of several other useful formulations that express the connection between a semiconducting material's radiative efficiency and its ability to generate photovoltage and current, such as Rau's reciprocity relation [45], Würfel's description of the chemical potential of photons [47], and Ross's treatment of the work that can be extracted from a photochemical system [46,48]. Importantly, this relationship dictates that radiative recombination is an unavoidable energetic loss. The emitting photon flux due to unavoidable radiative recombination can be expressed in terms of the material's properties, such as its absorption strength and energy landscape, according to the generalized form of Planck's law:

$$I(E) = \frac{\alpha(E)}{4\pi^2\hbar^3c^2}\frac{E^2}{\exp\left(\frac{E-\Delta\mu}{kT}\right)-1}, \tag{6.1}$$

where E is energy, α is the absorption coefficient, c is the speed of light, $\hbar$ is Planck's constant, k is the Boltzmann constant, T is temperature, and $\Delta\mu$ is the quasi-Fermi-level splitting. In the case where the Boltzmann approximation is valid ($E - \Delta\mu >> kT$), (6.1) can be simplified to

$$I(E) = \alpha(E)\phi_{bb}(E)\exp\left(\frac{\Delta\mu}{kT}\right), \tag{6.2}$$

where ϕ_{bb} is the blackbody spectrum.

Equation (6.2) highlights the important exponential relationship between photon emission and the quasi-Fermi-level splitting, where small changes in $\Delta\mu$ will strongly affect the emitted photon flux. This fundamental exponential relationship makes luminescence detection a facile and sensitive method for tracking variations in the local electrochemical potential energy, which can also be thought of as the PV cell's internal voltage. During luminescence imaging with a CCD camera, the spectrally integrated luminescence intensity is resolved across the sample area, and the images of band-to-band luminescence are used to generate implied-voltage maps using a similar exponential form to (6.2), according to:

$$I_{xy} \propto C_{xy}\exp\left(\frac{qV_{xy}}{kT}\right), \tag{6.3}$$

where I_{xy} is the spatially varying luminescence intensity, V_{xy} is the spatially varying internal voltage, and C_{xy} is a calibration constant that depends on the measured cell voltage, transmission of the optics, and quantum efficiency of the detector. The images are calibrated using the known average voltage across the cell, given by the photoinduced voltage (for PL imaging) or the forward bias across the electrical contacts (for EL imaging). In the case of PL, it may also be necessary to subtract the background PL under short-circuit conditions because these voltage-independent carriers to do not contribute to device operation [49].

Luminescence imaging and the associated spatial mapping of the cell voltage are useful for locating nonradiative recombination centers as well as spatial variations in electrical contact. Nonradiative recombination centers are apparent during luminescence imaging because they represent loss pathways that decrease the cell voltage. That is, the maximum possible quasi-Fermi-level splitting for a given material is equal to its bandgap; however, there will always be some recombination that depopulates the material's excited states and causes the quasi-Fermi levels to move away from the band edges deeper into the bandgap. In a perfect material where all the carriers surviving radiative recombination are collected and perform work, the open-circuit voltage will be at a maximum possible value of $E_g - qV_{\text{rad}}$, where q is the elemental charge and V_{rad} is the voltage loss from radiative recombination. This radiative recombination loss is an unavoidable material property, but additional nonradiative recombination pathways will further decrease the voltage by $E_g - qV_{\text{rad}} - qV_{\text{nonrad}}$. Identifying and suppressing such nonradiative recombination pathways is essential for improving cell efficiency.

Identifying issues with electrical contacts, such as characterizing the series resistance or shunt resistance, is also a strength of luminescence imaging, especially when it is possible to compare EL and PL images. For example, the local luminescence intensity can vary in shunted areas when thin-film PV cells become locally short-circuited at defects such as scribe-line failures, isolation scribe problems, pits or nodules that instigate reverse-bias breakdown under partial shading, or development of conductive precipitates during operation or fabrication. In addition, the cell's series resistance can hinder lateral carrier spreading across the cell from the point of injection and will create increasing contrast in the spatial distribution of luminescence intensity when imaging under higher injection conditions with increasing charge densities.

Several methods for spatially mapping series resistance in PV cells have been explored [2,50–56]. The most straightforward understanding of local series resistance is to visualize each volume element of the cell as a separate photodiode that is connected to the cell terminals over a local series resistance representing the path for current flow into and out of that volume element. The local series resistance can then be thought of as the voltage drop between the cell terminals and the local volume element divided by the current flowing into and out of that volume element. This logic can be expressed as a series resistance map using measurable current and voltage values along with the luminescence images according to

$$R_{s,xy} = \frac{\Delta V}{\Delta J} = \frac{V_{\text{terminal}} - V_T \ln\left(I_{xy}/C_{xy}\right)}{J_L - J_{0,xy} * I_{xy}/C_{xy}}, \tag{6.4}$$

where $R_{S,xy}$ is the spatially varying series resistance, V_{terminal} is the voltage at the cell terminals, J_L is the photocurrent density, and $J_{0,xy}$ is the local dark saturation current density. The dark saturation current can either be assumed as a constant value (J_0) or as spatially resolved ($J_{0,xy}$). The spatial resolution of $J_{0,xy}$ can be obtained either through luminescence imaging at varying current injection/extraction or through additional measurement techniques (such as LIT or light-beam-induced current [LBIC]) [55,57].

6.3.2 Lock-in thermography

Thermal images of PV devices represent areas where power is lost to heat dissipation instead of being converted to useful work. Combining maps of power loss (from thermal images) with the voltage and resistance information attainable from luminescence images enables detailed modeling of pixel-by-pixel local current–voltage (I–V) curves, as described by the works of Otwin Breitenstein *et al.* [15]. In this section, we give some brief theoretical background on thermal imaging and the lock-in acquisition method.

Planck's blackbody law mentioned in the previous section was originally applied to absorption and emission of thermal photons, resulting in the temperature dependence of MWIR and LWIR radiation described in Figure 6.5(a). Integrating the blackbody curve over all wavelengths gives the Stefan–Boltzmann law, which can be used to describe the total power radiated from an object in terms of its temperature as

$$P = A\varepsilon\sigma T^4, \tag{6.5}$$

where A is the surface area, ε is the emissivity, and σ is the Stefan–Boltzmann constant. The object is a blackbody emitter when $\varepsilon = 1$, although most real objects are not perfect blackbodies and exhibit $\varepsilon < 1$. The actual value of ε can vary depending on wavelength, temperature, surface roughness, and viewing angle, and this presents a challenge for thermal imaging of thin-film PV materials and devices. For example, shiny metal surfaces can have a low thermal emissivity of 0.03–0.3, whereas matte black surfaces can have emissivities of 0.96 [15]. The variations in surface thermal emissivity may thus be correlated with the presence of different materials across the cell surface, which can result in artifacts that must be appropriately interpreted.

Thermal imaging can be carried out in both steady-state and lock-in acquisition modes. Lock-in acquisition is desirable for improving the sensitivity for detecting heat generation that is related to a particular oscillating stimulus, while removing heat signals related to the DC background. DLIT is when the heat-inducing stimulus is an external bias, and illuminated LIT is when heating is induced by a pulsed light source. In both cases, the DC background is separated from the AC signal by integrating the product of the oscillating signal of interest with a correlation function. The correlation function can be thought of as a series of weighting factors related to the time dependence of the applied stimulus. In practice, this correlation function is a sine wave generated at the reference frequency of the AC stimulus. Assuming that the detected signal also oscillates as a sine wave, the output amplitude (A) would be the product of these functions,

$$A = A_{\text{sig}}\sin\left(\omega_{\text{sig}}t + \theta_{sig}\right) * A_{\text{ref}}\sin\left(\omega_{\text{ref}}t + \theta_{\text{ref}}\right), \tag{6.6}$$

where $A_{\text{sig/ref}}$, $\omega_{\text{sig/ref}}$, and $\theta_{\text{sig/ref}}$ are the amplitude, frequency, and phase of the signal or reference functions. Using the rule of addition for sine functions, when the reference frequency is equal to the signal frequency, the output becomes a DC amplitude dependent on the phase difference between the signal and the reference, $\theta = \theta_{\text{sig}} - \theta_{\text{ref}}$, given as

$$A = \frac{1}{2}A_{\text{sig}}A_{\text{ref}}\cos\left(\theta_{\text{sig}} - \theta_{\text{ref}}\right) \sim A_{\text{sig}}\cos(\theta). \tag{6.7}$$

The result of (6.7) implies a maximum signal amplitude when $\theta_{\text{sig}} = \theta_{\text{ref}}$, resulting in a 0° shift or in-phase measurement. The signal amplitude may then decrease when the phase is not matched. Interestingly, a signal that is 180° shifted from the reference would appear as a negative signal, corresponding to a long-lived phenomenon with respect to the modulation frequency or heating that occurs when the stimulus is turned off. Thus, the phase dependence could be used to produce time-dependent information related to heat diffusion or the spatial location (i.e., depth) of the heat-generating defect [58].

Usually, LIT is recorded as a two-channel measurement where a second component is 90° out of phase with the reference, called the quadrature component, resulting in two measured amplitudes given as

$$X = A_{\mathrm{sig}}\cos(\theta);\, Y = A_{\mathrm{sig}}\sin(\theta), \tag{6.8}$$

where the in-phase (X) and quadrature (Y) components are most often combined to produce thermal images of the spatially varying total amplitude (R) and phase as

$$R = \sqrt{X^2 + Y^2};\, \theta = \arctan\left(\frac{Y}{X}\right). \tag{6.9}$$

Importantly, thus far, we have described the detected signal as an amplitude that varies continuously and smoothly as a sine wave over the course of the measurement. However, in reality, the signal is recorded by the camera at discrete points in time that are defined by the integration time and the framerate. Many measurements with short integration time with respect to the lock-in frequency (i.e., ω_{ref} from (6.6)) must be processed to build an image of the modulated signal. Although it is most typical to design lock-in measurements with over-sampling (framerate $> \omega_{\mathrm{ref}}$), it is also possible to use under-sampling (framerate $< \omega_{\mathrm{ref}}$), where the modulated signal is constructed by sampling different points over many periods.

In both the over-sampling and under-sampling cases, the camera's framerate can either be synchronized with the lock-in frequency, or it can be purposefully asynchronized with ω_{ref}. In the synchronized case, the same timer is used to trigger both the cell stimulus and the camera, but the camera is triggered at a multiple of the lock-in frequency. For typical over-sampling, a framerate of at least $4 \times \omega_{\mathrm{ref}}$ is needed for a two-phase lock-in measurement to capture a two-point correlation for each of the in-phase and quadrature components [15]. Clearly, increasing framerate will increase the signal resolution by providing a better approximation of the sinusoidal modulation. For asynchronized measurements, the framerate is arbitrary with respect to ω_{ref}, with the exception of forbidden frequencies at certain multiples of ω_{ref} [15]. In this case, ω_{ref} must be supplied for on-line calculation of weighting factors for each captured frame. The sum of the weighting factors will not be zero for the asynchronized measurement, so a DC background subtraction is necessary in this case.

6.4 Examples of PV material and cell-level imaging

High-resolution PL imaging is useful for developing greater spatial understanding of how different processing conditions affect the active materials employed in thin-film PV devices. Figure 6.6 shows an example of PL imaging on CdTe films, where various processing steps are investigated to understand the effect on carrier recombination and the passivation of grain boundaries [10]. Although typical grain sizes for thin-film solar cells are on the order of ~1 μm, the study in Figure 6.6 uses thick CdTe films supporting much larger grain sizes (~5–50 μm) to enable optical imaging of the grain boundaries. These films were polished with ion milling to minimize artifacts from surface roughness including shadowing and spectral reflections. The Cu-doped film shows higher and more uniform PL intensity compared with the undoped sample, consistent with the known effect of improved solar cell performance due to Cu acting as an acceptor when it occupies Cd sites [59]. The Cu-doped sample with a Te annealing treatment shows a bright perimeter around each grain boundary, while the center of the grain boundary remains dark, suggesting a gettering effect when excess Te segregates to the grain boundaries [60]. The P-doped films show overall higher PL intensity and regions of brighter PL along the grain boundaries, consistent with the longer lifetimes measured using time-resolved PL mapping. The Cd-annealed samples show even higher average PL intensity and longer carrier lifetimes, though the Cd annealing treatment decreases the relative luminescence intensity at the grain boundaries. This high-resolution PL imaging study enabled greater understanding of various doping and annealing treatments for CdTe films and showed the promise of applying PL imaging techniques to understand grain boundaries in other thin-film technologies.

In completed device structures, comparing PL images with EL and LIT images is a powerful tool to differentiate spatial variations in material quality from variations in the quality of electrical contacts. Figure 6.7 shows an example of combined PL, EL, and DLIT imaging on several thin-film CIGS devices that were stressed

Figure 6.6. PL images with 450 μm field of view of CdTe films under various processing conditions including (left to right) as-deposited, Cu-doped, Cu-doping followed by Te-annealing, P-doped, and P-doping followed by Cd annealing. Time-resolved PL maps are shown for the right three samples with 375 μm across each map. The image is modified with permission from Johnston et al. [10]

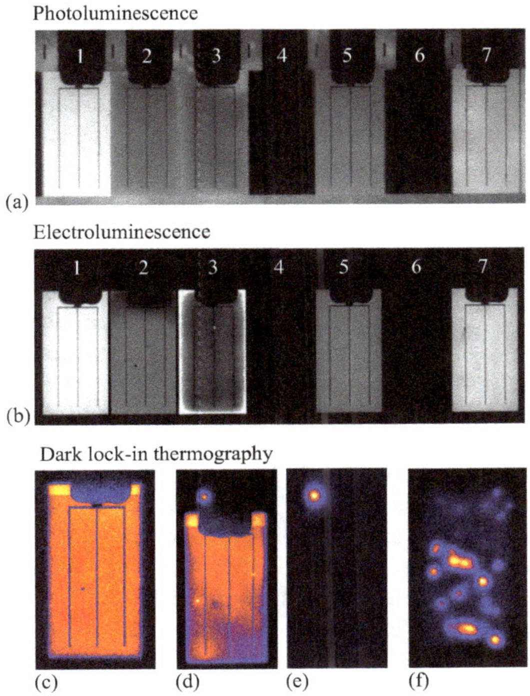

Figure 6.7. Example of CIGS cells that underwent degradation by reverse-bias stressing, investigated by (a) PL, (b) EL, and (c–e) DLIT. Dark lock-in thermography of a CIGS cell stressed for potential-induced degradation (PID) in (f) shows a different heating pattern due to a different degradation mechanism

under reverse bias with a goal of discovering ways to make the cells more resilient to reverse-bias breakdown. Mediating possible reverse-bias breakdown has received considerable attention in the continued development of thin-film modules because these modules are not most often equipped with bypass diodes, so partial shading places cells in reverse bias, causing high breakdown current densities that concentrate at defects and result in cell shunting [9,61–63].

The cells in Figure 6.7(a) and (b) show various degrees of degradation that occurred during the reverse-bias stress. Cells 1 and 7 are control cells that were not stressed, and these cells show overall brighter PL and EL intensity compared with the stressed cells. Cells 4 and 6 appear dark in both the PL and EL images likely because a severe short circuit does not allow current injection for EL and drains photogenerated carriers during PL measurement. Interestingly, DLIT imaging of cells 4 and 6 showed that the short circuit for both of these devices developed in the electrode area in a region away from the active CIGS material. Cells 2, 3, and

5 have decreased PL intensity compared with the control cells, which could be caused either by the less severe shunting compared with cells 4 and 6 (where shunting in this case developed within the active CIGS area) or by degradation of the active material. Interestingly, cell 3 shows bright edges during EL measurement, which could be caused by partial delamination of the transparent conductive oxide (TCO) contact near the center of the cell.

The DLIT images in Figure 6.7(c)–(f) demonstrate the identification of shunted areas either within the cell area or along the electrode using thermal imaging. Figure 6.7(c) shows an unstressed cell in comparison with a cell with a short circuit in the electrode area (Figure 6.7(d) and (e)). Figure 6.7(d) shows the shorted cell under forward bias, where a hot spot indicates a short-circuited area in the electrode, although some current is still injected into the active cell area causing mostly uniform heating across the cell. Under reverse bias, Figure 6.7(e) shows that all current flows through the short circuit in the electrode, and this is the spot where the cell eventually breaks down under reverse-bias stressing.

Figure 6.7(f) shows an example of a CIGS cell from a different experiment that was stressed under conditions for PID. The mechanism of PID is still not widely understood for thin-film modules, but it suggested the shunts formation due to sodium migration along grain boundaries [64]—similar to the mechanism of PID shunting in silicon PV, where shunts are formed by sodium migration, causing conductive pathways along stacking faults [65,66]. Consistent with this expected degradation mode for PID, Figure 6.7(f) shows the PID-affected cell degraded by the formation of many shunts within the active area.

Thermal imaging of series-connected thin-film cells is particularly well-equipped to study reverse-bias breakdown caused by partial shading in real time. Figure 6.8 shows an example of thermal imaging carried out in video mode to investigate the worm-like defects that form in CIGS cells and modules under partial-shading conditions. This study showed that breakdown under reverse bias initiates at localized defects either within the cell or at scribe lines, leading to further melting of the active material that can be seen as hot spot migration in a worm-like pattern [9,62,67]. LIT with limited current can be used to study the initial defects that may lead to reverse-bias breakdown, and defects have been

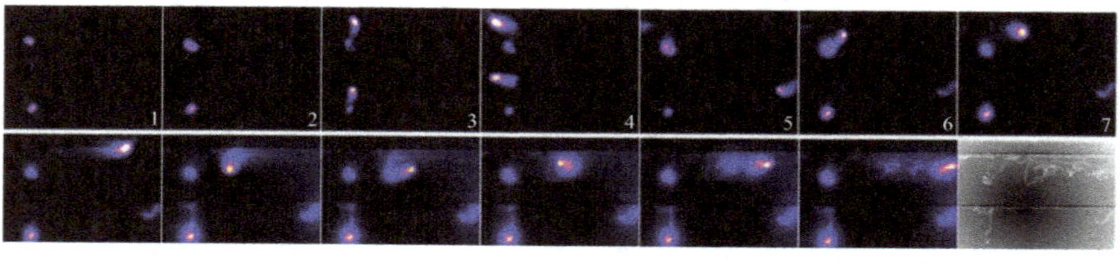

Figure 6.8. Selected frames from a thermal camera video of worm-like defects formed under reverse-bias breakdown in a CIGS mini-module along with a final image showing the visible damage in the cell. This image is reproduced with permission from Johnston et al. [9]

identified as craters, voids, and nodules, in addition to scribe-line imperfections [9,62,68]. Collecting thermal videos of reverse-bias breakdown is not a lock-in technique, and the high current densities that pass through the cell lead to permanent damage, similar to that which occurs under partial shading in the field. The images in Figure 6.8 from left to right are selected frames from a thermal video that spans a region across two cells that were placed under reverse bias. Frame 1 of the video shows that breakdown originates at mid-cell defects, but under 30 mA/cm^2 reverse-bias current the defects burn out after random propagation for a few seconds. Frames 2–3 show that the hot spots begin to propagate again when the current density is increased to 60 mA/cm^2, and frames 4–13 show how breakdown eventually reaches the scribe lines and propagates along the scribe line under the high current density of 150 mA/cm^2. Importantly, these current densities are calculated based on the entire cell area, and the actual current densities through the point-like breakdown sites are very high.

The hot spots that eventually lead to the formation of worm-like defects or shunting in thin-film solar cells have also been investigated by combining thermal imaging with laser micromachining, as demonstrated in Figure 6.9 [38]. Figure 6.9 (a) is a thermal image of a thin-film cell that was stressed under reverse bias to simulate partial-shading conditions, where the stressing caused development of one hot spot at the scribe line and one hot spot within the interior of the cell. Figure 6.9 (d) shows that the reverse-bias stress caused the cell to become shunted. That is, the dark *I–V* curve after stressing (red trace) resembles resistor characteristics in comparison with the rectifying diode characteristics before stressing (gray trace). Figure 6.9(b) shows a steady-state image where the defect in the cell interior was electrically isolated from the cell by scribing with a pulsed femtosecond laser. Figure 6.9(c) shows the thermal image after laser scribing, where the isolated defect is no longer connected to the cell and does not heat up. The *I–V* curve after laser scribing in Figure 6.9(d) confirms that the defect in the cell interior was primarily responsible for the cell shunting because the final *I–V* curve very closely resembles the initial *I–V* curve before stressing.

Recently, thermal and luminescence imaging has also become increasingly popular for developing perovskite devices as the cell areas are increasing and larger-scale homogeneity is becoming more important [2,69–75]. Figure 6.10 shows an example of using PL, EL, and DLIT imaging to study the effects of accelerated stress testing in perovskite solar cells. An unstressed perovskite solar cell (a–c) is compared with one that degraded under an accelerated stressing procedure (d–f). The PL image for the control cell is largely uniform, aside from some dark points that could either be pinholes, shunts, or variations in the material composition. Interestingly, many of these dark points appear enlarged in the EL image, with several of them having a bright halo. This result suggests that the enlarged points in the EL image are current sinks, such as shunts or nonradiative recombination centers, although some of the spatial variation may additionally be related to inhomogeneity in the material composition. The degraded cell shows significantly larger dark spots in the PL and EL images. Interestingly, further EL and PL imaging with a different set of band-pass filters showed that the dark spots

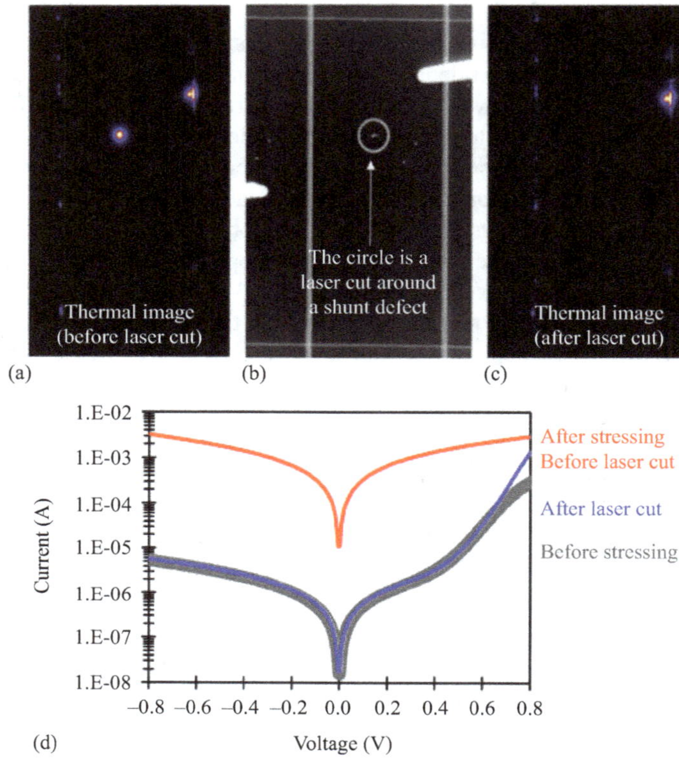

Figure 6.9. Example of laser micromachining to remove shunt-type defects in a thin-film solar cell. (a) DLIT image of cell with two shunts before micromachining; (b) steady-state image showing micromachining used to cut the layers of the cell and isolate the shunt; (c) DLIT after laser scribing; (d) I–V curves of the solar cell before and after laser scribing. The figure is modified with permission from Johnston et al. [38]

in Figure 6.10(d) and (e) are regions with different material composition and lower-energy emission. The DLIT images show more heat dissipation in the degraded cell, where the heat dissipation is likely co-incident with the dark spots in the luminescence images. However, the LIT images for both cells are dominated by a dark outline of the metal contact. This dark outline is caused by the low thermal emissivity of shiny metal surfaces, and this study demonstrates that further optimization of the surface thermal emissivity is necessary to use LIT on cells of this architecture.

While the study in Figure 6.10 used different sets of band-pass filters to discover qualitative changes in the material composition and emission wavelength

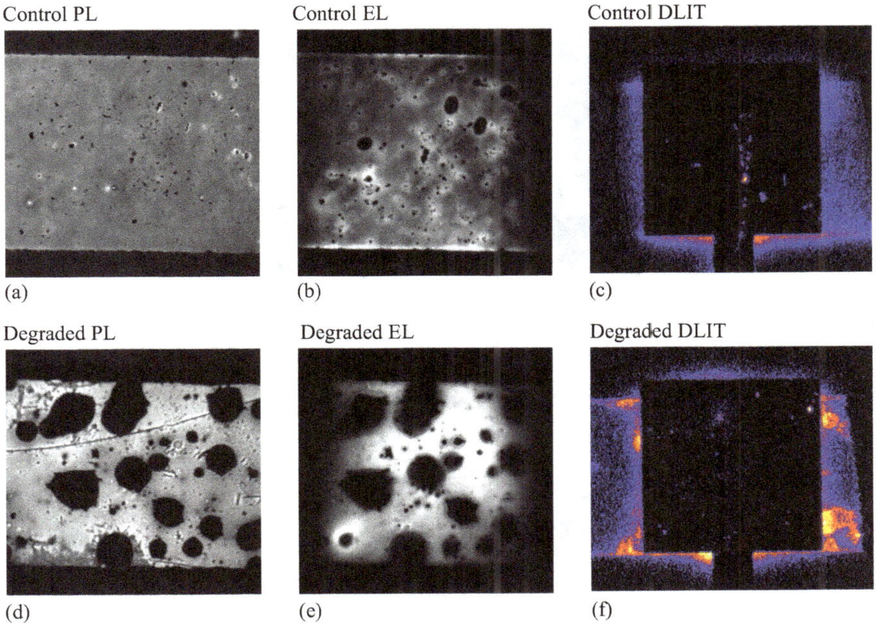

Figure 6.10. *Example of two perovskite solar cells, including a control cell (a–c) and a degraded cell that has undergone accelerated stress testing (d–f). Cells are imaged by PL (a and d), EL (b and e), and DLIT (c and f)*

caused by stressing the perovskite cells, the study in Figure 6.11 demonstrates a systematic method that has been used to quantitatively map the shift in EL peak position with sample degradation [71]. In this study, the authors noted that perovskites tend to have Gaussian-shaped luminescence spectra, where emission occurs at the optical bandgap and there are no significant defect bands. Because this spectral shape can be assumed, the only parameters necessary to extract spatial variations in the optical bandgap were the Gaussian peak position and full width at half maximum. The authors were able to find these two unknown values using two luminescence ratios. That is, three luminescence images were collected with different optical filters, and then two distinct ratio images along with the wavelength-dependent responsivity of the imaging system were used to map the spatially varying optical bandgap. This technique can be applied to different PV systems with Gaussian-shaped luminescence profiles, provided that the correct filters are chosen and there are no defect emission bands within the camera's detectivity range. This can serve especially as a useful screening tool for thin-film PV because spatial variations in chemical and morphological properties that evolve during degradation must be continuously monitored and controlled.

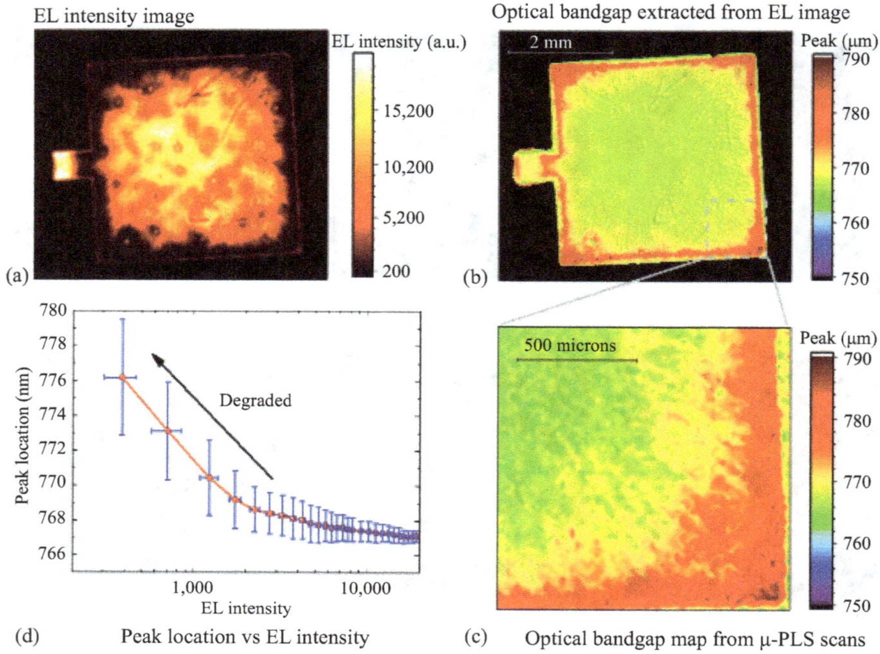

Figure 6.11. Imaging of a perovskite solar cell that was degraded by remaining for >2 months in nitrogen atmosphere with no humidity control, including (a) integrated EL intensity image, (b) optical bandgap image extracted from the ratio of filtered images, (c) optical bandgap image from spectral PL mapping, and (d) peak location as a function of luminescence intensity. This figure was reproduced with permission from Chen et al. [71]

Given the emphasis on characterizing degradation of perovskite cells in the studies represented by Figures 6.10 and 6.11, it is apparent that long-term stability of perovskite thin-film PV has been a particularly important concern, especially as efficiencies have increased and discussions of commercialization progress. Figure 6.12 further demonstrates a practical use of EL and PL imaging to teach a lesson about perovskite device stability in the context of common lab practices where a nominally nondestructive and routine characterization method (measuring illuminated *I–V* curves) can irreversibly degrade a perovskite cell [73]. Figure 6.12 shows EL images before illuminated *I–V* measurements compared with the images either directly after or a day after the measurements. Using EL imaging, the authors observed brighter luminescence in the cell area defined by the aperture of the shadow mask, which is used to define the illuminated area during *I–V* scans. The increased EL was found to correlate with improvements in the *I–V* curves over several repeating illuminated scans. The authors concluded that the act of

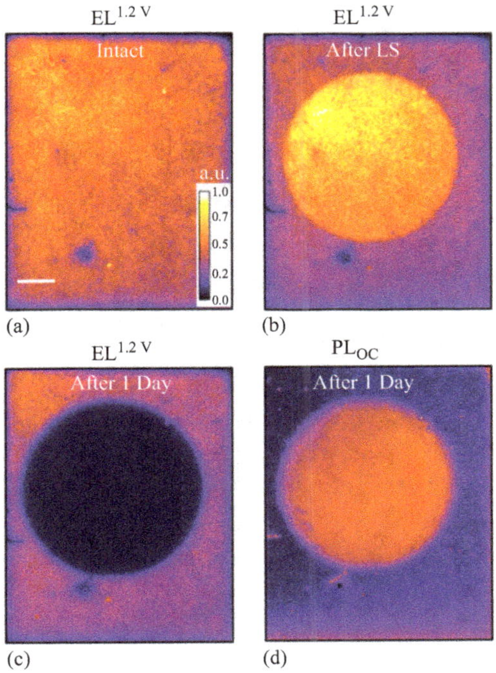

Figure 6.12. EL images of a perovskite solar cell (a) before measurement, (b) immediately after illuminated I–V measurement, and (c) after 1 day, along with (d) the PL image after 1 day. This figure is reproduced with permission from Soufiani et al. [73]

measuring the illuminated *I–V* curve caused a decrease in contact resistance at the TiO_2/perovskite interface due to ion migration in the direction opposing the photoinduced voltage. Interestingly, after a day of storing the device in a controlled atmosphere, the EL intensity in the aperture area decreased significantly relative to the area of the cell that was not illuminated, and this correlated with a relative increase in the open-circuit PL intensity. The authors found that this pattern was caused by delamination of the perovskite from the TiO_2 contact, which decreased the EL amplitude due to poor carrier injection, but increased the PL amplitude because of less nonradiative recombination at interfacial defects.

Although these studies have highlighted instability of perovskite device structures, it is important to note that their longevity continues to increase [76], and several frameworks for commercialization are actively developed even before realizing the target of 10–20-year stability [76]. The development of tandem perovskite devices deposited onto high-efficiency silicon heterojunction cells has become one of the leading candidates for commercializing perovskite technology [77]. Figure 6.13 shows the use of EL imaging with different filter combinations along with DLIT imaging to investigate a section of a large-area tandem

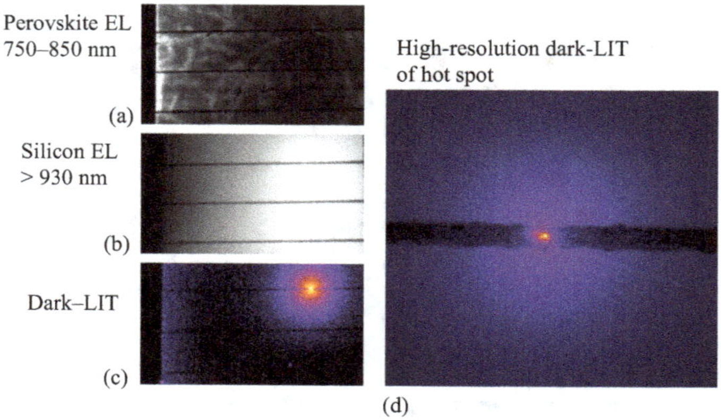

Figure 6.13. Example of imaging a tandem perovskite-silicon cell using EL with (a) 750–850-nm band-pass filter and (b) 930-nm long-pass filter along with DLIT of (c) the full cell and (d) a grid-finger defect

perovskite/silicon solar cell. In this example, nonuniformities in the unoptimized perovskite layer are imaged using a 750–850-nm band-pass filter, demonstrating scratches in the perovskite layer, particulates, and more intense luminescence at the cell edge. Imaging of the silicon luminescence with a 930-nm long-pass filter cuts out the perovskite signal and confirms that these spatial variations are indeed related to the perovskite, while the silicon cell shows mostly uniform luminescence aside from greater recombination at the outer cell regions caused by unpassivated edges formed during cell cleavage. DLIT imaging shows a mid-cell hot spot, along with heating at the unpassivated cell edge. Higher-resolution LIT of the hot spot shows that the defective area is a scratch that has caused a break in the screen-printed finger. Figure 6.13 shows an important example of separating defects or nonuniformities in particular layers of tandem devices, enabling informed cell optimization.

6.5 Examples of module-level imaging

Large-scale imaging at the PV module level is useful for developing greater spatial understanding of outdoor degradation mechanisms, understanding the impacts of accelerated stress conditions, developing new module fabrication processes, and monitoring manufacturing quality control. In this section, we provide some examples of how thin-film module-scale imaging has been used to accomplish these goals.

Figure 6.14 shows PL, EL, and DLIT images for two thin-film modules that degraded under different types of stress. The CdTe module (upper row) underwent accelerated stress testing for PID including −1,000 V applied to the shorted leads of the module at elevated temperature and humidity [78]. PID is caused by

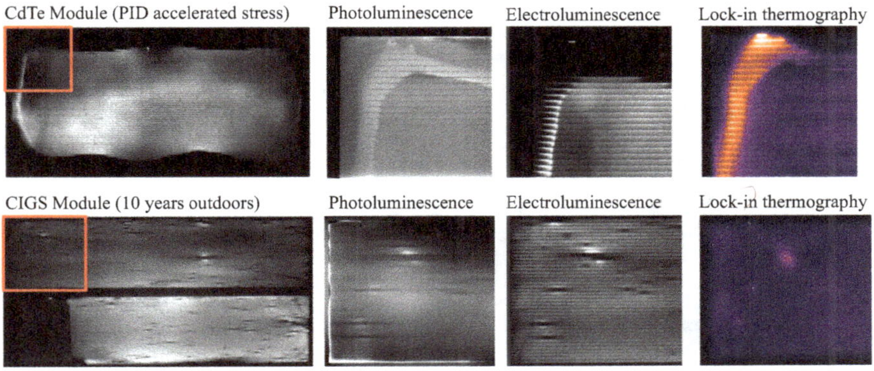

Figure 6.14. Imaging of degraded CdTe (top row) and CIGS (bottom row) PV modules. The image on the left is a full-module EL image, where the red box indicates the region corresponding to higher-resolution images on the right

migration of positive sodium atoms toward the cells when cells are under relative negative bias compared with the module package, and power loss in the field can occur by both shunting and increases in series resistance caused by TCO degradation and changes in local chemistry or microstructure in the active material.

The PID degradation example in Figure 6.14 shows a dark perimeter around the module in the EL images, which is caused by increased series resistance that prevents current injection into the outer module regions and effectively removes these regions from the circuit. The PL image shows a brighter band encircling the edge of the module, where PL intensity is higher because the series resistance prevents photoexcited carriers from leaving this region. The DLIT image shows heat dissipation in the high-resistance area. Importantly, visible inspection alone of this CdTe module did not show significant signs of degradation, underscoring the importance of using these spatially resolved imaging methods to understand PID degradation mechanisms.

The CIGS module in the second row of Figure 6.14 degraded after outdoor operation for 10 years. The EL for this module shows dark spots throughout the module corresponding to shunts, where some of the most severe shunts have bright regions nearby caused by current crowding. The lower left corner in the module is dark due to a crack in the glass. The PL image shows a similar pattern to the EL images, where the shunted cells are darker due to a lower voltage in these cells. The DLIT images show point-like hot spots where high current densities through the shunts cause resistive heating. Importantly, these localized shunts heat up in both forward and reverse bias (as expected for a resistor), and they overlay with the darker regions in the PL and EL images.

The shunted defects in the CIGS module resemble previous examples of breakdown caused by partial shading, where worm-like defects originate either at scribe lines or within cells [9,67,79,80]. Similar shunting patterns have also been

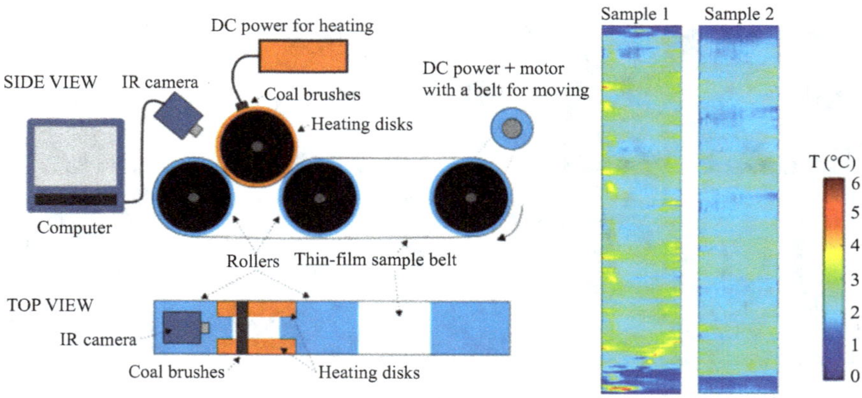

Figure 6.15. Schematic for in-line thermal imaging during roll-to-roll processing, along with thermal images ITO on polyethylene terephthalate films, 1.5 m long and 5 cm wide. Adapted with permission from Remes et al. [82]. Copyright from The Optical Society

observed in as-fabricated modules, where a combination of EL and DLIT imaging has been used to show that the relative degrees of mid-cell shunting versus scribe-line shunting depend on the amount of sodium segregated at the surface of the absorber layer as well as the use of different buffer-layer deposition methods [68]. Combined analysis of EL and LIT images using a SPICE model has also enabled quantitative characterization of shunt resistance (from LIT) versus series resistance of the front and back contacts (from EL) [1,81].

Finally, we noted that imaging methods have been increasing in popularity for quality control in manufacturing, and Figure 6.15 gives an example of using thermal imaging for in-line defect identification during roll-to-roll processing of thin films by synchronizing image acquisition with the sample motion [82]. The schematic in Figure 6.15 depicts how DC power is applied to metal disks using coal brushes, in which the TCO film is then heated by electrical injection as the edges of the sample are pressed against the metal disks with rollers. The top view of the schematic shows that the camera captures thermal images of the sample area between the metal disks. In recent reports, a similar in-line thermal imaging method has been implemented as a contactless technique [83]. That is, instead of inducing heating with current injection under direct contact with the sample, the new contactless technique causes heating in the sample with eddy currents produced by an induction coil [83].

The thermal images on the right-hand side of Figure 6.15 show in-line thermal images of two indium tin oxide (ITO) samples on flexible substrates, in which Sample 1 was mechanically stressed by twisting and bending and Sample 2 was unstressed and handled carefully. The impact of mechanical stress on brittle ITO is clearly shown with the in-line imaging as a higher overall thermal signal and greater inhomogeneity. These results show both the promise of using thermal imaging for quality control during roll-to-roll processing and demonstrating the interest at the manufacturing level for imaging TCO layers in addition to the active absorber layers in thin-film devices.

6.6 Concluding remarks

In this chapter, we discussed the utility of luminescence and thermal imaging for characterizing thin-film PV materials and devices. Wide-field camera-based imaging has proven to be a time-efficient method for characterization across varied length scales—from studying microscale materials properties such as bandgap inhomogeneities and grain-boundary passivation to investigating module-scale degradation mechanisms that occur in fielded commercial devices. Especially for these developing thin-film technologies, imaging of lab-scale solar cells has been particularly useful for targeted optimization. At early stages of device development, in which understanding of new materials or fabrication methods may be incomplete, imaging can play a crucial role in quickly identifying defects and expediting progress.

At the manufacturing level, although luminescence imaging has become widely applied in silicon PV quality control, there are still additional challenges for imaging thin-film devices at the large-scale fabrication stage. In particular, thin-film modules are directly deposited onto module-scale substrates and cannot be tested cell-by-cell before assembly in analogy to silicon. For this reason, new camera-based methods and experimental geometries must be developed for in-line screening of thin-film PV. The progress in this area is exciting, with new understanding of contactless thermal and luminescence methods potentially enabling mobile quality control of roll-to-roll devices in addition to conventional module structures. The ongoing advancement and adaptation of imaging techniques for thin-film PV characterization reflects the widespread need for fast, reliable, and quantitative screening methods that can be applied to a broad range of continuously evolving PV materials.

References

[1] Gerber A., Huhn V., Tran T.M.H., *et al.* “Advanced large area characterization of thin-film solar modules by electroluminescence and thermography imaging techniques.” *Sol. Energy Mater. Sol. Cells* 2015, vol. 135, pp. 35–42.

[2] Walter D., Wu Y., Duong T., *et al.* “On the use of luminescence intensity images for quantified characterization of perovskite solar cells: Spatial distribution of series resistance.” *Adv. Energy Mater.* 2018, vol. 8(2), p. 1701522.

[3] Lisco F., Bowers J., Walls J., *et al.* “Photoluminescence imaging analysis of doping in thin film CdS and CdS/CdTe devices.” *IEEE 44th Photovoltaic Specialists Conference (PVSC)*, Washington, DC, 2017.

[4] Johnston S., Unold T., Repins I., *et al.* "Correlations of Cu(In, Ga)Se2 imaging with device performance, defects, and microstructural properties." *J. Vac. Sci. Technol. A* 2012, vol. 30(4), p. 04D111.
[5] Johnston S., Unold T., Repins I., *et al.* "Imaging characterization techniques applied to Cu(In,Ga)Se2 solar cells." *J. Vac. Sci. Technol. A* 2010, vol. 28 (4), pp. 665–670.
[6] Runai F.R., Schwäble F., Walter T., *et al.* "Imaging and performance of CIGS thin film modules." *2011 37th IEEE Photovoltaic Specialists Conference*, 2011, pp. 003399–003403.
[7] Bokalič M., Gerber A., Pieters B.E., Rau U. and Topič M. "Bandgap fluctuations observed by EL in various Cu(In,Ga)(Se,S)2 PV modules". *IEEE J. Photovolt.* 2018, vol. 8(1), pp. 272–277.
[8] Bokalič M., Pieters B.E., Gerber A., Rau U. and Topič M. "Bandgap imaging in Cu(In,Ga)Se2 photovoltaic modules by electroluminescence." *Prog. Photovolt. Res. Appl.* 2017, vol. 25(2), pp. 184–191.
[9] Johnston S., Sulas D., Palmiotti E., *et al.* "Thin-film module reverse-bias breakdown sites identified by thermal imaging." *IEEE 7th World Conference on Photovoltaic Energy Conversion (WCPEC)*, Waikoloa, HI, 2018, pp. 1897–1901.
[10] Johnston S., Motz A.A., Reese M.O., Burst J.M. and Metzger W.K. "Photoluminescence imaging of large-grain CdTe for grain boundary characterization." *IEEE 42nd Photovoltaic Specialist Conference (PVSC)*, New Orleans, LA, 2015, pp. 1–4.
[11] Uchida I. "Observation of the recombination radiation from silicon p-n junction." *Jpn. J. Appl. Phys.* 1963, vol. 2(9), pp. 561–564.
[12] Fuyuki T., Kondo H., Yamazaki T., Takahashi Y. and Uraoka Y. "Photographic surveying of minority carrier diffusion length in polycrystalline silicon solar cells by electroluminescence." *Appl. Phys. Lett.* 2005, vol. 86(26), p. 262108.
[13] Fuyuki T., Kondo H., Kaji Y., Yamazaki T., Takahashi Y. and Uraoka Y. "One shot mapping of minority carrier diffusion length in polycrystalline silicon solar cells using electroluminescence." *IEEE 31st Photovoltaic Specialists Conference*, 2005, pp. 1343–1345.
[14] Breitenstein O., Langenkamp M., Lang O. and Schirrmacher A. "Shunts due to laser scribing of solar cells evaluated by highly sensitive lock-in thermography." *Sol. Energy Mater. Sol. Cells* 2001, vol. 65(1), pp. 55–62.
[15] Breitenstein O., Wilhelm W. and Langenkamp M. *Lock-in thermography*, second edn. Berlin Heidelberg: Springer-Verlag; 2010.
[16] Ramspeck K., Bothe K., Hinken D., Fischer B., Schmidt J. and Brendel R. "Recombination current and series resistance imaging of solar cells by combined luminescence and lock-in thermography." *Appl. Phys. Lett.* 2007, vol. 90(15), p. 153502.
[17] Jordan D.C. and Kurtz S.R. "Photovoltaic degradation rates—an analytical review." *Prog. Photovolt. Res. Appl.* 2013, vol. 21(1), pp. 12–29.

[18] Janssen G. J. M., Slooff L. H. and Bende E.E. "2D-Finite Element Model of a CIGS Module." *IEEE 38th Photovoltaic Specialists Conference (PVSC)*, Austin, TX, 2012

[19] Ott T., Runai F.R., Schwäble F. and Walter T. "2D network simulation and luminescence characterization of Cu(In,Ga)Se2 thin film modules." *Prog. Photovolt. Res. Appl.* 2012, vol. 20(5), pp. 600–605.

[20] Rau U. and Werner J.H. "Radiative efficiency limits of solar cells with lateral band-gap fluctuations." *Appl. Phys. Lett.* 2004, vol. 84(19), pp. 3735–3737.

[21] Werner J.H., Mattheis J. and Rau U. "Efficiency limitations of polycrystalline thin film solar cells: Case of Cu(In,Ga)Se2." *Thin Solid Films* 2005, vol. 480–481, pp. 399–409.

[22] Noel N.K., Stranks S.D., Abate A., *et al.* "Lead-free organic–inorganic tin halide perovskites for photovoltaic applications." *Energy Environ. Sci.* 2014, vol. 7(9), pp. 3061–3068.

[23] Wilkes T.C., McGonigle A.J.S., Willmott J.R., Pering T.D. and Cook J.M. "Low-cost 3D printed 1 nm resolution smartphone sensor-based spectrometer: Instrument design and application in ultraviolet spectroscopy." *Opt. Lett.* 2017, vol. 42(21), pp. 4323–4326.

[24] Wilkes C.T., McGonigle J.A., Pering D.T., *et al.* "Ultraviolet imaging with low cost smartphone sensors: Development and application of a raspberry pi-based UV camera." *Sensors* 2016, vol. 16(10), p. 1649.

[25] Wilkes C.T., Pering D.T., McGonigle J.A., Tamburello G. and Willmott R.J. "A low-cost smartphone sensor-based UV camera for volcanic SO2 emission measurements." *Remote Sensing* 2017, vol. 9(1), p. 27.

[26] Owen-Bellini M., Sulas-Kern D.B., Spataru S., North H. and Hacke P. "In-situ performance characterization of photovoltaic modules during combined-accelerated stress testing." *IEEE 46th Photovoltaic Specialists Conference (PVSC)*, Chicago, IL, 2019.

[27] Sinha A., Bliss M., Wu X., Roy S., Gottschalg R. and Gupta R. "Cross-characterization for imaging parasitic resistive losses in thin-film photovoltaic modules." *J. Imaging* 2016, vol. 2(3), p. 23.

[28] Helbig A., Kirchartz T., Schaeffler R., Werner J.H. and Rau U. "Quantitative electroluminescence analysis of resistive losses in Cu(In, Ga)Se2 thin-film modules." *Sol. Energy Mater. Sol. Cells* 2010, vol. 94(6), pp. 979–984.

[29] deQuilettes D.W., Jariwala S., Burke S., *et al.* "Tracking photoexcited carriers in hybrid perovskite semiconductors: Trap-dominated spatial heterogeneity and diffusion." *ACS Nano* 2017, vol. 11(11), pp. 11488–11496.

[30] Delamarre A., Paire M., Guillemoles J.-F. and Lombez L. "Quantitative luminescence mapping of Cu(In, Ga)Se2 thin-film solar cells." *Prog. Photovolt. Res. Appl.* 2015, vol. 23(10), pp. 1305–1312.

[31] Sinton R.A. "Contactless electroluminescence for shunt-value measurement in solar cells." *23rd European Photovoltaic Solar Energy Conference and Exhibition*, Valencia, 2008, pp. 1157–1159.

[32] Johnston S. "Contactless electroluminescence imaging for cell and module characterization." *IEEE 42nd Photovoltaic Specialist Conference (PVSC)*, New Orleans, 2015, pp. 1–6.

[33] Sulas D.B., Johnston S. and Jordan D.C. "Comparison of photovoltaic module luminescence imaging techniques: Assessing the influence of lateral currents in high-efficiency device structures." *Sol. Energy Mater. Sol. Cells* 2019, vol. 192, pp. 81–87.
[34] Sulas D.B., Johnston S. and Jordan D.C. "Imaging lateral drift kinetics to understand causes of outdoor degradation in silicon heterojunction photovoltaic modules." *Solar RRL* 2019, vol. 3, p. 1900102.
[35] Razani M., Parkhimchyk A. and Tabatabaei N. "Lock-in thermography using a cellphone attachment infrared camera." *AIP Advances* 2018, vol. 8 (3), p. 035305.
[36] Sulas D.B., Johnston S., Seitzman N., Platt H., Al-Jassim M. and Guthrey H. "Defect detection in solid-state battery electrolytes using lock-in thermal imaging." *J. Electrochem. Soc.* 2018, vol. 165(13), pp. A3205–A3211.
[37] Johnston S., Albin D., Hacke P., *et al*. "Imaging, microscopic analysis, and modeling of a CdTe module degraded by heat and light." *Sol. Energy Mater. Sol. Cells* 2018, vol. 178, pp. 46–51.
[38] Johnston S., Sulas-Kern D.B. and Kroeger G.F. "Laser cutting and micromachining for localized and targeted solar cell characterization." *IEEE 46th Photovoltaic Specialists Conference (PVSC)*, Chicago, IL, 2019.
[39] Kendig D., Alers G.B. and Shakouri A. "Thermoreflectance imaging of defects in thin-film solar cells." *IEEE International Reliability Physics Symposium*, 2010, pp. 499–502.
[40] Katz N., Patterson M., Zaunbrecher K., Johnston S. and Hudgings J. "High-resolution imaging of defects in CdTe solar cells using thermoreflectance." *Electron. Lett.* 2013, vol. 49(24), pp. 1559–1561.
[41] Matatagui E., Thompson A.G. and Cardona M. "Thermoreflectance in semiconductors." *Phys. Rev.* 1968, vol. 176(3), pp. 950–960.
[42] Demirtaş M., Tamyürek B., Kurt E., Çetinbaş İ. and Öztürk M.K. "Effects of aging and environmental factors on performance of CdTe and CIS thin-film photovoltaic modules." *J. Electron. Mater.* 2019, vol. 48, pp. 6890–6900.
[43] Nelson J. *The physics of solar cells*. London: Imperial College Press; 2003.
[44] Sze S.M. and Ng K.K. *Physics of semiconductor devices*, second edn. New York: Wiley; 2006.
[45] Rau U. "Reciprocity relation between photovoltaic quantum efficiency and electroluminescent emission of solar cells." *Phys. Rev. B* 2007, vol. 76(8), p. 085303.
[46] Ross R.T. "Some thermodynamics of photochemical systems." *J. Chem. Phys.* 1967, vol. 46(12), pp. 4590–4593.
[47] Wurfel P. "The chemical potential of radiation." *J. Phys. C: Solid State Phys.* 1982, vol. 15(18), pp. 3967–3985.
[48] Ross R.T. "Thermodynamic limitations on the conversion of radiant energy into work." *J. Chem. Phys.* 1966, vol. 45(1), pp. 1–7.
[49] Juhl M.K. and Trupke T. "The impact of voltage independent carriers on implied voltage measurements on silicon devices." *J. Appl. Phys.* 2016, vol. 120(16), p. 165702.

[50] Kasemann M., Reindl L.M., Michl B., Warta W., Schütt A. and Carstensen J. "Contactless qualitative series resistance imaging on solar cells." *IEEE J. Photovolt.* 2012, vol. 2(2), pp. 181–183.

[51] Potthoff T., Bothe K., Eitner U., Hinken D. and Köntges M. "Detection of the voltage distribution in photovoltaic modules by electroluminescence imaging." *Prog. Photovolt. Res. Appl.* 2010, vol. 18(2), pp. 100–106.

[52] Li B., Stokes A. and Doble D.M.J. "Evaluation of two-dimensional electrical properties of photovoltaic modules using bias-dependent electroluminescence." *Prog. Photovolt. Res. Appl.* 2011, vol. 20(8), pp. 936–944.

[53] Haunschild J., Glatthaar M., Kasemann M., Rein S. and Weber E.R. "Fast series resistance imaging for silicon solar cells using electroluminescence." *Phys. Status Solidi RRL* 2009, vol. 3(7–8), pp. 227–229.

[54] Ren A., Lai H., Liu C., *et al.* "A luminescence-based interpolation method for series resistance imaging in thin film solar cells." *Jpn. J. Appl. Phys.* 2019, vol. 58(5), p. 050908.

[55] Breitenstein O., Bauer J., Hinken D. and Bothe K. "The reliability of thermography- and luminescence-based series resistance and saturation current density imaging." *Sol. Energy Mater. Sol. Cells* 2015, vol. 137, pp. 50–60.

[56] Bandhauer T.M., Garimella S. and Fuller T.F. "A critical review of thermal issues in lithium-ion batteries." *J. Electrochem. Soc.* 2011, vol. 158(3), pp. R1–R25.

[57] Glatthaar M., Giesecke J., Kasemann M., *et al.* "Spatially resolved determination of the dark saturation current of silicon solar cells from electroluminescence images." *J. Appl. Phys.* 2009, vol. 105(11), p. 113110.

[58] Breitenstein O., Straube H. and Iwig K. "Lock-in thermography with depth resolution on silicon solar cells." *Sol. Energy Mater. Sol. Cells* 2018, vol. 185, pp. 66–74.

[59] Perrenoud J., Kranz L., Gretener C., *et al.* "A comprehensive picture of Cu doping in CdTe solar cells." *J. Appl. Phys.* 2013, vol. 114(17), p. 174505.

[60] Kumar S.G. and Rao K.S.R.K. "Physics and chemistry of CdTe/CdS thin film heterojunction photovoltaic devices: Fundamental and critical aspects." *Energy Environ. Sci.* 2014, vol. 7(1), pp. 45–102.

[61] Silverman T.J., Mansfield L., Repins I. and Kurtz S. "Damage in monolithic thin-film photovoltaic modules due to partial shade." *IEEE J. Photovolt.* 2016, vol. 6(5), pp. 1333–1338.

[62] Palmiotti E., Johnston S., Gerber A., *et al.* "Identification and analysis of partial shading breakdown sites in CuInxGa(1−x)Se2 modules." *Sol. Energy* 2018, vol. 161, pp. 1–5.

[63] Silverman T.J., Deceglie M.G., Sun X., *et al.* "Thermal and electrical effects of partial shade in monolithic thin-film photovoltaic modules." *IEEE J. Photovolt.* 2015, vol. 5(6), pp. 1742–1747.

[64] Harvey S.P., Guthrey H., Muzzillo C.P., *et al.* "Investigating PID shunting in polycrystalline CIGS devices via multi-scale, multi-technique characterization." *IEEE J. Photovolt.* 2019, vol. 9(2), pp. 559–564.

[65] Harvey S.P., Moseley J., Norman A., *et al.* "Investigating PID shunting in polycrystalline silicon modules via multiscale, multitechnique characterization." *Prog. Photovolt. Res. Appl.* 2018, vol. 26(6), pp. 377–384.

[66] Harvey S.P., Sulas D., Hacke P., Johnston S. and Al-Jassim M. "Comparison of PID shunting in polycrystalline and single-crystal silicon modules via multi-scale, multi-technique characterization." *IEEE 46th Photovoltaic Specialists Conference (PVSC)*, Chicago, IL, 2019.

[67] Lee J.E., Bae S., Oh W., *et al.* "Investigation of damage caused by partial shading of CuInxGa(1−x)Se2 photovoltaic modules with bypass diodes." *Prog. Photovolt. Res. Appl.* 2016, vol. 24(8), pp. 1035–1043.

[68] Bin Mo C., Park S.J., Bae S., *et al.* "Impact of buffer layer process and Na on shunt paths of monolithic series-connected CIGSSe thin film solar cells." *Sci. Rep.* 2019, vol. 9(1), p. 3666.

[69] Moerman D., Eperon G.E., Precht J.T. and Ginger D.S. "Correlating photoluminescence heterogeneity with local electronic properties in methylammonium lead tribromide perovskite thin films." *Chem. Mater.* 2017, vol. 29(13), pp. 5484–5492.

[70] Soufiani A.M., Tayebjee M.J.Y., Meyer S., *et al.* "Electro- and photoluminescence imaging as fast screening technique of the layer uniformity and device degradation in planar perovskite solar cells." *J. Appl. Phys.* 2016, vol. 120(3), p. 035702.

[71] Chen B., Peng J., Shen H., *et al.* "Imaging spatial variations of optical bandgaps in perovskite solar cells." *Adv. Energy Mater.* 2019, vol. 9(7), p. 1802790.

[72] Bercegol A., Ramos F.J., Rebai A., Ory D., Rousset J. and Lombez L. "Investigation of in-depth transport and absorption properties of various perovskite materials using luminescence imaging." *World Conference on Photovoltaic Energy Conversion (WCPEC-7)*, Waikoloa, HI, 2018.

[73] Soufiani A.M., Hameiri Z., Meyer S., *et al.* "Lessons learnt from spatially resolved electro- and photoluminescence imaging: Interfacial delamination in ch3nh3pbi3 planar perovskite solar cells upon illumination." *Adv. Energy Mater.* 2017, vol. 7(9), p. 1602111.

[74] Hameiri Z., Mahboubi Soufiani A., Juhl M.K., *et al.* "Photoluminescence and electroluminescence imaging of perovskite solar cells." *Prog. Photovolt. Res. Appl.* 2015, vol. 23(12), pp. 1697–1705.

[75] El-Hajje G., Momblona C., Gil-Escrig L., *et al.* "Quantification of spatial inhomogeneity in perovskite solar cells by hyperspectral luminescence imaging." *Energy Environ. Sci.* 2016, vol. 9(7), pp. 2286–2294.

[76] Wang R., Mujahid M., Duan Y., Wang Z.-K., Xue J. and Yang Y. "A review of perovskites solar cell stability." *Adv. Funct. Mater.* 2019, p. 1808843.

[77] Sahli F., Werner J., Kamino B.A., *et al.* "Fully textured monolithic perovskite/silicon tandem solar cells with 25.2% power conversion efficiency." *Nat. Mater.* 2018, vol. 17(9), pp. 820–826.

[78] Hacke P., Spataru S., Johnston S., *et al.* "Elucidating PID degradation mechanisms and in situ dark i-v monitoring for modeling degradation rate in

CdTe thin-film modules." *IEEE J. Photovolt.* 2016, vol. 6(6), pp. 1635–1640.

[79] Westin P.-O., Zimmermann U., Stolt L. and Edoff M. "Reverse bias damage in CIGS modules." *24th European Photovoltaic Solar Energy Conference*, Hamburg, Germany, 2009, pp. 2967–2970.

[80] Wendlandt S. and Podlowski L. "Influence of near field shadowing on the performance ratio of thin film modules." *35th European Photovoltaic Solar Energy Conference*, Brussels, Belgium, 2018, pp. 1230–1235.

[81] Tran T.M.H., Pieters B.E., Siegloch M., *et al.* "Characterization of shunts in Cu(In, Ga)Se2 solar modules via combined electroluminescence and dark lock-in thermography analysis." *26th European Photovoltaic Solar Energy Conference*, Hamburg, Germany, 2011, pp. 2981–2985.

[82] Remes K., Leppänen K. and Fabritius T. "Thermography based online characterization of conductive thin films in large-scale electronics fabrication." *Opt. Express* 2018, vol. 26(2), pp. 1219–1229.

[83] Remes K., Järvenpää A. and Fabritius T. "Contactless online characterization of large-area conductive thin films by thermography and induction." *Opt. Lett.* 2019, vol. 44(10), pp. 2574–2577.

Chapter 7

Application of spatially resolved spectroscopy characterization techniques on $Cu_2ZnSnSe_4$ solar cells

Qiong Chen[1] and Yong Zhang[1]

7.1 Raman spectroscopy and CZTSe solar cells

7.1.1 Raman spectroscopy

Raman spectroscopy is named after C.V. Raman who observed the Raman effect for the first time in 1928. Today, Raman spectroscopy is widely used in various fields—from fundamental research to applied solutions. This spectroscopic technique takes advantage of light interaction with vibrational and rotational states of materials.

When light shines on a sample and interacts with the material, the dominant scattering is elastic scattering that does not involve energy change and is referred to as Rayleigh scattering. However, it is also possible that the incident light interacts with the material in such a way that the energy is either transferred to or received from the sample, resulting in an energy change of scattered photons or a frequency shift of the scattered light. Such inelastic scattering is the so-called Raman scattering. Similar to Rayleigh scattering, the material in Raman scattering also needs to be polarizable because the incident photon interacts with the electric dipoles in the material. In classic terms, the interaction can be viewed as a perturbation to the material's polarization field [1]. From a quantum mechanics perspective, the incident photons can excite the vibrational modes of the material; in this way, photons give up energy and are scattered with a lower (red-shifted) frequency than that of Rayleigh-scattered light. A spectral analysis of the scattered light can reveal light signals at frequencies different from the incident light, and the lower-frequency lines are referred to as Stokes Raman scattering. Alternatively, if the system is already in a vibrational excited state, it is possible that the vibrational energy can be transferred to the incident photon, leading to a blue-shifted frequency that is greater than the incident frequency, referred to as anti-Stokes Raman scattering. Figure 7.1

[1]Department of Electrical and Computer Engineering, The University of North Carolina at Charlotte, Charlotte, North Carolina, USA

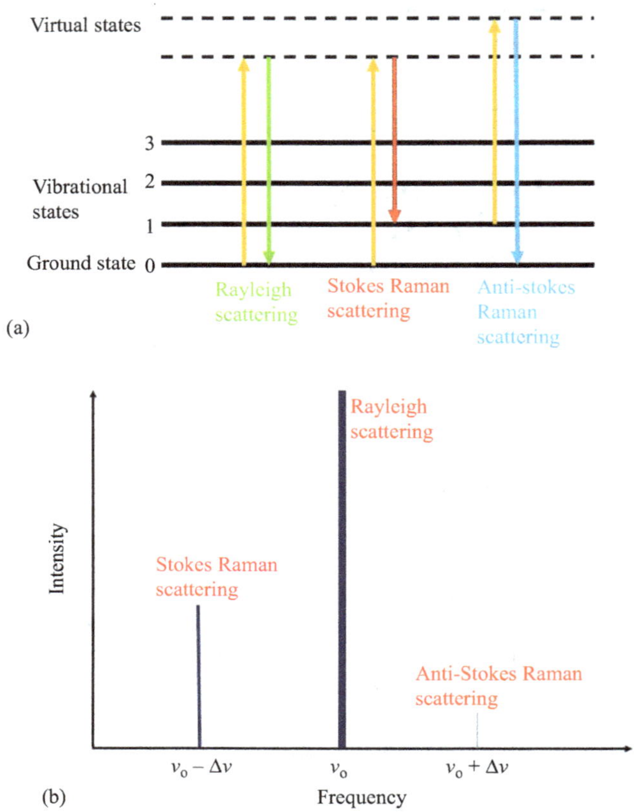

Figure 7.1 Energy-level diagram and relative locations in frequency of Rayleigh, Stokes, and anti-Stokes scattering: (a) energy diagram and (b) Raman and Rayleigh lines

(a) shows the energy-level diagram involved in Rayleigh and Raman scattering, and Figure 7.1(b) shows the relative locations of three different scattering lines in frequency domain. In most cases, the probability of a system in the vibrational excited state is much lower than in the ground state. As a result, the probability of the anti-Stokes scattering is usually much less than that of the Stokes scattering. Usually, Stokes scattering is studied in most applications.

The frequency shift or Raman shift is typically measured in wavenumber, with a unit of inverse length. Photon energy is inversely proportional to wavelength, so the energy change can be converted to a shift in wavenumber in Raman spectrum using the following equation:

$$\Delta\omega = \frac{1}{\lambda_0} - \frac{1}{\lambda}$$

where $\Delta\omega$ is the Raman shift in wavenumber, λ_0 is the incident wavelength, and λ is the scattered wavelength. Usually, the unit chosen for the Raman shift is inverse centimeter, cm^{-1}, and the wavelength in the above equation is in centimeter.

Raman scattering originates from a change in the polarizability of the material due to an interaction of light with the material, which corresponds to specific energy transitions. As a result, Raman spectroscopy is a form of vibrational spectroscopy. The energy of the vibrational mode is determined by many aspects: crystal structure, atomic mass, iconicity, bonding, and so on, all of which are directly related to the specific characteristics of the material. By probing individual vibration modes, Raman spectroscopy can be used to identify the structure and composition because it provides a "fingerprint" of the material being measured. Raman spectroscopy is a very important tool in semiconductor research and characterization.

By using light interaction with a sample, Raman spectroscopy involves only a laser illuminating a sample and then collecting the scattered photons, which offers advantages such as being nondestructive, requiring no sample preparation, and providing a fast measurement. Since Raman spectroscopy is noncontact and nondestructive (if the power is sufficiently low), it is possible to make repeated measurement as well as correlative analyses by applying other techniques.

7.1.2 CZTSe solar cells

The quaternary compounds Cu_2ZnSnS_4 (CZTS) and $Cu_2ZnSnSe_4$ (CZTSe) are promising thin-film absorber materials by substituting zinc and tin for indium and gallium in a $CuInGaS_2$ (CIGS) absorber. The quaternary compound has three types of crystal structures—kesterite, stannite, and premixed Cu–Au (PMCA)—with kesterite being the main structure used for photovoltaic (PV) application. Although first made in 1967 [2], scientists did discover the PV effect of CZTS [3] until 1988. CZTS, as a p-type semiconductor, has a high absorption coefficient for visible light. The direct optical bandgap is 1.45 eV, which matches the best semiconductor bandgap value for PV application according to the Shockley–Queisser limit [4]. Thereafter, major effort was devoted to fabricate solar devices with these materials, and the efficiencies of CZT(S,Se) cells have continuously increased over the past 20 years. The first CZTS cell was fabricated in 1996 with only 0.66% efficiency and open-circuit voltage (V_{oc}) of 400 mV [5]. The efficiency was improved to 2.3% and V_{oc} to 470 mV for both CZTS and CZTSe cells the next year [6]. By 2008, CZTS cell efficiency improved to 6.7% [7]. In 2010, 7.2% and 9.6% CZTSSe cells were fabricated by different methods [8,9]. Also the first CZTSSe cell with over 10% efficiency was demonstrated in 2011 [10]. In 2013, the highest energy efficiency obtained from a CZTSSe solar cell was 12.6% [11]. Unfortunately, the record efficiency remains the same till today. This situation is largely due to the emergence of a new PV material $CH_3NH_3PbI_3$ that has shown unprecedented rapid improvement in efficiency, currently reaching 23.7%, which has led to the diversion of the resources from areas like CZT(S,Se) to the new area.

Although CZT(S,Se) is not a PV material of broad interest currently, in this chapter we use it as a prototype system to illustrate how multiple spatially resolved optoelectronic characterization techniques when applied correlatively can lead to more comprehensive insights to the material properties and device performance than applied them independently. The methodologies and practices can be easily adopted for other PV materials and devices and in general (opto)electronic materials and devices. With this finding, we hope that the unique insights obtained through our studies could benefit those researchers who have not given up on CZT(S,Se).

7.2 Spatially resolved Raman spectroscopy in conjunction with laser-beam-induced current/photoluminescence/reflectance/scanning electron microscope/atomic force microscopy

Laser-beam-induced current (LBIC) is another high-resolution, nondestructive optoelectronic characterization technique. It is used to spatially map device photo-response. In the technique, a laser beam is focused on the solar cell surface and the short-circuit current induced by the laser beam in the solar cell is measured. By scanning the laser beam across the sample or moving the sample with a XY stage, the current values at different spatial points can be translated to a color value and shown as a two-dimensional (2D) LBIC mapping image representing the spatial distribution of the photo-response in the scanned semiconductor region. As a result, electrical defects can be easily visualized. Depending on the laser spot size and scanning step size, the spatial resolution can range from submicron to several hundred micrometers. It is well known that the spatial inhomogeneity in a semiconductor can have a significant effect on the uniformity in the performance of the final optoelectronic device. With micro-LBIC, inhomogeneity at the micron scale can be identified.

Reflection signals are categorized as specular reflection and diffuse reflection according to the surface condition. In a microscopic-scale measurement, a major portion of diffuse reflection as well as specular reflection can be captured simultaneously with the help of an objective lens with high numerical aperture (NA) focused on the sample surface [12].

Integrating and correlating multiple nondestructive spatially resolved techniques is a powerful approach in studying optoelectronic properties of thin-film solar cell materials and devices [12]. μ-Raman spectroscopy, probing the "fingerprints" of the material structures and compositions, can be used to analyze the chemistry in the sample. Micro-photoluminescence (μ-PL) can be used to estimate the bandgap and detect possible defects. μ-LBIC can probe the photo-response of the solar cell under different illumination density conditions and visualize the structural defects and the impact of microscale film inhomogeneity at the same time. μ-Reflectance offers a more accurate method to estimate the reflectance loss of a rough sample surface, especially comparing the general area with nonuniform regions.

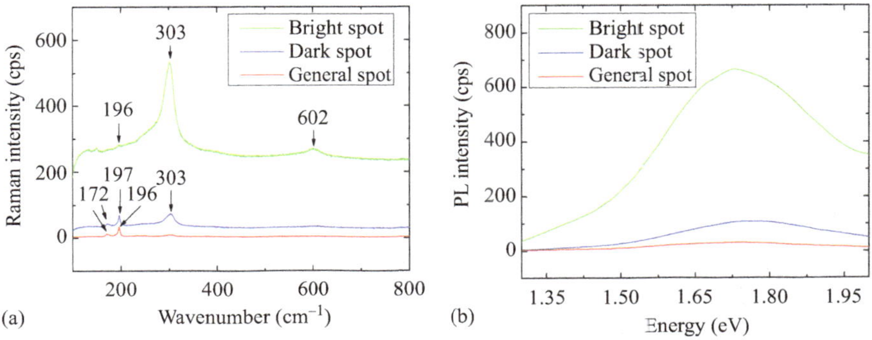

Figure 7.2 Raman and PL spectra from three types of regions: (a) Raman and (b) PL

On a microscopic scale, three distinctly different types of regions—referred to as general area, dark spot, and bright spot—can be observed on a typical CZTSe solar cell under optical microscope with white-light illumination. Figure 7.2(a) and (b) compares typical Raman and PL spectra, respectively, from the three regions measured with a 532-nm laser. In Figure 7.2(a), in addition to the two typical CZTSe Raman modes at 172 and 196 cm^{-1}, another Raman mode at 303 cm^{-1} is also observed from all the three regions. This 303 cm^{-1} peak is strongly enhanced at both dark and bright spots, in particular, with the appearance of another peak of 602 cm^{-1} at the bright spot. These two peaks match, respectively, the first- and second-order longitudinal optical (LO) phonon Raman peaks of CdS. The observation of 1LO and 2LO peaks can be understood as the resonant Raman effect of CdS because the excitation energy of the 532-nm laser (~2.33 eV) is very close to that of the CdS bandgap (2.42 eV). Furthermore, as shown in Figure 7.2(b), both bright and dark spots give rise to a broad PL band at 1.73 eV, which is far above the bandgap of CZTSe (~0.96 eV), presumably associated with defects in the polycrystalline CdS. Therefore, the CdS layer is the thickest at the bright spot and thinnest at the general area. To further study the effect of the microscopic-scale thickness fluctuation of CdS on the device performance, one can perform an array of mutually correlated microscopic characterizations from the same area of the device.

Figure 7.3 shows 2D imaging or mapping of optical microscope, scanning electron microscope (SEM), reflectance, Raman, PL, and LBIC—all acquired approximately from a single 20 μm × 20 μm area on a CZTSe device with a 100× objective lens (NA = 0.9). Reflectance (Figure 7.3(c)) was obtained with ~18 pW power (~4.4×10^{-3} W/cm^2), LBIC (Figure 7.3(f)) with ~2 μW power (~4.9×10^2 W/cm^2), and Raman (Figure 7.3(d)) and PL (Figure 7.3(e)) with 200 μW power (~4.9×10^4 W/cm^2), all with the 532-nm laser. The optical image (Figure 7.3(a)) reveals a bright region of a few micrometer size near the center and some scattered dark spots, which are identified to be CdS-rich according to its Raman and PL spectroscopy, as shown in Figure 7.2. The SEM image (Figure 7.3(b)) further confirms the variation in surface morphology. The extra height of the bright regions

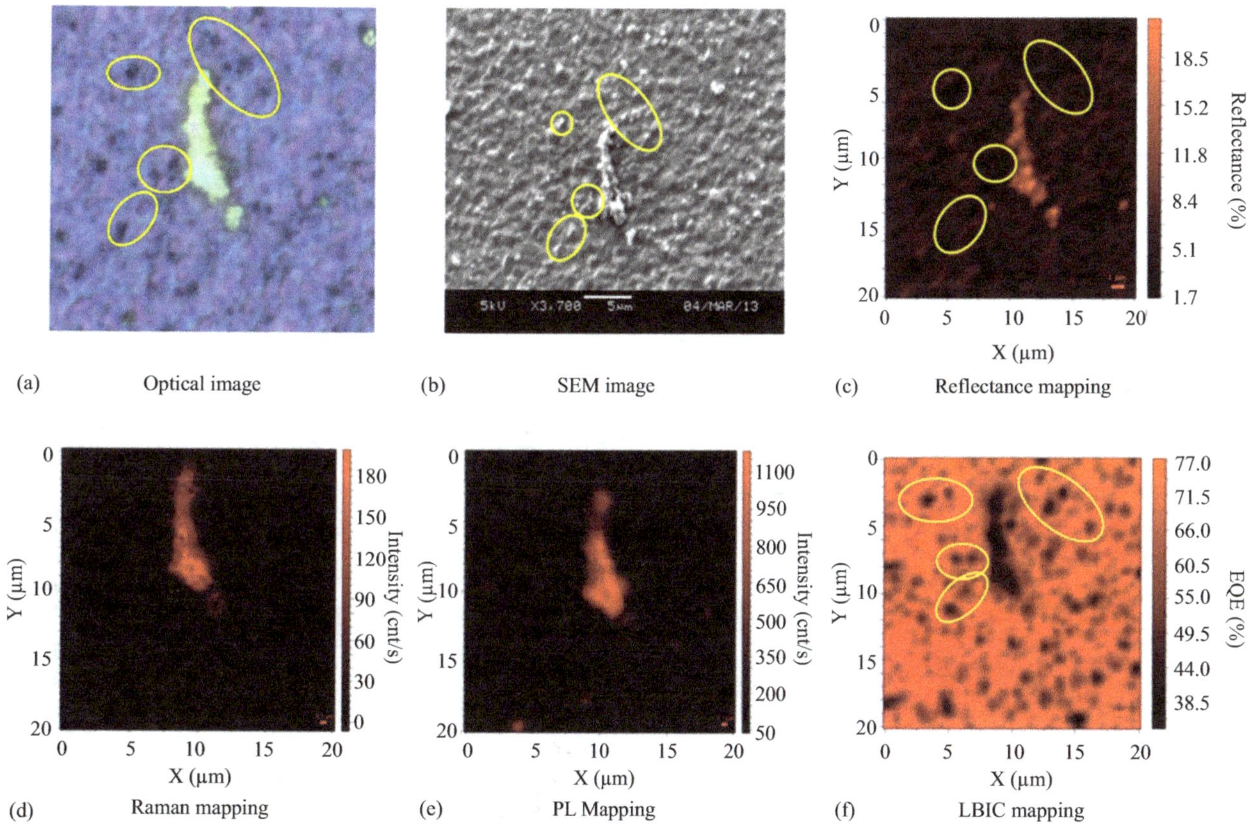

Figure 7.3 Correlative 2D imaging and mapping results of a CZTSe cell

is estimated to be 1–1.5 μm using atomic force microscopy (AFM). An aluminum mirror with its reflectance data provided by the vendor was used as a reference to obtain the reflectance mapping shown in Figure 7.3(c). The reflectance mapping shows that the bright CdS-rich region is much more reflective than the general area, whereas the dark spots are less reflective than the rest of the area—in good agreement with the optical image because both of them result from reflection. Both Raman (monitored at 303 cm^{-1}) and PL (monitored at 1.73 eV) clearly reveal stronger signals from the bright region than from the general area, as shown in Figure 7.3(d) and (e). As expected (and observed in Figure 7.3(f)), the CdS-rich regions, including the large bright island and scattered black spots, yield much smaller photocurrent—reduced by as much as a factor of 3—than the general area. Yellow circles in LBIC mapping correlate with the dark spots in the optical images. By comparing Figure 7.3(c) and (f), one can find the general anti-correlation between the reflectance and LBIC mappings. Lower current from the CdS-rich region indicates the potential to optimize the thickness of the CdS layer to further improve device performance.

7.3 Secondary-phase identification

Similar to its parent CIGS, CZT(S,Se) also has a narrow region of phase stability in which the device-quality material can be synthesized without major adverse effects of secondary phases. The formation of binary and ternary secondary phases—including $Zn_xS(Se)$, $Cu_xS(Se)$, $Sn_xS(Se)$, and $Cu_xSnS(Se)_y$—are often observed in CZT(S,Se) films. Usually, film quality, defect formation, and secondary-phase formation are coupled. The presence of these secondary phases has been assumed to contribute to lowering the energy conversion efficiency of the solar cell. For example, Cu-Sn-S clusters could act as recombination centers, resulting in a lower V_{OC}; $Cu_{2-x}S$ could even short the device; and ZnS may reduce the volume of the useful material and absorb the shorter wavelengths and lower external quantum efficiencies (EQEs) when present at the film surface. Achieving a high-quality film requires a higher degree of understanding and control of the CZT(S,Se) phase space. Detection of secondary phases will guide how to improve the growth method for the CZT(S,Se) thin film. Identifying secondary phases using only X-ray diffraction (XRD) in CZT(S,Se) is not as easy as in CIGS because kesterite CZTS shares multiple peaks with cubic ZnS and Cu_2SnS_3. Raman spectroscopy is a powerful characterization method to reveal vibrational signatures of CZT(S,Se) and especially of secondary phases.

Two CZTSe samples—one prepared by a sputtering method (S1) and the other by a co-evaporation method (S2)—were measured and results are shown in Figure 7.4. Raman spectroscopy was performed at room temperature with a 147 μW (3.5×10^4 W/cm^2) 532-nm laser and 100× lens. Both samples exhibit rather similar Raman features: two intense Raman peaks at 195–196 cm^{-1} and 172–173 cm^{-1} and a weak peak at 232–233 cm^{-1}, which are very close to single-crystal CZTSe Raman peaks at 196, 173, and 231 cm^{-1}. Except for the primary CZTSe Raman scattering peaks, additional peaks at 222, 244, and 251 cm^{-1} are also observed.

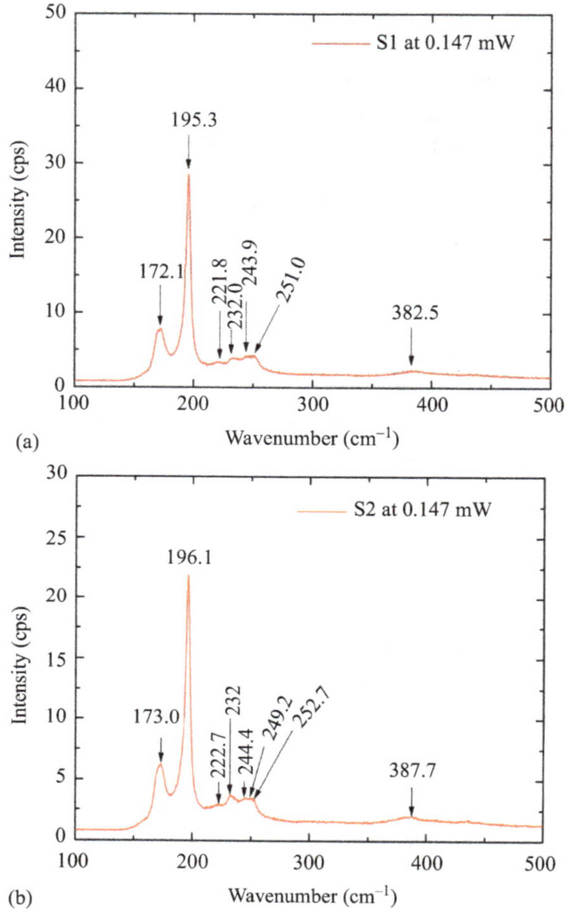

Figure 7.4 Raman spectra of two CZTSe samples measured at 0.147 mW from front surface: (a) S1 and (b) S2

Based on the theoretical calculations, the peak of ~222 cm^{-1} is believed to be a transverse-optical (TO) Raman mode in E symmetry from kesterite structure [13–15]. The peak of ~244 cm^{-1} was also reported both experimentally and theoretically to be CZTSe-related [16–18]. The peak of ~251.5 cm^{-1} is close to the Raman modes of these possible secondary phases: 253 cm^{-1} of ZnSe, 251 cm^{-1} of Cu_2SnSe_3, and ~250 cm^{-1} of amorphous selenium (a-Se) [19–23]. The origin of this peak is shown to be a-Se in Section 7.4.

Raman spectroscopy can be performed from the front surface and also from a cleaved edge. By comparing Raman spectra from different positions of the cleaved edge, one can achieve a better understanding of the composition depth profile and the subtle difference between films prepared by different methods. Raman spectra are measured with the laser beam focused to different depths of the sample: Mo layer,

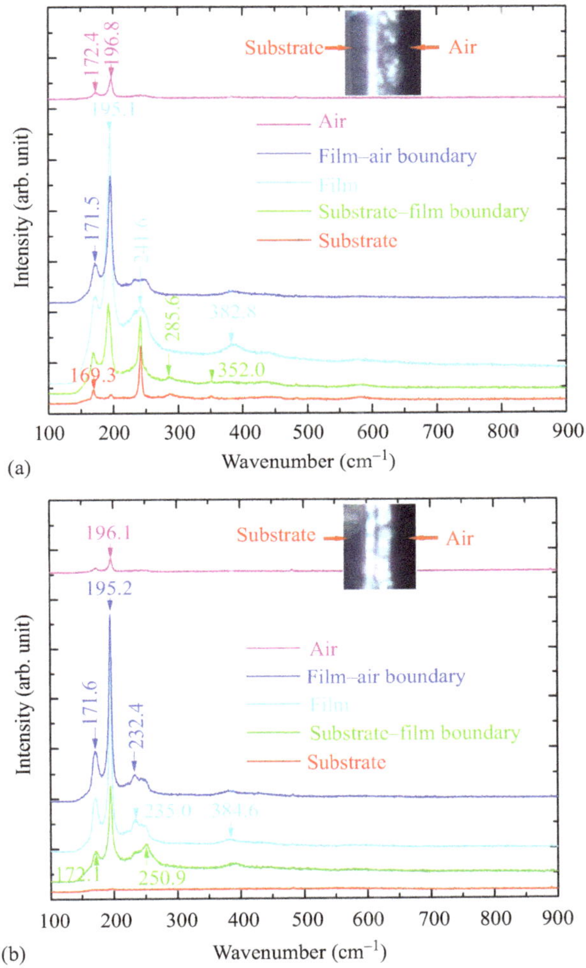

Figure 7.5 Raman spectra of two CZTSe samples measured from cleaved edge: (a) S1 and (b) S2

Mo/CZTSe boundary, body of film, film/air boundary, and air (when the beam is just moved off the edge of the front surface). Due to the limitation of the laser spot size, these measurements can only reveal the qualitative change. In Figure 7.5, measured at 0.147 mW (3.5 × 10^4 W/cm^2) from the cleaved edge, one can clearly see the variation of the spectral feature with film depth. The spectra reveal the evolution of the dominant components from (Mo) substrate to the absorber film. In Figure 7.5(a), only very weak peaks at 169, 196, 241, 286, and 351 cm^{-1} are detected on the substrate. Observed close to the molybdenum layer, $MoSe_2$ is the dominant component at the substrate/film boundary with Raman modes at 169, 242, 286, and 351 cm^{-1} in S1. The mode at ~196 cm^{-1} appears in both Mo and CZTSe layers, although substantially weaker in the Mo region. When moving toward the front surface, the $MoSe_2$-related peaks gradually diminish, although they remain throughout the film thickness, whereas the CZTSe peaks at 196 and 172 cm^{-1} grow. Compared with the results measured from the front surface, one can conclude that $MoSe_2$ concentrates at the substrate/film boundary and exists throughout the film thickness in this sputtered sample. In contrast, no Raman signals from $MoSe_2$ are observed near the substrate of S2, as shown in Figure 7.5(b). Only CZTSe Raman modes at 172, 196, and 232 cm^{-1} as well as secondary phases such as ternary compound Cu_xSnSe_y or a-Se at 250.9 and 235 cm^{-1} are observed from S2.

7.4 Laser-induced-modification Raman spectroscopy

Transmission electron microscopy (TEM) is generally considered to be a direct probe for the material structure. However, the probing volume of TEM tends to be small—typically on the order of a few tens of nanometers—which places a severe limit on the characterization efficiency when one needs to have macroscopic-scale structural information of an inhomogeneous material. Two better-known laser-based material analysis techniques—laser-induced-breakdown spectroscopy (LIBS) and femtosecond-laser tomography—are destructive and do not offer information on chemical bonding. In contrast, "high power" (HP) Raman spectroscopy, as a special form of laser-induced-modification spectroscopy (LIMS) [24], uses a tightly focused continuous-wave laser with a power density that is just high enough to induce a local structural change but usually without causing major material ablation; it measures the change in Raman features compared with the spectrum before the illumination under "low power" (LP). Performing spatially resolved Raman mapping on an as-grown material can already reveal composition and/or structural variations in the sample. However, the HP Raman spectroscopy can provide additional information beyond that obtainable from the conventional technique. For instance, some structural or chemical fluctuations are too weak to be detected or distinguished in the as-grown sample or between different ones, but they are magnified after being modified by the HP illumination. The application of the HP Raman spectroscopy can reveal some subtle but important structural differences in two samples that might otherwise appear to be indistinguishable if they were only subjected to conventional probes, such as Raman spectroscopy, PL, XRD, and device characterization.

S1, S2, and another co-evaporated CZTSe device, S3, are compared. Raman spectra of S1 and S2 are obtained with two powers—32.8 μW (7.9×10^3 W/cm^2) and 146 μW (3.5×10^4 W/cm^2)—as shown in Figure 7.6, from the same location with no grating movement between the two measurements. The primary CZTSe Raman modes of S1 experience a 0.4 cm^{-1} redshift when the laser power is increased. When remeasured with the lower power, the observed redshift is reversible, indicating an elastic change. For S2, no shift is observed with the power change. The slight CZTSe Raman mode redshifts indicate that a small heating effect is introduced for the sputtered sample under 146 μW illumination, but the

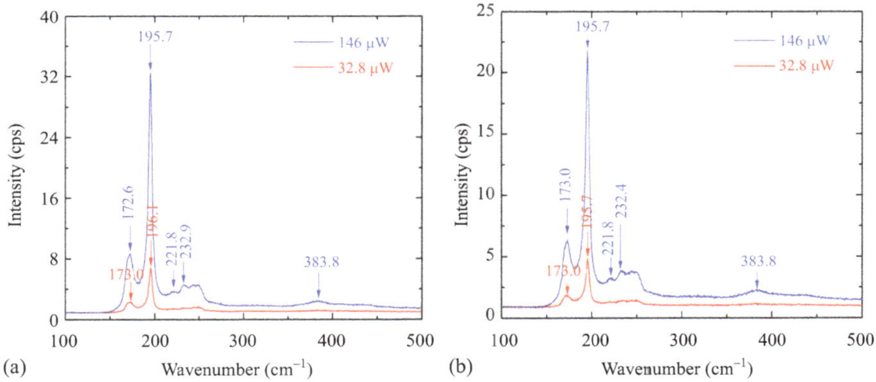

Figure 7.6 Raman spectra of the bare CZTSe films measured at 32.8 and 146 μW: (a) S1 and (b) S2

power level is not high enough to cause irreversible material modification. The comparison suggests some subtle structural difference between the two samples manifested in the difference in thermal conductivity.

Figure 7.7 examines the effects of HP illumination by comparing the LP Raman spectra from the same location at 0.146 mW before and after 100 s illumination at 2.47 mW (5.9×10^5 W/cm^2) and a further 36 s at 4.5 mW (1.1×10^6 W/cm^2). After the 2.47-mW illumination, the illuminated spot shows some color change under optical microscope but no apparent ablation. However, the 4.5-mW illumination typically results in some local material ablation. However, for a bulk Si shown in Figure 7.7(f), even illuminated with the full power (about 20 mW, 4.9×10^6 W/cm^2), there is practically no change when returned to the LP condition.

The red curves in Figure 7.7(a)–(c) are typical Raman spectra from the three CZTSe samples measured first at 0.146 mW, and the green and blue curves are the corresponding spectra remeasured after the two higher-power illuminations. All three samples exhibit rather similar Raman features in their initial states. The "mesa-like" band to the right of the 196 cm^{-1} peak with multiple weak peaks is compared directly in Figure 7.7(d) among the three samples in their initial states, indicating that there is no significant or distinct difference between the samples prepared differently. As a matter of fact, more variations can be found within one sample, due to composition inhomogeneity [25], as shown in Figure 7.7(e) between different locations in S2. Therefore, it is impractical to use conventional Raman spectroscopy to reliably reveal potential structural variations from sample to sample. The peak of ~251.5 cm^{-1} is close to the Raman modes of these possible secondary phases: ZnSe, Cu_2SnSe_3, and a-Se. It is possible that extra selenium is present in CZTSe films because they are fabricated in a Se-rich condition. a-Se may also have a weak feature at ~235 cm^{-1} due to the presence of crystalline-phase trigonal Se (t-Se) [23].

After the 2.47-mW illumination, the two main CZTSe peaks at 196 and 172 cm^{-1} exhibit both intensity reduction and redshift in three samples, as shown

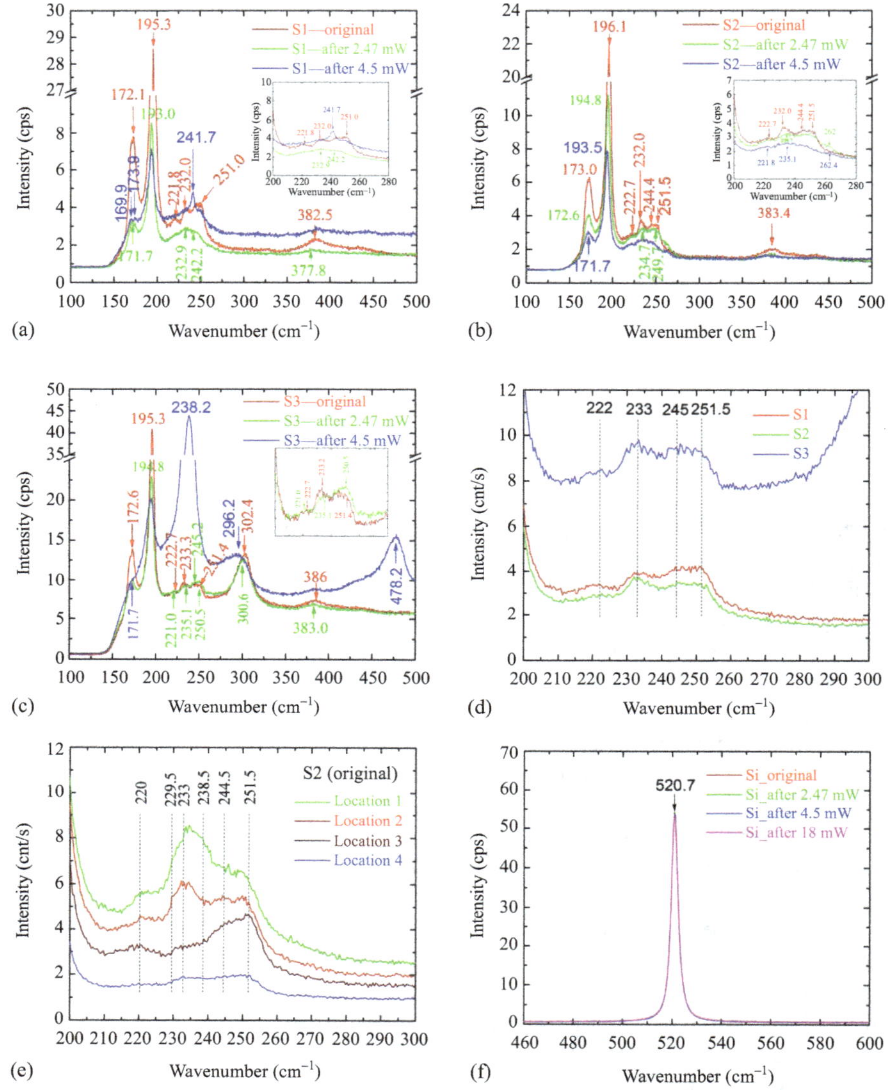

Figure 7.7 Raman spectra of CZTSe and Si samples at 0.146 mW before and after being illuminated by high powers: (a) S1; (b) S2; (c) S3; (d) comparison of S1, S2, and S3; (e) comparison of different locations of S2; and (f) Si

in the spectra in Figure 7.7(a)–(c) (in green). In S1, the CZTSe main peak at 195.3 cm^{-1} shifts to 193.0 cm^{-1}, and 172.1 cm^{-1} shifts to 171.7 cm^{-1}. The redshift of the 195–196 cm^{-1} is the largest for S1 at 2.3 cm^{-1}, compared with 1.3 cm^{-1} for S2 and 0.5 cm^{-1} for S3. These changes seem to suggest that the atomic bonds at the illuminated site are thermally expanded irreversibly as a result of the local heating if no atomic rearrangement occurs. However, it is possible that the redshift is caused by the cation sublattice disordering, as suggested for CZTS [26], although the magnitude in CZTSe is much smaller (despite the higher excitation density in our case) and is sample-dependent. The exact mechanism remains to be examined by a direct structural probe (e.g., TEM), but it is likely not as simple as the proposed disordering effect. The weak features in the mesa-like band appear to smear out and weaken in both S1 and S2, as shown in Figure 7.7(a) and (b). In S2, a new feature at ~262 cm^{-1} also emerges, whereas in S3 those weak features remain, but their relative intensities change—for instance, the peaks at ~235 and 250.5 cm^{-1} become more apparent, as shown in Figure 7.7(c).

After illuminating further at 4.5 mW, with the spectra given in Figure 7.7(a)–(c) (in blue), the intensity of the strongest CZTSe peak initially at 195–196 cm^{-1} decreases further for all three samples. But the peak position remains nearly the same as that after the first HP illumination, except for S2; it shifts further from 194.8 to 193.5 cm^{-1}, resulting in the largest redshift of 2.6 cm^{-1} among all. For S1, as shown in Figure 7.7(a), the intensity of the 193 cm^{-1} peak is reduced to only 25% of its original value before any HP illumination, and the 172 cm^{-1} mode becomes a weak shoulder from 168 to 178 cm^{-1}. Interestingly, for S2, the main CZTSe peak intensity decreases by a factor of 3, and the CZTSe peak of ~172 cm^{-1} remains sharp. For the spectral region of the mesa-like band, a sharp peak emerges at 241.7 cm^{-1} as the most prominent feature in S1 but not in S2 and S3. No significant further change occurs in S2 except that the feature of ~262 cm^{-1} is slightly better resolved. In S3, the main peak intensity reduces only by about a factor of 2, but 172 cm^{-1} becomes nearly invisible. Also, a very strong sharp peak at 238.2 cm^{-1} appears, which is accompanied by a second-order peak at 478.2 cm^{-1} (and a third-order peak at ~714 cm^{-1}, data not shown) and a broad band at ~296 cm^{-1}. Obviously, the second HP illumination brings more qualitatively different changes to these samples. The contrast between S1 and S2 suggests that the CZTSe structure prepared by co-evaporation seems to be more robust than that prepared by sputtering against HP-illumination-induced structural modifications.

Changes in Raman spectra from all three CZTSe samples after HP illumination indicate partial decomposition and plastic changes of the materials. However, different samples respond rather differently to HP illumination, indicating that seemingly similar CZTSe materials may differ significantly in their microscopic structures. The highest power applied to Si was, in general, safe for an epitaxial GaAs sample, but it could induce a structural change when a dislocation defect was illuminated [27]. Apparently, quaternary CZTSe is structurally not as robust as an elemental or binary semiconductor such as Si and GaAs. The contrast is mostly because of the difference in chemical bonding strength, but it may also because of the structural defects and lower thermal conductivity of the polycrystalline film.

2D Raman mapping is performed to examine the spatial extension of the local heating effect caused by HP illumination on the CZTSe samples, as shown in Figure 7.8 for 2.47-mW illumination. One random spot from each sample surface is first illuminated by 2.47 mW for 100 s. Raman mapping is then acquired with 0.146-mW laser power for a 5 × 5 μm^2 or 8 × 8 μm^2 area centered at the illuminated spot. For all the three samples, the illuminated spot appears darker after being illuminated, as shown in the optical images of Figure 7.8(a), (d), and (g). For each sample, Raman intensity mapping reveals a dark circle for the 196 cm^{-1} peak, as shown in

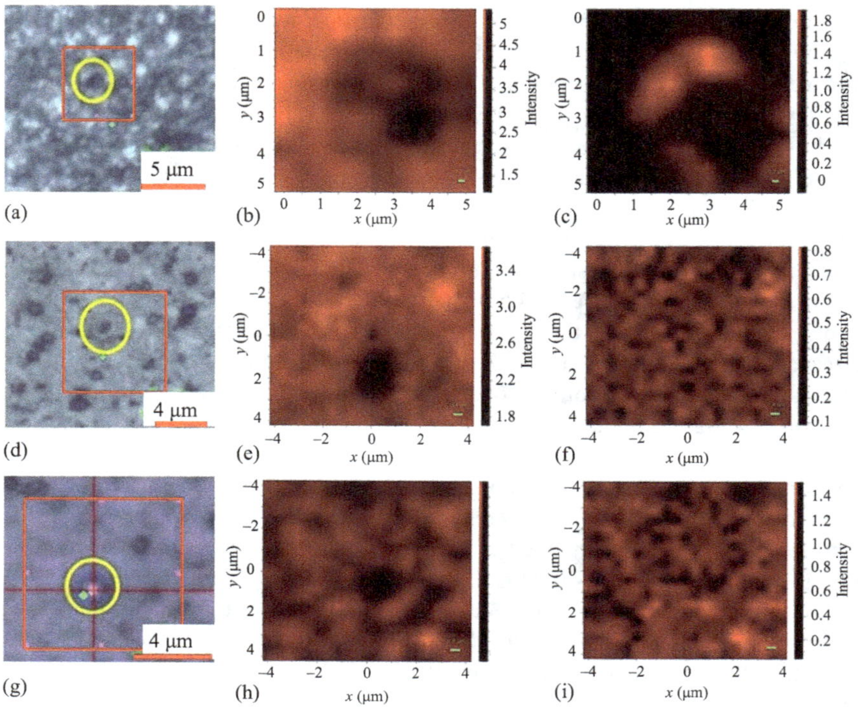

Figure 7.8 S1, S2, and S3 Raman mapping after one spot being illuminated by 2.47 mW for 100 s: (a), (d), and (g) optical images of the illuminated spots (the red square indicates the area of Raman mapping); (b), (e), and (h) Raman mapping of the primary CZTSe mode at 196 cm^{-1}; (c), (f), and (i) Raman mapping of the Cu_xSe_y mode at 261 cm^{-1}

Figure 7.8(b), (e), and (h), which confirms the intensity reduction of this mode shown in Figure 7.7. In the Raman mapping of S1, a Raman peak at ~261 cm^{-1} is observed in a ring outside of the illumination site (with ~4-μm diameter), as evident in Figure 7.8(c). However, in the mapping results of S2 and S3, the 261 cm^{-1} peak is not observed, and the intensity maps of 220–270 cm^{-1} show no significant spatial variation, as shown in Figure 7.8(f) and (i). The appearance of the 261 cm^{-1} peak in S1 suggests the formation of Cu_xSe_y [28–31] occurring mostly at some distance from the illumination site, which likely results from a specific temperature profile caused by the laser heating. The formation of the ring structure also explains when measured at the illumination site, why the Cu_xSe_y feature is not resolved in S1 and S3, and is only very weak in S2, as shown in Figure 7.7(b). The Raman mapping of the primary CZTSe mode at 196 cm^{-1} reveals most directly the spatial extension and that the material is being affected by the local heating. For S1, as shown in Figure 7.8(a), beyond the darkest 1-μm circle (comparable with the laser spot size of ~0.76 μm) at the illumination site, a dark area is observed with a diameter of ~3 μm comparable with the size of the 261 cm^{-1} Cu_xSe_y ring. On the other hand, the illumination affects

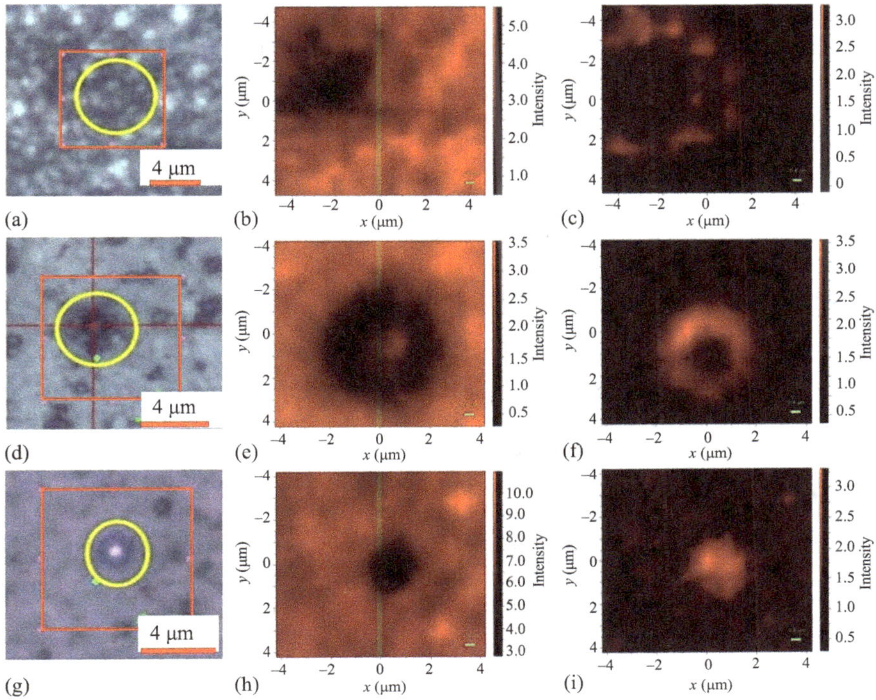

Figure 7.9 *S1, S2, and S3 Raman mapping after one spot being illuminated by 4.5 mW for 36 s: (a), (d), and (g) optical images of the illuminated spots (the red square indicates the area of Raman mapping); (b), (e), and (h) Raman mapping of the primary CZTSe mode at 196 cm^{-1}; (c) and (f) Raman mapping of Cu_xSe_y mode at 261 cm^{-1}; and (i) Raman mapping of the t-Se mode at 238 cm^{-1}*

only an area of ~1.5 μm diameter in S2 and less than 1 μm diameter in S3. The difference between S1 and S2 seems to suggest that the two nominally similar films have rather different thermal conductivities, which might be partially responsible for the different responses in the structural change that we mentioned earlier. The extra layers above CZTSe in device S3 might improve the thermal conductivity of the structure as a whole; thus, it shows the least spatial extension. The contrast between S1 and S2 suggests that the CZTSe absorber layer prepared by the sputtering method (S1) is more sensitive to HP illumination than the film fabricated by the co-evaporation method (S2).

Figure 7.9 shows the mapping results with 0.146 mW after 4.5-mW illumination. Compared with the results of 2.47-mW illumination, larger affected regions are seen in the optical images in Figure 7.9(a), (d), and (g). Raman mapping in Figure 7.9(b) shows that the 196 cm^{-1} peak intensity of S1 reduces in a circular region with a diameter larger than 6 μm. The affected area of S2 in Figure 7.9(e) also enlarges to about 5 μm in diameter. For S3, the affected area in Figure 7.9(h) has a diameter only about 2 μm. The extra layers in the CZTSe device again offer some protection to the CZTSe absorber. In the intensity map of the 261 cm^{-1} peak, a Cu_xSe_y-rich ring still exists in S1 as shown in Figure 7.9(c), but now a similar Cu_xSe_y ring also appears in S2, although smaller, as shown in Figure 7.9(f). The size of the Cu_xSe_y-rich ring reflects the size of the heated region that reaches the optimal phase-transition temperature, which is comparable with the affected CZTSe area in S1 but smaller in S2. For S3, again, no Cu_xSe_y ring is observed. However, consistent with the observation of a strong Raman peak at ~238 cm^{-1} shown in Figure 7.7(c)—due to the formation of t-Se at the illuminated site under the same HP illumination—the Raman map of 238 cm^{-1} (Figure 7.9(i)) shows a bright region with a comparable size of the dark region for the CZTSe mode in Figure 7.9(h). The affected CZTSe area and the size of the Cu_xSe_y ring both depend on the thermal conductivity of the material, which differs significantly among the three samples.

It has been shown through TEM and other elemental studies that polycrystalline CZTSe films tend to exhibit various sizes of voids, domains of the secondary phases, and elemental segregation at the domain or interface boundaries, and these defects depend sensitively on the growth and postgrowth treatment conditions [32–34]. It is well known that grain boundaries and defects can have a major impact on the thermal conductivity, which in turn can significantly affect the size and shape of a new microstructures generated by local laser-heating-induced structural modification [35]. The structural modifications observed in the vicinity of the illuminated sites clearly indicate that S1 has a lower thermal conductivity than S2 and S3, implying that the film produced by sputtering is likely more defective. Furthermore, all these samples are likely to have a-Se as grown, and the sputtered sample has more inter-diffused Mo in the film, which is supported by the cleaved-edge Raman probe [36]. It is generally known that structural imperfections such as dislocations and grain boundaries reduce material thermal conductivity [37], but the substrate may also affect the thermal conductivity of a thin film [38]. In these CZTSe films, the Mo layer and glass substrate can, in principle, improve heat dissipation of the laser-induced heating, which yields a higher effective thermal conductivity. However, due to the low thermal conductivity of CZTSe [39] and short absorption length below the film thickness, the effect of the layers underneath is expected to be less significant. In fact, if it were significant, one would find the highest thermal conductivity in S1, which is apparently contrary to the experimental findings. Therefore, the lower crystallinity of S1 is likely the primary reason for its low thermal conductivity. In addition to revealing the microscopic-scale structural differences between samples, the HP illumination provides a way to generate new microscopic structures on an as-grown material for various possible applications, such as Cu_xSe_y rings in samples S1 and S2, and t-Se disks in sample S3. A similar method was used to generate graphene disks on an epitaxially deposited SiC [35].

7.5 Other applications

In Section 7.2, a CdS-rich region was found to yield lower EQEs and energy conversion efficiencies than the general area. However, these regions show much less

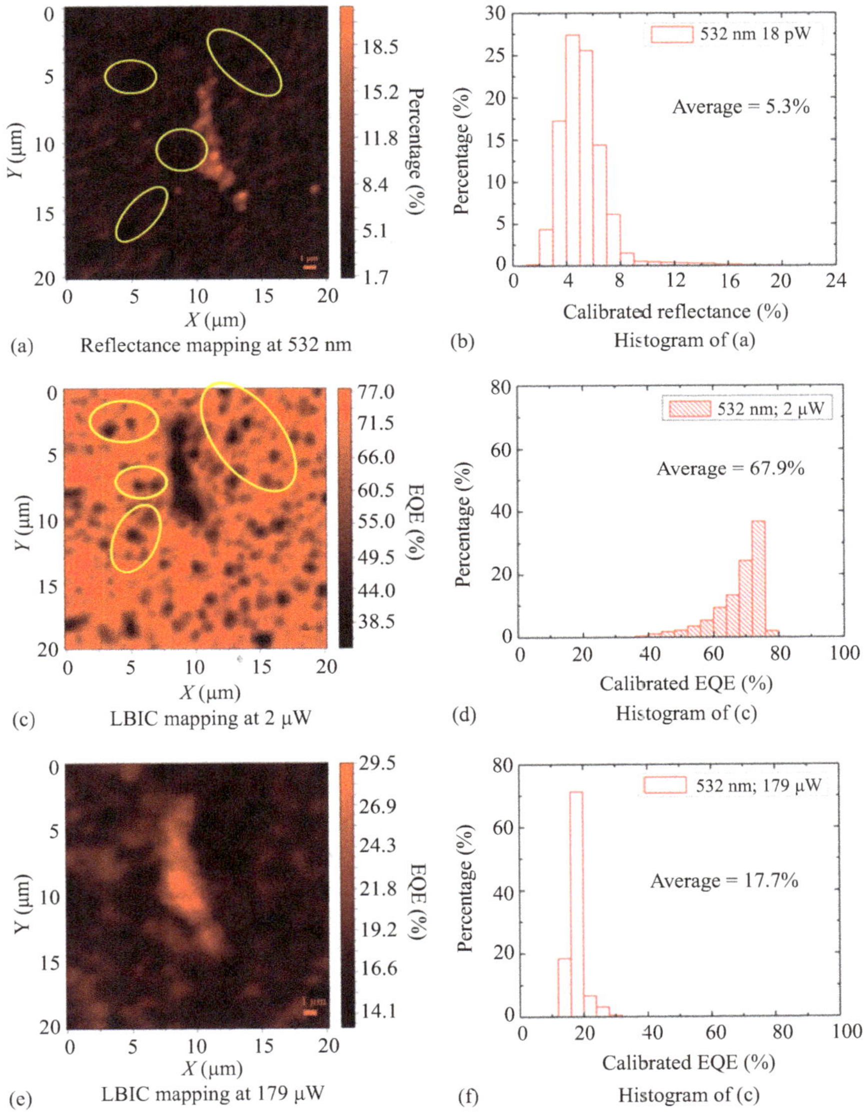

Figure 7.10 Reflectance and LBIC mapping with 532-nm laser at low and high power levels: (a) reflectance mapping at 532 nm; (b) histogram of (a); (c) LBIC mapping at 2 μW; (d) histogram of (c); (e) LBIC mapping of 179 μW; and (f) histogram of (e)

efficiency degradation at high illumination intensity, leading to an inversion of LBIC contrast in the area mapping as shown in Figure 7.10, which compares the LBIC mappings of the same area under two representative low and high laser powers. Here, "LP" means a level that can yield an average EQE comparable with the macroscopic probe, whereas "HP" means the EQE of the general area that shows significant degradation, but the power is not as high as to cause permanent damage to the material. At the LP level, the device shows rather high average EQE for the 532-nm laser: 67.9% ± 0.3%, varying from 34.3% to 77.6%, which is consistent with the ~81% macroscopic EQE of a similar device. However, at the HP level, the average EQE reduces drastically down to 17.7% ± 2.3%. Interestingly, the reduction at the CdS-rich region is much smaller, from ~45% to ~25%, which leads to the reversal of the LBIC contrast, as evident from Figure 7.10(c) and (e). Clearly, the device region with a thicker CdS layer is more immune to the EQE droop with increasing illumination power, although the thicker region has a lower initial EQE. One possible mechanism could be that the electronic structure of the CZTSe/CdS heterojunction, such as the effective barrier height, depends sensitively on the CdS layer thickness. The low initial EQE of the CdS-rich region is likely related to the properties of CdS: the high reflectivity and residual absorption near the bandgap, and the poor carrier transport of the polycrystalline phase. These issues could, in principle, be mitigated or improved to allow the adoption of a thicker CdS layer—in particular, for high-illumination applications in either concentrated PV or photo-detection.

It is well-recognized that elementary ratios of Cu, Zn, Sn, and S(Se) within the CZTS(Se) absorber layer can greatly impact the CZTS(Se) solar cell performance. Empirically, the Cu-poor and Zn-rich condition tends to offer better performance than stoichiometry when the stoichiometry condition prompts point-defect formation [9,11,40,41]. However, it is also reported that the cell with elemental ratios closer to stoichiometry is more immune to EQE droop than that more off-stoichiometry. The variation in elemental ratios results not only in fewer or more defects but also in changes to the band alignment at the absorber/back-contact interface, absorber/window-layer interface, or both [42].

Figure 7.11 shows LBIC mappings obtained from the same area of a CZTSe solar cell (S4) at LP and HP illumination. The average EQE of S4 at LP level is 58.4% from Figure 7.11(a). However, at HP level, as shown in Figure 7.11(b), the average EQE of S4 is 55.5%. Comparing the histograms in Figure 7.11(c) and (d), the result at HP is similar to that at LP, i.e., there is almost no EQE droop. Raman spectra obtained from S4 and the CZTSe cell measured in Figure 7.2 show very similar CZTSe and CdS Raman modes and intensities, as shown in Figure 7.12. In other words, it is not the variation in CdS layer thickness that leads to the different EQE droops. The main difference between the two cells is the Cu/(Zn+Sn) and Zn/Sn ratios in the absorber. Therefore, the difference in the elemental ratios is likely to be the main cause of the different illumination density dependencies. CdS/CZTSe is mostly found to have a "spike"-like alignment, in which the conduction band minimum (CBM) of CdS is higher than that of CZTSe. Type I alignment benefits in V_{OC} but becomes a barrier for carriers moving across the heterojunction. Increasing the Zn/Sn ratio is expected to increase the CBM of CZTSe, which is beneficial for electron collection. Thus, CZTSe devices with higher Zn/Sn and lower Cu/(Zn+Sn) ratios usually yield better photo-response. However, when illuminated with HP density, raising of the electron Fermi level might lower the

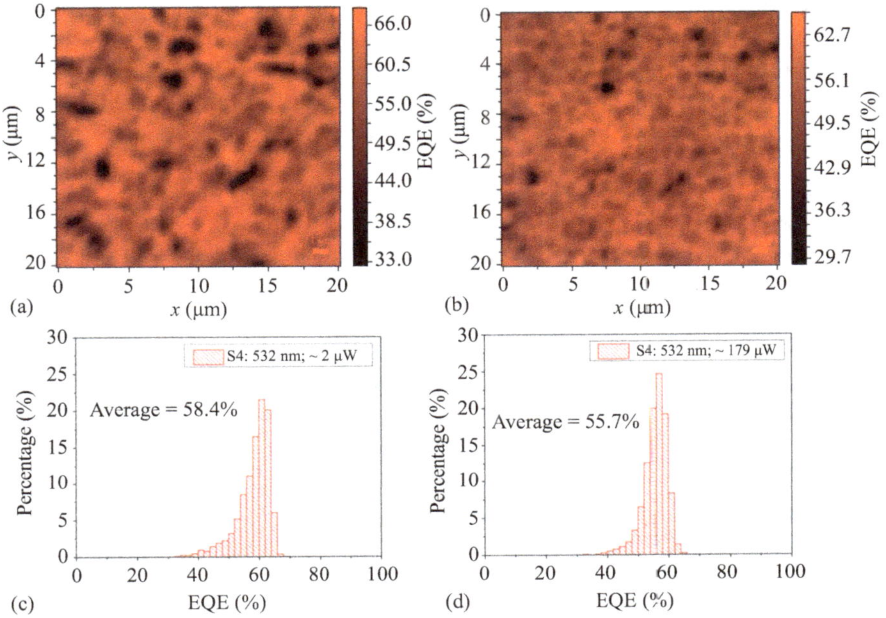

Figure 7.11 *S4 LBIC mapping with 532-nm laser at low and high power levels: (a) at ~2 μW; (b) at ~179 μW; (c) histogram of (a); and (d) histogram of (b)*

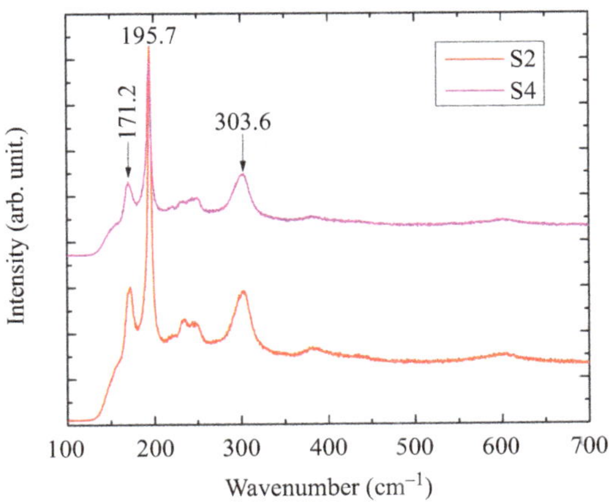

Figure 7.12 *Raman spectra of S2 and S4 from a random spot*

electron barrier at the back contact, which might allow electrons to leak out through the back contact. Thus, materials with higher Cu and lower Zn content or their ratio closer to stoichiometry, which are expected to have a higher Mo back electron-blocking barrier, could still generate a relatively high EQE at HP illumination, despite that the initial short-circuit current is lower at low illumination density because of unfavorable band alignment between CZTSe and CdS. In addition, an CZTSe absorber with closer-to-stoichiometry elemental ratios might be less defective, which could be beneficial to the device performance. As demonstrated in a recent comparative study of inorganic and organic–inorganic hybrid PV materials, specifically, GaAs and CdTe vs. $CH_3NH_3PbI_3$, the most important reason for the impressive performance of the hybrid perovskite is that it tends to have much less detrimental point defects than its inorganic counterparts [43]. The structural defects lead to lower PL efficiencies in the inorganic materials at the low illumination density (e.g., close to 1 Sun). The defects could very well be the reason for the low open circuit voltage in the CZT(S,Se) cells, which is one of the key challenges to improve their efficiencies. To conclude, there might be alternative approaches to optimize the device performance other than altering the stoichiometry, which also depends on the device operation condition.

7.6 Summary

Integration and correlation of multiple nondestructive spatially resolved techniques is a powerful characterization approach to study the optoelectronic properties of thin-film solar cell materials and devices, as demonstrated for CZT(S,Se) in this chapter. μ-Raman spectroscopy, revealing the "fingerprints" of the material structures and compositions, was used to analyze the chemistry in CZT(S,Se) films. μ-PL was used to estimate the bandgap of CZT(S,Se) and detect possible defects within the film. μ-LBIC was used to probe the photo-response of CZT(S,Se) solar cells in different illumination densities and to visualize the structural defects and the impact of microscale film inhomogeneity at the same time. μ-Reflectance offers a more accurate method to estimate the reflectance loss of a rough sample surface—especially to compare the general CZT(S,Se) area with nonuniform regions. SEM and AFM were used to characterize the surface morphology. By applying this approach, in conjunction with high-temperature and high-excitation-power optical spectroscopy, one can probe the microscopic-scale variations between samples and devices that appear to be very similar from macroscopic material and device characterizations. Thus, the approach serves as a very useful tool to understand the underlying microscopic material structures, provides hints for future improvement in material quality and device design, and predicts the potential of improvement in device performance.

Acknowledgments

We are very grateful to Dr. Ingrid Repins and Dr. Sergio Bernardi for providing the CZTSe samples and helpful discussions. Y.Z. acknowledges the support of Bissell Distinguished Professorship.

References

[1] Y. Peter and M. Cardona, *Fundamentals of Semiconductors: Physics and Materials Properties*, Springer-Verlag Berlin Heidelberg, Springer Science & Business Media, 2010.

[2] R. Nitsche, D. Sargent, and P. Wild, Crystal growth of quaternary chalcogenides by iodine vapor transport, *J. Crystal Growth* **1**, 52–53 (1967).

[3] K. Ito and T. Nakazawa, Electrical and optical properties of stannite-type quaternary semiconductor thin films, *Jpn. J. Appl. Phys.* **27**, 2094 (1988).

[4] W. Shockley and H. J. Queisser, Detailed balance limit of efficiency of p-n junction solar cells, *J Appl Phys* **32**, 510–519 (1961).

[5] H. Katagiri, N. Sasaguchi, S. Hando, S. Hoshino, J. Ohashi, and T. Yokota, Preparation and evaluation of Cu_2ZnSnS_4 thin films by sulfurization of E-B evaporated precursors, *Solar Energy Mater Solar Cells* **49**, 407–414 (1997).

[6] T. M. Friedlmeier, N. Wieser, T. Walter, H. Dittrich, and H. Schock, Heterojunctions based on Cu_2ZnSnS_4 and $Cu_2ZnSnSe_4$ thin films, in *14th European PVSEC* (1997).

[7] H. Katagiri, K. Jimbo, S. Yamada, *et al.*, Enhanced conversion efficiencies of Cu_2ZnSnS_4-based thin film solar cells by using preferential etching technique, *Appl. Phys. Exp.* **1**, 041201 (2008).

[8] Q. Guo, G. M. Ford, W.-C. Yang, *et al.*, Fabrication of 7.2% efficient CZTSSe solar cells using CZTS nanocrystals, *J. Amer. Chem. Soc.* **132**, 17384–17386 (2010).

[9] T. K. Todorov, K. B. Reuter, and D. B. Mitzi, High-efficiency solar cell with earth-abundant liquid-processed absorber, *Adv. Mater.* **22**, E156–E159 (2010).

[10] D. A. R. Barkhouse, O. Gunawan, T. Gokmen, T. K. Todorov, and D. B. Mitzi, Device characteristics of a 10.1% hydrazine-processed $Cu_2ZnSn(Se, S)_4$ solar cell, *Prog. Photovoltaics Res. Appl.* **20**, 6–11 (2012).

[11] W. Wang, M. T. Winkler, O. Gunawan, *et al.*, Device characteristics of CZTSSe thin-film solar cells with 12.6% efficiency, *Adv. Energy Mater.* **4** (2014).

[12] Q. Chen and Y. Zhang, The reversal of the laser-beam-induced-current contrast with varying illumination density in a $Cu_2ZnSnSe_4$ thin-film solar cell, *Appl. Phys. Lett.* **103** (2013).

[13] A. Khare, B. Himmetoglu, M. Johnson, D. J. Norris, M. Cococcioni, and E. S. Aydil, Calculation of the lattice dynamics and Raman spectra of copper zinc tin chalcogenides and comparison to experiments, *J. Appl. Phys.* **111**, 083707 (2012).

[14] M. Dimitrievska, A. Fairbrother, E. Saucedo, A. Pérez-Rodríguez, and V. Izquierdo-Roca, Influence of compositionally induced defects on the vibrational properties of device grade $Cu_2ZnSnSe_4$ absorbers for kesterite based solar cells, *Appl. Phys. Lett.* **106**, 073903 (2015).

[15] T. Gürel, C. Sevik, and T. Çağın, Characterization of vibrational, and mechanical properties of quaternary compounds Cu_2ZnSnS_4, and

$Cu_2ZnSnSe_4$ in kesterite, and stannite structures, *Phys. Rev. B* **84**, 205201 (2011).

[16] M. Grossberg, J. Krustok, K. Timmo, and M. Altosaar, Radiative recombination in $Cu_2ZnSnSe_4$ monograins studied by photoluminescence spectroscopy, *Thin Solid Films* **517**, 2489 (2009).

[17] M. Grossberg, J. Krustok, J. Raudoja, K. Timmo, M. Altosaar, and T. Raadik, Photoluminescence and Raman study of $Cu_2ZnSneSe_xS_{1-x}$ monograins for photovoltaic applications, *Thin Solid Films* **519**, 7403 (2011).

[18] A. Redinger, K. Hönes, X. Fontané, *et al.*, Detection of a ZnSe secondary phase in coevaporated $Cu_2ZnSnSe_4$ thin films, *Appl. Phys. Lett.* **98**, 101907 (2011).

[19] H. Li, B. Wang, and L. Li, Study on Raman spectra of zinc selenide nanopowders synthesized by hydrothermal method, *J. Alloys Compd.* **506**, 327 (2010).

[20] J. Scott, T. Damen, W. Silfvast, R. Leite, and L. Cheesman, Resonant Raman scattering in ZnS and ZnSe with the cadmium laser, *Opt. Commun.* **1**, 397 (1970).

[21] C. Shan, Z. Liu, X. Zhang, C. Wong, and S. Hark, Wurtzite ZnSe nanowires: Growth, photoluminescence, and single wire Raman properties, *Nanotechnology* **17**, 5561 (2006).

[22] M. Altosaar, J. Raudoja, K. Timmo, *et al.*, $Cu_2Zn_{1-x}Cd_xSneSe_{1-y}S_y$ solid solutions as absorber materials for solar cells, *Phys. Status Solidi A* **205**, 167 (2008).

[23] V. V. Poborchii, A. V. Kolobov, and K. Tanaka, An in situ Raman study of polarization-dependent photocrystallization in amorphous selenium films, *Appl. Phys. Lett.* **72**, 1167 (1998).

[24] Q. Chen, S. Bernadi, and Y. Zhang, Spatially-resolved laser induced modification Raman spectroscopy for probing the microscopic structure variations in quaternary alloy CZTSe, *Phys. Rev. Appl.* **8**, 034008 (2017).

[25] P. Schöppe, G. Gurieva, S. Giraldo, *et al.*, Discrepancy between integral and local composition in off-stoichiometric $Cu_2ZnSnSe_4$ kesterites: A pitfall for classification, *Appl. Phys. Lett.* **110**, 043901 (2017).

[26] M. Y. Valakh, V. Dzhagan, I. Babichuk, X. Fontane, A. Perez-Rodriquez, and S. Schorr, Optically induced structural transformation in disordered kesterite $Cu_2ZnSnSe_4$, *JETP Lett.* **98**, 255 (2013).

[27] T. Gfroerer, Y. Zhang, and M. Wanlass, An extended defect as a sensor for free carrier diffusion in a semiconductor, *Appl. Phys. Lett.* **102**, 012114 (2013).

[28] B. Minceva-Sukarova, M. Najdoski, I. Grozdanov, and C. Chunnilall, Raman spectra of thin solid films of some metal sulfides, *J. Mol. Struct.* **410**, 267 (1997).

[29] K.-H. Kim and I. Amal, Growth of $Cu_2ZnSnSe_4$ thin films by selenization of sputtered single-layered Cu-Zn-Sn metallic precursors from a Cu-Zn-Sn alloy target, *Electron. Mater. Lett.* **7**, 225 (2011).

[30] F. Luckert, D. I. Hamilton, M. V. Yakushev, N. S. Beattie, G. Zoppi, M. Moynihan, I. Forbes, A. V. Karotki, A. V. Mudryi, M. Grossberg, J. Krustok,

and R.W. Martin, Optical properties of high quality $Cu_2ZnSnSe_4$ thin films, *Appl. Phys. Lett.* **99**, 062104 (2011).

[31] C. M. Fella, A. R. Uhl, C. Hammond, I. Hermans, Y. E. Romanyuk, and A. N. Tiwari, Formation mechanism of $Cu_2ZnSnSe_4$ absorber layers during selenization of solution deposited metal precursors, *J. Alloys Compd.* **567**, 102 862 (2013).

[32] M. Buffiere, G. Brammertz, M. Batuk, *et al.*, Microstructural analysis of 9.7% efficient $Cu_2ZnSnSe_4$ thin film solar cells, *Appl. Phys. Lett.* **105**, 183903 (2014).

[33] T. Schwarz, O. Cojocaru-Mirédin, P. Choi, *et al.*, Atom probe tomography study of internal interfaces in $Cu_2ZnSnSe_4$ thin-films, *J. Appl. Phys.* **118**, 095302 (2015).

[34] M. Neuschitzer, Y. Sanchez, T. Olar, *et al.*, Complex surface chemistry of kesterites: Cu/Zn reordering after low temperature postdeposition annealing and its role in high performance devices, *Chem. Mater.* **27**, 5279 (2015).

[35] N. Yue, Y. Zhang, and R. Tsu, Ambient condition laser writing of graphene structures on polycrystalline SiC thin film deposited on Si wafer, *Appl. Phys. Lett.* **102**, 071912 (2013).

[36] Q. Chen, S. Bernardi, and Y. Zhang, Nondestructive high-power-high-temperature Raman spectroscopy for probing microscopic structural variations in CZTSe alloys, *MRS Adv.* **2**, 3187–3198 (2017).

[37] J. He, S. N. Girard, M. G. Kanatzidis, and V. P. Dravid, Microstructure-lattice thermal conductivity correlation in nanostructured $PbTe_{0.7}S_{0.3}$ thermoelectric materials, *Adv. Funct. Mater.* **20**, 764 (2010).

[38] L. Su and Y. Zhang, Temperature coefficients of phonon frequencies and thermal conductivity in thin black phosphorus layers, *Appl. Phys. Lett.* **107**, 071905 (2015).

[39] Y. Dong, H. Wang, and G. S. Nolas, Synthesis and thermoelectric properties of Cu excess $Cu_2ZnSnSe_4$, *Phys. Status Solidi RRL* **8**, 61 (2014).

[40] I. Repins, C. Beall, N. Vora, C. DeHart, D. Kuciauskas, P. Dippo, *et al.*, "Co-evaporated $Cu_2ZnSnSe_4$ films and devices," *Solar Energy Materials and Solar Cells*, **101**, 154 (2012).

[41] S. Chen, L.-W. Wang, A. Walsh, *et al*, "Abundance of CuZn+ SnZn and 2CuZn+ SnZn defect clusters in kesterite solar cells," *Appl. Phys. Lett.* **101**, 223901 (2012).

[42] Q. Chen and Y. Zhang, Effect of elemental ratios in CZTSe film on CZTSe/CdS band alignment and inhomogeneity in absorber thermal conductivity, *Proceedings of the 43nd IEEE Photovoltaic Specialists Conference (PVSC)*, Portland, Oregon, USA, June 5–10, 2016.

[43] F. Zhang, J. F. Castaneda, S. Chen, *et al.*, Comparative studies of optoelectrical properties of prominent PV materials: Halide perovskite, CdTe, and GaAs, Materials today. DOI: 10.1016/j.mattod.2020.01.001.

Chapter 8

Time-resolved photoluminescence characterization of polycrystalline thin-film solar cells

Darius Kuciauskas[1]

8.1 Electro-optical characteristics of thin-film solar cells

Some semiconductor structures are fabricated to eliminate most defects, but this is usually not possible for solar cells, where fast growth of large-area semiconductor films is required. Therefore, photovoltaics (PV) characterization research focuses on identifying, analyzing, and eliminating the *most detrimental semiconductor defects*. When efficiency-limiting defects are identified, their impact can be reduced using several approaches. First, defect concentration can be reduced based on thermodynamics. For example, the density of Te antisites (Te_{Cd}) in CdTe can be reduced at higher growth temperatures or by increasing Cd partial pressure during the growth or annealing [1]. The second approach is based on changing semiconductor composition. This approach was used when ternary CdSeTe absorbers were introduced for CdTe solar cells [2]. The third approach is based on modifying solar cell device architectures, such as using different buffer and contact or passivation layers. This approach was used when the CdS buffer in CdTe solar cells was replaced by a MgZnO buffer [3]. Finally, the fourth approach is based on the concept of "defect-tolerant" semiconductors, and it can be important when developing new solar cell materials such as perovskites. Nevertheless, defects are present in all solar cells, which lead to losses in voltage, fill factor, current, and efficiency.

To identify the most detrimental semiconductor defects and their location in devices (such as interfaces, bulk, space-charge region, or back contact), we refer to the thin-film solar cell model in Figure 8.1. The minimal set of electro-optical (EO) characteristics required to evaluate defect impact typically includes interface recombination velocity (S_{int}), grain-boundary (GB) recombination velocity (S_{GB}), back-contact recombination velocity (S_{back}), and minority-carrier lifetime (τ_B). Combined with the knowledge of doping (net acceptor density, N_A), absorption coefficients (α), and band alignment and band offsets at interfaces, a solar cell device model can be developed.

[1]National Renewable Energy Laboratory (NREL), Golden, CO, USA

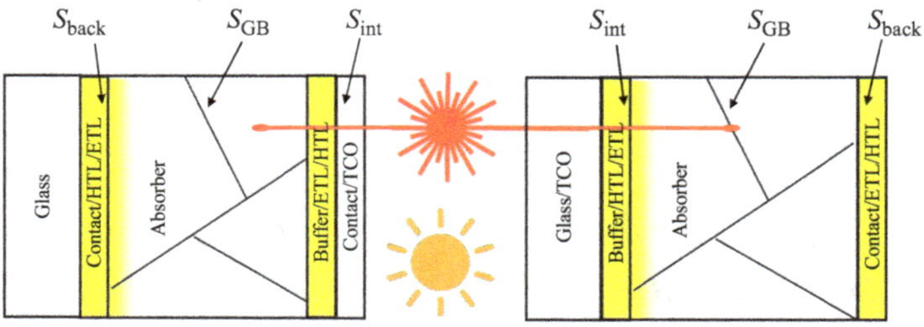

Figure 8.1 Electro-optical model for substrate (left) and superstrate (right) thin-film solar cells. Recombination velocities at the back contact (S_{back}), grain boundary (S_{GB}), and interface (S_{int}) describe losses at PV interfaces. ETL/HTL are electron and hole transport layers, respectively. The laser used in photoluminescence characterization can provide excitation similar in wavelength and intensity to solar radiation

We do not attempt to describe specific electronic defects because such analysis would be different for each semiconductor. For CdTe, we provided a brief summary [4]. The model based on recombination velocities and lifetimes is more general and, to some degree, allows comparison of different absorbers and solar cells. To express EO semiconductor characteristics using time-resolved photoluminescence (TRPL) measurement results, we refer to (8.1)–(8.4):

$$\frac{1}{\tau_{\text{TRPL}}} = \frac{1}{\tau_{\text{R}}} + \frac{1}{\tau_{\text{SRH}}} + \frac{1}{\tau_{\text{int}}} = \frac{1}{\tau_{\text{B}}} + \frac{1}{\tau_{\text{int}}} \tag{8.1}$$

$$\frac{1}{\tau_{\text{R}}} = B(N_{\text{A}} + n) \approx BN_A \tag{8.2}$$

$$\frac{1}{\tau_{\text{SRH}}} = v_{\text{th}}\, \sigma N_{\text{SRH}} \tag{8.3}$$

$$S_{\text{int}} = v_{\text{th}}\, \sigma N_{\text{int}} \tag{8.4}$$

where $\tau_{\text{TRPL}}{}^{-1}$ is the experimentally measured recombination rate (calculated from TRPL lifetime τ_{TRPL}), $\tau_{\text{R}}{}^{-1}$ is the radiative recombination rate (calculated from radiative lifetime τ_{R}), $\tau_{\text{SRH}}{}^{-1}$ is the Shockley–Read–Hall (SRH) recombination rate (calculated from SRH lifetime, τ_{SRH}), $\tau_{\text{int}}{}^{-1}$ is the interface recombination rate, and $\tau_{\text{B}}{}^{-1}$ is the bulk recombination rate that combines radiative and SRH recombination. B is the radiative recombination coefficient, n is injection, σ is electron or hole capture cross section, v_{th} is the thermal velocity, N_{SRH} is the SRH recombination center density in the semiconductor bulk (measured in cm^{-3}), and N_{int} is the interface defect density (measured in cm^{-2}).

Radiative recombination has more impact in highly doped materials. (When $B = 2 \times 10^{-10}$ cm^3/s, $\tau_{\text{R}} \approx 50$ μs/500 ns/5 ns for $N_{\text{A}} = 10^{14}/10^{16}/10^{18}$ cm^{-3}, respectively.) When doping is increased, SRH recombination is reduced, TRPL lifetimes approach radiative limit, and optimization of radiative processes enables additional efficiency improvements. Currently, this is applied only in III–V solar cells [5].

The simplified form of (8.2) can be used when $n < N_A$. At such low-injection conditions, bulk and interface SRH recombination rates can be determined more accurately.

Equation (8.4) for S_{int} can be applied to different interfaces, such as front, back, or grain boundaries. S_{int} can range from ~1 cm/s for single-crystal PV interfaces with the best passivation [6] to ~10^6 cm/s. The high limit is similar to the thermal velocity $v_{th} = \sqrt{3k_B T/m}$, where k_B is Boltzmann's constant, T is temperature, and m is effective mass.

Additional processes (such as charge-carrier trapping, de-trapping, diffusion, drift, or thermionic emission) sometimes need to be included to the TRPL data analysis models. The key concept is that measured lifetime τ_{TRPL} depends on many processes. This provides a challenge in data interpretation and also an opportunity to identify and evaluate different loss mechanisms and transport processes in solar cells.

PL/TRPL measurement conditions can usually be chosen so that a simpler data analysis model is sufficient. Experimental variables are injection, temperature, excitation wavelength (one-photon excitation [1PE] or two-photon excitation [2PE]), or additional (light or electrical) bias. It is desirable to apply at least some such variables to obtain more comprehensive datasets. Detailed modeling, when applied to a limited dataset, is usually not very predictive, because modeling assumptions need to be verified experimentally.

We describe some photoluminescence (PL) approaches to characterize thin-film solar cells with emphasis on time-resolved methods. Spectral PL analysis is complementary and was recently reviewed [7]. Since TRPL is not very commonly used in PV characterization, in Section 8.2 we describe some experimental aspects. We then consider interface and bulk recombination in the test structures (Section 8.3) and in devices (Section 8.4). In Section 8.5, we briefly describe charge-carrier transport and recombination microscopy. Finally, in Section 8.6, we summarize and compare some CdTe, CdSeTe, CIGS, kesterite, and perovskite EO characteristics.

8.2 Experimental aspects of thin-film characterization with TRPL

Contactless and nondestructive EO characterization can be applied to semiconductor films, incomplete device stacks, and devices. For metrology, PL measurements can be rapid and implemented on research and development (R&D) or fabrication lines.

Semiconductors differ in bandgaps, absorption coefficients, doping, recombination center density, and other properties. PL/TRPL instrumentation needs to be adapted for such different measurements; therefore, many researchers develop and use custom instrumentation. For example, low injection is usually required for the SRH recombination center analysis. Many PV semiconductors are doped only 10^{14}–10^{15} cm^{-3}, so high measurement sensitivity is required to achieve low-injection conditions. In this section, we consider some experimental aspects that increase measurement speed and sensitivity.

8.2.1 Lasers for TRPL

A pulsed laser is required for time-resolved measurements, and relevant considerations are energy per pulse, laser pulse repetition rate, and wavelength tunability. For measurements in low injection, 1–10 nJ energy per pulse is usually sufficient, and this is commonly available from simple and inexpensive semiconductor lasers [8]. Amplified laser systems with $\geq$ 1 μJ energy per pulse also allow high-injection and enable experiments with non-focused excitation, which is especially useful when carrier diffusion length is greater than several micrometers. In addition, amplified femtosecond laser systems can provide tunable excitation and peak power sufficient for nonlinear 2PE [9]. The regeneratively amplified femtosecond laser system and an optical parametric amplifier (OPA) provide the most flexible excitation source for 1PE/2PE, low or high injection, and wavelength tunability from 350 nm to 3 μm. Regeneratively amplified femtosecond systems based on Ti:sapphire lasers provide the shortest pulses (typically 100 fs), but they are complex instruments requiring regular maintenance. Amplified Yb:KGW laser systems can be more compact and reliable for use without a dedicated ultrafast laser lab. Pulse length is typically 300 fs.

The laser repetition rate is an important parameter. To accurately measure lifetime τ_{TRPL}, the separation between the laser pulses needs to be at least $3\tau_{TRPL}$. Thus, to measure 1,000-ns lifetimes, laser repetition rate needs to be 300 kHz or less. When ~10-ns lifetimes are measured, a 30-MHz repetition rate laser would make experiments faster. Therefore, a laser with adjustable repetition rate (ideal range 100 kHz–30 MHz) enables the most flexible characterization. Changing repetition rate is usually possible with Yb:KGW laser systems. However, currently, it is not possible to have 1-μJ pulse energy for laser systems working above 1 MHz, and a compromise needs to be made between the energy per pulse and repetition rate.

8.2.2 Time-correlated single-photon counting

Different approaches can be used to measure time-dependent optical signals. As the simplest choice, an analog (not single-photon counting) photomultiplier (PMT) can be used for detection. The voltage from the PMT can be recorded by a digital oscilloscope, perhaps already available in the lab [8]. This simple method has somewhat reduced time resolution and sensitivity, both of which are less important for higher-lifetime and higher-doped semiconductors.

A time-correlated single-photon counting (TCSPC) detection method provides the highest sensitivity (single photons are counted) and high time resolution (down to 20 ps). Since a photon registration is a binary yes/no event, TCSPC provides excellent signal-to-noise characteristics. Introductions to TCSPC are published by manufacturers of such instrumentation [10,11]. To avoid photon pileup, the single-photon counting rate needs to be below 2% of the excitation laser repetition rate, which is the most significant limitation of the TCSPC method. The number of photons that need to be recorded in TCSPC measurement depends on the fitting model and allowable uncertainly of the fitting parameters. Exponential fitting models are commonly used, and statistical estimates suggest that 20,000 photons per TRPL decay are sufficient for two-exponential data fitting [12]. This is illustrated in Figure 8.2, which shows lifetimes derived from one- and two-exponential fits to TRPL data for CdTe solar cells [13]. Laser repetition rate was 1 MHz and photon count rate 2%, or 20,000 photons/s. Lifetimes derived from one- and two-exponential fits fall into the acceptable range after 1 s, improve slightly after 2 s, and do not change when measurement time is increased to 100 s.

Data acquisition time per pixel is very important for TRPL microscopy, where—for a 100 × 100-pixel image—10,000 TRPL decays need to be recorded. With parameters illustrated in Figure 8.2, measurement time would be at least 10,000 s, or almost 3 h. If, for example, a 10-MHz laser could be used, the measurement time would be only 1,000 s (less than 20 min).

To partially overcome this data-acquisition-time limitation in TRPL microscopy, we use a CMOS camera for wide-field imaging and time-resolved detectors

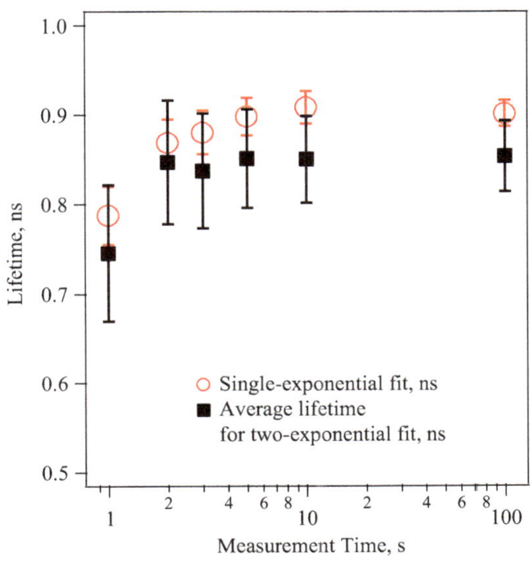

Figure 8.2 Values and uncertainty for TRPL lifetimes derived from one- (open circles) and two-exponential (squares, average lifetime $\tau_{av} = (A_1\tau_1 + A_2\tau_2)/(A_1 + A_2)$ is shown) fitting when counting rate was 20,000 photons/s and integration time was 1–100 s

for pixel-by-pixel data acquisition [14]. With camera imaging it is possible to quickly identify the region of interest (ROI), and with suitable ROI selection it is possible to obtain sufficient resolution in TRPL microscopy when the image size is 100 × 100 or 200 × 200 pixels (Section 8.5).

8.2.3 Photon-counting detectors and electronics

Improvement of red-sensitive PMTs and single-photon avalanche diodes (SPADs) has made single-photon-counting measurements very practical. In particular, SPAD detectors are now more durable (e.g., not damaged by accidental exposure to room light) and less expensive. For Si SPADs, typical quantum efficiencies in the 700–1,050-nm spectral range (important for most PV materials) are 10%–30%, time responses are as short as 50 ps (can be improved to ~20 ps with deconvolution), counting rates are up to 10 MHz, and detector diameters are 100 μm and larger. High-sensitivity TRPL measurements at 900–1,800 nm are possible with InGaAs SPADs, but with some compromise in the detector's characteristics, such as active area or time response.

TCSPC electronics are typically provided with software that can be used to perform TRPL experiments. Electronics typically have a time resolution of 4 ps and better; thus, overall system response (laser, detector, and electronics) is limited by the detector and can be about 20–100 ps. This resolution is sufficient to measure recombination velocities $>10^5$ cm/s and lifetimes <100 ps, which is sufficient for all typical PV absorbers.

8.2.4 PL measurement setups

Figure 8.3 illustrates two PL/TRPL experimental setups. In Figure 8.3(a), different optics are used for laser excitation and for PL signal collection. The advantage of this approach is that the excitation spot can be large (e.g., by moving the excitation lens out of focus), which aids achieving low injection. This approach requires careful alignment, which is simplified by observing light scattered from the sample to the entrance slit of the spectrograph or the detector. The approach illustrated in Figure 8.3(a) is usually implemented on the optical table where sample mount, collection, and excitation optics can be aligned accurately.

In Figure 8.3(b), optical paths for excitation and signal collection are the same. This simplifies alignment and, for example, enables measurements inside fabrication or processing chambers. When ~100-μm-core-diameter multimode optical fiber is used to deliver excitation pulses using the setup illustrated in Figure 8.3(b), the diameter of the excitation spot is 20–300 μm even when the numerical aperture (NA) of the objective focusing optics is relatively high. This allows achieving low injection and efficient collection using the same compact and simple optics [15]. Figure 8.3(c) shows the TRPL setup in NREL's Process Development and Integration Laboratory, implementing an optical scheme in Figure 8.3(b). This setup allows TRPL characterization in an ultrahigh-vacuum (UHV) environment.

In PL/TRPL microscopy, the collinear arrangement is also used (see Figure 8.8). Multimode fibers are not used in microscopy, so excitation can be

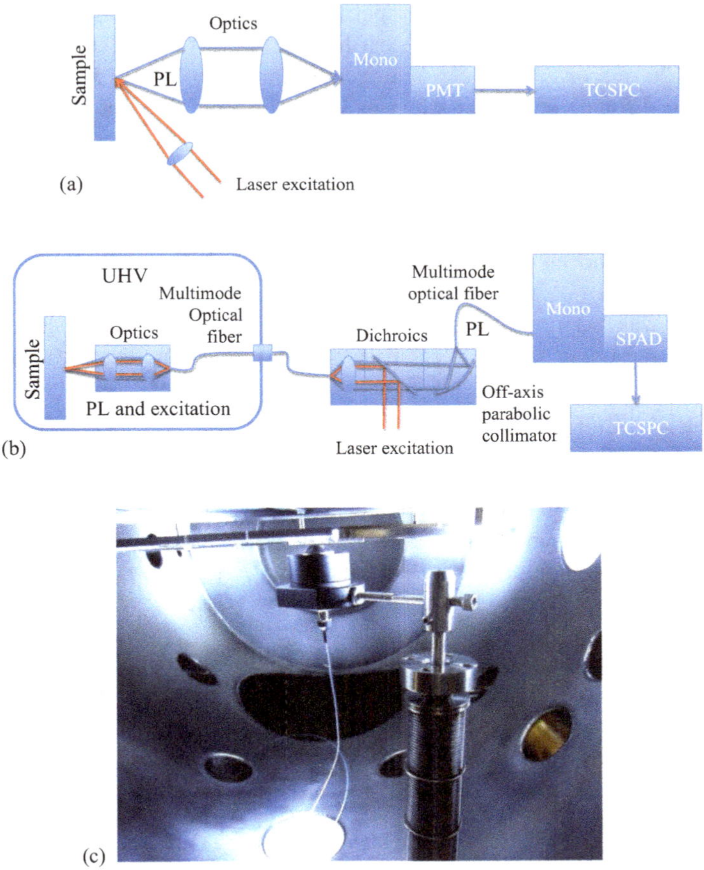

Figure 8.3 *(a) Off-axis PL measurement geometry. (b) Collinear PL measurement where the same multimode optical fiber is used to deliver excitation pulses and to collect PL/TRPL signals. Mono is a monochromator or a spectrograph. (c) Photograph of the optical-fiber-based TRPL spectrometer developed at NREL. As illustrated in (b), a multimode optical fiber is used to deliver excitation pulses and to collect PL in the UHV environment*

focused to a diffraction-limited spot. This provides better than 1-μm lateral resolution, but usually results in higher injection. In addition, as described in Section 8.5, TRPL microscopy data can be affected by carrier transport, which could be a limitation or advantage for a particular study.

8.2.5 PL collection efficiency

PL spectrometer is a relatively simple optical system aligned according to geometric optics. With proper alignment, PL collection efficiency is proportional to

$(NA)^2$, where NA is the numerical aperture of the optics used to collected PL emission. When the working distance (distance from the optics to the sample) needs to be 20 mm or greater, NA is typically limited to NA 0.15. With shorter 5-mm working distance (e.g., still allowing measurements through 3-mm glass substrates), NA can be increased to NA 0.55. In microscopy, where the working distance can be as short as 0.1 mm, NA for measurements in air can be increased to NA 0.9.

Table 8.1 illustrates PL measurement sensitivity with different collection optics. We used the New Focus 5720 aspheric lens kit, which provides several lenses with similar clear apertures but different working distances and NAs. Comparison was performed for two samples: an InGaP double heterostructure (DH), where the optics can be placed very close to the sample, and a CdTe solar cell, where the excitation and measurement were performed through the 2-mm-thick glass. Collinear measurement geometry (Figure 8.3(b) and (c)) was used.

Data in Table 8.1 show that higher NA optics collect PL signals more efficiently. For example, for the InGaP DH measurement, counts per second decrease from 13,000 with NA 0.65 to 2,200 with NA 0.25. When normalized as counts/NA^2, collection efficiency is about the same (30,800 vs. 35,000). When collection and focusing optics have different diameters (e.g., larger NA 0.50 lens is used for collection), counts/NA^2 = 18,000 due to size (also called etendue, which includes both diameter and NA) mismatch.

Data in Table 8.1 also show that when measurements are taken through glass—such as for CdTe solar cells illuminated through the glass substrate—signals are weaker and only certain optics can be used. However, TRPL analysis can still be performed without sample preparation or damage. In comparison with sample preparation requirements for analytical microscopy and surface analysis, contactless and nondestructive EO characterization offers distinct advantages. Therefore, EO methods are commonly used for initial assessments of material and device properties as well as for inline characterization.

Table 8.1 PL measurement sensitivity with different collection optics

Numerical aperture for measurement optics, NA	Clear aperture, mm	Working distance, mm	Counts, InGaP DH[a]	Counts/ NA^2	Counts, CdTe solar cell[b]	Counts/ NA^2
–[c]	–	0	10,000	–	No signal	
0.65	3.6	1.6	13,000	30,800	No signal	
0.55	5.0	2.9	8,600	28,400	1,680	5,500
0.40	5.0	3.5	5,600	35,000	960	6,000
0.50	8.0	5.9	4,500	18,000	725	2,900
0.25	5.5	8.0	2,200	35,000	425	6,800

[a]Sample InGaP double heterostructure, τ_{TRPL} = 50 ns, PL measured at 670 nm.
[b]Sample CdTe device, τ_{TRPL} = 2 ns, PL measured at 840 nm through 2-mm glass substrate.
[c]Optical fiber is touching the sample; PL collection optics is not used.

8.3 Interface and bulk recombination in thin-film absorbers

Defects in semiconductor bulk and at the interfaces can have different impacts on solar cells [16]. Therefore, one of the first questions is to determine whether dominant recombination losses occur near the interfaces or in the bulk. If such identification can be made, materials science efforts can be selectively applied to improve interface passivation or to reduce bulk defect density. This question is most easily answered if samples with the same interfaces but different thickness, d, can be fabricated. Such samples are often called double heterostructures. For DHs, measured carrier lifetime τ_{TRPL} can be expressed as (see (8.1)) [17]:

$$\frac{1}{\tau_{TRPL}} = \frac{1}{\tau_B} + \frac{1}{\tau_{int}} = \frac{1}{\tau_B} + \frac{2S_{int}}{d} \tag{8.5}$$

Recombination velocity S_{int} is assumed to be identical at both interfaces. An additional assumption used in deriving (8.5) from the drift-diffusion model is that carrier density is uniform in the absorber. (Carriers are generated near the front interface; density becomes uniform when diffusion is faster than recombination.) This criterion can be expressed as $L_d = \text{sqrt}(\tau_B D) > d$, where L_d is carrier diffusion length and D is carrier diffusion coefficient.

The model of (8.5) predicts a linear dependence for τ_{TRPL}^{-1} vs. d^{-1} with the slope $2S_{int}$ and intercept $1/\tau_B$. Therefore, from a simple linear regression, S_{int} and τ_B are easily determined. The impact of recombination is described by rates, where the first term in (8.5) describes the bulk recombination rate and the second term describes the interface recombination rate. For example, if $\tau_B = 1{,}000$ ns, $S_{int} = 100$ cm/s, and $d = 2$ μm, then bulk and interface recombination rates $\tau_B^{-1} = 2S_{int}/d = 10^6\ \text{s}^{-1}$. In this hypothetical case, interface and bulk defects make the same impact on overall recombination. Even if defect density were the same, interface recombination rate would be greater for a thinner sample ($2S_{int}/d > 10^6\ \text{s}^{-1}$ when $d < 2\,\mu\text{m}$), and the bulk recombination rate would be higher for a thicker sample ($2S_{int}/d < 10^6\ \text{s}^{-1}$ when $d > 2\,\mu\text{m}$). Thus, the relative importance of bulk vs. interface recombination depends on the absorber thickness and other aspects of device design, and not only on the defect cross sections and densities.

If S_{int} is very high, then recombination occurs immediately after the absorption, and defects at the distance larger than L_d from the front interface do not significantly impact recombination. The simplified expression for evaluating S_{int} in the high interface recombination case is [18]:

$$\frac{1}{\tau_{TRPL}} \approx \frac{S_{int}}{d}, \tag{8.6}$$

where the distance parameter d is not sample thickness, but rather the absorption depth; $d \approx \alpha^{-1}$ (α is the absorption coefficient at the excitation wavelength). For example, when $\alpha = 10^5\ \text{cm}^{-1}$ and $S_{int} = 10^5$ cm/s, the interface recombination rate is $10^{10}\ \text{s}^{-1}$ and $\tau_{TRPL} = 0.1$ ns. This set of parameters is typical for some as-grown direct-bandgap semiconductors (e.g., GaAs or CdTe). This high limit for the interface recombination rate is relevant when selecting instrumentation; for example, to analyze recombination for non-passivated interfaces, the instrument's time response needs to be better than 0.1 ns. Section 8.2 describes lasers, detectors, and TCSPC electronics that enable such analysis.

Equations (8.5) and (8.6) present two limiting cases. For example, we used (8.5) for CdSeTe DHs passivated with Al_2O_3 ($S < 100$ cm/s, $\tau_B > 200$ ns) [19], and (8.6) for as-grown single-crystal and polycrystalline CdTe ($S > 10^5$ cm/s) [9].

When DH samples are not available, or when a pn junction is present in the sample, separating interface and bulk recombination is more difficult and usually requires numerical modeling of the carrier dynamics after pulsed excitation [20–22]. In Section 8.4, we describe some additional experimental aspects that aid interface, space-charge region, and bulk recombination analysis.

8.3.1 Characterization of buried semiconductor interfaces with 2PE

Some PV interfaces are buried (not accessible to direct optical excitation), but recombination at buried interfaces can be significant for overall recombination losses. Examples of such high-recombination-velocity buried interfaces are Cu(In, Ga)(S,Se)$_2$/Mo, CdTe/ZnTe, or CdTe/Te, which all have $S_{int} > 10^5$ cm/s. Despite high recombination velocity, such interfaces are used in solar cells due to favorable band alignments that avoid barriers for hole collection. To limit recombination losses at buried PV interfaces with high S_{int}, the device structures called "electron reflectors" can be used [23]. The most successful electron reflectors are fabricated using Ga/(In+Ga) grading for CIGS solar cells, and more recently, CdSeTe/CdTe grading for CdTe solar cells. In both cases, grading of the conduction band energy reduces electron recombination near the back contact. For such graded absorbers, the impact of the buried interface recombination is reduced, but not eliminated. Direct measurements of the buried interface recombination velocity could be informative when evaluating and optimizing electron reflectors.

Buried interface characterization requires generating carriers near the buried interface, 2–3 μm (and sometimes deeper) in the absorber. In direct-bandgap semiconductors, this is not possible with excitation in the visible spectral range because when $\alpha > 10^4$ cm^{-1}, excitation is attenuated to <1% of the initial intensity at the depth $2/\alpha < 2$ μm.

We evaluated whether nonlinear 2PE can be applied to analyze recombination at II–VI and III–V semiconductor-buried interfaces [24]. An illustration of this analysis for a buried homoepitaxial CdTe interface is shown in Figure 8.4. To enable this study, 15-μm-thick epitaxial CdTe absorbers were grown by molecular beam epitaxy (MBE). Surface passivation was not applied; therefore, recombination velocity at the front surface $S_{surface} = (1–2) \times 10^5$ cm/s. Figure 8.4(b) shows integrated PL intensity measured during the axial (depth) scan. Figure 8.4(c) shows a linear intensity profile, where we observe the strongest PL intensity from the middle of the epilayer and reduced intensity near the top interface and the buried interface. The fit in Figure 8.4(c) shows that the width of the high-recombination region is 1.3 ± 0.2 μm, and the buried interface with high defect density is clearly resolved. Application of (8.6) to fit 2PE TRPL decays directly measured at the buried interface yielded $S_{\text{int}} \approx 1 \times 10^5$ cm/s [24].

One limitation of the buried interface analysis illustrated in Figure 8.4 is uncertainty of the axial coordinate: optical beams change direction when propagating in the semiconductor, and translation of the sample by some micrometers does not equal the analysis depth in the sample. To accurately locate interfaces, optical diffraction and refraction modeling is required. For buried interface analysis with PL-based methods, we developed a simpler approach based on spectral shifts due to PL reabsorption. The apparent red shift in the PL emission spectrum increases with depth, and as shown in Figure 8.4(d) for CdTe we resolve the PL shift at 0–30-μm depth. This shift can be modeled using a near-interface PL emission spectrum (measured with above-bandgap excitation) and the absorption

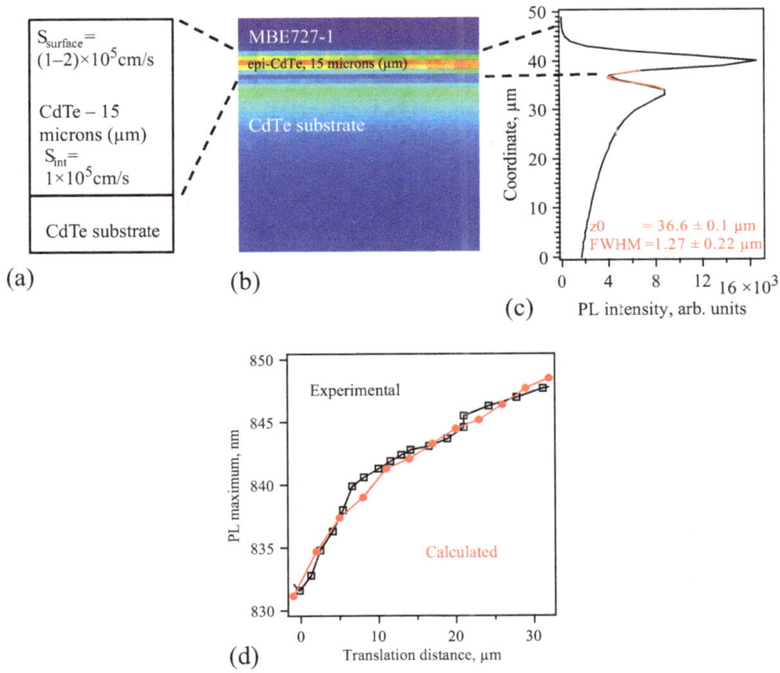

Figure 8.4 Buried homoepitaxial interface characterization with 2PE TRPL. (a) Sample structure. (b) Integrated PL intensity shown in cross section (scan area 50 × 50 μm²). (c) Axial PL intensity profile for the data in (b). (d) Experimental (open squares) and calculated (full circles) shift of the PL emission maximum vs. the axial coordinate. Red PL emission shift occurs due to reabsorption and can be used to accurately evaluate the depth coordinate of PL/TRPL analysis

spectrum. As shown in Figure 8.4(d), this simple model reproduces the experimental data. Therefore, the axial coordinate is determined accurately from the apparent PL maximum. For CdTe, this analysis is especially simple for up to 7-μm-thick films (the thickness range used in solar cells), and the calibration curve in Figure 8.4(d) is the same for single-crystal, epitaxial, and polycrystalline samples. We call these data the "spectroscopic ruler" [24]. At >7 μm depth (for CdTe), PL emission spectral shifts depend on the band-tail absorption; thus, at >7 μm the calibration curves would be different for the electronically disordered absorbers characterized by the band tails in the optical spectra.

Buried interface characterization illustrated in Figure 8.4 was applied to a 15-μm-thick epilayer, but solar cell absorbers are typically $\leq$3 μm thick. By using higher NA optics, the resolution can be improved, and buried interfaces of thinner semiconductor layers can be studied. Practical limits are NA 0.9 in air and NA 1.3 with solid immersion lenses (SIL) [25], predicting up to 4× improvement in measurement resolution. Thus, buried interface analysis for 3-μm-thick absorbers should be possible, but SIL microscopy has not yet been applied to thin-film solar cells.

8.4 Time-resolved emission spectroscopy of thin films and solar cells with graded absorbers

When the bandgap E_g is changing with depth (i.e., the absorber is "graded"), PL maximum indicates the depth coordinate in the absorber. When the PL lifetime is measured at that energy, the recombination rate is determined at a specific depth coordinate; in this way, it is possible to "map out" recombination and drift rate distribution. This analysis requires that TRPL decays are measured at the range of energies that span full E_g variation. Data combine time and spectral information, and the method is called time-resolved emission spectroscopy (TRES). To impact electron diffusion in solar cells, bandgap grading significantly exceeds thermal energy k_BT and is typically at least 200–300 meV. Therefore, high spectral resolution is not needed for TRES analysis of graded absorbers. The spatial resolution in TRES measurements mostly depends on the absorber grading profile. For CdS/CdTe solar cells and graded Cu(In,Ga)Se_2 absorbers, resolution is sufficient for detailed characterization of the first 0.5 μm, which corresponds to the buffer/absorber interface and the space-charge region. Interestingly, such axial/depth resolution exceeds the optical diffraction limit; yet, it can be simply applied to solar cells with graded absorbers.

8.4.1 *TRES of CdS/CdTe solar cells*

Figure 8.5 illustrates TRES data for the CdS/CdTe device [26]. This device structure results in substantial alloying at the interface, and secondary-ion mass spectrometry (SIMS) finds that sulfur diffuses about 0.5 μm into CdTe. Therefore, the bandgap is changing with depth, and this is evident as two PL emission peaks attributed to CdSTe ($E_g \approx 1.46$ eV) and CdTe ($E_g \approx 1.50$ eV). Depending on

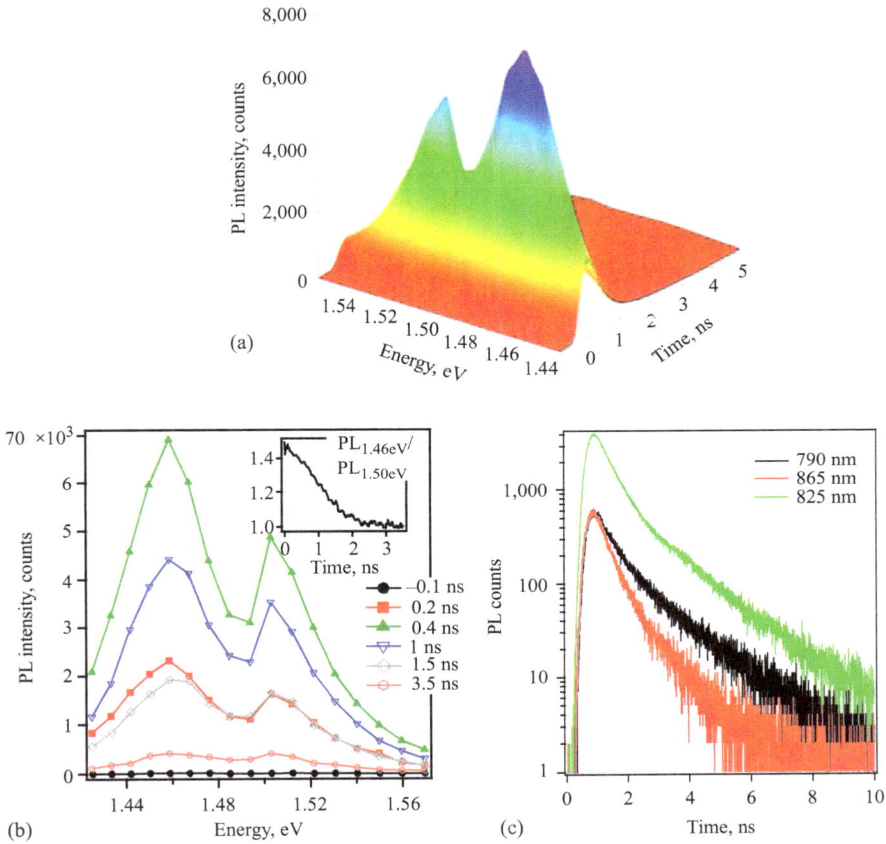

Figure 8.5 (a) Three-dimensional representation of TRES data for CdS/CdTe solar cell. (b) Time-resolved PL emission spectra at different time delays indicated in the legend. Inset shows the ratio of CdSTe (1.46 eV) and CdTe (1.50 eV) emission peaks. (c) TRPL decays 157 eV (790 nm), 1.43 eV (865 nm), and 1.50 eV (825 nm)

device fabrication, CdSTe and CdTe peaks can be less distinct [13], and in that case CdSTe and CdTe contributions can be separated by comparing them with CdTe PL emission (e.g., measured from the back side of the same solar cell). The three-dimensional TRES data in Figure 8.5(a) provides spectral and time-domain information. Cross sections of these data in spectral (shown in Figure 8.5(b)) and time (Figure 8.5(c)) domains are easier to analyze.

After absorption, carriers are generated in the CdSTe region, and at $t = 0.2$–0.4 ns emission from CdSTe is stronger, with $PL_{CdSTe}/PL_{CdTe} = 1.45$ (inset in Figure 8.5(b)). After 2 ns, $PL_{CdSTe}/PV_{CdTe} \approx 1.0$, and this change is attributed to electron drift from the interface to the CdTe bulk on the time scale <2 ns. These dynamics are also seen in

Figure 8.5(c), where the 865-nm decay (1.43 eV, CdSTe) is faster than the 790-nm decay (1.57 eV, CdTe). The faster dynamics in CdSTe occurs because recombination and drift rates are additive. Therefore, measurement at <1.46 eV will reflect interface processes (which includes drift and recombination), whereas at >1.5 eV we can investigate bulk carrier lifetime. Such measurement of carrier redistribution from the near-interface to the bulk region also allows estimating carrier mobility, which in this case was found to be 15–45 $cm^2/(Vs)$ [26]. Recombination lifetimes are relatively short, $\tau_B \approx 2$ ns.

Since TRPL lifetimes positively correlate with CdTe solar cell voltage [27,28], TRPL analysis is widely applied in CdTe PV research. The illustration in Figure 8.5 shows that, in addition to the bulk recombination lifetime, TRES measurements provide information about the interface and the space-charge region in CdS/CdTe solar cells.

More recently, alloying absorbers with Se (CdSeTe) and using bilayer CdSeTe/CdTe absorbers was one of the factors that enabled >19% efficient CdSeTe solar cells [2,29]. It is generally observed that carrier lifetimes are ≥ 10 times longer in polycrystalline CdSeTe (in comparison with CdTe) [29,30], but the origin of such improvement is not fully understood. These results also have not been reproduced in CdSeTe single crystals. Carrier dynamics in such more-complex "CdTe" absorbers is currently actively investigated. Among the factors that are considered are bandgap grading, different interface and bulk defect properties, nonuniform Se distribution in crystalline grains that could affect GB potentials and passivation [31], doping [32], band offsets and bowing [33], and other effects.

8.4.2 *Interface and space-charge region recombination in $Cu(In,Ga)Se_2$ thin films*

Some $Cu(In,Ga)Se_2$ (CIGS) solar cells use double-graded absorbers, where bandgap is increased at the front and back interfaces by changing absorber composition, such as Ga/(In+Ga) ratio or Se concentration. This device architecture helps to minimize interface recombination losses even when the interface defect density remains the same, enabling >20% efficiency with $S_{front} > 10^3$ cm/s [34] and $S_{back} > 10^5$ cm/s [35]. Since the CIGS/Mo interface is not transparent, lift-off methods need to be used to study electronic properties of the back-interface region [35]. The front-interface and space-charge region can be studied nondestructively by taking advantage of combined time and spectral information available from TRES. This analysis is illustrated in Figure 8.6 for CIGS-graded absorbers fabricated in NREL's three-stage process [36].

Auger electron spectroscopy (AES) data in Figure 8.6(a) indicates $E_g = 1.20$–1.24 eV at the front interface and $E_g = 1.10$–1.11 eV at a 0.5–1-μm depth. As a result, minority carriers drift from the interface, thus reducing the impact of the interface recombination. This can be quantified experimentally by recording TRPL decays at 1.24 eV (interface, sample 1) and 1.10 eV (E_g minimum). As shown in Figure 8.6(b), 1.24-eV decay has a fast 0.3-ns component,

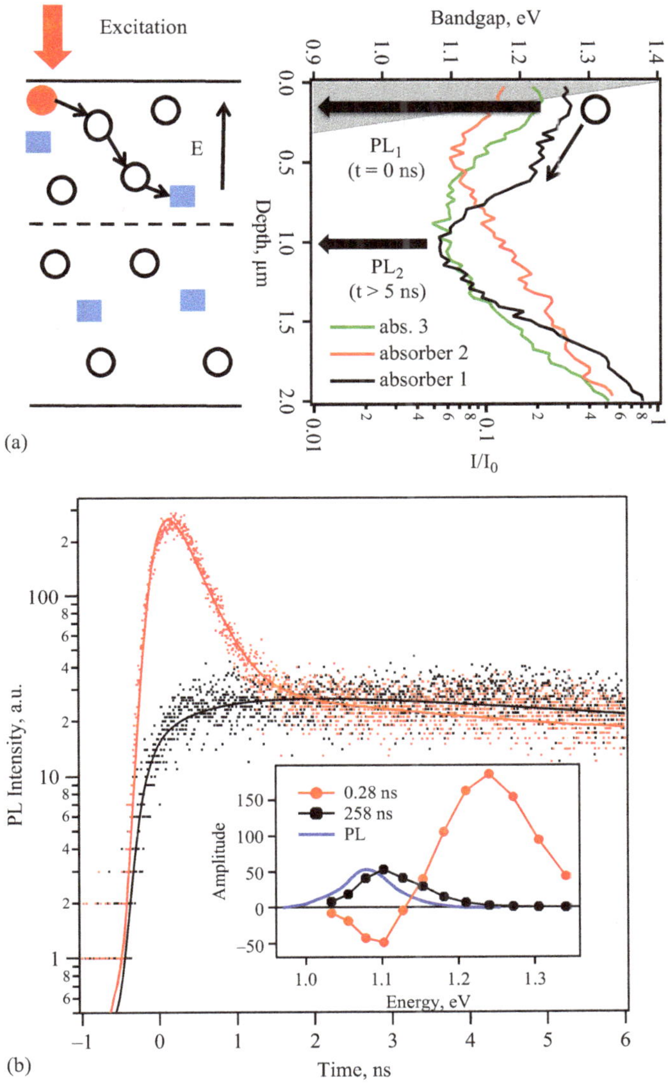

Figure 8.6 *(a) Depth-dependent E_g for three solar cells fabricated using NREL's three-stage process. Gray-shaded area illustrates Beer absorption profile at 640 nm, the excitation wavelength used in TRPL measurements. The open circle illustrates minority-carrier (electron) drift from the near-interface region, when the PL emission maximum changes from 1.2–1.3 eV to 1.1 eV. (b) TRPL decays at 1.24 eV (red) and 1.10 eV (black). Inset shows decay-associated spectra for lifetime components τ_1 (circles) and τ_2 (squares) and PL emission spectrum measured with continuous excitation (solid line)*

whereas the 1.10-eV decay has a rise (increasing PL intensity) component with the same lifetime. Therefore, rate $(0.3\ \text{ns})^{-1} = 3.3 \times 10^9\ \text{s}^{-1}$ predominantly indicates carrier drift from the interface to the E_g minimum. Since drift rate is larger than the interface recombination rate (for $S_{front} = 10^3$ cm/s, interface recombination rate is $\approx 10^7$ cm/s), the interface recombination impact is reduced.

When the experiment is repeated at several wavelengths and all data are fit together using a two-exponential decay model $A_1\exp(-t/\tau_1) + A_2\exp(-t/\tau_2)$, where A_1/A_2 are amplitudes of corresponding lifetime components, we obtain τ_1 and τ_2 spectra (also called decay-associated spectra) shown in the inset of Figure 8.6(b). The spectrum for τ_1 has a maximum that corresponds to the near-interface $E_g = 1.24$ eV and a negative amplitude (indicating increase in signal intensity) at 1.10 eV. Spectrum for the component τ_2 is similar to the PL emission spectrum measured with continuous excitation, and it predominantly occurs from the space-charge region.

TRES data for front-graded CIGS provide TRPL lifetime (which, in this case, $\tau_2 = 258$ ns), and also bandgap at the interface (1.24 eV), bandgap at the E_g minimum (1.10 eV), and the drift rate ($3.3 \times 10^9\ \text{s}^{-1}$), which can be used to evaluate the impact of the interface recombination and to estimate mobility $\mu = 55$–$230\ \text{cm}^2/(\text{Vs})$ [36]. These characteristics are much more detailed than available from TRPL measured at a single emission wavelength, and TRES analysis could be used more broadly in absorber-grading optimization and front-interface passivation research.

8.4.3 Recombination in solar cell devices

Carrier lifetime analysis is important not only in partial device stacks but also in devices. One experimental complication is that in a device with a pn junction drift will change carrier dynamics and TRPL signals. When doping is $<10^{15}\ \text{cm}^{-3}$, typical for CdTe solar cells, higher injection can be used to partially screen the pn junction field, and recombination lifetime can be measured in devices [26–28]. When device doping is $\geq 10^{16}\ \text{cm}^{-3}$, SRH recombination lifetimes cannot be usually measured accurately with higher injection; at the injection level required to screen the space-charge field, radiative recombination will begin to dominate.

To measure carrier lifetimes in devices, one can remove the junction (e.g., using etching or liftoff). This approach is not reversible and may change absorber properties. Alternatively, electrical bias can be applied to reduce the space-charge field. This is evident from the bias-dependent band diagrams for ZnO/CdS/CIGS solar cells (Figure 8.7(a)) [37]. At the forward bias approaching the open-circuit voltage, the space-charge field strength is reduced, and recombination lifetime can be measured more accurately. Data in Figure 8.7(b) indicate TRPL decay changes with the applied bias. The drift rate is increased at the negative bias, and the drift rate is reduced when the bias is positive. Figure 8.7(c) shows that only the slower decay component (τ_2) depends on the applied bias, whereas the faster decay component (τ_1) is mostly unaffected. This is in agreement with τ_1 attribution to the interface properties for front-graded CIGS absorbers (Figure 8.6).

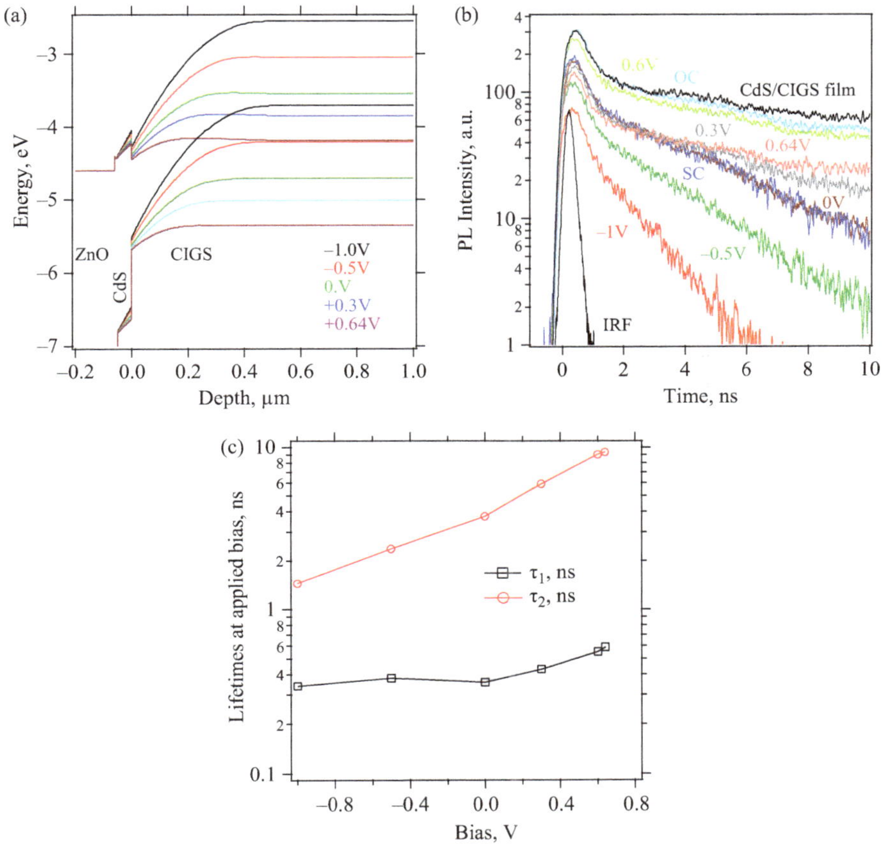

Figure 8.7 Band diagrams and TRPL analysis for ZnO/CdS/CIGS solar cell. (a) Device band diagrams when applied bias is changed from −1 to +0.64 V. (b) Bias-dependent TRPL decays. Two-exponential fitting model was applied to the data in (b), and lifetimes τ_1/τ_2 as a function of applied bias are shown in (c)

8.5 2PE TRPL and second-harmonics generation microscopy for recombination and charge-carrier transport analysis

Absorbers in thin-film solar cells have crystalline grains that can range from tens of nanometers to tens of micrometers. The key property for time-resolved microscopy in such films and devices is charge-carrier diffusion length, L_d; because charge carriers are mobile, with optical resolution better than L_d, we will be able to time-resolved recombination and also charge-carrier transport. Analysis will be simplified when the grain size is $\geq 2L_d$, because in that case carriers generated in the grain interior (GI) will effectively be in the single-crystal environment, and carriers generated near GBs will have reduced lifetimes due to GB recombination.

In this section, we give examples of charge-carrier transport and recombination measurements in CdTe and perovskites. We also briefly describe correlative second-harmonics generation (SHG) microscopy. The experimental scheme illustrated in Figure 8.8(a) is similar to a non-microscopic setup in Figure 8.3, except

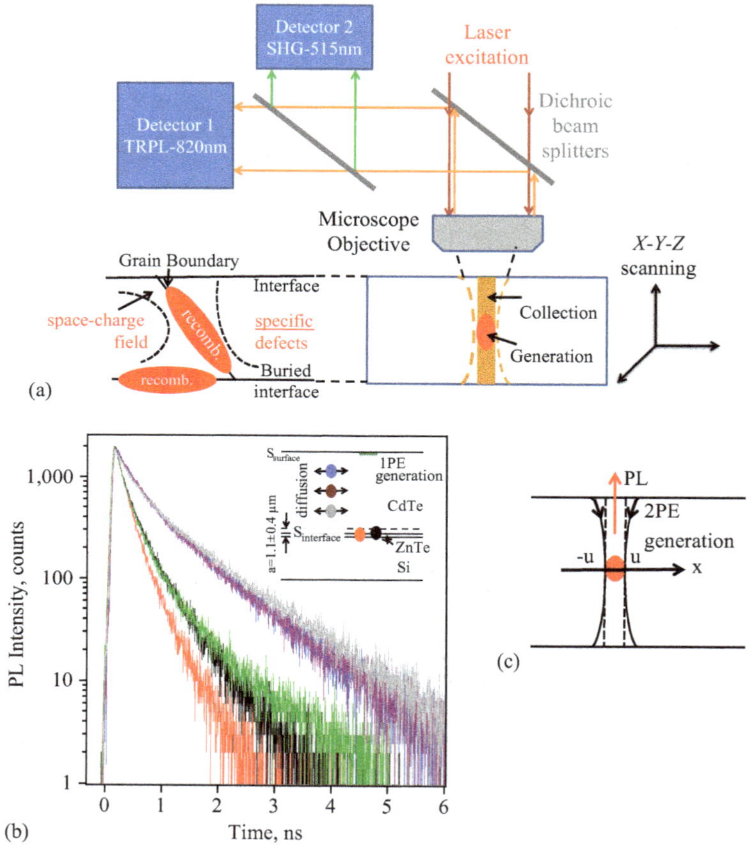

Figure 8.8 *(a) Illustration of correlative EO microscopy. 2PE is focused to the relevant area of the sample, and carrier lifetime (TRPL) and space-charge field strength (second-harmonics generation (SHG)) are measured at different detection wavelengths. (b) 1PE (green) and 2PE (other colors) TRPL decays are measured at different depths in CdTe/ZnTe/Si heterostructure. Inset indicates approximate axial coordinates where measurements are taken. (c) Illustration of the impact of the charge-carrier diffusion on TRPL measurement. Radius of the excitation spot, u, is assumed to define generation and PL signal collection region*

that laser excitation is focused to a diffraction-limited spot using a microscope objective. When 2PE is used, sample scanning in the *xyz* directions can provide nondestructive lateral (*xy*, planar) and axial (*xz*, vertical, e.g., Figure 8.4(b)) imaging, which is a unique feature of 2PE analysis.

8.5.1 Charge-carrier transport in CdTe and perovskites

In thin-film solar cells, carrier lifetimes primarily indicate SRH recombination center density ((8.1)–(8.4)). Recombination lifetime is also one of the factors which determines carrier diffusion length. To more accurately characterize carrier dynamics, it is desirable to measure both τ_B and L_d under the same excitation conditions. This can be realized if we determine charge-carrier diffusion coefficient $D = e/k_B T \mu$, where μ is mobility and e is elementary charge. When $L_d > u$ (u is the radius of the generation spot, Figure 8.8(c)), time-dependent PL intensity $I_{PL}(t)$ can be described as [38]:

$$I_{PL}(t) = I_0 \mathrm{erf}\left(\frac{u}{2\sqrt{Dt}}\right)e^{-t/\tau_B}, \tag{8.7}$$

where I_0 is PL intensity at $t = 0$ and t is time. Data in Figure 8.8(b) illustrate such measurements in 17.5-μm-thick heteroepitaxial CdTe. As predicted in (8.7), TRPL decays are not single exponential. The initial part of the decay (described by the error function erf) depends mostly on mobility, and the later part of the decay depends mostly on the bulk lifetime τ_B. Fits of three decays measured in the middle of the epilayer indicate $\tau_B = 2.2$ ns and $D = 17$ cm^2/s, or $\mu = 650$ cm^2/(Vs). Good structural quality of the epilayer (as evaluated by the full width at the half maximum [FWHM] of the X-ray double-crystal rocking curve [DCRC]) results in high mobility, even with relatively short SRH recombination lifetime. The diffusion length is $L_d = 1.9$ μm.

The model of (8.7) is convenient for data analysis, but the limits of its validity need to be verified. Such analysis was presented by Gaury and Haney [39] using analytical methods and by Kanevce *et al.* [20] using technology-aided computer design (TCAD) simulations. For the data in Figure 8.8(b), three approaches produced comparable results. Analytical and TCAD models are more detailed and also allow evaluating impact of the surface/interface/GB recombination, which is not included in deriving (8.7). However, analytical and TCAD methods cannot be easily applied to fit large datasets generated with TRPL microscopy.

Application of recombination and transport analysis to $CH_3NH_3PbI_3$ (or $MAPbI_3$) perovskite thin films is summarized in Figure 8.9 and Table 8.2. Data were measured every 0.25 μm, resulting in 1,000 TRPL decays over the 5 × 12.5 μm^2 area. Time-integrated PL in Figure 8.9(a) shows two crystalline grains with 3–4 times stronger PL emission intensity. A linear intensity profile in Figure 8.9(b) was used to estimate grain diameters and to select several pixels (A–E) for recombination lifetime and mobility analysis. TRPL decays at pixels A–E in Figure 8.9(c) indicate that, at these microscopic locations, the sample is heterogeneous in both mobility and recombination lifetimes. Fitting results in

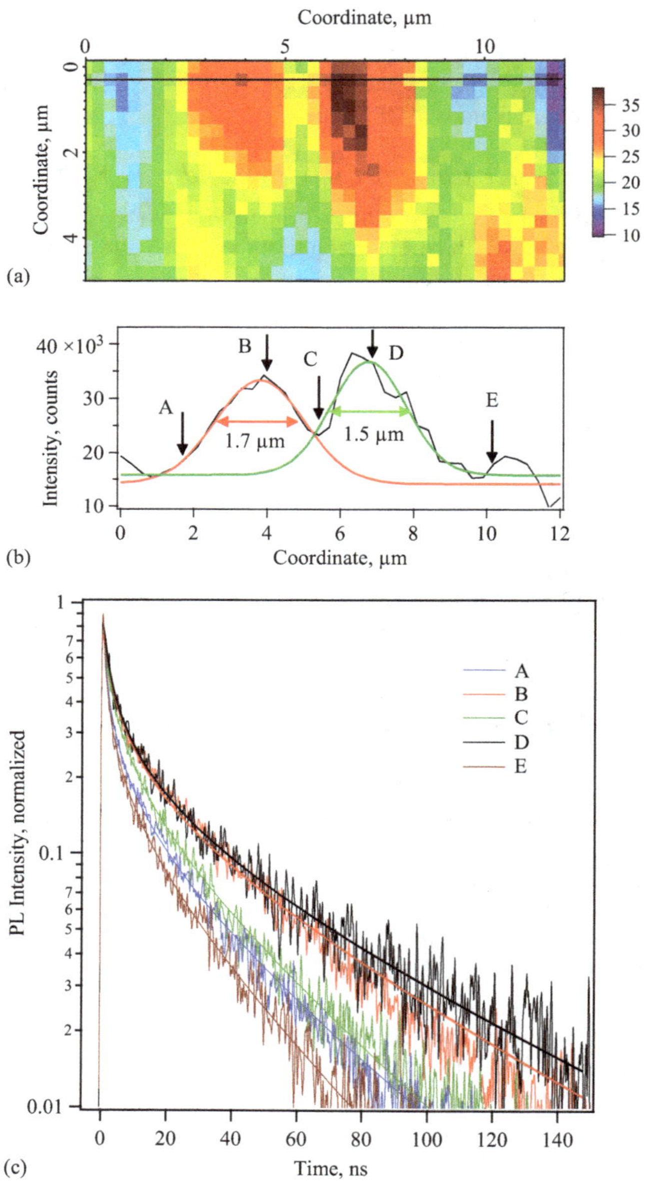

Figure 8.9 2PE TRPL microscopy results for $MAPbI_3$ thin films. (a) Integrated PL intensity. (b) Linear intensity profile for the horizontal line shown in (a). Gaussian fits are used to estimate grain diameters of 1.5 and 1.7 µm. (c) 2PE TRPL decays at pixels A–E. Solid lines in (c) indicate fits to (8.7)

Table 8.2 Microscopic variation of charge-carrier diffusion coefficients, mobilities, and recombination lifetimes in $MAPbI_3$ thin films determined with (8.7) for pixels A–E in Figure 8.9(c)

Location	D, cm²/s	μ, cm²/(Vs)	τ_B, ns	L_d, μm
A	1.79 ± 0.03	69 ± 4	50.2 ± 0.4	3.0
B	0.48 ± 0.02	18 ± 1	74.9 ± 0.9	1.9
C	0.83 ± 0.02	32 ± 2	46.6 ± 0.4	2.0
D	0.56 ± 0.02	22 ± 1	84.6 ± 1.0	2.2
E	2.60 ± 0.10	98 ± 3	39.3 ± 0.3	3.2

Table 8.2 indicate that in the GI with stronger PL (pixels B and D), lifetimes are longer but mobilities are smaller. The opposite is observed near GBs (pixels A, C, and E). Overall, this results in lower L_d for GIs. A similar result was earlier reported for $MAPbI_3$ films fabricated at NREL: GBs do not lead to increased recombination, and recombination primarily occurs in non-GB regions [40]. Based on such data, it was suggested that potential barriers could exist at GBs [40]. For $MAPbI_3$ and other perovskites, this finding is still debated [41]. In Section 8.5.3, we describe correlative SHG microscopy, which could be applied to directly measure GB potentials in such samples.

We also observed microscopic mobility distributions in epitaxial CdTe, where extended defects affecting carrier transport were not GBs but threading dislocations [42]. Therefore, for absorbers with high extended defect density, it is desirable to microscopically analyze not only recombination but also transport. This is because the impact of the extended defects on recombination depends on the probability that minority carriers diffuse to such defects, and this, in turn, depends on the diffusion length.

In some cases, recombination lifetime alone provides an incomplete description of carrier dynamics in solar cells. For example, data in Figure 8.8 for CdTe and in Figure 8.9 for $MAPbI_3$ show samples with comparable diffusion lengths (L_d = 1.9 vs. 1.9–3.2 μm) despite having recombination lifetimes that differ by an order of magnitude (τ_B = 2.2 ns for CdTe vs. τ_B = 47–85 ns for $MAPbI_3$).

The largest diffusion length we measured with TRPL microscopy was $L_d \approx$ 20 μm (in passivated epitaxial CdTe with high mobility and τ_B = 560 ns) [43]. Earlier, Fluegel *et al.* obtained similar results using gated-camera imaging [44]. These results illustrate that single-crystal II–VI semiconductors can have excellent transport characteristics, but this aspect is less clear for polycrystalline II–VI absorbers [45].

8.5.2 GB recombination

Here, we consider GB effects on charge-carrier recombination. In comparison with cathodoluminescence (CL) spectrum imaging [46,47], models for such analysis are less-developed for TRPL microscopy. However, since lifetimes at GBs and in GIs are easily separated (e.g., Figure 8.9), the ability to directly analyze lifetime heterogeneity is very useful. Since carriers are mobile, the concept of "microscopic carrier lifetimes" should not be overused—carriers generated in one location can drift or diffuse to a different region of the solar cell. Instead, local recombination rates (described by (8.1)–(8.4)) can microscopically map bulk or interface defect distributions. TCAD simulations guide such data interpretation. An example is given in Figure 8.10(a), which shows simulated TRPL decays as a function of the distance from the GB. In this simulation, Kanevce *et al.* [20] assumed grain diameter = 7 μm, $\tau_B = 5$ ns, $L_d = 2$ μm, $S_{GB} = 10^5$ cm/s, and $S_{int} = 0$ cm/s. Decay at $x = 0$ μm (black, directly at GB) is fast, with the amplitude decreasing to <1% in 2 ns. Decays further away from the GB are progressively slower, until simulated TRPL traces at $x = 2.5$–3.5 μm (distance larger than L_d from the GB) correspond to the GI not affected by the GB recombination. Simulated TRPL data in the GI ($x = 2.5$ and 3.5 μm) still have some initial decay due to diffusion, a feature also described by (8.7).

Experimental 2PE TRPL data for As-doped polycrystalline CdTe solar cells [48] are in good agreement with the TCAD simulations. Three representative 2PE TRPL decays are shown in Figure 8.10(b): GI corresponds to the GI, and GB1/GB2 denote decays at two GBs. In good agreement with the modeling predictions, TRPL decays GB1 and GB2 have a larger initial decay component due to GB recombination. Based on the TCAD simulations, GB recombination velocity in As-doped CdTe solar cells is in the $S_{GB} \approx 10^5$ cm/s range.

In addition to TCAD analysis, we used an experimental approach to derive similar values of $S_{GB} = (1–4) \times 10^5$ cm/s near GBs in As-doped polycrystalline CdTe [48]. GB recombination with $S_{GB} = (1–4) \times 10^5$ cm/s and bulk

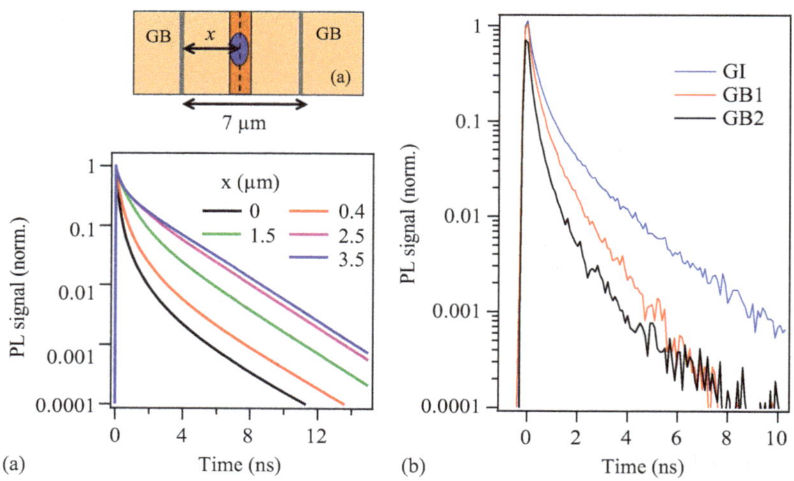

Figure 8.10 (a) Simulated TRPL decays for polycrystalline CdTe assuming parameters described in the text. (b) Experimental 2PE TRPL decays for As-doped polycrystalline CdTe solar cells

recombination with $\tau_B = 1.0$–2.4 ns were attributed to incomplete As activation (1%–4%), resulting in the formation of As-related SRH recombination centers. First-principles studies indicate that As_2 dimers and other As complexes can be recombination centers [49]. Research efforts are applied to increase As activation [50] with the expectation that this will result in higher doping and in lower bulk and GB SRH recombination center density.

8.5.3 Electric-field-induced second-harmonics microscopy for space-charge fields

Space-charge fields perform different roles in thin-film solar cells. First, a pn junction is intentionally created to assist charge-carrier separation and collection. Second, interface and GB defects can be charged, which leads to band banding near interfaces and GBs. For CdTe, band bending can exceed 0.5 V [51], and this significantly affects interface band alignment, carrier dynamics, and collection. Since space-charge field measurements near GBs require microscopy, fewer data are available, but it is typically found or assumed that GBs are also charged [52]. Simulations predict charged GB effects on solar cell voltage and efficiency [53]. Space-charge fields undoubtedly affect carrier dynamics, and thus TRPL data. But TRPL decays are already used to characterize transport and recombination, and additional parametrization would result in models that could easily be overfit.

We proposed to use SHG as a contactless microscopic optical probe for space-charge fields in II–VI heterostructures [43] and thin-film solar cells [48]. SHG microscopy can be correlative with TRPL (see Figure 8.8(a)), where the same fs laser excitation is used to generate electron–hole pairs (observed via PL) and nonlinear optical SHG signals. Measurements on CdTe heterostructures showed that SHG is generated due to electric-field-induced second harmonics (EFISH), and EFISH sensitivity is better than 1,000 V/cm [43]. For comparison, 0.5-V band bending over the distance of 1 μm (typical depletion width for 10^{14} cm^{-3} doping) would result in the space-charge field strength $E_{SCF} \approx 5{,}000$ V/cm. The field would be stronger in more highly doped samples with shorter depletion width. Therefore, SHG/EFISH measurements have sufficient sensitivity for thin-film solar cell analysis. SHG occurs only during the fs laser excitation. Due to this property, diffraction-limited optical resolution is easier to achieve with SHG/EFISH than with TRPL microscopy [43].

In correlative 2PE TRPL and SHG/EFISH measurements, we use 1,030-nm excitation with 300-fs pulses. This results in SHG at 515 nm, and space-charge field measurement is performed using 515-nm optical signals. In such a configuration, SHG/EFISH is a near-interface contactless and nondestructive optical probe: absorption coefficient (for CdTe) $\alpha_{515\text{nm}} = 7.7 \times 10^4$ cm^{-1} and SHG analysis depth $1/\alpha_{515\text{ nm}} = 0.13$ μm. Such near-interface EFISH characterization can be used to differentiate defect and field-effect passivation mechanisms [19].

Data from correlative 2PE TRPL and SHG/EFISH measurements on CdTe heterostructures are illustrated in Figure 8.11. Such CdTe/ZnTe/Si heterostructures have high threading-dislocation density (10^7 cm^{-2}), and CL can resolve individual

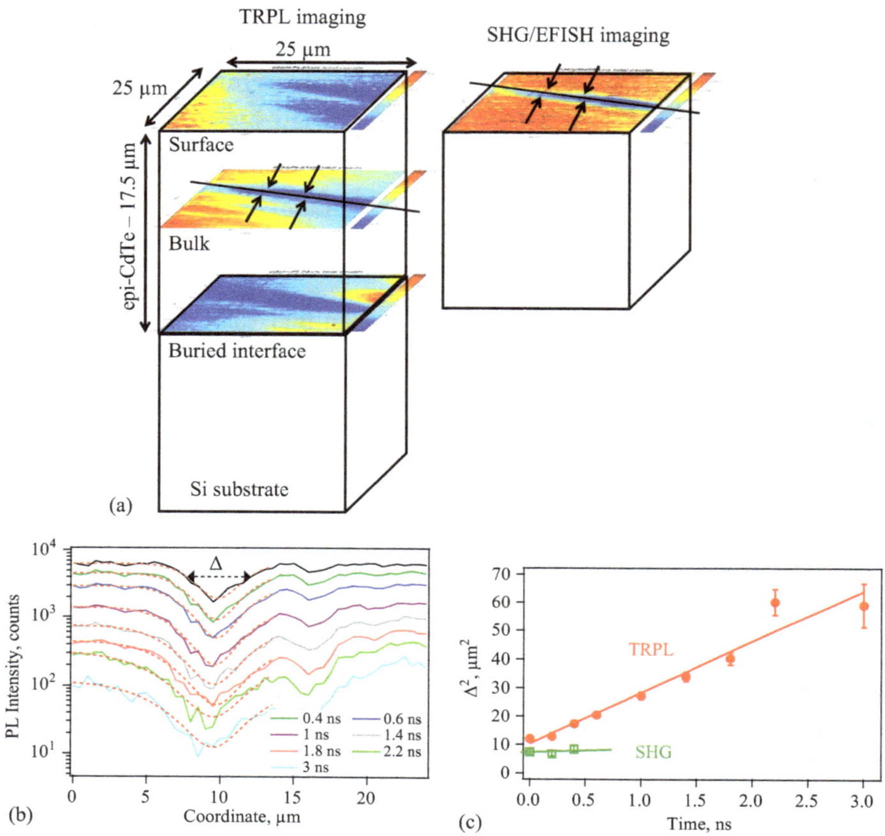

Figure 8.11 (a) Correlative 2PE TRPL and SHG/EFISH imaging for CdTe/ZnTe/Si heterostructure. (b) Time-dependent 2PE TRPL intensity profiles for the defect feature indicated by arrows in (a). Gaussian fits are used to estimate time-dependent changes in the FWHM, Δ. (c) The PL width Δ plotted vs. time indicates diffusion with $D = 16.3 \pm 1.5$ cm²/s. The width of the SHG/EFISH signal is constant because it does not depend on the carrier dynamics

extended defects [38]. In addition to extended defects that lead to short bulk lifetime, PL and SHG data in Figure 8.11(a) indicate some epilayer damage. This damage can be more clearly seen in the PL data measured in the middle of the epilayer. Carrier diffusion in the directions indicated by arrows increases the impact of recombination. This can be described using Gaussian fits to the time-dependent PL intensity profiles in Figure 8.11(b). The FWHM for the Gaussian function, Δ, increases over time. For diffusion, a linear Δ^2 vs. time dependence is expected, and it is observed in Figure 8.11(c). The fit for the TRPL data shows $D = 16.3 \pm 1.5\ \text{cm}^2/\text{s}$ and $\mu = 630 \pm 60\ \text{cm}^2/(\text{Vs})$. Such transport characteristics are in good agreement with the data in Figure 8.8(b), where the model of (8.7) was applied to the same samples. The FWHM for the EFISH signal $\Delta^2 = 9\ \mu m^2$ is similar to that for PL at $t = 0$ ns. Since SHG is generated only during the fs laser excitation, Δ(SHG) does not increase over time.

SHG/EFISH intensity for as-grown (not passivated) CdTe surface is similar with PL intensity (300-fs pulses were used, and SHG depends on the laser peak power, which is the function of the laser pulse length). In contrast, for CdMgTe/CdTe heterostructures with interface defect passivation, EFISH signals are not generated [19]. This is attributed to substantially reduced strength of the near-interface space-charge field. Therefore, correlative TRPL/EFISH microscopy provides interface recombination velocities from lifetime data and the field strength from the SHG data—a comprehensive dataset for PV interface passivation studies.

8.6 Carrier lifetimes, recombination velocities, and diffusion lengths for CdTe, CdSeTe, CIGS, kesterites, and perovskites

To compare thin-film PV absorbers, Table 8.3 summarizes some EO characteristics. Since carrier lifetimes and mobilities depend on injection and other experimental conditions, it is easier to evaluate results from similar measurements. Therefore, this summary is not comprehensive and includes only results from our laboratory. Absorbers include single-crystal and polycrystalline CdTe, polycrystalline CdSeTe, CIGS, kesterites, and perovskites. Semiconductors in Table 8.3 have similar doping, $N_A \approx 10^{14}$ cm^{-3}, with exceptions being CIGS and kesterites with $N_A \geq 10^{15}$ cm^{-3}.

Comparison of the small-area solar cell efficiency and voltage deficit from the study by Green *et al.* [56] shows that short carrier lifetimes lead to larger voltage deficits (e.g., kesterites). Therefore, SRH recombination lifetime is perhaps the single most important EO characteristic for thin-film PV absorbers. Positive correlations between τ_{TRPL} and device performance are observed for kesterites [8], CIGS [8], and CdTe [27,28]. An additional important metric is radiative efficiency, which is still low for most thin-film PV absorbers [54]. High τ_{TRPL} and L_d do not necessarily predict high radiative efficiency (see later). The implication of this difference could be more fully explored in fundamental device analysis models [55].

For perovskites and passivated CdSeTe, both of which are not intentionally doped, TRPL lifetimes can be in the 100–1,000-ns range. For direct-bandgap semiconductors in Table 8.3 (and also GaAs), radiative recombination coefficient $B = (1\text{–}2) \times 10^{-10}$ cm^{-3}, and with $N_A < 10^{14}$ cm^{-3}, the radiative lifetime $\tau_R >$ 50–100 μs (8.2). This means that $\tau_{TRPL} \approx 1{,}000$ ns is limited by SRH recombination due to bulk and interface defects, and internal radiative efficiency is <1%. It is perhaps unlikely that SRH center density in polycrystalline materials can be reduced to obtain 50-μs free-carrier lifetimes; therefore, higher radiative efficiency could perhaps be more easily obtained with higher doping. For example, for CIGS

Table 8.3 Some EO characteristics of thin-film solar cell absorbers

	Single-crystal CdTe	**Polycrystalline CdTe**	**CdSeTe**	**Cu(In,Ga)Se$_2$**	**Kesterites**	**Perovskites**
Efficiency and voltage deficit, V[a]		22.1% 0.53		22.9% 0.36 (1.1 eV)	12.6% 0.62	23.7% 0.42
Charge-carrier lifetime τ_{TRPL}, ns	Radiative	30	800	400 (E_g = 1.1 eV) 2 (E_g = 1.5 eV)	1–10	>1,000
Mobility μ, cm^2/(Vs)	600–800	50–100	20–100	20–200		20–100
Diffusion length L_d, μm	20	2.5	10	6 (E_g = 1.1 eV) 0.1 (E_g = 1.5 eV)		15
S_{int}, cm/s (passivation)	<100		<100	10^3		<10
S_{int}, cm/s (as-grown)	>10^5	>10^5	>10^5	10^3		<100

[a]From Green *et al.* [56]. Voltage deficit is estimated as $E_g/e - V_{OC}$, where V_{OC} is open-circuit voltage. EO characteristics are representative of the "best" samples, but they do not include world-record solar cells.

with $\tau_{TRPL} = 400$ ns and $N_A \approx 10^{15}$ cm^{-3}, internal radiative efficiency has the highest value of about 8% (Table 8.3). TRPL lifetimes >100 ns are only observed for CIGS with front- and back-graded absorbers, suggesting lower bulk SRH recombination-center density than the interface defect density.

An exception to carrier lifetimes limited by SRH recombination is P-doped single-crystal CdTe, where (despite $S_{int} > 10^5$ cm/s) $\tau_{TRPL} \approx \tau_{rad}$ in the crystal bulk [57]. As-doped single-crystal CdTe also has radiative efficiency that can exceed 10% [50]. These results are enabled by intentional doping and high dopant activation [50] because inactivated dopants can be SRH recombination centers [49].

For semiconductors listed in Table 8.3, mobility is the highest in undoped single-crystal and epitaxial CdTe. This includes samples with TRPL lifetimes 560 [43] and 2.2 ns [38]. For compensated single-crystal CdTe with $\tau_B = 670$ ns, mobility is somewhat lower [58]. Therefore, a high SRH recombination-center density does not necessarily reduce mobility, and high τ_B does not directly indicate high mobility. In polycrystalline absorbers, mobility is reduced from the single-crystal values due to defect scattering. It is likely that GB scattering and potentials reduce mobility. Since in some samples grain diameter is comparable with L_d (e.g., perovskites in Figure 8.9), there are other scattering centers in addition to GBs. Mobility is comparable for all polycrystalline absorbers in Table 8.3, despite bulk carrier lifetime difference by an order of magnitude. Exception is kesterites, where lower mobility can be attributed to cation disorder and potential fluctuations [59].

Diffusion length L_d combines lifetime and mobility, with possible improvements in both characteristics. Since scattering and recombination centers are likely different, increased lifetimes will not necessarily lead to higher mobilities.

In as-grown semiconductors in Table 8.3, S_{int} for the free surface differs by three orders of magnitude. Surface recombination velocity is the lowest in perovskites (10^2 cm/s), has intermediate values for CIGS (10^3 cm/s), and is high for CdTe/CdSeTe (>10^5 cm/s). In this aspect, CdTe is similar to GaAs [60]. (Only the free CIGS surface has some passivation, $S_{int} > 10^5$ at the CIGS/metal interface [35].) More research is needed to determine SRH recombination-center defects at PV absorber surfaces and interfaces. In addition to defects, it is likely that near-interface space-charge fields are important for interface recombination and also for passivation.

Interface recombination velocity in devices can be different from that for free absorber surfaces. Such passivation occurs due to buffers used in device structures, and also due to intentionally applied passivation. For single-crystal CdTe, near-lattice-matched CdMgTe provides defect passivation similar to that for GaAs [6,43], but this approach was not transferred to polycrystalline absorbers. For polycrystalline CdSeTe, Al_2O_3 passivation enables $S_{int} < 100$ cm/s [19]. SHG/EFISH data suggest field-effect passivation, but more analysis is needed. It is likely that polycrystalline CdTe passivation with Al_2O_3 is similar, but in this case absorbers do not have sufficient τ_B to evaluate S_{int} [61]. Alumina in DHs is insulating, so different device architectures are required if Al_2O_3 passivation will be used in solar cells.

It appears that S_{int} is similar for the free CIGS surface and for CdS/CIGS interface [34,62]. Instead of defect passivation, absorber grading provides effective passivation in CIGS films and solar cells. For perovskites, surface passivation enables internal radiative efficiencies approaching and exceeding 90% [63]. However, such passivation can lead to carrier trapping and low L_d [64], which illustrates complex interdependence of absorber EO characteristics.

In summary, PV material and device research has progressed from a search of empirical correlations that lead to efficiency improvements to advances based on detailed modeling and analysis [32,49]. Advanced EO measurements enable the development and verification of such models.

Acknowledgments

We thank collaborators acknowledged in the references, especially Patricia C. Dippo, Ana Kanevce, Jian V. Li, and John Moseley. This work is funded by the U. S. Department of Energy's Office of Energy Efficiency and Renewable Energy (EERE) under Solar Energy Technologies Office (SETO) Agreement Number 34350 and supported by the U.S. Department of Energy under Contract No. DE-AC36-08-GO28308 with the National Renewable Energy Laboratory.

References

[1] Ma, J., Kuciauskas, D., Albin, D., *et al.*, "Dependence of the minority-carrier lifetime on the stoichiometry of CdTe using time-resolved photo-luminescence and first-principles calculations," *Phys. Rev. Lett.* **111**(6), 067402 (2013).

[2] Duggal, A. R., Shiang, J. J., Huber, W. H., and Halverson, A. F. (First Solar Inc.), "Photovoltaic devices and method of making," US Patent Application 13/923,644 (2013).

[3] Kephart, J. M., "Optimization of the front contact to minimize short-circuit current losses in CdTe thin-film solar cells," Ph.D. Thesis, Colorado State University, Fort Collins (2015).

[4] Kuciauskas, D. and Krasikov, D., "Spectroscopic and microscopic defect and carrier-lifetime analysis in cadmium telluride," *IEEE J. Photovoltaics* **8** (6), 1754–1760 (2018).

[5] Steiner, M. A., Geisz, J. F., Garcia, I., *et al.*, "Effects of internal luminescence and internal optics on V_{oc} and J_{sc} of III–V solar cells," *IEEE J. Photovoltaics* **3**(4), 1437–1442 (2013).

[6] Olson, J. M., Ahrenkiel, R. K., Dunlavy, D. J., Keyes, B., and Kibbler, A. E., "Ultralow recombination velocity at $Ga_{0.5}In_{0.5}P$/GaAs heterointerfaces," *Appl. Phys. Lett.* **55**(12), 1208–1210 (1989).

[7] Unold, T. and Gutay, L., "Photoluminescence analysis of thin film solar cells," in *Advanced Characterization Techniques for Thin Film Solar Cells*,

Editors Abou-Ras, D., Kirchartz, T., and Rau, U., Weinheim, Germany, Wiley-VCH, pp. 151–175 (2011).

[8] Repins, I. L., Egaas, B., Mansfield, L. M., *et al.*, "Fiber-fed time-resolved photoluminescence for reduced process feedback time on thin-film photovoltaics," *Rev. Sci. Instrum.* **86**(1), 013907 (2015).

[9] Kuciauskas, D., Kanevce, A., Burst, J. M., *et al.*, "Minority carrier lifetime analysis in the bulk of thin-film absorbers using subbandgap (two-photon) excitation," *IEEE J. Photovoltaics* **3**(4), 1319–1324 (2013).

[10] Becker, W. *Advanced Time-Correlated Single Photon Counting Techniques*, Springer (2005).

[11] Wahl, M. "Modern TCSPC electronics: Principles and acquisition modes," in *Advanced Photon Counting, Applications, Methods, Instrumentation*, Editors Kapusta, P., Wahl, M., and Erdmann, R., Heidelberg, Germany, Springer (2015).

[12] Kollner, M. and Wolfrum, J., "How many photons are necessary for fluorescence lifetime measurements?" *Chem. Phys. Lett.* **200**(1–2), 199–204 (1992).

[13] Kuciauskas, D., Duenow, J. N., Li, J. V, Young, M. R., Dippo, P., and Levi, D. H., "Time resolved photoluminescence spectrometer for thin-film absorber characterization and analysis of TRPL data for CdS/CdTe interface," *Conf. Rec. 38th IEEE Photovolt. Spec. Conf.* (June), 001721–001726 (2012).

[14] Motz, A. A., Squier, J., Kuciauskas, D., Johnston, S., Kanevce, A., and Levi, D., "Development of two-photon excitation time-resolved photoluminescence microscopy for lifetime and defect imaging in thin film photovoltaic materials and devices," *Conf. Rec. 42nd IEEE Photovolt. Spec. Conf.*, DOI: 10.1109/PVSC.2015.7356388 (2015).

[15] Repins, I. and Kuciauskas, D., "Photoluminescence-based quality control for thin film absorber layers of photovoltaic devices," US Patent 9,075,012 (2015).

[16] Kanevce, A., Reese, M. O., Barnes, T. M., Jensen, S. A., and Metzger, W. K., "The roles of carrier concentration and interface, bulk, and grain-boundary recombination for 25% efficient CdTe solar cells," *J. Appl. Phys.* **121**(21), 214506 (2017).

[17] R. K. Ahrenkiel, "Minority carrier lifetime in III–V semiconductors," in *Semiconductors and Semimetals*, vol. 39, New York, NY: Academic Press, pp. 39–150 (1993).

[18] Ahrenkiel, R.K. and Johnston, S.W., "An optical technique for measuring surface recombination velocity," *Solar Energy Mat. Solar Cells*, **93**, 645–649 (2009).

[19] Kuciauskas, D., Kephart, J.M., Moseley, J., Metzger, W.K., Sampath, W.S., and Dippo, P., "Recombination velocity less than 100 cm/s at polycrystalline Al_2O_3/CdSeTe interfaces," *Appl. Phys. Lett.* **112**(26), 263901 (2018).

[20] Kanevce, A., Levi, D. H., and Kuciauskas, D., "The role of drift, diffusion, and recombination in time-resolved photoluminescence of CdTe solar cells determined through numerical simulation," *Prog. Photovoltaics Res. Appl.* **22**(11), 1138–1146 (2014).

[21] Kanevce, A., Kuciauskas, D., Levi, D. H., Allende Motz, A. M., and Johnston, S. W., "Two dimensional numerical simulations of carrier dynamics during time-resolved photoluminescence decays in two-photon microscopy measurements in semiconductors," *J. Appl. Phys.* **118**(4), 045709 (2015).

[22] Maiberg, M., "Numerical and analytical study of time-resolved luminescence in thin-film semiconductors," Dissertation, Martin Luther-Universität, Halle-Wittenberg (2016).

[23] Hsiao, K.-J, "Electron-reflector strategy for CdTe thin-film solar cells," Ph. D. Thesis, Colorado State University, Fort Collins (2010).

[24] Kuciauskas, D., Malik, R., and Steiner, M. A., "Buried interface and luminescent coupling analysis with time-resolved two-photon excitation microscopy in II–VI and III–V semiconductor heterostructures," *Physics, Simulation, Photonic Eng. Photovolt. Devices VIII 1091307(February)*, Editors Freundlich, A., Sugiyama, M., and Lombez, L., vol. 9, SPIE, Bellingham, WA, USA (2019).

[25] Serrels, K. A., Ramsay, E., Dalgarno, P. A., *et al.*, "Solid immersion lens applications for nanophotonic devices," *J. Nanophotonics* **2**(1), 021854 (2008).

[26] Kuciauskas, D., Kanevce, A., Duenow, J. N., *et al.*, "Spectrally and time resolved photoluminescence analysis of the CdS/CdTe interface in thin-film photovoltaic solar cells," *Appl. Phys. Lett.* **102**(17), 173902 (2013).

[27] Metzger, W. K., Albin, D., Levi, D., *et al.*, "Time-resolved photoluminescence studies of CdTe solar cells," *J. Appl. Phys.* **94**(5), 3549–3555 (2003).

[28] Kuciauskas, D., Dippo, P., Zhao, Z., *et al.*, "Recombination analysis in cadmium telluride photovoltaic solar cells with photoluminescence spectroscopy," *IEEE J. Photovoltaics* **6**(1), 313–318 (2016).

[29] Munshi, A. H., Kephart, J., Abbas, A., *et al.*, "Polycrystalline CdSeTe/CdTe absorber cells with 28 mA/cm^2 short-circuit current," *IEEE J. Photovoltaics* **8**(1), 310–314 (2018).

[30] Amarasinghe, M., Colegrove, E., Moseley, J., *et al.*, "Obtaining large columnar CdTe grains and long lifetime on nanocrystalline CdSe, MgZnO, or CdS layers," *Adv. Energy Mater.* **8**(11), 1702666 (2018).

[31] Fiducia, T. A. M., Mendis, B. G., Li, K., *et al.*, "Understanding the role of selenium in defect passivation for highly efficient selenium-alloyed cadmium telluride solar cells," *Nat. Energy* **4**, 504–511 (2019).

[32] Sankin, I. and Krasikov, D., "Kinetic simulations of Cu doping in chlorinated CdSeTe PV absorbers," *Phys. Status Solidi* **216**(15), 1800887 (2019).

[33] Yang, J. and Wei, S.-H, "Enhancing CdTe solar cell performance by reducing the 'ideal' bandgap of CdTe through $CdTe_{1-x}Se_x$ alloying," *Chinese Physics B* **28**, 08606 (2019)

[34] Li, J. V, Mansfield, L. M., Egaas, B., and Ramanathan, K., "Electrical properties of CdS/Cu(In,Ga)Se_2 interface," *Jpn. J. Appl. Phys.* **57**(8), 085701 (2018).

[35] Weiss, T. P., Bissig, B., Feurer, T., Carron, R., Buecheler, S., and Tiwari, A. N., "Bulk and surface recombination properties in thin film semiconductors with different surface treatments from time-resolved photoluminescence measurements," *Sci. Rep.* **9**(1), 5385 (2019).

[36] Kuciauskas, D., Li, J. V, Contreras, M. A., *et al.*, "Charge carrier dynamics and recombination in graded band gap $CuIn_{1-x}Ga_xSe_2$ polycrystalline thin-film photovoltaic solar cell absorbers," *J. Appl. Phys.* **114**(15), 154505 (2013).

[37] Kuciauskas, D., Li, J. V, Kanevce, A., *et al.*, "Charge-carrier dynamics in polycrystalline thin-film $CuIn_{1-x}Ga_xSe_2$ photovoltaic devices after pulsed laser excitation: Interface and space-charge region analysis," *J. Appl. Phys.* **117**(May), 185102 (2015).

[38] Kuciauskas, D., Farrell, S., Dippo, P., *et al.*, "Charge-carrier transport and recombination in heteroepitaxial CdTe," *J. Appl. Phys.* **116**(12), 123108 (2014).

[39] Gaury, B. and Haney, P. M., "Probing surface recombination velocities in semiconductors using two-photon microscopy," *J. Appl. Phys.* **119**(12), 125105 (2016).

[40] Yang, M., Zeng, Y., Li, Z., *et al.*, "Do grain boundaries dominate non-radiative recombination in $CH_3NH_3PbI_3$ perovskite thin films?" *Phys. Chem. Chem. Phys.* **19**, 5043–5050 (2017).

[41] Adhyaksa, G. W. P., Brittman, S., Āboliņš, H., *et al.*, "Understanding detrimental and beneficial grain boundary effects in halide perovskites," *Adv. Mater.* **30**(52), 1804792 (2018).

[42] Kuciauskas, D., Wernsing, K., Jensen, S. A., *et al.*, "Analysis of recombination in CdTe heterostructures with time-resolved two-photon excitation microscopy," *IEEE J. Photovoltaics* **6**(6), 1581–1586 (2016).

[43] Kuciauskas, D., Myers, T. H., Barnes, T. M., Jensen, S. A., and Allende Motz, A. M., "Time-resolved correlative optical microscopy of charge-carrier transport, recombination, and space-charge fields in CdTe heterostructures," *Appl. Phys. Lett.* **110**(8), 083905 (2017).

[44] Fluegel, B., Alberi, K., DiNezza, M. J., Liu, S., Zhang, Y.-H., and Mascarenhas, A., "Carrier decay and diffusion dynamics in single-crystalline CdTe as seen via microphotoluminescence," *Phys. Rev. Appl.* **2**(3), 034010 (2014).

[45] Alberi, K., Fluegel, B., Moutinho, H., Dhere, R. G., Li, J. V., and Mascarenhas, A., "Measuring long-range carrier diffusion across multiple grains in polycrystalline semiconductors by photoluminescence imaging," *Nat. Commun.* **4**(6), 2699 (2013).

[46] Moseley, J., Rale, P., Collin, S., *et al.*, "Luminescence methodology to determine grain-boundary, grain-interior, and surface recombination in thin-film solar cells," *J. Appl. Phys.* **124**(11), 113104 (2018).

[47] Moseley, J., Metzger, W. K., Moutinho, H. R., *et al.*, "Recombination by grain-boundary type in CdTe," *J. Appl. Phys.* **118**(2), 025702 (2015).

[48] Kuciauskas, D., Lu, D., Grover, S., Xiong, G., and Gloeckler, M., "Separating grain-boundary and bulk recombination with time-resolved photoluminescence microscopy," *Appl. Phys. Lett.* **111**(23), 233902 (2017).

[49] Krasikov, D. and Sankin, I., "Beyond thermodynamic defect models: A kinetic simulation of arsenic activation in CdTe," *Phys. Rev. Mater.* **2**(10), 103803 (2018).
[50] A. Nagaoka, K. Nishioka, K. Yoshinho, D. Kuciauskas, and M. A. Scarpulla, "Arsenic doped Cd-rich CdTe: Equilibrium doping limit and long lifetime for high open-circuit voltage solar cells greater than 900 mV," *Appl. Phys. Exp.*, **12**(8), 089401 (2019).
[51] Niles, D. W., Li, X., Sheldon, P., and Höchst, H., "A photoemission determination of the band diagram of the Te/CdTe interface," *J. Appl. Phys.* **77** (9), 4489 (1995).
[52] Li, C., Wu, Y., Pennycook, T. J., *et al.*, "Carrier separation at dislocation pairs in CdTe," *Phys. Rev. Lett.* **111**(9), 096403 (2013).
[53] Gaury, B. and Haney, P. M., "Quantitative theory of the grain boundary impact on the open-circuit voltage of polycrystalline solar cells," *ACS Appl. Energy Mater.* **2**(1), 144–151 (2019).
[54] Green, M. A., "Radiative efficiency of state-of-the-art photovoltaic cells," *Prog. Photovoltaics Res. Appl.* **20**(4), 472–476 (2012).
[55] Kirchartz, T. and Rau, U., "What makes a good solar cell?" *Adv. Energy Mater.* **8**(28), 1703385 (2018).
[56] Green, M. A., Hishikawa, Y., Dunlop, E. D., *et al.*, "Solar cell efficiency tables (Version 53)," *Prog. Photovoltaics Res. Appl.* **27**(1), 3–12 (2019).
[57] Burst, J. M., Duenow, J. N., Albin, D. S., *et al.*, "CdTe solar cells with open-circuit voltage breaking the 1 V barrier," *Nat. Energy* **1**(3), 16015 (2016).
[58] Ščajev, P., Miasojedovas, S., Mekys, A., *et al.*, "Excitation-dependent carrier lifetime and diffusion length in bulk CdTe determined by time-resolved optical pump-probe techniques," *J. Appl. Phys.* **123**(2), 025704 (2018).
[59] M. Grossberg, J. Krustok, C. H. Hages, *et al.*, "The electrical and optical properties of kesterites", *J. Phys. Energy*, **1**(4), 044002 (2019)
[60] Cadiz, F., Paget, D., Rowe, A. C. H., *et al.*, "Surface recombination in doped semiconductors: Effect of light excitation power and of surface passivation," *J. Appl. Phys.* **114**(10), 103711 (2013).
[61] Kephart, J. M., Kindvall, A., Williams, D., *et al.*, "Sputter-deposited oxides for interface passivation of CdTe photovoltaics," *IEEE J. Photovoltaics* **8**(2), 587–593 (2018).
[62] Metzger, W. K., Repins, I. L., and Contreras, M. A., "Long lifetimes in high-efficiency Cu(In,Ga)Se_2 solar cells," *Appl. Phys. Lett.* **93**(2), 022110 (2008).
[63] Braly, I. L., Dequilettes, D. W., Pazos-Outón, L. M., *et al.*, "Hybrid perovskite films approaching the radiative limit with over 90% photoluminescence quantum efficiency," *Nat. Photonics* **12**(6), 355–361 (2018).
[64] Stoddard, R. J., Eickemeyer, F. T., Katahara, J. K., and Hillhouse, H. W., "Correlation between photoluminescence and carrier transport and a simple in situ passivation method for high-bandgap hybrid perovskites," *J. Phys. Chem. Lett.* **8**(14), 3289–3298 (2017).

Chapter 9

Fundamentals of electrical material and device spectroscopies applied to thin-film polycrystalline chalcogenide solar cells

Michael A. Scarpulla[1]

9.1 Introduction

This chapter examines the fundamentals of characterization methods using AC-modulated electrical excitation and measurement of the complex-valued response as applied to chalcogenide polycrystalline thin-film solar cells. Capacitance voltage (CV), drive-level capacitance profiling (DLCP), admittance spectroscopy (AS), deep-level transient spectroscopy (DLTS) and its variants, so-called transient photocurrent (TPI) and transient photocapacitance (TPC), and deep-level optical spectroscopy (DLOS) are the members of this class sometimes referred to as capacitance-based techniques. The primary desirable trait of capacitance measurements compared to others is their ability, in ideal cases, to accurately quantify charge and thus densities of states as functions of space and energy. However, other material and device features may produce similar signals; and only in very restricted cases (such as an epitaxial pn homojunction with two Ohmic contacts) are traps or defects in the depletion width the only possible origin of a defect-like signature. The applicability of assumptions of capacitance spectroscopies and interpretations developed for single-crystalline and epitaxial materials and devices are examined in the context of today's chalcogenide thin-film photovoltaic (PV) technologies, which present both material and band-diagram nonidealities in this regard.

This is by no means the first review of such techniques [1], nor of their application to chalcogenide thin-film solar materials and devices. The differentiating foci of this chapter are (1) a centralized overview of the fundamentals of such techniques and their relations to capacitance concepts from other semiconductor device and solar cell fields, (2) an examination of interpretations of results in the context of thin-film polycrystalline chalcogenide devices, which do not satisfy the same assumptions appropriate for single-crystalline or homojunction

[1]Departments of Materials Science & Engineering and Electrical & Computer Engineering, University of Utah, Salt Lake City, UT, USA

devices. As a result, it is suggested that the thin-film PV research community should reconceptualize these techniques as "material and device spectroscopies" rather than "junction spectroscopies" or "defect spectroscopies." Such a change in conceptualization may, it is hoped, bring these powerful techniques back to their roots as real experiments yielding results requiring rigorous elimination of alterative hypotheses before conclusions are drawn—as opposed to routine characterization techniques guaranteed to measure parameters of point defects no matter the situation. In the same process of broadening our understanding of these techniques applied to thin-film polycrystalline solar cells, it is hoped that opportunities for broadening the scope of information and insights into materials and devices extractable from these techniques will be realized.

9.2 Fundamentals of current, voltage, admittance, and impedance

The primary topology of the current path in any solar cell is the series connection of the absorber layer and its carrier-separating contacts. In polycrystalline thin-film devices, this is at least true within each absorber grain of a polycrystalline device if it spans the full thickness. Contact and spreading resistances and effects of cabling (especially resistance and inductance) are further added in series. The propagation of the signal through a semiconductor device and response of each subregion can be represented by the usual time-dependent partial-differential continuity and Poisson equations for the quasi-Fermi levels (QFLs) and electric potential. The highest level of computation of responses comes from explicitly solving these time-dependent equations [2]. One step of simplification uses expressions for small signal linear response derived from them and computing the numerical simulation of device responses [3,4]. Also, automatic networks of lumped elements can be generated [5,6]. Simpler models can capture essential features—first-multiple conductor transmission-line representations of these equations with each segment representing an infinitesimal spatial unit are perhaps the most direct and exact models [4,7–10]. Even further simplification is achieved by assumed lumped-element equivalent-circuit models, which is perhaps the most widely used approach [11].

Physically, the primary difference between transmission-line representations and lumped-element equivalent-circuit models is that the latter neglects propagation delays, which may or may not be significant depending on frequency, system-response frequency, and system size. In larger cells, full modules, and arrays, such effects are critical to understanding, and time-domain techniques can allow spatial identification of the elements connected in series [12]. The through-thickness direction of solar cells is sufficiently small that, for frequencies up to about 1 MHz, propagation delays and transmission-line effects are negligible and analysis within lumped-element equivalent circuits is sufficient. At typical transmission-line propagation phase velocities near $2/3c$ (with c as vacuum light speed), a sine-wave signal will advance 200 m within 1 μs of one cycle of a 1 MHz signal; thus, propagation across a 1 cm device would appear to impart a phase delay of only 50 ps, or $\pi/10^4$ radians. However, the lateral voltage spreading velocity may be slowed by high capacitance or spreading resistance. Thus, consideration of transmission-line effects in the lateral direction may be required, especially for devices with large lateral dimensions. A study of the effects of cell area on capacitance–frequency (*C–f*) data from $CuInSe_2$ solar cells by Kneisel *et al.* showed that the spreading resistance in their ZnO layers imparted changes in the cutoff frequency of the data (and thus, smaller cells could yield data over a larger dynamic range of frequencies) [13]. It is unclear, however, whether or not the transmission-line approach could have been replaced in this study by a lumped spreading resistance to obtain the same conclusions. Nardone and Karpov introduced a lateral decay length in thin-film solar cell diodes to determine what portions of a device are being probed [14].

For sinusoidal voltage input $\tilde{V} = V_{AC}e^{-i\omega t}$ with frequency ω, the linear response is given by the transfer function's impedance Z or admittance Y:

$$\tilde{Z}(\omega) = \tilde{V}(\omega)/\tilde{I}(\omega) = 1/\tilde{Y}(\omega) \qquad \tilde{Y}(\omega) = \tilde{I}(\omega)/\tilde{V}(\omega) = 1/\tilde{Z}(\omega), \qquad (9.1)$$

which are the complex generalizations, respectively, of resistance and conductance and which allow for the possibility of out-of-phase current response $\tilde{I} = I_{AC}e^{-i\omega t+\delta}$. For a single capacitor, positive capacitance appears as $\text{Im}(\tilde{Z}) < 0$ or $\text{Im}(\tilde{Y}) > 0$. Incidentally, the nonlinear second harmonic signal from a diode may also be used in an alternative depth-profiling method for doping [15], and the nonlinear response to V_{AC} at the fundamental frequency provides the basis for DLCP [16–20].

A solid foundation in admittance/impedance/capacitance measurement techniques and analysis can be found in books and handbooks [11,21–25] and references considering the response of inorganic semiconductor diodes and solar cells [26–34]. The resistance R or conductance G (=1/R) relate the in-phase, real, instantaneous current response to $\tilde{V}$. Capacitance C or inductance L are reactive loads, and the impedances and admittances R, C, and L passive circuit elements are given by:

$$\begin{aligned} \tilde{Z}_R(\omega) &= R & \tilde{Y}_R(\omega) &= G = 1/R \\ \tilde{Z}_C(\omega) &= X_C = -j/\omega C & \tilde{Y}_C(\omega) &= 1/X_C = j\omega C \\ \tilde{Z}_L(\omega) &= X_L = j\omega L & \tilde{Y}_L(\omega) &= 1/X_L = -j/\omega L \end{aligned} \qquad (9.2)$$

Impedances in series or parallel add like resistors, whereas admittances add like capacitors. In the absence of negative differential resistance, both Z and Y are constrained to the first and fourth quadrants of the Argand plane. A capacitive Z response appears in the fourth quadrant with negative imaginary component whereas inductors have positive imaginary Z component in the first quadrant. The quadrants are swapped for Y responses. Cole–Cole plots display the real and imaginary parts of $Y(\omega)$ or $Z(\omega)$ parametrically, whereas Bode plots are semi-log plots of a quantity such as capacitance vs. frequency. Representative Cole–Cole plots of

Z and Y for thin-film cells at room temperature with low series resistance are shown as functions of applied bias in Figure 9.1. The curves are very close to semi-circles, which indicates that a simplified equivalent circuit with series resistance (R_s), shunt resistance (R_{sh}), and depletion capacitance (C_{depl}) can capture most of the Y or Z behavior. However, this does not imply that fitting experimental data to a simplified model can accurately capture all details of the capacitance because small changes in Z or Y correspond to larger fractional changes in C. More in-depth discussion is contained in later sections, but these typical material and device parameter values are such that the dominant effect of the diffusion admittance is to add an effective shunt across the depletion capacitance, shrinking the semicircles and slightly distorting them.

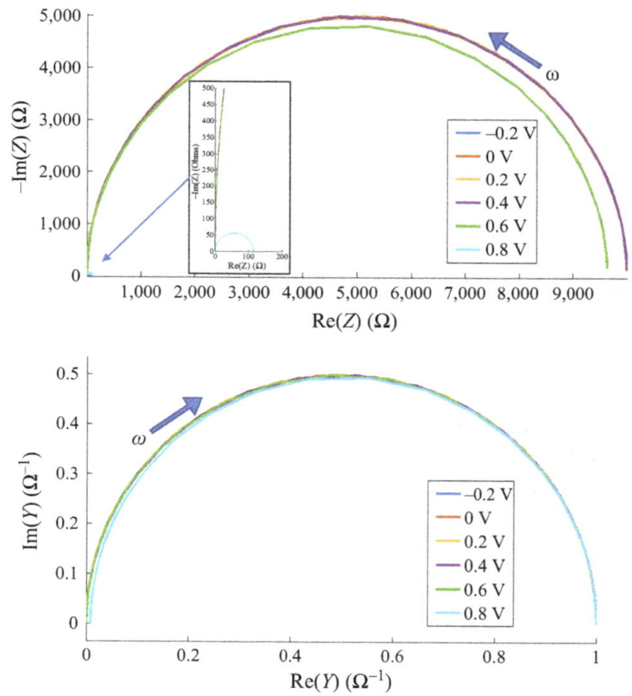

Figure 9.1 *Calculated CdTe solar cell impedance Z (top) and admittance Y (bottom) presented as Cole–Cole plots. Frequency is from 1–10^{12} Hz and increases in the directions shown by arrows. The cell was assumed to have $R_s = 1\ \Omega$, $R_{sh} = 10^4\ \Omega$ and the junction capacitance including depletion capacitance and diffusion admittance was computed for a 1-sided n^+-p junction from material parameters. The doping and lifetime on the p side were taken as 10^{16} /cm^3 and 10 ns, while on the n+ side these were 10^{18}/cm^3 and 0.1 ns, respectively*

Apparent negative capacitance has been documented in PV cells of many types, and some consensus is emerging that it may occur when both ionic and electronic processes coexist [35–38]. This highlights an observation that ionic transport and defect reactions are more the rule than the exception in many solar cells, which commonly occurs in high defect concentrations and grain boundaries (GBs), perhaps only excluding epitaxially grown ones.

In a linear system, only two linearly independent parameters can simultaneously be determined for each frequency because only the real and imaginary parts of Z or Y are independent. Example pairs include Re(Z) and Im(Z) or $|Z|$ and $\measuredangle Z$, or other derivative quantities [21,22]. Commonly, LCR meters offer many choices of output pairs common for different applications, whereas capacitance meters will typically offer to report the measured $\widetilde{Z}(\omega, T, V_{DC}, V_{AC})$ projected onto either the series or parallel RC models given by:

$$\widetilde{Z}_s = R_s + 1/j\omega C_s \qquad C_s = -1/\omega \mathrm{Im}(1/\widetilde{Z}_s)$$

$$\widetilde{Y}_p = 1/R_p + j\omega C_p = G_p + j\omega C_p \quad C_p = 1/\omega \mathrm{Im}(\widetilde{Y}_p) \tag{9.3}$$

Projection of any measured data consisting of real and imaginary components onto any model with two independent free elements can be accomplished point by point. Projection of higher-order models containing more than two elements onto 2-element models can be accomplished analytically by equating the real and imaginary parts of the impedance or admittance [33]. The result, however, is that the values of the circuit elements in the 2-element model become frequency-dependent (see (9.18)). Models with more than two free parameters must be fit to measured data over significant frequency ranges in order to determine parameter values accurately. Simultaneous fitting of the real and imaginary parts of the data is necessary, and different results and sensitivity to different model parameters will be obtained depending on the data representation in the objective function minimized (e.g., Re(Z) and Im(Z) vs. Re(Y) and Im(Y), etc.) [11].

Capacitance is related to charging and discharging of reservoirs for carriers, and phase delays are related to the imaginary component of $\widetilde{Z}(\omega)$ or $\widetilde{Y}(\omega)$. In all cases—whether at the interface between a depletion width and quasi-neutral region (QNR), at a heterointerface band offset, or at a location where a trap's energy level crosses a QFL—a capacitance arises from the flow of real and displacement currents to change the occupation of mobile carriers in some available density of states (trap-related or band-related). At low frequencies, ions and charged point defects—which are present in all thin-film PV devices (e.g., Na, Cu, Ag, Au in CdTe, CIGSe, and CZTSSe) and are especially mobile in GBs—can also add capacitance features. These phenomena are critical to understanding the response of halide perovskite devices in which emergent effects such as negative capacitance have been observed [38,39]. In chalcogenide thin-film devices, such point-defect transport- and reaction-related transients will affect low-frequency impedances as well as time-domain measurements such as slow current-voltage (I-V) sweeps and long DLTS transients. These are also implicated in many modes of long-term degradation of both thin-film and Si PV devices—for example, potential-induced degradation (PID), where Na from the module's glass becomes mobile in devices. Mobile and dynamic charged point defects and complexes that respond to light and voltage bias are the rule rather than the exception in all PV technologies.

The most fundamental measurement in the class of frequency-domain material and device spectroscopy experiments is thus the measurement of impedance or admittance at a single frequency. The response of the device in terms of changing the occupation of charge reservoirs and Ohmic carrier drift is measured at the excitation frequency. In principle, the same information can also be obtained over a bandwidth of frequencies from time-domain measurements such as transient responses to impulse or step changes in voltage or light. In the time domain, the impulse response is arguably the most fundamental; however, the response to step functions of bias or light are typically used for reasons of signal amplitude (i.e., a finite pulse time is sometimes required to fill states). Most characterization techniques add modulation of other variables to the unit measurement of $\widetilde{Y}$ or $\widetilde{Z}$ in order to determine paths or surfaces within the more general $\widetilde{Z}(\omega, t, T(t), V_{DC}(t), V_{AC}(t), \Gamma(\lambda, t))$ space defined by the response to variation of frequency (ω), time (t), temperature (T), DC bias (V_{DC}), AC amplitude (V_{AC}), and illumination spectrum $\Gamma(\lambda)$. Capacitance–voltage profiling (C–V), capacitance–frequency (C–f), and impedance or AS represent such lower-order cuts through this higher-dimensional response space.

The other variables such as T and V_{DC} must be varied at rates producing responses slower than $1/\omega$. But many orders of magnitude are available and may be used to include, exclude, and characterize slower processes such as trap capture and emission, emission over energy barriers, or other transport processes such as ionic conduction. For doped semiconductors at room temperature, the slew rate of V_{DC} is usually quasi-static compared to the signal propagation speed in the QNRs (limited by the dielectric relaxation (DR) time of the doped region)—meaning that the occupations of electrons and holes in the conduction and valance bands follow the applied bias. However, in the presence of localized states in the device, including deep dopants or band tails, large voltage slew rates may produce hysteretic responses or even suppress the contributions of these states to capacitance, thus allowing the separation of traps and shallow centers [32,33,40]. The characteristic voltage slew rate at which behaviors change will be about $k_B T \cdot e(T)$, in which $e(T)$ is the temperature-dependent trap emission rate (s^{-1}). $e(T)$ typically depends exponentially on T, so trap states can "freeze out" of the junction response, which is the basis of their characterization by many techniques. Prior to the introduction of DLTS, temperature-scanning techniques dominated the field of electrical defect spectroscopies [30,31,41]. In addition to temperature, the occupation of localized states within the bandgap can be modulated by sub-gap optical illumination with energies sufficient to cause localized-to-band transitions, which results in the changes in junction capacitance. Experiments using photoemission from trap states are discussed in a later section and include optical DLTS (ODLTS), DLOS, and the TPC/TPI used by Cohen *et al.*

9.3 Displacement/Depletion and chemical capacitances

Capacitance can be defined in a general thermodynamic sense as the storage of energy in charging reservoirs or in the electric field F between reservoirs in response to a change in bias or chemical potential of that charge carrier. The total current j appearing in Maxwell's equations within a semiconductor device can be subdivided into displacement current (j_D) and band-transport currents (j_n, j_p): $j = j_n + j_p + j_D$. The total capacitance has components relating to displacement fields—usually termed depletion capacitance—and components related to charge carriers' occupation of states—chemical capacitance.

The most familiar depletion capacitance is a mutual capacitance arising from the electric field in the region between two oppositely charged objects. For a parallel-plate capacitor of area A and gap h filled with material of dielectric constant $\varepsilon = \varepsilon_r \varepsilon_o$, the capacitance is constant with voltage and equal to

$$C_D = \frac{dQ}{dV} = \frac{\varepsilon_r \varepsilon_o A}{h}, \tag{9.4}$$

in which dQ is the magnitude of charge on each plate modulated by the voltage increment dV applied between the plates. The capacitance is constant vs. distance because no charge density exists in the space between the plates. The field exists only between the plates because they each perfectly screen it beyond. The energy can be viewed as being stored in the displacement field $\vec{D} = \varepsilon \vec{F}$ [42,43], although through the Poisson equation the fields and charges are intimately linked; thus, energy storage could also be viewed as being stored in the changes in charge on the plates. The displacement current is given by the time derivative of the displacement field

$$j_D = \frac{\partial D}{\partial t} = \frac{\partial}{\partial t}(\varepsilon F) = \frac{\partial}{\partial t}\left(\varepsilon \frac{\partial \phi}{\partial x}\right), \tag{9.5}$$

in which F is the electric field, ϕ is the electric potential, and we have allowed for a frequency-dependent dielectric constant.† In inorganic semiconductors, the low- and high-frequency limits of the dielectric constant in the regime (about 10^0–10^{12} Hz) are governed, respectively, by the defect processes and DR of the majority transport in the QNRs. The defect component can include formation, annihilation, complexation, and transport. The DR or freezeout frequency of a uniform slab $\omega_{DR,QNR} = \sigma_{QNR}/\varepsilon_{QNR}$ controls the capacitance step of the geometric capacitance. This step occurs at about 10^8–10^{12} Hz for doping in the 10^{13}–$10^{17}/\text{cm}^3$ range assuming $\varepsilon_r = 10$ and $\mu = 100\ \text{cm}^2/\text{V}\cdot\text{s}$. Section 9.8 discusses the freezeout of the junction capacitance step because of changes in QNR conductivity. For the

†In [4], the assumed low-frequency dielectric relaxation represents ionic transport and/or rotations of metal halide octahedra, leading to colossal orders-of-magnitude changes in dielectric constant that have been documented in halide perovskite materials, especially with illumination. In traditional inorganic semiconductors, including polycrystalline chalcogenides, ionic conductivity may be present but will not be as significant.

parallel-plate capacitor with constant D field, integration of j_{D} by time yields $\delta Q_{\mathrm{D}} = \delta \mathrm{D}$ and thus demonstrates that $C_D = \delta D/\delta\phi$, illustrating its intimate tie to the electric field and potential.

In the case of a depletion width in a semiconductor, the capacitance is a function of potential change across the depletion width because of the nonzero charge density between the plates and can be related to the net charge distribution using the Poisson equation [42,43]. Again, the capacitance is related to the increments of charge $\delta \mathrm{Q}$ on either side of the depletion width modulated by a small change in potential across the depletion width $\delta\phi$. The depletion capacitance becomes voltage dependent but is given by $C_{Wd}(V) = \varepsilon A/W_d(V)$. In wafer-based solar cells (i.e., Si cells), the depletion width is much smaller than the photocarrier collection length in the base QNR or the total thickness. However, in thin-film cells, W_{d} is frequently a substantial fraction of the absorber thickness h; thus, the geometric capacitance $C_{\mathrm{geo}} = \varepsilon A/h$ between the front and back contacts may not be negligible.

A prototypical one-dimensional band diagram is shown for a thin-film PV device in the top panel of Figure 9.2. A device of n^+ fluorinated tin oxide (FTO)/n MZO/p-CdTe was chosen; however, the main band-diagram features are also common to ZnO:Al/CdS/CIGSe and ZnO:Al/CdS/CZTSSe. In the bottom panel, the depletion capacitance arises between the sheets of charge density change in the n-type regions and the region at the interface between the QNR and depletion regions where the free holes change their concentration with applied V_{AC} (in this case, at about 2.5–3 μm for 100 mV V_{AC}). The relative charge densities and screening of the two differential charge sheets on either side of the MZO layer determine whether the dielectric constant and thickness of the MZO must be included in the calculation of capacitance; this physics is very similar to an MOS capacitor [44,45]. We will use C_{Wd} to denote the capacitance of a semiconductor depletion width and C_{j} to denote the total junction capacitance when other components are added.

The chemical capacitance can be considered a self-capacitance and relates to the occupation of carriers in electronic densities of states. The current carried in the conduction band by mobile electrons is driven by the spatial gradients in the electron QFL or chemical potential for conduction-band electrons

$$j_n = n\mu_n \frac{\partial E_{Fn}}{dx}, \tag{9.6}$$

where n is the electron density and μ_n is the mobility (similarly for hole density p and E_{Fp}). The carrier continuity equation

$$\frac{\partial n}{\partial t} = G - R - \frac{1}{q}\nabla \cdot \vec{j}_n, \tag{9.7}$$

in which G and R are the generation and recombination rates, respectively, allows for time-dependent local accumulation (depletion) of excess electrons above (below) the thermal equilibrium values. Similar equations hold for valence-band holes.

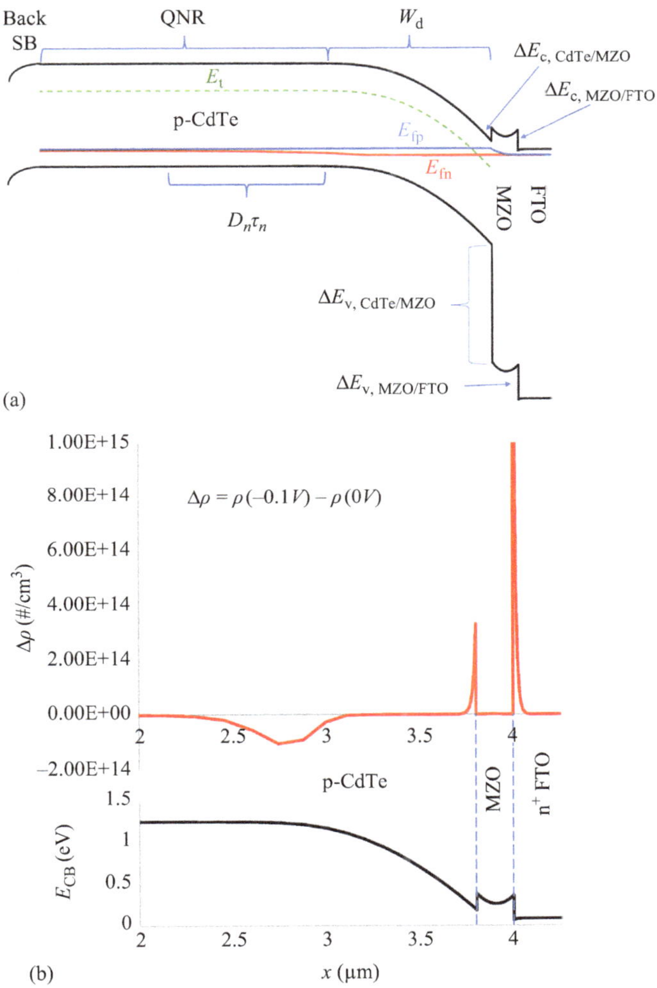

Figure 9.2 *(Top) Schematic band diagram for a thin polycrystalline PV cell represented by n^+-FTO/n-MZO/p-CdTe. The device is shown at a reverse bias of −0.1 V, so E_{Fn} is below E_{Fp} in W_d. A trap level with energy E_t is indicated in green. A back Schottky barrier (SB), QNR, and depletion width (W_d) are shown in the CdTe. The electron minority-carrier diffusion length measured from the W_d edge is also shown. The various band offsets between layers are also noted. (Bottom) Charge density differences vs. distance in the same band diagram between −0.1 and 0 V showing the wide area at the edge of the depletion width over which the free hole density changes as well as two electron charge sheets at the conduction band offsets on either side of the magnesium zinc oxide (MZO)*

The chemical capacitance describes such local charge storage in an electronic densities of states $g(E)$ [4,8,9,46,47]. It is given by

$$C_{\phi i} = \frac{\partial Q}{\partial \phi_i} = \frac{\partial}{\partial \phi_i} q \int g(E) f(E - q\phi_i) dE, \tag{9.8}$$

where $i = \mathrm{n,p}$; ϕ_i is the carrier electrochemical potential given by $q\phi_\mathrm{I} = E_\mathrm{Fn}$ or E_Fp; $g(E)$ is the density of states ($\mathrm{cm}^{-3}\ \mathrm{eV}^{-1}$), and $f(E)$ is the carrier occupation distribution. For example, the charge stored per unit volume by a density of electrons n in the conduction band's density of states is $Q = qn$. For nondegenerate cases,

$$Q = qn = q \int g(E) f(E - E_{Fn}) dE = qN_c \exp\left(-\frac{E_c - E_{Fn}}{k_B T}\right), \tag{9.9}$$

and thus,

$$C_{\mu,n} = \frac{q^2 N_c}{k_B T} \exp\left(-\frac{E_c - E_{Fn}}{k_B T}\right) = \frac{q^2 n(E_{Fn})}{k_B T}. \tag{9.10}$$

In thermal equilibrium when $E_\mathrm{Fn} = E_\mathrm{Fp} = E_\mathrm{F}$ even for doped semiconductors, this is very small. The diffusion capacitance C_diff arising at the edges of a pn junction depletion width and which increases exponentially with applied forward bias from minority carriers injected across the junction is an example of chemical capacitance. In an equivalent-circuit representation of a pn diode, it appears in parallel with the depletion capacitance. Additionally, a uniformly illuminated or infinite mobility absorber layer at open-circuit voltage (V_oc) will have a uniform distribution of excess carriers that stores energy, just as for a tank pressurized with an ideal gas. The ability of this energy to be extracted and converted to work depends on differences in carrier energy, and thus, the thermal equilibrium carrier distributions serve as the reference state for DC excitation.

The capacitance arising from a density of trap states in a depletion width, as shown in Figure 9.2 (top), is a chemical capacitance, and for low densities compared to the doping, it will appear in parallel with the junction capacitance. At high density or when there are no shallow dopants, the traps themselves can govern the junction capacitance. Under AC excitation, this capacitance arises from the change in occupation of the trap states around the point in space and energy where the dominant QFL crosses the trap level [42,48,49]. This is the case that gives junction spectroscopies their ability to measure the energetic density of states in the case of uniform spatial distribution. Experiments are arranged in which the QFLs are perturbed by light, applied bias, and temperature, and the resulting changes in capacitance are converted into density of states, usually within the assumption of uniform spatial density.

A special term "quantum capacitance" was introduced to describe capacitance arising from two-dimensional (2D) charge sheets (2DEGs) in semiconductor devices, either in the form of quantum wells (QWs) or an inversion layer that arises from filling quantized densities of states [50]. This is a specific case of the more general chemical capacitance. The term "quantum capacitance" typically refers to charge sheets occurring inside a depletion width but incapable of fully screening the electric field because of insufficient sheet charge density. 2DEGs capacitance usually occurs in series with the depletion capacitance because of charge sharing between the three or more plates of the equivalent ideal capacitors. In the context of thin-film PV devices, this general concept is relevant because inversion and accumulation sheet charges can arise at band offsets under different bias and illumination conditions—for example, at the n-CdS/p-CIGSe or even at the n^+ ZnO:Al/n-CdS interfaces as in Figure 9.2 (bottom)—and can lead to changes in measured capacitance (usually interpreted as arising from depletion in the absorber alone) because of the more complex mutual capacitances. An enumeration of the possible situations is given by Scheer in [45].

9.4 Capacitance from defect states crossing QFLs in a depletion width

Defect levels that cross a QFL in the depletion width and change their occupation in response to the AC excitation produce an additional capacitive response in the junction. The physics of optical and thermal transitions underpinning most space-charge spectroscopy techniques was laid out by Sah *et al.* and other valuable references [27–29,31,33,43,48,49]. The density of states presented by the defect level or distribution of levels can only respond within the period of the AC signal and add to the junction capacitance if its limiting capture or emission rate is sufficiently rapid (in the usual case, thermal emission to a band is the limiting rate). For steady-state DC bias in a nondegenerate region, the capture and emission (c, e) rates of electrons and holes (n,p) from a Schottky–Sah single charge-transition level appropriate for a defect with insignificant structural relaxations are given by:

$$\begin{aligned} r_{c,n} &= v_n\sigma_n N_C \exp[-(E_C - E_{Fn})/k_BT]N_t(1-f_t) \\ r_{c,p} &= v_p\sigma_p N_V \exp\left[-(E_{Fp} - E_V)/k_BT\right]N_tf_t \\ r_{e,n} &= v_n\sigma_n N_C \exp[-(E_C - E_t)/k_BT]N_tf_t \\ r_{e,p} &= v_p\sigma_p N_V \exp[-(E_t - E_V)/k_BT]N_t(1-f_t) \end{aligned}, \tag{9.11}$$

in which the carrier velocities v_n and v_p are taken as the thermal velocity—thus absorbing any field dependence of the velocity (e.g., that would be present in differently doped junctions or at different biases) into the capture cross sections σ_n and σ_p because these terms always appear as a product. N_C, N_V are the bands' effective densities of states; E_C, E_V are the band edge energies; E_t is the trap charge-transition level; E_{Fn}, E_{Fp}, are the QFLs; N_t is the number density of traps; and f_t is the electron occupation function for the trap. A defect or flaw with charge-transition level E_t acts as a trap if it re-emits a carrier before capturing one of the other type, thus leading to defect-assisted Shockley–Read–Hall recombination.

Sometimes, a demarcation energy is defined separating recombination centers from traps, but this is of limited utility because the behavior changes with trap cross sections and with bias or illumination through the QFLs. In a depletion width, all defects with E_t between the two QFLs contribute to recombination and generation. This is an important point—the case of maximum possible recombination/generation rate in a depletion width occurs for $E_t = E_i$ and $\sigma_n = \sigma_p$, both large. However, any defect with E_t between the QFLs will contribute, even if it is removed from E_i [51].

In the case of spontaneous capture of band carriers, detailed balance determines f_t to be:

$$f_t = \frac{v_n\sigma_n N_C e^{\frac{-(E_C-E_{Fn})}{k_BT}} + v_p\sigma_p N_V e^{\frac{-(E_t+E_V)}{k_BT}}}{v_n\sigma_n N_C\left(e^{\frac{-(E_C-E_{Fn})}{k_BT}} + e^{\frac{-(E_C-E_t)}{k_BT}}\right) + v_p\sigma_p N_V\left(e^{\frac{-(E_{Fp}+E_t)}{k_BT}} + e^{\frac{-(E_t+E_V)}{k_BT}}\right)} = \frac{1}{e^{\frac{(E_t-E_{Ft})}{k_BT}} + 1} \tag{9.12}$$

This occupation function differs from the normal Fermi–Dirac occupation function applicable for thermal equilibrium. It can be shown that it consists of two superimposed Fermi–Dirac distributions vs. energy centered very near E_{Fp} and E_{Fn} with amplitudes, respectively, $R/(1+R)$ and $1/(1+R)$, with $R \equiv v_p\sigma_p N_V/v_n\sigma_n N_C$. Thus, at 0 K, $f = 1$ for E_t below about E_{Fp}, $f = 1/(1+R)$ for E_t between about E_{Fp} and E_{Fn}, and $f = 0$ for E_t above about E_{Fn}. An interesting implication is that f_t collapses to the Fermi–Dirac distribution when $E_{Fn} = E_{Fp}$, thus implying that a relation exists between the degeneracy factor and capture cross sections. It has further been shown that the same description can be used for energetic distributions of traps sharing the same capture cross sections for electrons and holes [52]. A trap QFL E_{Ft} has been defined by Lutz using the second equality in (9.12), although its role as a chemical potential has not been fully evaluated theoretically [53–56]. If a capture energy barrier ζ is present, an additional factor $\exp(-\zeta/k_BT)$ multiplies the capture rates. If the same transition state applies for the emission processes, the factor would also multiply the emission process. Numerical device-physics simulations can provide more nuanced insight into trap occupations and QFLs [57], and thus, the interpretation of observed signals. It allows the local QFLs to be used to determine the occupation of the trap, for example, in the case of the filling pulse used for DLTS experiments.

During a capacitance experiment, one thermal emission rate will dominate and the trap will show a cutoff frequency ω_o that depends exponentially on temperature according to the emission rate equations above. Useful analytical results are provided by Blood [42] for the total junction capacitance C_j as a function of frequency for a one-sided p^+n or n-type Schottky junction having depletion width x_d, doping on the lower-doped n-type side N, and a density of electron trap states N_t with energy E_t such that E_t crosses the dominant QFL at a position x_1 from the interface:

$$C_j(\omega) = \frac{\varepsilon A[u(e_n) \cdot N_t(x_1) + N(x_d)]}{[u(e_n) \cdot x_1 \cdot N_t(x_1) + x_d \cdot N(x_d)]}. \tag{9.13}$$

In (9.13), A is the cross-sectional area, and the factor $u(e_n)$ is a step function for $T = 0$ smeared by Boltzmann factors for $T > 0$ that determines whether or not the trap capacitance will contribute to C_j based on the conditions on measurement frequency ω and temperature,

$$u(e_n) = \begin{cases} 0, & \omega \cdot e_n(T) > 1 \\ 1, & \omega \cdot e_n(T) < 1 \end{cases}. \tag{9.14}$$

Equation (9.13) shows that the total capacitance is related to the center of mass of the responsive portion of the charge distribution. When $u(e_n) = 0$, $C_j = C_{\mathrm{Wd}}$. The criteria of (9.14) govern whether or not a trap's contribution will be present in experiments such as CV, DLCP, and AS, which measure the trap's capacitance directly on an absolute scale. A reasonable threshold on the detectability in these techniques is thus that the trap density would be at least about 10^{-2} to 10^{-3} times the doping N. Interestingly though, DLTS experiments are capable of much higher detection sensitivities (below 10^{-6} times N) because: (a) they are differential (in the sense that they measure the change in capacitance of the depletion width resulting from a filling pulse) and (b) the trap capacitance is not measured directly (in the limit of $N_t << N$), but rather, via its influence on the depletion capacitance. In the case where N_t is significant compared to N (or even dominant in the case of non-shallow doping), constant capacitance DLTS must be used [58]. This case is highlighted because in many of the thin-film materials used as polycrystalline absorber layers (such as CdTe, CZTS, and perhaps CIGSe), dopant activation energies are rather large; and as temperature decreases, the junction capacitance can no longer be observed because of the dielectric freezeout of the junction capacitance driven by the increase in series resistance of the QNRs and contacts as the dopant ionization decreases. This case is discussed in Section 9.8.

Returning to the observability of a trap capacitance let us examine the case of AS in the typical range of frequencies from approximately 1 Hz to 1 MHz and temperatures from approximately room temperature to nitrogen or helium temperatures. A case like the one in Figure 9.2 is considered in which the band bending is such that E_t crosses E_{Fn} or E_{Fp} inside the depletion width. The crossing could also occur within the depletion width at a back SB (not shown in Figure 9.2). The total junction capacitance can fairly straightforwardly be approximated by (9.13) and (9.14). As is seen in Figure 9.3 (which was generated by representing C_{Wd}, C_{trap}, and C_{geo} in an equivalent-circuit model), a step in the junction capacitance on plots of C vs. f or C vs. T with magnitude related to N_t and E_t is added to C_{Wd} and C_{geo} [28]. The junction capacitance step will downshift in frequency with decreasing temperature according to the activation energy for e_n given by $E_c - E_t$. This is also the basis of DLTS detection of traps in the depletion width—only the

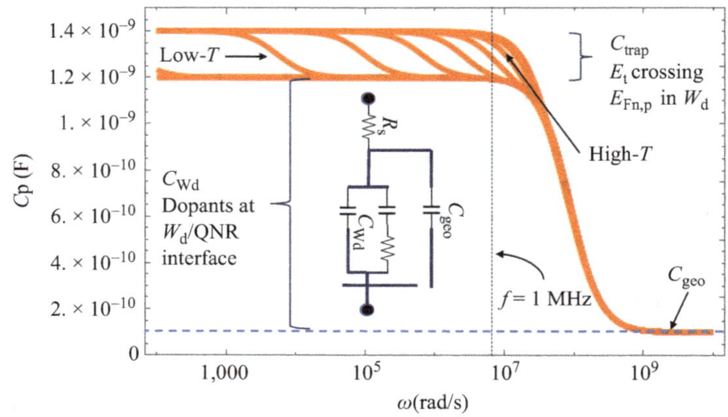

Figure 9.3 Calculated capacitance vs. frequency for a diode showing freezeout of a trap-related step with decreasing temperature. C_{trap} is added to C_{Wd} and C_{geo} as shown in the inset. Note that only the step related to C_{trap} exhibits freezeout in this case, whereas in Section 9.8, the case of dopant freezeout is considered. Also note that this trap response appears nearly identical to the measured data in Figure 9.4 in which the response originates from a band offset

frequencies associated with transients measured from ~100 μs to ~10 s span 10^{-1} to 10^4 Hz.

In terms of equivalent-circuit response, as long as x_1 does not vary greatly, the trap contribution may be represented by the series connection of R_T and C_T placed in parallel with the depletion and diffusion components. C_T must quantitatively account for the density of traps; so, for simple modeling, it is useful to determine C_T and ω_o and then compute the needed R_T from $\omega_o = 1/R_T C_T$. A simplified model assumes that ω_o follows Arrhenius behavior (with a prefactor having T^2 arising from V_{th} and N_c or N_v from capture/emission physics):

$$\omega_o = \omega_o^{'} T^n \exp\left(\frac{-\Delta E}{k_B T}\right), \tag{9.15}$$

in which ω_0' is the high-T limit and ΔE is $E_c - E_t$ for electron emission or $E_t - E_v$ for hole emission. A single monoenergetic trap will yield a sharp capacitance step consistent with a single Debye relaxation process (transition within about 1 decade of frequency), whereas an energetic distribution can yield broader steps. Figure 9.3 illustrates how a single trap level uniform vs. distance in W_d should manifest in AS as a capacitance step that downshifts in frequency as temperature decreases. It is common to define the inflection point of a capacitance step as the characteristic frequency $\omega_o(T)$ of the relevant process for Arrhenius analysis. Frequently, it is found as the maximum point of a peak in the logarithmic derivative for the data taken at different temperatures:

$$\frac{-\partial C(\omega, T)}{\partial \ln(\omega)} = \frac{-1}{\omega}\frac{\partial C(\omega, T)}{\partial \omega} = 0. \tag{9.16}$$

An energetic distribution of traps will induce a distribution of time constants, and thus, the slope of any trap capacitance step will decrease. Constant-phase elements (CPEs) are sometimes used to capture such dispersion effects. Using the CPEs to replace the trap contribution only, rather than that of the entire C_j, is more justified. If at some temperature this trap step crosses the measurement frequency (e.g., 1 MHz) used in CV, DLCP, DLTS, or other techniques, then a change in results should also be expected at that crossing temperature. It is important to note that the trap response is added to C_{Wd} and that this does not change strongly with decreasing temperature. This is the case when neither DR in the QNR nor equivalent-circuit effects from the total series resistance R_s combined with C_j affect the measurements.

It is noted that differentiation erases information about constant capacitances, and thus, *which capacitance step* $\omega_o(T)$ corresponds to. This erasing of information can lead to misinterpretation, which is rampant in at least the CZTSSe literature, where freezeout of the C_j step (which leaves C_{geo} at low temperatures and high frequencies) as discussed in Section 9.8 has been widely mislabeled as a "trap feature." To be able to differentiate between C_t and C_j showing Arrhenius behavior, it is critical that both $C_{Wd}(V_{DC})$ and C_{geo} be estimated and that the low-temperature, high-frequency constant-valued capacitance be compared. The complete trap response as a function of bias, temperature, and frequency and extraction of the trap density of states was computed by Walter *et al.* [59]. Caution is urged here: the analysis of Walter can be applied to any AS dataset and activation energies and densities of states extracted—but again, the interpretation of the results as a trap density of states depends on first eliminating other possible origins.

9.5 Capacitance from excess carriers (Diffusion capacitance)

We now discuss the effects of injected minority carriers in forward bias, which leads to the diffusion admittance Y_{diff}. The injected minority-carrier currents that diffuse and recombine adjacent to the depletion width edges in forward bias give rise to both capacitance and conductance (i.e., a low-resistivity voltage-dependent shunt) in parallel with C_{Wd}. In a small-signal model (justified when $V_{th} < V_{AC} << V_{DC}$), the diffusion admittance and its conductance and capacitance components within a parallel model are given by:

$$Y_{Diff} = \frac{I_{DC}(V_{DC})}{V_{th}}\left[\sqrt{1 + j\omega\tau_p} + \sqrt{1 + j\omega\tau_n}\right] = G_{p,Diff} + j\omega C_{p,Diff}, \tag{9.17}$$

in which $I_{DC}(V_{DC})$ is the DC diode diffusion current at V_{DC} and τ_n, τ_p are the minority-carrier lifetimes or transit times on the n and p sides of the junction. For $\omega << 1/\tau$, in which τ is the limiting minority-carrier lifetime for long diodes or the transit time in short ones with high interface recombination velocities, the diffusion capacitance acts like a G and C in parallel. At higher frequencies approaching and exceeding the inverse minority-carrier lifetimes, the impedance becomes a CPE with phase angle of 45° because of the square roots. This represents behavior in some sense half-way between resistors and capacitors. To observe such linear behavior on a Cole–Cole plot, it is also necessary that the semicircle determined by C_j with R_s and R_{sh} (e.g., Figure 9.1) has not reached the R_s limit at the relevant ω. Both components of Y_{diff} scale exponentially with voltage (the total admittance increases in magnitude); so, at sufficient forward bias, the measured apparent capacitance drops precipitously, essentially because of the diode's enormous on-state conductance which acts as an AC shunt. At high frequency and high forward bias, the diode's impedance approaches its series limits $R_s + L_s$. This is the origin of the apparent maximum in measured capacitance near the diode turn-on voltage, followed by rapid decrease in CV profiling into forward bias. In doping vs. W_d plots extracted from such CV data, this can form a divergence and apparent W_d increasing with forward bias.

The diffusion capacitance is linked to the bias across a diode; thus, in work on solar cells, it is inexorably linked to the open-circuit voltage V_{oc} and chemical potential difference of minority carriers. The same type of chemical capacitance caused by excess carriers is present in both illuminated and dark solar cells for all forward biases. The topic has been examined by several authors, including in thin-film CdTe devices [4,60,61]. The lifetime of minority carriers in Si can be in the μs to ms range, meaning that diffusion admittance may approach its high-frequency limit even for frequencies of kHz to MHz. For direct-bandgap thin-film PV materials in which lifetimes are in the ns range, it is unlikely that these effects would be seen below ~100 MHz–1 GHz, unless the effective lifetime (including grain-boundary recombination) approaches the radiative limit. High-efficiency solar cells have long lifetimes; thus, diffusion admittance should not be ignored. Similarly, the lifetime in polycrystalline thin-film cells has been very short (<10 ns) for most of their history. But nowadays, in cells with efficiency >20% for hybrid perovskites, CIGSe, and CdTe materials, bulk lifetimes can approach 1 μs [62–67]. In long-lifetime cells, it is also critical to pay attention to the V_{DC} slew rate in *I–V* and *C–V* experiments to ensure that the cell is in steady state.

9.6 Capacitance at band offsets, metal/semiconductor contacts, horizontal grain boundaries, and other defects in series

In capacitance or the more generalized impedance/admittance experiments, time-varying voltage or current signals are imposed between the contact terminals to investigate the time- or frequency-dependent changes in current and capacitance of the device. The goal is usually to determine the density of shallow dopants or deeper states in or near a depletion width. In ideal cases, this can be accomplished, and more than 50 years of research and application in Si and other conventional electronic devices speaks to the power and utility of these techniques. However, it is not necessarily simple to apply the interpretations—developed for Schottky junctions and pn junctions made of near-ideal single-crystalline and epitaxial semiconductors such as Si or GaAs—to devices such as thin-film PV, which use materials with no hydrogenic dopants, have significant band tails and defect densities, always contain heterojunctions, and also contain horizontal and vertical GBs and other structural defects. Clearly, it is necessary to examine the effects of these attributes on experimental results and the subsequent interpretations.

In addition to the depletion and chemical capacitances discussed above, capacitance responses can arise from semiconductor heterojunctions (isotype or heterotype) and metal/semiconductor contacts. The case of Schottky junctions is well documented and the primary origin is a depletion capacitance [43]. Insulator layers and trap states at the interface or crossing a QFL will also induce contributions. Therefore, this section concentrates on semiconductor heterojunctions, which are ubiquitous in thin-film solar cells and at horizontal GBs that are present in all except the most perfect thin-film devices. Importantly, it has been recently shown that many of the "defect" and even "back contact" features especially reported in the CIGSe literature probably arise from other layers in thin-film devices and not from trap states crossing Fermi levels [45,68–71]. The band structures of the major classes of thin-film polycrystalline chalcogenide devices across CdSeTe, CIGSe, and CZTSSe are similar to Figure 9.2. So, this hints that it might be wise to evaluate carefully the interpretations of at least some of the existing literature for AS on thin-film solar cells.

The overall structure of an ideal solar cell operating at the maximum power point near V_{oc} (as opposed to one using field-assisted photocarrier collection operating near short-circuit current density (J_{sc})) is a light-absorbing layer with long carrier lifetime and two carrier-selective contacts allowing extraction of photocarriers at their QFLs. In organic solar cells, energy-level offsets at both contacts play this role; but in inorganic semiconductors, capable of supporting space charges in depletion widths, a pn junction commonly serves as one of the carrier-separating one-way valves for photocarriers. Energy-level or band offsets may be such that DC transport is always "downhill" in forward bias, but this cannot be the case for AC transport—any band offset introduces asymmetry for the AC signal because of energy irreversibility. For example, in a typical thin-film device, during one half of an AC cycle, the signal must propagate through the front contact, window layer, any window/buffer/absorber heterointerfaces, depletion width, QNR, back contact, and finally, exit the device. On the other half of the AC signal, the signal's path is reversed; therefore, the effects of any band offsets or built-in voltages are reversed. This implies that whenever energy offsets are larger than about several k_BT, the AC response will in general be different from the DC one—and thus, more detail regarding energy-level offsets can be determined. Practically, at room temperature, offsets smaller than about 75–100 meV or ~3–4 k_BT usually are not the current-limiting resistance in the device; thus, they are difficult to resolve using temperature-dependent admittance techniques.

Considering an isotype heterojunction (e.g., the canonical n-GaAs/n-AlGaAs) at equilibrium, two sheet charges appear on either side of the heterojunction—an accumulation layer on the GaAs side and a partial depletion layer on the AlGaAs side. There is no depletion region, yet a capacitance still arises because of the changes in electron concentration on either side. Additional capacitance can arise from trap states in the heterojunction interface itself. A temperature-dependent RC delay time can arise from the processes of filling and draining these charge reservoirs via a thermally activated process such as thermionic emission.

AS data measured on an InGaP pn junction solar cell usually used in 3-junction cells epitaxially grown on n-GaAs provides an example in which an isotype InGaP/GaAs heterojunction produces a signature that is outwardly indistinguishable from a trap in the depletion width, as shown for experimental data presented in Figure 9.4. Misattribution of band offsets as trap signatures has recently been explained in the thin-film PV context. Recent exemplary work has shown that conduction-band offsets in the typical n-type ZnO:Al/i-ZnO/CdS stack, commonly used as the front contact in CIGSe and CZTS devices, results in an Arrhenius-activated capacitance step in AS that may explain the N1 "defect" reported widely in the CIGSe literature [45,68,71]. Window/buffer/absorber stacks with similar thicknesses and band offsets are used in many thin-film solar cell designs (e.g., ZnO/CdS/CZTS or MZO/CdTe). Therefore, practitioners must be cognizant that similar "trap" signatures may have a similar origin.

An RC time constant can arise from the chemical capacitance C of the regions on either side of the heterointerface and the $R = 1/G$ arising from the limiting (smallest) conductance through which C is charged and discharged. Majority carrier band transport through most of the device will only play this role in the carrier freezeout regime. The thermionic emission and/or tunneling currents over/through the band offset can become this limiting conductance at sufficiently low temperature or high frequency if it becomes too small to bring the device into steady state for all times during the sinusoidal signal. This leads to a capacitance step that will behave similarly to the band-offset case shown above. Similarly, AS, CV, and other AC techniques have been used to characterize QWs [33,72]. In some sense, localized defects and QWs share the same physics of spontaneous trapping and thermally activated emission of carriers (but not for details of recombination). These processes are shown in Figure 9.5. Also, an attractively charged GB or other structural defect can trap and emit carriers as long as thermal emission is faster than recombination. The kinetic energy prefactor including temperature dependence will change between emission from localized states compared to thermionic emission from delocalized states, but the dominant Arrhenius behavior remains. This points out a possible pitfall in interpreting data from imperfect thin polycrystalline film devices containing small-grained regions or other horizontal planar defects.

The stated aim of undertaking capacitance or impedance techniques is frequently to detect and characterize localized states, whether dopants or deep states, in terms of electronic transitions between those states and the delocalized

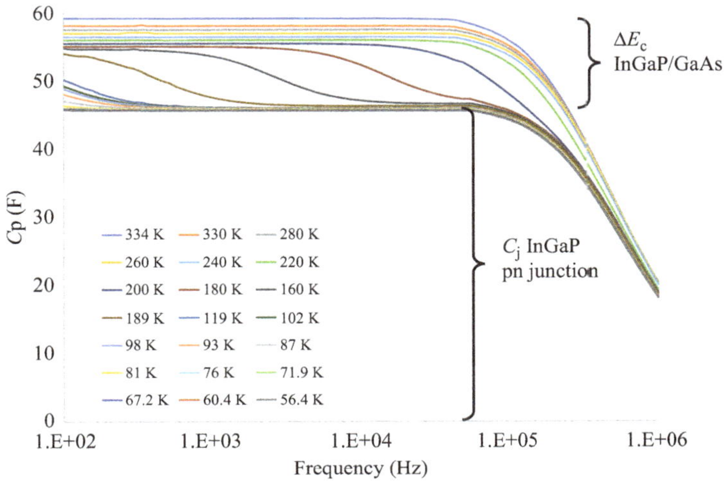

Figure 9.4 *Capacitance-frequency data from an epitaxial InGaP solar cell grown by MBE on n^+-GaAs for selected temperatures from 334 to 56.4 K. The trap-like signature with activation energy ~100 meV arises from thermionic emission at the n^+-InGaP/n^+-GaAs conduction-band offset, illustrating the importance of considering all possible origins of signatures exhibiting Arrhenius behavior before assigning them to a defect in the depletion width. Sample courtesy of Dr. R. Jones-Albertus, formerly of Solar Junction, Inc*

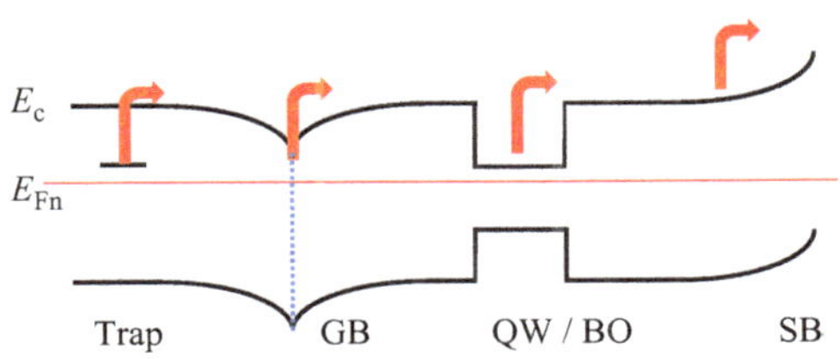

Figure 9.5 *Schematic band diagram showing thermally activated emission processes from (left to right): a localized state, a positively charged GB, a QW with localized states, one band offset of a wide double-heterostructure with delocalized states (BO), or a SB*

conducting states and to extract an activation energy and other parameters. However, any process that involves thermal emission over an energy barrier can give nearly identical responses, which has led to many incorrectly interpreted results. This should not be considered a weakness of the techniques per se, but rather, the result of confirmation bias stemming from inadequately examining assumptions and possible alternative hypotheses on the part of practitioners treating as routine these powerful experiments as applied to complex polycrystalline heterostructure-based semiconductor devices. These other sources of features manifested in capacitance techniques are enumerated here because their presence must be eliminated before any signal is declared to be a localized defect state in the bandgap of the absorber layer.

9.7 Equivalent-circuit representations of thin-film PV devices

It is important to reiterate that (1) ultimately all measurements referred to as "capacitance techniques" always measure both the imaginary and real parts of the impedance and (2) the apparent capacitance depends on both portions. The aim of capacitance techniques is typically to quantify the reservoirs for charge inside the device (doping or trap density of states), yet the values of Ohmic conductances or thermally activated transport processes in parallel or series with the reservoir may modify the observable value. Another view is that the measured or apparent capacitance is the result of a capacitance meter projecting the admittance or impedance onto (typically) the parallel C_p, R_p circuit model. It is not the true junction capacitance that is measured, but rather, the $C_{p,meas}$, which can be significantly modified by large parasitic series resistance (R_s) or shunt conductance (G_{sh}). If these are large, a measurement will be made of how much charge can be moved in and out of an available reservoir within the period of the signal instead of a measurement of the total capacity of the reservoir.

A diode is a nonlinear circuit element, and there is no way to represent it as a single equivalent circuit composed of passive linear components with constant values over all ranges of AC and DC voltages, frequency, and illumination or other stimuli. Thus, equivalent circuits composed of such elements can be used as a framework for analysis, but it should be remembered that this imposes regimes of applicability. Good solid-state diodes with small R_s and G_p are typically closer to the parallel model; but if R_s is very large, the device may be better described by the series model. For variable-frequency experiments or those where the device response changes considerably with temperature, light, or voltage (e.g., AS), the minimal model for a PV device at low frequencies is that of Schofield consisting of L_s and R_s in series, with R_p and C_p in parallel, as shown in Figure 9.6(a) (also sometimes called the Randles circuit in electrochemical and dielectric fields) [33,73].

For reverse to zero bias, only C_{Wd}, diode reverse saturation conductance (usually small), and shunts past the diode are significant, whereas in forward bias, standard small-signal analysis results in diffusion admittance (capacitance and conductance) being added. The capacitance of the pn junction includes effects from the depletion region plus about three minority diffusion lengths on either side (for the diffusion capacitance). Real shunting around or through the diode will dominate the conductance, whereas barrier lowering in the case of SBs and enhanced tunnel conductance through band offsets in heterojunctions may also contribute to differential AC conductance. For DC bias smaller than ~$3V_{th}$ (with $V_{th} = k_BT/q$), the

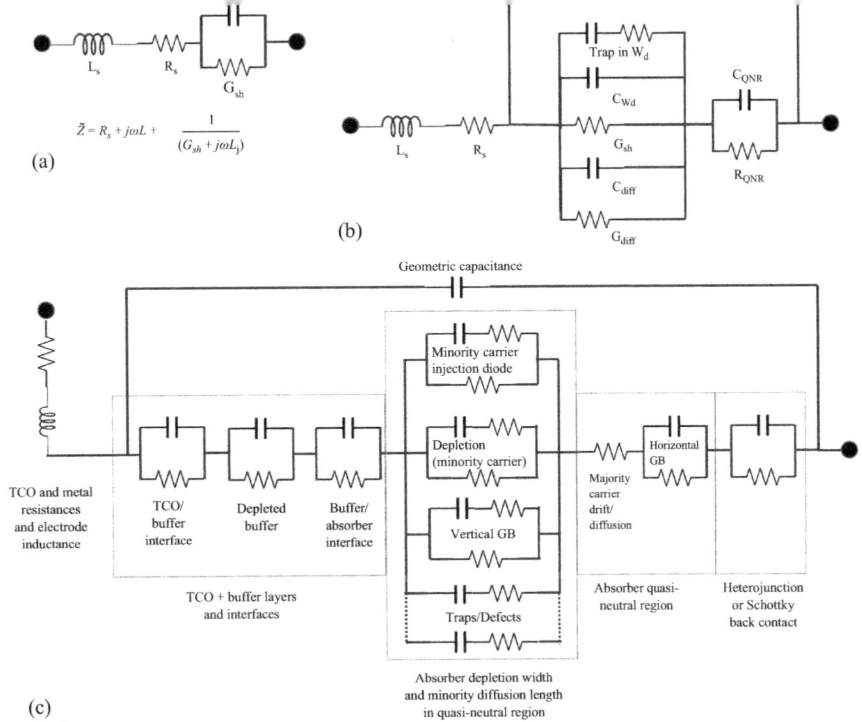

Figure 9.6 (a) Very minimal Schofield circuit applicable for reverse bias and solar cells where $W_d << h$, along with its impedance. (b) Semi-elaborated circuit capable of capturing phenomena observed in thin-film solar cells including traps, forward bias, geometric capacitance, and high frequencies. Note that the lumped, constant R_s from contact and spreading resistances is external to C_{geo}. (c) Further-elaborated model enumerating additional sources of capacitive response in thin-film solar cells and allowing for dopant freezeout and from which the others can be derived. In (b) and (c), note that many effects are in series and are thus inseparable from electrical measurements alone (i.e., determining whether a signal arises from the buffer layers or back contact requires changing at least one of them with light, bias, or a separate structure)

conductance of the nonideal diode equation yields a differential conductance $G = J_oA/mV_{th}$, with m as the ideality factor. The depletion capacitance prevails in reverse bias where the differential conductance is zero for an ideal diode for $|V| > 3V_{th}$. For diodes with significant trap-assisted generation current, this current increases with reverse bias because of the expanding depletion width and the reverse bias current will not saturate. This is intimately tied to voltage-dependent carrier collection under illumination.

In addition to C_{Wd} and C_{diff}, C_{geo} is always present, arising from the entire device thickness (h) acting as a dielectric between high-conductivity contacts observed at high frequencies or in conditions of punch-through (when $W_d = h$). C_{geo} must be in parallel with C_j, but it must also span any features inside the contacts. This topology is shown in Figure 9.6(b) and reduced forms of it are referred to as the Debye circuit. The QNR is inside the voltage nodes connected by C_{geo}, whereas other sources of series resistance occur outside of these nodes. So, only in the case where C_{geo} is completely negligible (e.g., in Si cells) is it proper to lump the QNR resistance with other series resistances in the device or measurement circuit. This condition is usually not strictly met for CIGSe and CZTS devices with thicknesses near 2 μm, but it may be met for CdTe:Cu devices with an absorber thickness near 4 μm. Thinner CdTe devices at fixed doping would necessitate the inclusion of C_{geo}, whereas recent advances in group-V doping will increase the depletion capacitance relative to C_{geo}. It would certainly be met for Si cells. Note that in Figure 9.6, L_s is placed outside of C_{geo}; but if necessary, this too could be split into parts inside and outside of the semiconductor [44].

The standard DC equivalent circuit of a solar cell includes both lumped series (R_s) and shunt (R_{sh}) parasitic resistances, as does the minimal AC model. Differences arise because this minimal AC model represents the differential response for vanishingly small sinusoidal AC signal and averages over displacement and real currents of both polarities. The R_s and R_{sh} values will be identical only when all elements are Ohmic and energy levels are symmetric across all material interfaces. An example in which the AC R_s differs from the DC R_s is when any band offsets, including heterojunction and Schottky contacts, are present in the device. For one direction of DC forward bias, current does not face any energy barrier; for the reverse direction, the thermionic emission process is sufficiently rapid so as not to impede transport at higher temperatures. However, as the temperature is lowered, the conductance of the thermionic emission process decreases and may limit at least one direction of the AC signal.

Assembling all of the physical effects discussed leads to the proposed "parent" equivalent circuit in Figure 9.6(b) and further elaborated in Figure 9.6(c). Many possible features have been added especially in series with the main junction—but not all are expected to be significant in any one type of device. Nevertheless, they are all included to illustrate the multitude of alternative hypotheses that must be eliminated before interpreting any data in junction spectroscopy experiments as related to defects crossing a QFL inside W_d. Simpler models can be derived in particular cases by eliminating insignificant elements. For example, in the case of $W_d << h$, the Schofield circuit of Figure 9.6(a) is recovered [69,73]. The inclusion of capacitance of the QNR may seem somewhat questionable; however, it is necessary at high frequencies, low doping, low mobility, or large QNR thickness where dielectric freezeout may occur. For example, a high-frequency study of CIGSe impedance accounted for the freezeout of the junction capacitance step-down to the geometric capacitance by including both R and C components for the QNR [7]. It can be shown through much tedious algebraic manipulation that this is equivalent to adding C_{geo} in parallel with C_{j} as for the Debye circuit; so, both C_{geo} and C_{QNR} should not be implemented simultaneously (it is shown in Figure 9.6 for clarity of the topology being discussed). Also, such a series capacitance has been used to represent the QNR of CdTe thin-film devices [74]. Most of the other literature on admittance and capacitance techniques applied to thin-film chalcogenide PV devices has (explicitly or implicitly) used the simplified Schofield circuit with reasonable accuracy.

Interpretation of any impedance- or capacitance-based technique must be mindful of the effects of parasitic resistances and conductances on the measured capacitance, which make it differ from the true C_{j} (thus, in principle, preventing accurate quantification of the shallow or trap states). The Schofield circuit in Figure 9.6(a) gives a good basis for discussion. Herein, R_{s} and L_{s} represent the contributions of the measurement fixturing as well as any internal series resistance for the device, and G_{sh} can account for shunting past the diode as well as conductance through the diode. The inductance of thin-film cells and modules is expected to be small, whereas for Si-based cells, these inductances can be significant for cells with extensive front-contact grids and AC loops arising from each string in the module and its bypass diode. This framework can be used in reverse bias, where the dominant capacitances are the depletion and geometric capacitances (depending on frequency) or forward DC bias where the diffusion capacitance and conductance dominate.

Schroeder [33] discusses nicely the projection of higher-order equivalent circuits onto the RC series or parallel models—a procedure which can allow the extraction of true capacitance from measured or apparent capacitance if (1) the parasitic values are known or (2) the complete admittance or impedance data (not just capacitance) is fit to a model over large frequency bandwidth. For example, by equating the real and imaginary parts of the Schofield model with those of the RC parallel model and adding the geometric capacitance (the high-frequency limit), we obtain:

$$\begin{aligned} C_{p,meas} &= C_{geo} + \frac{C_j - L_s\left(G^2 - C_j^2\omega^2\right)}{(1+R_sG)^2 + \left(G^2L_s^2 - 2C_jL_s + R_s^2C_j^2\right)\omega^2 + C_j^2L_s^2\omega^4} \\ &\approx C_{geo} + \frac{C_j}{(1+R_sG)^2 + R_s^2C_j^2\omega^2} \end{aligned} \tag{9.18}$$

The final form of (9.18) applies for negligible L_{s}, which must be carefully checked—especially as high frequencies are approached—because stray inductances are commonly present from wiring and probe fixturing (e.g., inside a cryostat) and are not removed if they occur after the 4-wire to 2-wire conversion points. For a thin-film PV device with capacitance of 1 nF measured at 1 MHz and stray inductance L below 0.1 μH, the apparent capacitance will be within 5% of the true junction capacitance for series resistance r below 2, 20, and 200 Ω for G_{sh} less than 10^{-2}, 10^{-3}, and 10^{-4} Siemens (R_{shunt} greater than 10^2, 10^3, and 10^4 Ω), respectively. These conditions are fairly easily met for reasonable efficiency devices measured on probe stations at room temperature. However, the series resistance of many cells with imperfect Ohmic contacts or heterojunction band offsets increases exponentially with decreasing temperature, thus introducing non-negligible differences between the true junction capacitance and the apparent $C_{p,meas}$. As shown in Figure 9.7(a), the effects of parasitic R_s of increasing value manifest first as the step at $\omega = 1/R_sC_j$ in the capacitance between C_j and C_{geo} downshifting in frequency; and for very large values, the low-frequency limit of $C_{p,meas}$ also decreases. Additionally, the G_{sh} must be low (R_{sh} high). In Figure 9.7(b), the effect of G_p increasing (i.e., R_{sh} decreasing) is to decrease $C_{p,meas}$. The lessening of the apparent capacitance step with G_{sh} decreasing is a result of the current division between the C and G: when $G < i\omega C$, most of the current flows instead through G. The downshifting of the apparent capacitance step with R_s is, in essence, the circuit's filtering effect—when the R_s becomes too high, C_j cannot be charged and discharged within one cycle for higher frequencies, and thus, it cannot fully respond to the applied V_{AC}.

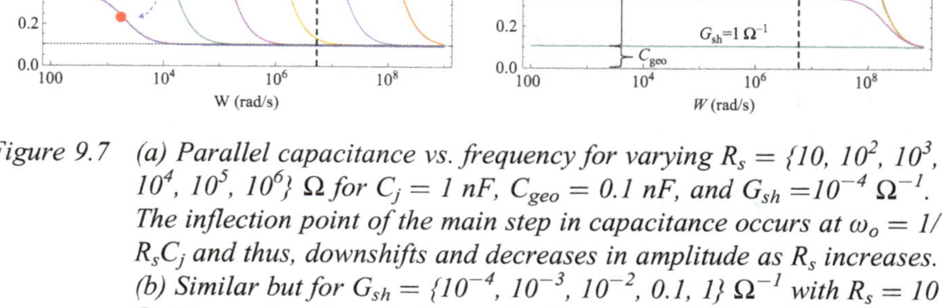

Figure 9.7 *(a) Parallel capacitance vs. frequency for varying* $R_s = \{10, 10^2, 10^3, 10^4, 10^5, 10^6\}$ *Ω for* $C_j = 1$ *nF,* $C_{geo} = 0.1$ *nF, and* $G_{sh} = 10^{-4}\ \Omega^{-1}$*. The inflection point of the main step in capacitance occurs at* $\omega_o = 1/R_sC_j$ *and thus, downshifts and decreases in amplitude as* R_s *increases. (b) Similar but for* $G_{sh} = \{10^{-4}, 10^{-3}, 10^{-2}, 0.1, 1\}\ \Omega^{-1}$ *with* $R_s = 10$ *Ω. Decreasing* G_{sh} *only lowers the* $C_{apparent}$*. Both* C_j *and* C_{geo} *remain constant at 1 nF and 0.1 nF, respectively, but the measured, apparent, or observable* C_p *changes with* R_s *and* G_{sh}*—for example, at the common measurement frequency of 1 MHz. This leads, for example, to over-estimation of* W_d *if it is assumed that* $C_{meas}(1\ MHz) = C_j(1\ MHz)$ *for* $R_s > 10$ *Ω.*

If both the real and imaginary parts of the impedance are measured with accuracy and precision over large frequency ranges, C_j and any other desired parameters can be extracted from model fitting. It should be noted that computing C_p within the parallel model from Z is insensitive compared to from Y because Y is directly proportional to C_p. Extensive discussion of impedance modeling is found in [11]. Examining (9.18) carefully, to accurately measure C_j directly as the measured C_p, point-by-point without fitting to a model requires that the following conditions are met: (1) $R_sG_{sh} << 1$ and (2) $\omega < R_sC_j$. Inequality 1 can be broken in forward bias of a diode, where the diffusion conductance increases exponentially with bias; but it is commonly known that CV data should not be taken far into forward bias to avoid unphysical artifacts. Inequality (2) may not be fully satisfied in experiments such as CV profiling or DLTS of thin-film solar cells, especially as T is decreased. If both equalities are not met, then model fitting of the complex admittance data is required rather than assuming $C_j = C_p$ measured.

This can have profound implications for experiments such as CV profiling, DLCP, and DLTS. In many instruments, the AC excitation frequency is fixed at 1 MHz because that frequency is appropriate for high-performance Si and GaAs devices with small R_s and shallow dopants that do not freeze out at low temperature. For experiments such as CV or DLCP profiling attempting to measure only dopant states, it is desirable to maximize the frequency to reduce the contributions of trap states. On the other hand, above about 1 MHz, practical RF challenges become more significant. Thus, 1 MHz has been an acceptable compromise for single-frequency instruments. However, if one seeks to measure the true C_j directly, an excitation frequency below the R_sC_j step (for all temperatures of interest) is required. Frequencies as low as 1 kHz may be required if R_s is too high. This means that before making and interpreting any CV, DLCP, or DLTS measurements, the *C–f* behavior should be examined for all voltages and temperatures in the experiment and either (1) the series resistance decreased by using 4-wire measurements, providing larger and closer contact pads, reducing contact and probing resistances, (2) choosing a different frequency for which the inequalities can be satisfied (i.e., choose a frequency below the R_sC_j step), or (3) bounds on the varied experimental parameters such as temperature or voltage should be identified that keep both inequalities satisfied. Otherwise, the (typically fixed) AC frequency may fall below the junction capacitance step during some subset of the experiment. For some solar cells, 1 MHz may actually be on or above the capacitance step from C_j to C_{geo}, which means that the measured capacitance will be lower than the actual junction capacitance. This is especially true when significant band tailing is present as for CZTSSe, which smears capacitance steps over many decades of frequency.

For example, our common experience with CZTSSe PV cells of many types using two wires carried to the sample within in a cryostat was that the data showed the onset of the step below 1 MHz, in many cases at room temperature. Underestimation of C_j would lead to overestimation of the apparent depletion width and inaccurate estimation of the doping. Note that although reverse bias will increase the depletion width and thus decrease C_j, this will not cause appreciable (e.g., >1 order of magnitude) shifts in the capacitance step unless the reverse bias is

>100 V (because of the square-root V dependence of C_{Wd}). In DLCP and DLTS, the same considerations apply. In DLCP or CV, it is relatively common to measure at a few different frequencies and interpret the increasingly lower apparent net doping vs. $<x>$ with increasing frequency as a sign that increasingly fewer trap states are participating. In the case of DLCP, compared to CV, these would be bulk traps because of the insensitivity of DLCP to direct capacitance contributions from the Fermi levels moving through an interface density of trap states [17,18,20]. However, it is also possible that the R_sC_j step has been exceeded with the measurement frequency, thus leading to apparently smaller capacitance, larger depletion width, and smaller apparent doping.

9.8 Dielectric freezeout driven by dopant freezeout: the case of non-shallow dopants

These arguments demonstrate that even the typical depletion-width capacitance has a characteristic cutoff frequency. For the case of truly hydrogenic shallow dopants, this frequency might only be observed at liquid-helium temperatures <10 K because its R_sC_j time constant is always very short. However, in the case of doping with a deeper dopant or distribution of dopant states such that the band carrier density freezes out as temperature is lowered [75,76], the ability of the QNR to deliver carriers to the edge of the depletion width is restricted and the dopant freezeout will induce dielectric freezeout. In this case, the interpretation of, for example, AS of p-CdTe, p-CZTSSe, some samples of CIGSSe, and other such materials lacking shallow doping must be reconsidered somewhat. Consider findings of "trap" states in which the low-temperature, high-frequency limiting capacitance associated with the "trap" capacitance step is the geometric capacitance C_{geo} rather than the sum of C_{geo} and the depletion capacitance C_{Wd}. Such a step indicates that the freezeout of the dominant acceptor states in the absorber have been measured and not a "trap" per se. Whenever the dopant ionization energy is rather high compared to the thermal energy during the experiment—perhaps >50 meV for common cases using nitrogen and helium cooling. This is a common case not only in emerging PV materials such as CZTSSe doped by native V_{Cu} and Cu_{Zn}, but also in established materials like CdTe doped by and GaN doped with Mg.

Here, a combined device-physics and equivalent-circuit model is developed to capture effects of the freezeout of the dominant dopant in the lower-doped side of a pn junction. Some example material systems for which this would be observed at typical temperatures for junction spectroscopy include CdTe:Cu, Si:In, CZTSSe, and GaN:Mg; but any dopant will eventually enter this regime when $k_BT < E_a$ for the dopant. It is found that, despite the complex device physics, Arrhenius analysis [59] will yield fairly accurate values for the dopant ionization energy, which appears as the largest activation energy. However, one or more spurious "trap" signatures may also be induced within the common measurement bandwidth, especially for larger ionization energies.

We first consider the propagation of an AC signal through the majority carrier gas in a homogenously doped slab of semiconductor—e.g., in a QNR of a pn junction diode alone, not including the depletion width. By combining the Poisson and continuity equations, a RC time constant $\tau_{DR} = \varepsilon/\sigma$ arises in sufficiently long-doped regions that defines a frequency $\omega_{DR} = en\mu/\varepsilon$ above which a signal cannot propagate through the majority carrier gas. This phenomenon is known as the DR or dielectric freezeout of the majority carriers [11,25,34,77]. At very high frequencies ($>10^8$ Hz approximately for $\mu \approx 100$ cm$^2/V_s$, $p > 10^{14}$ /cm^3, and $\varepsilon_r \approx 10$), the QNR alone can act as a low-pass filter if the resistivity and/or dielectric constant are sufficiently high. Below ω_{DR}, the geometric capacitance of the doped slab will be measurable, whereas above it, this capacitance step will freeze out and not be measurable. Note that both the capacitance and resistance considered in this case come from the same physical region (a doped slab or the QNR itself).

A distinct but related phenomenon involves an RC constant formed from the QNR resistance R_{QNR} and the junction capacitance C_j [69,74,77]. In the normal case of W_d significantly smaller than the layer thickness, or if the junction is forward biased so $C_{diff} > C_{Wd}$, then C_j is much larger than C_{geo}. Thus, if the resistance of the QNR (R_{QNR}) combines with C_j, then a lower cutoff frequency is established that may move into the measurement frequency range in the case of non-shallow dopants, resulting in the C_j step freezing out. The effect can be seen in [27] as the collapse of the capacitance at low temperature. Other series resistances such as contact resistances may be added to R_{QNR} to determine the cutoff frequency. The signature of this R_sC_j effect is that C_j (perhaps depreciated by R_s and R_{sh} effects per Figure 9.7) is measured below the cutoff in frequency (or above it in temperature), and C_{geo} is observed above the cutoff in frequency (or below it in temperature). We reiterate here that resistive effects other than the QNR may contribute—from the example of the recent CIGSe literature—decreases in conductance of thermionic currents over band offsets or SBs as in Figure 9.5 (which would be inside the lateral spreading on both electrodes in terms of the equivalent circuit).

If the absorber layer of a PV device is doped by defects having non-hydrogenic ionization energy, then the entire physics of the pn junction—and thus, its capacitance response—changes strongly with temperature. This causes somewhat puzzling responses which—if the standard analysis of Walter [59] are applied over limited frequency bandwidth (i.e. paying attention only to the derivative of the capacitance and not to the value of the capacitance steps in question)—may lead to the conclusion that multiple (up to three) traps are present. However, in reality only one physical process related to defects—dopant freezeout of the primary dopant—is occurring with temperature. The complexities of the device physics including bandgap variation with temperature and the equivalent circuit give rise to multiple features in the capacitance signal that must be carefully interpreted.

We believe that much of the analysis of CZTSSe devices claiming "trap" responses (from a trap crossing the Fermi level) in the literature may in fact be other effects. We reiterate that a trap in W_d will yield a capacitance step in reverse to zero bias where the high-frequency limit is C_{Wd}. Many of the reported signatures in the CZTSSe literature arise from capacitance steps where the high-frequency limit is C_{geo} instead.

To model the effects of non-shallow dopants in junction spectroscopy experiments, we define the relevant equivalent circuit starting with the structure of Figure 9.6(b). We include only C_{Wd} and C_{geo}—i.e., there are no trap or diffusion impedances. A small, fixed series resistance (R_s) is assumed to represent lateral spreading in the TCO and electrodes. L_s and C_{QNR} are also eliminated. C_{geo}, R_{sh}, and R_s are temperature independent, whereas C_{Wd} and R_{QNR} vary with the ionized acceptor concentration vs. temperature. The temperature-dependent device physics of the junction that causes variation in C_{Wd} and R_{QNR} includes the temperature dependence of the built-in voltage (V_{bi}), which causes (1) some variation of the depletion width W_{d} (which would be accentuated at lower doping), and (2) the freezeout of the acceptor in the QNR, which exponentially increases its resistivity as temperature decreases. Additionally, dopant freezeout may also increase contact resistances although this effect was not included herein.

The relevant device physics are next detailed here. For simplicity, we assume a pn homojunction with Ohmic contacts in the absence of degeneracy, but this could easily be extended to the case of a heterojunction providing the band offset(s) are also included. The built-in voltage V_{bi} and intrinsic carrier density n_i are given by

$$V_{bi}(T) = \frac{k_B T}{q} \ln\left(\frac{n_{no}(T) p_{po}(T)}{n_i^2(T)}\right) \tag{9.19}$$

$$n_i^2(T) = N_c(T) N_v(T) \exp\left(-E_g(T)/k_B T\right) \tag{9.20}$$

in which n_{no} and p_{po} are the free-carrier densities in the bands on the two sides of the pn junction, which we must carefully differentiate from the ionized (N_{d}^+,N_{a}^-) and total dopant concentrations (N_{d}, N_{a}) in both the absorber and oppositely doped window/buffer. In heterojunctions, the appropriate band offsets and bandgaps must also be included. It is commonly assumed that the depletion capacitance does not change with temperature, even in the absence of dopant freezeout (i.e., for $E_{\mathrm{a}} = 0$). However, a roughly 20% decrease in W_{d} and thus C_{Wd} between room temperature and low temperature will be present for common PV absorbers. This is primarily because of the temperature dependence of V_{bi}, which, in turn, comes primarily from the dependence of E_{g}. For narrower-gap or wider-gap semiconductors, the proportional difference will vary because the temperature dependence of the bandgap is roughly 100 meV between 0 and 300 K for a wide range of semiconductors, independent of E_{g} (300 K).

In the case that incomplete ionization of the dopants—in our example, the acceptor in the absorber—cannot be assumed, then p_{po} must be determined from [78,79].

$$p_{po} = \frac{\sqrt{\left(N_{d,comp} + N_v'\right)^2 + 4N_v'\left(N_a - N_{d,comp}\right)} - \left(N_{d,comp} + N_v'\right)}{2} \tag{9.21}$$

$$N_v' = \frac{N_v}{2} \exp(-E_a/k_B T) \tag{9.22}$$

where $N_{d,comp}$ represents the density of compensating donors in the absorber (to differentiate between N_d in the other side of the pn junction); N_a is the total density of acceptors; N_v is the temperature-dependent valence-band density of states; and E_a is the acceptor ionization energy. We have neglected the degeneracy factor for acceptor occupation for simplicity. At higher temperatures or intermediate temperatures for small N_a or large E_a, p_{po} becomes n_i and the contribution of the absorber doping to V_{bi} disappears entirely. Through complete numerical coincidence, a rule of thumb exists that dopant freezeout will become significant roughly at the temperature in K as the dopant ionization energy in meV.

In the absence of traps in significant concentrations, the depletion width W_d is set by the electrostatics of the net doping densities and V_{bi}, which represents the differences in Fermi levels at distances far from the depletion width. Thus, importantly, the depletion width depends both on (N_a − $N_{d,comp}$), which is temperature-independent, and on p_{po}, which depends exponentially on temperature. We state here that the charge in the depletion width, which sets its dimension for a given total bias, is determined by the net doping, irrespective of temperature and freezeout. The built-in voltage depends on the equilibrium Fermi levels, and thus, the actual band carrier densities n_{no} and p_{po}, and it thus depends on dopant freezeout. In prior versions of this model, a mistaken assumption was included that the depletion width also depended on the free-carrier density on the low-doped side (as opposed to net doping $N_a - N_d$) [80–82]. This led to the incorrect conclusion that the ionization energy of the dopant was significantly different than the results of Arrhenius analyzing the C_{Wd} capacitance step. This error has been corrected herein and the conclusion is that the Arrhenius behavior of the C_{Wd} capacitance step fairly accurately reflects the dopant ionization energy (because the main effect is the series connection of R_{QNR} with C_{Wd}).

For the assumed n^+p homojunction, W_d is given by

$$W_d = \sqrt{\frac{2\epsilon_r\epsilon_o\left(N_{d,n} + \left(N_a - N_{\mathrm{d,comp}}\right)\right)\left(V_{bi} - V - 2k_BT/q\right)}{qN_{d,n}\left(N_a - N_{\mathrm{d,comp}}\right)}} \tag{9.23}$$

where we introduce $N_{d,n}$ as the donor concentration on the n-type side of the junction to differentiate from the compensating donors $N_{d,comp}$ on the p-type side. The two factors of $2k_BT/q$ arise from the Debye tails and are included for completeness [33]. For $N_{d,n} >> (N_a - N_{d,comp})$, the junction is one-sided and the depletion width is effectively all in the absorber layer.

The device physics is now used to determine the circuit-element values. As stated earlier, R_s, C_{geo}, and R_{sh} can for now be assumed to be temperature independent. R_{sh} will arise from both Ohmic shunts and from the diode currents. The former shows up in both the DC and AC (differential) equivalent-circuit models, whereas, to first approximation, a diode has infinite differential resistance at reverse to zero bias because I_o is constant. In detail, the diode saturation current dominated by generation in the depletion width depends on voltage as it changes the depletion width. C_{Wd} and C_{geo} are defined in the usual way as

$$C_{Wd} = \frac{\varepsilon_r \varepsilon_o A}{W_d} \tag{9.24}$$

$$C_{geo} = \frac{\varepsilon_r \varepsilon_o A}{h} \tag{9.25}$$

where A is the device area and h is the total thickness of the QNR plus W_{d} which always equals h, the thickness of the total low-doped side. R_{QNR} is given as

$$R_{QNR} = \frac{(h - W_d)}{A} \frac{1}{q\mu p_{po}} \tag{9.26}$$

where the QNR thickness is $(h - W_{\mathrm{d}})$, mobility is μ (assumed constant herein), and the primary temperature dependence enters from p_{po}. The main effects arise from the exponential dependence of R_{QNR} on temperature, with some changes also in C_{j}.

The apparent C_{p} vs. f for a 1-cm^2 cell at different temperatures computed from the described model as well as the circuit model used (a reduced form of Figure 9.6 (b) or (c)) are shown in Figure 9.8 over a simulated range of 350 to 30 K. Here, a n+p junction was assumed with $N_{\mathrm{d}} = n_{\mathrm{no}} = 10^{18}$ /cm^3 on the high-doped side. On the low-doped side, h was assumed to be about 10 times W_{d} at 300 K. N_{a} was 10^{16} /cm^3 with 10^{15} /cm^3 compensating donors. Some characteristic features appear. First, at high temperatures, the measured C_{p} begins to decrease without any downshifting of the capacitance step, indicating that this is not caused by R_{QNR} but

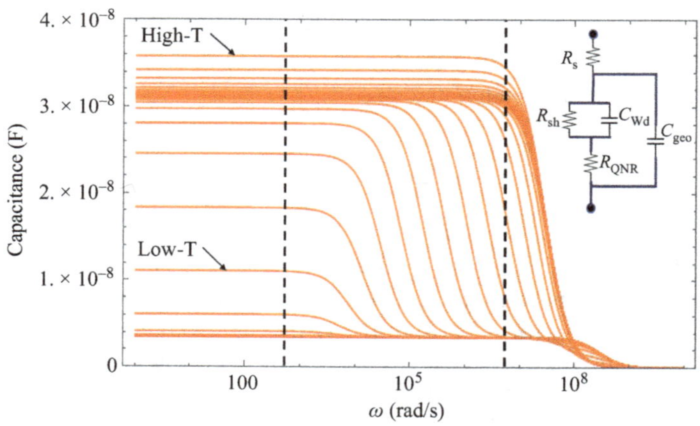

Figure 9.8 Simulation of C_p vs. frequency for the circuit shown combined with device physics for a n^+p homojunction with shallow doping on n^+ side and $(N_a - N_d) = 9 \times 10^{15}$ /cm^3 with $E_a = 150$ meV on the 3-μm-thick p side. The dashed lines indicate 100 Hz to 1 MHz, which is a common bandwidth for impedance spectroscopy measurements

rather by the mild temperature dependencies in C_{Wd} itself. Then, over a range of temperatures, the C_p then "bunches up" or stays fairly constant. Then, at lower temperatures where p_{po} begins to change exponentially, the effects of R_{QNR}'s exponential increase with minimal change in C_{Wd} become apparent as the C_{Wd} capacitance step rapidly downshifts in frequency. Arrhenius analysis of the inflection points of the simulated data yields activation energies very close to the ionization energy of the dopant (99 meV when 100 meV was used and 191 meV for 200 meV). In this modeled case with 10% compensation (as opposed to 1% or less), the "half-slope" dopant freezeout regime is traversed over a relatively narrow range of temperatures, so is not apparent. In the case of very low compensation, this "half-slope" regime would be apparent and the activation energies would be expected to follow $E_a/2$ [78,79]. At the lowest temperatures, the capacitance reverts to C_{geo}. At very high frequencies from the temperatures of dopant freezeout, this model predicts that the C_{geo} capacitance step may show some shifts vs. temperature. The other activation energies (two additional) detected by inflection-point analysis occur in this step and arise from the entire circuit response because C_{geo} does not change vs. temperature. The detected activation energies are 59 and 4 meV for an assumed 100-meV dopant, and 110 and 38 meV for an assumed 200-meV dopant. These two additional activation energies would not usually be measured because they would be at frequencies higher than the typical experimental bandwidth.

The salient features indicating that the physics may be caused by the freezeout of the dominant dopant in the absorber as temperature decreases include (1) the bunching of capacitance at an intermediate temperature and (2) for temperatures below this bunching temperature, the simultaneous decrease in low-frequency capacitance, and down-shifting in frequency of the capacitance step corresponding to C_j caused by the exponential increase in R_{QNR}, leaving C_{geo} as the high-frequency limit at low temperatures. A capacitive response from a trap-level crossing a QFL within W_d (as for E_t in Figure 9.2 and the response depicted in Figure 9.3) will manifest "on top of" the main C_j step. Thus, as temperature is lowered, the trap step will downshift exponentially in frequency and eventually disappear, leaving C_{Wd} (as in Figure 9.3). In the case of freezeout of the main dopant in the QNR, the C_j step itself downshifts in frequency as R_{QNR} increases, leaving only C_{geo} observable at low temperatures.

As temperature decreases, the physics required to cause the C_j step to freeze out is an exponentially increasing resistance in series with C_j and inside the nodes defining C_{geo}. A few mechanisms involving energy barriers could also provide such increasing R_s: thermionic emission over a band offset, thermionic emission over the potential well or band offset induced by horizontal GBs or other planar defects, and thermionic emission over a heterojunction or Schottky back-contact. It is expected that these band-diagram features would also manifest in ways nearly identical to QNR dopant freezeout driving the C_j capacitance step. In fact, multiple mechanisms could simultaneously operate and superimpose in series. Thus, the microscopic origin of Arrhenius-activated R_s should not be assumed to be QNR dopant freezeout until other causes are excluded. Unfortunately, the common practice is to blindly apply the standard analysis and label any Arrhenius-activated capacitance steps as a "trap" in the depletion width because that is ostensibly what most researchers are hoping to find when undertaking AS of thin-film solar cells. If confirmation bias can be minimized, perhaps other insights into possible device improvements can be made such as opportunities for eliminating horizontal GBs from small-grained nucleation layers and improving non-Ohmic back contacts.

Here, we clearly state the main points of this section: AS data were computed for the case of temperature-dependent ionization of the primary dopant in the QNR of a one-sided junction. No traps or band-offset physics were included; yet, if the conventional Arrhenius analysis of the inflection points in the AS data is carried out blindly, up to three "trap" features would be found (the two at higher frequencies will be smaller in magnitude and correspond to small changes in the C_{geo}-related step). The results may be considered as coming from the equivalent circuit's response [69], although it does correspond to a physical effect wherein charge cannot be moved into and out of the edge of the depletion width sufficiently fast to keep up with high-frequency signals. The applicability of the analysis of the trap density of states [59] from the modeled AS data was not assessed—only the simple Arrhenius analysis of the inflection point.

Additionally, from the perspective of a band diagram, one would not expect to be able to measure the density of states for a shallow acceptor as a chemical capacitance, because its energy level would not be expected to cross its QFL like E_t does in Figure 9.2 (top). However, this examination (and prior literature [69,77]) shows that it is possible to use AS to measure the ionization energy of the dominant dopant in a case where it freezes out and causes exponential changes in R_{QNR}. The model indicates that the freezeout of the main C_{Wd} capacitance step can be used with good accuracy to measure the ionization energy of the dopant in the low-doped side of the n^+p junction, as has been done for example in CZTS [77,83–85]. In recent years, the combination of photoluminescence (PL) and AS have differentiated deep dopant traps and other effects with higher ionization energies [86,87]. The contacts were assumed to be Ohmic at all temperatures, which may not be realistic—because the net ionized doping decreases the effects of any SB and may become significant, thus adding another source of exponentially changing series resistance and possibly even capacitance. This would produce modified response details, so there is no guarantee that in a real physical system the dopant ionization energy would be exactly reflected in Arrhenius analysis.

The presence of this effect in AS caused by freezeout of the primary QNR dopant also implies that related effects will be found in all other capacitance-based experiments such as CV, DLCP, and DLTS; therefore, extreme care is required in interpretations whenever a shallow, hydrogenic dopant is not present. This raises questions about what exactly is being measured, for example, in CV or DLCP profiling undertaken at temperatures and frequencies such that the excitation frequency is higher than the main C_j capacitance step. Thus, it is important to examine the C–f–T behavior before selecting a frequency at which to do CV and DLCP, and to estimate both C_{geo} and C_{Wd} to give perspective.

9.9 Descriptions of some capacitance-based measurement techniques: CV, DLCP, AS, DLTS, MCDLTS, ODLTS, DLOS, and TPC/TPI

Many capacitance-based characterization techniques have been applied to thin-film polycrystalline chalcogenide solar cells over decades. Rather than reporting on results from each technique and material technology, here, we give a short description meant only to disambiguate and begin to explain the physics behind each technique.

AS is considered here as the base technique for understanding the cell's AC response. The cell's admittance is measured as functions of frequency and temperature, and the capacitance extracted from it using (9.3) for cases of inconsequential parasitic impedances (see Figure 9.7 and (9.18)) or fitting to a more complex model (Figure 9.6). Steps in C–f or C–T indicate the freezeout of a physical process in the device, and Arrhenius analysis of the step can yield the activation energy. In principle, DC bias can be applied to determine depth profiles of trap states, to forward bias a back-contact SB [27] to increase its conductance and eliminate its influence, and to determine if the device may be in punch-through. The technique has a long history in semiconductor device analysis including thin-film polycrystalline chalcogenides [7,13,14,59,68,69,71,73,74,77,84,85,87–110]. Examination of AS data, or at least measurements at a few different f and T values, is important before undertaking most other techniques. For example, the default 1-MHz excitation frequency standard on many capacitance instruments may be completely unsuitable for measuring the dominant dopant concentration in CV in the presence of large R_S. A related technique, modulated photocurrent, in which excitation is provided by a light source driven with a sine wave instead of electrically exciting the cell, has also been applied [111–118].

Capacitance-voltage profiling is an extremely well-established technique for analysis of semiconductor junctions, and standard references discuss it in detail [33]. Thus, only a few features are mentioned here. Thin-film polycrystalline solar cells frequently present characteristic U-shaped CV curves, the origins of which are not entirely settled [119]. It is not uncommon to see roll-over of Mott–Schottky plots of $1/C^2$ vs. V that seem to indicate two doping densities vs. depth and preposterous built-in voltage larger than the bandgap. These features frequently point to the possible presence of a significant trap density in addition to the shallow doping or a large density of traps at an interface [33]. Finally, if the CV profiles are repeated for different frequencies at room temperature, it is common to see higher apparent doping and smaller apparent W_d for high frequencies and the opposite for low. This can be caused by the presence of dopants significantly distributed in energy, wide band tails overlapping with the dopants, traps crossing E_f as in Figure 9.2, and interface states.

The other method for profiling doping in the thin-film PV community is DLCP [16,17,19,120]. DLCP is like CV in that it seeks to measure the net charge density at the edge of the depletion width. However, each DLCP measurement for a given depletion width requires at least three (3–10 typically) measurements taken at coordinated values of V_{DC} and V_{AC} such that the least-negative total bias remains constant. It is important that the peak-to-peak and RMS V_{AC} are not confused (check which the instrument specifies). Hypothetical waveforms used to measure the net doping (N_{DLCP}) at two different values of W_d are shown in Figure 9.9. In the quasi-static case at low frequency, this implies that W_d is modulated from a minimum value to successively larger values. The capacitance vs. V_{AC} is then fit with a second-order polynomial and the constant and linear terms used to determine the minimum value of W_d and the net doping just beyond this value in the region sampled by the different V_{AC}. Then, a different minimum W_d is chosen and the process is repeated. An improved mathematical method for extracting the net charge density was recently introduced by realizing that only a few of the terms in the Taylor expansion of C in powers of V_{AC} are independent [121]. It has been argued that DLCP is less sensitive to capacitance induced by QFLs traversing interface densities of states than CV, which has made it a popular alternative to CV [18,19]. The influence of charged GBs on CV and DLCP (let alone DLTS and related), especially in the case where the Fermi level is pinned by large defect densities, has not been fully elucidated.

DLTS and related emission transient techniques are mixed time- and frequency-domain techniques [26,122]. In DLTS, for example, the transient response of the junction capacitance to a step change in voltage is measured at frequency ω for times approximately $10/\omega$ to 10 s, equating to the frequency bandwidth 0.1 to $\omega/10$ rad/s (with ω typically but not always 2π MHz). If the junction contains trap states that change their charge by a thermally activated capture or emission process, then the depletion capacitance reacts to their changing charge state as a function of time. Thus, in essence, DLTS uses the depletion capacitance as a signal amplifier for the change in occupation of a small density of defect states, which is the secret of its sensitivity. Capacitance-based techniques

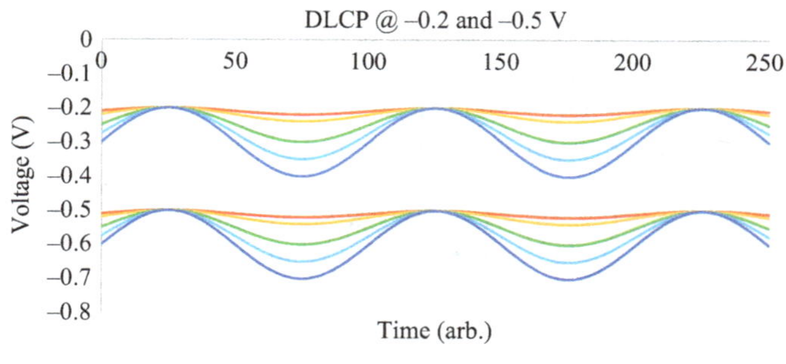

Figure 9.9 Total waveforms $V_{DC} + V_{AC}$ for DLCP at two different positions of W_d corresponding to -0.2 and $-0.5V_{DC}$. Five different V_{AC} values (peak-to-peak) of 10, 20, 50, 75, and 100 mV are shown for each measurement and would be used to determine the charge density assigned to the two values of W_d

such as DLTS are generally most useful for doped semiconductors, whereas thermal emission techniques using temperature scanning are well-suited for high-resistivity samples [30,123,124]. Here we mention especially the thermoelectric effect emission spectroscopy (TEES) introduced by Lynn at Washington State University. While developed originally for high resistivity CdTe radiation detectors, it has been shown to work extremely well for deep states in CdTe doped to $10^{17}/cm^3$.

Lang's original paper introduced many innovations in both the data acquisition and signal processing of transients simultaneously. The double-boxcar correlator method of transient analysis was designed for use with analog electronics and plotters, yet it continues to yield robust signals insensitive to temperature variations even in the age of digital signal processing. Variants of DLTS using voltage pulses to fill localized states include minority DLTS, in which the pulse takes the diode to forward bias to inject minority carriers; optical DLTS (O-DLTS), which denotes a DLTS experiment with constant sub-bandgap illumination; current DLTS (I-DLTS), wherein the current rather than capacitance is measured (for low-resistivity samples, similar sensitivity is possible) [125,126]; charge DLTS (q-DLTS), in which the transients are integrated (may be more appropriate for high-resistivity, low-mobility samples such as organic PV) [127]; and the TPC technique originated by Cohen's group at the University of Oregon (usually paired with TPI), which is in a sense a hybrid of DLTS and O-DLTS and is very similar to DLOS [128]. The Oregon TPC/TPI technique uses a lock-in amplifier thus both the in- and -out-of-phase transient components are measured simultaneously yielding both current and capacitance. The monochromatic illumination wavelength is scanned and for each wavelength a typical DLTS voltage pulsing sequence is carried out both with and without the shutter open. The purely optical emission transient is obtained by subtracting the "dark" transients from the "light" transients, ensuring that any thermal emission from traps is removed from the photocapacitance and photcurrent transient data. This technique is a generalized version of DLOS in that it records both current and capacitance. In this TPC/TPI experiment, the trap energy is identified from a step at a threshold photon energy. In O-DLTS, the activation energy is (at least in principle) still determined from thermal emission capacitance transients while the sub-gap light changes the occupation of defect states. Optical filling and emptying of localized states is of great utility in high-resistivity and wide-bandgap semiconductors. As opposed to many techniques for which the acronym denotes a fairly specific experimental procedure, the original paper introducing DLOS took an expansive view that almost any technique using photon energy to determine deep level parameters should be termed "DLOS" [132]. DLOS today as practiced by Ringel and Arehart at Ohio State University refers to a technique whereby voltage pulsing is used to fill trap states and long optical pulses of varying wavelength are used to empty them and the trap energy is identified from a step in the photon energy [129,130]. In the OSU implementation, after each voltage pulse, the DLTS transient is recorded and ensured to have ended. Then, a shutter is opened and the sample illuminated to begin the optical emission transient. Thus the Oregon TPC and the Ohio DLOS techniques are extremely similar with the main difference being the details of correcting for thermal emission transients. The term DLOS was first introduced to describe a wide range of experiments using light to help characterize deep levels (including, but not limited to, the modern DLOS) [131]. One variant that keeps voltage fixed, fills traps with light pulses, and uses thermal emission to characterize trap energy levels was denoted as optical DLTS [131]; but this use of the acronym O-DLTS was mostly eclipsed. Finally, in the age of digital signal processing, the original isothermal space-charge spectroscopy described by Sah and direct fitting of transients can be carried out; however, in practice, the requirements for temperature stability of the sample are limiting. If the temperature of the sample can be stabilized within fractions of a degree for long periods, then exquisite energy resolution and sensitivity to small trap concentrations are possible in Laplace DLTS (L-DLTS) [132,133].

9.10 Closing remarks

It is typical to focus attention on the pn junction and interpret results in terms of models of junction response—the most common goal being to characterize a device-limiting defect in the depletion width. In fact, most of these techniques, with the exception of CV profiling, are commonly referred to as "defect spectroscopies." But spectroscopy on defect states in the depletion width can only be unambiguously achieved if all other sources of similar responses have been understood or eliminated. In advanced single-crystalline Si and III–V homojunction devices with near-ideal characteristics, shallow dopants, and good Ohmic contact technologies, these conditions are met and it was appropriate to interpret responses as arising from "traps" or "defects" while these techniques were developed in the past 50 years. The complexities of heterojunctions and GBs ubiquitous in thin-film chalcogenide solar cells, as well as other material non-idealities such as large band tails, present challenges for interpretation but also opportunities for extending the application of techniques and gaining new insights. Direct simulation of device responses using device simulators [2,70,134,135] as opposed to analytical treatments include the self-consistent calculation of QFLs and their effects on capture and emission rates [57] and the complexities of 2D and 3D band structure induced by the microstructure of thin-film PV devices. Also, the impacts of any features such as defect states on device operation in illuminated forward bias can be assessed using the same physical model. The importance of quantifying the effects of deep levels with such modeling rather than speculating without quantification is reiterated here.

In the field of thin-film polycrystalline PV devices—where Ohmic contacts and shallow dopants are rare and homojunction devices are nonexistent—it unfortunately is common for the origin of signals to be automatically ascribed to "defects" or "traps" essentially because that is the phenomenon the experimenter set out to detect without careful elimination of alternative hypotheses by decisive, differentiating experiments. Early in the twentieth century, Karl Popper introduced the concept of empirical falsification (scientists should strive to disprove hypotheses to eliminate falsifiable ones, eventually leaving only the truth) in favor of inductivism; yet, one unfortunate comment on the field of capacitance spectroscopies on chalcogenide thin-film solar cells is that inductivism and confirmation bias remain fairly frequent. This is not to say that efforts to apply capacitance techniques to thin-film polycrystalline solar cells is hopeless or pointless. Great advances can be achieved by reframing these techniques as "material and device capacitance spectroscopies," acknowledging that multiple hypotheses for the physical origin of "trap signatures" always exist, and using carefully conceived experimental designs to isolate the physical origin of signals rather than assuming *a priori* that they arise from a defect in the depletion width. What could be more fun and intellectually stimulating for a scientist? Discoveries are still possible, but they just may not always be the lifetime-killing mid-gap defect one set out to find.

This chapter has attempted to survey some fundamentals of capacitance measurements and junction spectroscopies in inorganic semiconductor diodes, particularly thin-film polycrystalline chalcogenides. These techniques and their standard analyses and interpretations were developed for near-ideal devices such as Si and GaAs diodes. The devices and material properties found in thin-film chalcogenide photovoltaic devices contain many more features and thus the same interpretations and conclusions are not always warranted. Some main themes raised are that multiple physical features in the band structure and material properties of these devices may give rise to similar features. So, the assumption that every such feature arises from a defect state crossing a quasi-Fermi energy in the depletion width is unjustified until all other competing hypotheses have been eliminated. This chapter is not meant to replace all of the papers and books that came before it. Rather, it aims to fill some gaps and establish connections with other topics not usually covered therein. We hope that it proves to be of value and inspiration to push our field further.

Acknowledgments

I am indebted to the students and postdocs who have undertaken capacitance and admittance measurements in my lab and learned these topics with me: Dr. Ashish Bhatia, Dr. Liz Lund, Dr. Volodymyr Kosyak, Mr. Tom Wilenski, Ms. Anna Caruso, Ms. Kholoud Alajmi, Mr. Max Hansen, Mr. Sudhajit Misra, Dr. Rujun Sun, Dr. Yu Kee Ooi, Mr. Mashaduddin Saleh, Ms. Laura Treider, Mr. Jingxiang Zhou, Mr. Aleksandr Luchinovich, and Prof. Vipul Kheraj. Special thanks to Dr. Aaron Arehart, Dr. Jian V. Li, Dr. Steve Johnston, Dr. Thomas Weiss, Prof. Angus Rockett, Dr. D. Westley Miller, and Dr. Charles Warren for many insightful discussions on these topics. Thanks lastly to Don Gwinner for careful proofreading.

References

[1] Peaker AR, Markevich VP, Coutinho J (2018) Tutorial: junction spectroscopy techniques and deep-level defects in semiconductors. *Journal of Applied Physics*, 123(16):161559.

[2] Sozzi G, Di Napoli S, Menozzi R, Weiss TP, Buecheler S, Tiwari AN (2018) Interpretation of admittance signatures in Cu(In,Ga)Se_2 solar cells. *2018 IEEE 7th World Conference on Photovoltaic Energy Conversion, WCPEC 2018 – A Joint Conference of 45th IEEE PVSC, 28th PVSEC and 34th EU PVSEC*, 2515–2519.

[3] Burgelman M, Decock K, Niemegeers A, Verschraegen J, Degrave S (2019) SCAPS Manual.

[4] Bisquert J, Bertoluzzi L, Mora-Sero I, Garcia-Belmonte G (2014) Theory of impedance and capacitance spectroscopy of solar cells with dielectric relaxation, drift-diffusion transport, and recombination. *The Journal of Physical Chemistry C*, 118(33):18983–18991.

[5] Alam MA, Luryi S, Mastrapasqua M, Pacelli A (2000) Automatic generation of equivalent circuits from device simulation. *Technical Proceedings of the 2000 International Conference on Modeling and Simulation of Microsystems*, 337–340.

[6] Pacelli A, Mastrapasqua M, Luryi S (2000) Generation of equivalent circuits from physics-based device simulation. *IEEE Transactions on Computer-Aided Design of Integrated Circuits and Systems*, 19(11):1241–1250.

[7] Lee J, Cohen JD, Shafarman WN (2005) The determination of carrier mobilities in CIGS photovoltaic devices using high-frequency admittance measurements. *Thin Solid Films*, 480–481:336–340.

[8] Bisquert J (2002) Theory of the impedance of electron diffusion and recombination in a thin layer. *The Journal of Physical Chemistry B*, 106 (2):325–333.

[9] Bisquert J, Mora-Sero I, Fabregat-Santiago F (2014) Diffusion-recombination impedance model for solar cells with disorder and nonlinear recombination. *ChemElectroChem*, 1(1):289–296.

[10] Bisquert J (2008) Beyond the quasistatic approximation: impedance and capacitance of an exponential distribution of traps. *Physical Review B*, 77(23)

[11] Barsoukov E, Macdonald JR (2005) Impedance Spectroscopy: Theory, Experiment, and Applications.

[12] Saleh MU, LaCombe J, Jayakumar NKT, *et al.* (2019) Signal propagation through piecewise transmission lines for interpretation of reflectometry in photovoltaic systems. *IEEE Journal of Photovoltaics*, 9(2):506–512.

[13] Kneisel J, Siemer K, Luck I, Bräunig D (2000) Admittance spectroscopy of efficient CuInS_2 thin film solar cells. *Journal of Applied Physics*, 88 (9):5474–5481.

[14] Nardone M, Karpov VG (2008) Admittance characterization of semiconductor junctions. *Journal of Applied Physics*, 103(8):084508.

[15] Copeland JA (1969) A technique for directly plotting the inverse doping profile of semiconductor wafers. *IEEE Transactions on Electron Devices*, 16 (5):445–449.

[16] Michelson CE, Gelatos AV, Cohen JD (1985) Drive-level capacitance profiling: Its application to determining gap state densities in hydrogenated amorphous silicon films. *Applied Physics Letters*, 47(4):412–414.

[17] Heath JT, Cohen JD, Shafarman WN (2004) Bulk and metastable defects in CuIn1À$_x$Ga$_x$Se$_2$ thin films using drive-level capacitance profiling. *Journal of Applied Physics*, 95(3):11.

[18] Heath JT, Cohen JD, Shafarman WN (2003) Distinguishing metastable changes in bulk CIGS defect densities from interface effects. *Thin Solid Films*, 431–432:426–430.

[19] Heath JT (2018) Capacitance-voltage and drive-level-capacitance profiling. In *Capacitance Spectroscopy of Semiconductors*, edited by Li JV and Ferrari G. Pan-Stanford Publishing.

[20] Heath JT, Zabierowsky PW (2016) Capacitance spectroscopy of thin-film solar cells. In *Advanced Characterization Techniques for Thin-Film Solar Cells*, 2nd Edition, edited by Abou-Ras D, Kirchartz T, and Rau U. Weinheim, Germany: Wiley.

[21] Hioki (2018) Impedance Measurement Handbook, 1st Edition. Hioki E.E. Corporation.

[22] Keysight Technologies (2016) Keysight Technologies Impedance Measurement Handbook, 6th Edition. Keysight Technologies.

[23] Lee S (2011) C-V Measurement Tips, Tricks, and Traps. download.tek.com/document/CV%20MeasTips%20WP.pdf

[24] Macdonald JR (1992) Impedance spectroscopy. *Annals of Biomedical Engineering*, 20(3):289–305.

[25] Vadim F Lvovich (2012) *Impedance Spectroscopy: Applications to Electrochemical and Dielectric Phenomena*. Wiley, Hoboken NJ.

[26] Lang DV (1974) Deep-level transient spectroscopy: A new method to characterize traps in semiconductors. *Journal of Applied Physics*, 45(7):3023–3032.

[27] Lang DV, Cohen JD, Harbison JP (1982) Measurement of the density of gap states in hydrogenated amorphous silicon by space charge spectroscopy. *Physical Review B*, 25(8):5285–5320.

[28] Cohen JD, Lang DV (1982) Calculation of the dynamic response of Schottky barriers with a continuous distribution of gap states. *Physical Review B*, 25 (8):5321–5350.

[29] Miller GL, Lang DV, Kimerling LC (1977) Capacitance transient spectroscopy. *Annual Review of Materials Science*, 7(1):377–448.

[30] Sah CT, Chan WW, Fu HS, Walker JW (1972) Thermally stimulated capacitance (TSCAP) in *p-n* junctions. *Applied Physics Letters*, 20(5):193–195.

[31] Sah CT, Forbes L, Rosier LL, Tasch AF (1970) Thermal and optical emission and capture rates and cross sections of electrons and holes at imperfection centers in semiconductors from photo and dark junction current and capacitance experiments. *Solid-State Electronics*, 13(6):759–788.

[32] Kimerling LC (1974) Influence of deep traps on the measurement of free-carrier distributions in semiconductors by junction capacitance techniques. *Journal of Applied Physics*, 45(4):1839–1845.

[33] Schroeder D (2015) Semiconductor Material and Device Characterization, Third Edition. Tempe, AZ: Arizona State University.

[34] Hauff E von (2019) Impedance spectroscopy for emerging photovoltaics. *The Journal of Physical Chemistry C*, 123(18):11329–11346.

[35] Klotz D (2019) Negative capacitance or inductive loop? – A general assessment of a common low frequency impedance feature. *Electrochemistry Communications*, 98:58–62.

[36] Kovalenko A, Pospisil J, Zmeskal O, Krajcovic J, Weiter M (2017) Ionic origin of a negative capacitance in lead halide perovskites. *Physica Status Solidi (RRL) – Rapid Research Letters*, 11(3):1600418.

[37] Mora-Seró I, Bisquert J, Fabregat-Santiago F, *et al.* (2006) Implications of the negative capacitance observed at forward bias in nanocomposite and polycrystalline solar cells. *Nano Letters*, 6(4):640–650.

[38] Jacobs DA, Shen H, Pfeffer F, *et al.* (2018) The two faces of capacitance: new interpretations for electrical impedance measurements of perovskite solar cells and their relation to hysteresis. *Journal of Applied Physics*, 124(22):225702.

[39] Jacobs DL, Scarpulla MA, Wang C, Bunes BR, Zang L (2016) Voltage-induced transients in methylammonium lead triiodide probed by dynamic photoluminescence spectroscopy. *The Journal of Physical Chemistry C*, 120 (15):7893–7902.

[40] Paull PK, Bailey J, Zapalac G, Arehart AR (2017) Fast C-V method to mitigate effects of deep levels in CIGS doping profiles. *2017 IEEE 44th Photovoltaic Specialist Conference (PVSC)*, 2414–2418.

[41] Braunlich P (1979) Thermally Stimulated Relaxation in Solids: Topics in Applied Physics, 37, Springer-Verlag Berlin.

[42] Blood P (1986) Capacitance-voltage profiling and the characterisation of III-V semiconductors using electrolyte barriers. *Semiconductor Science and Technology*, 1(1):7.

[43] Rhoderick EH, Williams RH (1988) Metal-Semiconductor Contacts, 2nd Edition, Clarendon Press.

[44] Sze SM, Ng KK (2006) Physics of Semiconductor Devices, 3rd Edition, Wiley.

[45] Scheer R, Schock H-W (2011) Chalcogenide Photovoltaics: Physics, Technologies, and Thin Film Devices 1st Edition, Wiley.

[46] Buttiker M, Thomas H, Pretre A (1993) Mesoscopic capacitors. *Physica Letters A*, 180:364–369.

[47] Bisquert J (2003) Chemical capacitance of nanostructured semiconductors: its origin and significance for nanocomposite solar cells. *Physical Chemistry Chemical Physics*, 5(24):5360–5364.

[48] Blood P, Orton JW (1992) The Electrical Characterization of Semiconductors: Majority Carriers and Electron States. Book 14 in series "Techniques of Physics", Academic Press.

[49] Blood P, Orton JW (1990) The Electrical Characterization of Semiconductors: Measurement of Minority Carrier Properties. Part of Series "Techniques in Physics", Academic Press.

[50] Luryi S (1988) Quantum capacitance devices. *Applied Physics Letters*, 52(6): 501–503.

[51] Kuciauskas D, Dippo P, Zhao Z, *et al.* (2016) Recombination analysis in cadmium telluride photovoltaic solar cells with photoluminescence spectroscopy. *IEEE Journal of Photovoltaics*, 6(1):313–318.

[52] Simmons JG, Taylor GW (1971) Nonequilibrium steady-state statistics and associated effects for insulators and semiconductors containing an arbitrary distribution of traps. *Physical Review B*, 4(2):502–511.

[53] Lutz G (1996) Effects of deep level defects in semiconductor detectors. *Nuclear Instruments and Methods in Physics Research Section A: Accelerators, Spectrometers, Detectors and Associated Equipment*, 377 (2):234–243.

[54] Reddy P, Kaess F, Tweedie J, *et al.* (2017) Defect quasi Fermi level control-based CN reduction in GaN: Evidence for the role of minority carriers. *Applied Physics Letters*, 111(15):152101.

[55] Alberi K, Scarpulla MA (2016) Suppression of compensating native defect formation during semiconductor processing via excess carriers. *Scientific Reports*, 6:srep27954.

[56] Alberi K, Scarpulla MA (2018) Effects of excess carriers on charged defect concentrations in wide bandgap semiconductors. *Journal of Applied Physics*, 123(18):185702.

[57] Kheraj V, Lund EA, Caruso AE, *et al.* (2016) Minority carrier electron traps in CZTSSe solar cells characterized by DLTS and DLOS. *2016 IEEE 43rd Photovoltaic Specialists Conference (PVSC)*, 2195–2199.

[58] Johnson NM, Bartelink DJ, Gold RB, Gibbons JF (1979) Constant-capacitance DLTS measurement of defect-density profiles in semiconductors. *Journal of Applied Physics*, 50(7):4828–4833.

[59] Walter T, Herberholz R, Müller C, Schock HW (1996) Determination of defect distributions from admittance measurements and application to Cu(In,Ga)Se_2 based heterojunctions. *Journal of Applied Physics*, 80 (8):4411–4420.

[60] Mora-Seró I, Garcia-Belmonte G, Boix PP, Vázquez MA, Bisquert J (2009) Impedance spectroscopy characterisation of highly efficient silicon solar cells under different light illumination intensities. *Energy & Environmental Science*, 2(6):678–686.

[61] Proskuryakov YY, Durose K, Al Turkestani MK, *et al.* (2009) Impedance spectroscopy of thin-film CdTe/CdS solar cells under varied illumination. *Journal of Applied Physics*, 106(4):044507.

[62] Burst JM, Duenow JN, Albin DS, *et al.* (2016) CdTe solar cells with open-circuit voltage breaking the 1 V barrier. *Nature Energy*, 1(3):16015.

[63] Zhao Y, Boccard M, Liu S, *et al.* (2016) Monocrystalline CdTe solar cells with open-circuit voltage over 1 V and efficiency of 17%. *Nature Energy*, 1(6):16067.

[64] Nagaoka A, Han K-B, Misra S, Wilenski T, Sparks TD, Scarpulla MA (2017) Growth and characterization of arsenic doped CdTe single crystals grown by Cd-solvent traveling-heater method. *Journal of Crystal Growth*, 467:6–11.

[65] Nagaoka A, Kuciauskas D, McCoy J, Scarpulla MA (2018) High p-type doping, mobility, and photocarrier lifetime in arsenic-doped CdTe single crystals. *Applied Physics Letters*, 112(19):192101.

[66] Nagaoka A, Kuciauskas D, Scarpulla MA (2017) Doping properties of cadmium-rich arsenic-doped CdTe single crystals: evidence of metastable AX behavior. *Applied Physics Letters*, 111(23):232103.

[67] Nagaoka A, Nishioka K, Yoshino K, Kuciauskas D, Scarpulla MA (2019) Arsenic doped Cd-rich CdTe: equilibrium doping limit and long lifetime for high open-circuit voltage solar cells greater than 900 mV. *Applied Physics Express*, 12(8):081002.

[68] Eisenbarth T, Unold T, Caballero R, Kaufmann CA, Schock H-W (2010) Interpretation of admittance, capacitance-voltage, and current-voltage signatures in $Cu(In,Ga)Se_2$ thin film solar cells. *Journal of Applied Physics*, 107(3):034509.

[69] Paul Weiss T, Redinger A, Luckas J, Mousel M, Siebentritt S (2013) Admittance spectroscopy in kesterite solar cells: defect signal or circuit response. *Applied Physics Letters*, 102(20):202105.

[70] Sozzi G, Napoli SD, Menozzi R, *et al.* (2017) Influence of Conduction Band Offsets at Window/Buffer and Buffer/Absorber Interfaces on the Roll-Over of J-V Curves of CIGS Solar Cells. *2017 IEEE 44th Photovoltaic Specialist Conference (PVSC)*: 2205–2208.

[71] Werner F, Siebentritt S (2018) Buffer layers, defects, and the capacitance step in the admittance spectrum of a thin-film solar cell. *Physical Review Applied*, 9(5):054047.

[72] Rashidi A, Nami M, Monavarian M, *et al.* (2017) Differential carrier lifetime and transport effects in electrically injected III-nitride light-emitting diodes. *Journal of Applied Physics*, 122(3):035706.

[73] Scofield JH (1995) Effects of series resistance and inductance on solar cell admittance measurements. *Solar Energy Materials and Solar Cells*, 37(2): 217–233.

[74] Li JV, Li X, Albin DS, Levi DH (2010) A method to measure resistivity, mobility, and absorber thickness in thin-film solar cells with application to CdTe devices. *Solar Energy Materials and Solar Cells*, 94(12):2073–2077.

[75] Schroder DK, Braggins TT, Hobgood HM (1978) The doping concentrations of indium-doped silicon measured by Hall, *C–V* , and junction-breakdown techniques. *Journal of Applied Physics*, 49(10):5256–5259.

[76] Scarpulla MA, Cardozo BL, Farshchi R, *et al.* (2005) Ferromagnetism in $Ga_{1-x}Mn_xP$: Evidence for inter-Mn exchange mediated by localized holes within a detached impurity band. *Physical Review Letters*, 95(20):207204.

[77] Gunawan O, Gokmen T, Warren CW, *et al.* (2012) Electronic properties of the $Cu_2ZnSn(Se,S)_4$ absorber layer in solar cells as revealed by admittance spectroscopy and related methods. *Applied Physics Letters*, 100(25):253905.

[78] Blakemore JS (1987) Semiconductor Statistics, Dover.

[79] Smith RA (1959) Semiconductors, Cambridge University Press.

[80] Caruso AE, Lund EA, Kosyak V, *et al.* (2016) Comparison and interpretation of admittance spectroscopy and deep level transient spectroscopy from co-evaporated and solution-deposited Cu2ZnSn(Sx, Se1 #x2212;x)4 solar cells. *2016 IEEE 43rd Photovoltaic Specialists Conference (PVSC)*, 3018–3022.

[81] Caruso AE, Pruzan DS, Lund EA, Scarpulla MA (2015) Effects of back contact resistance, depletion width and relaxation time distributions in admittance spectroscopy of CZTSe devices. *2015 IEEE 42nd Photovoltaic Specialist Conference (PVSC)*, 1–5.

[82] Caruso AE, Pruzan DS, Kosyak V, *et al.* (2014) Temperature dependence of equivalent circuit parameters used to analyze admittance spectroscopy and application to CZTSe devices. *2014 IEEE 40th Photovoltaic Specialist Conference (PVSC)*, 0733–0736.

[83] Lund EA (2014) Defect Characterization and Phase Analyses of Copper Zinc Tin Sulfide/Selenide ($Cu_2ZnSnSSe$), University of Utah.

[84] Hages CJ, Carter NJ, Moore J, *et al.* (2013) Device comparison of champion nanocrystal-ink based CZTSSe and CIGSSe solar cells: Capacitance spectroscopy. *2013 IEEE 39th Photovoltaic Specialists Conference (PVSC)*, 1966–1971.

[85] Koeper MJ, Hages CJ, Li JV, Levi D, Agrawal R (2017) Metastable defect response in CZTSSe from admittance spectroscopy. *Applied Physics Letters*, 111(14):142105.

[86] Grossberg M, Krustok J, Hages CJ, *et al.* (2019) The electrical and optical properties of kesterites. *Journal of Physics: Energy*, 1, 044002

[87] Levcenko S, Just J, Redinger A, *et al.* (2016) Deep defects in Cu_2ZnSnS,Se_4 solar cells with varying Se content. *Physical Review Applied*, 5(2):024004.

[88] Bueno PR, Varela JA, Longo E (2007) Admittance and dielectric spectroscopy of polycrystalline semiconductors. *Journal of the European Ceramic Society*, 27(13–15):4313–4320.

[89] Burgelman M, Nollet P (2005) Admittance spectroscopy of thin film solar cells. *Solid State Ionics*, 176(25–28):2171–2175.

[90] Decock K, Khelifi S, Burgelman M (2011) Accuracy of defect distributions measured by bias dependent admittance spectroscopy on thin film solar cells. *2011 37th IEEE Photovoltaic Specialists Conference*, 001691–001694.

[91] Decock K, Khelifi S, Pianezzi F, Tiwari AN, Burgelman M (2011) Grading and metastable effects in admittance spectroscopy of CIGS-based solar cells. *2011 37th IEEE Photovoltaic Specialists Conference*, 001262–001266.

[92] Decock K, Khelifi S, Buecheler S, Pianezzi F, Tiwari AN, Burgelman M (2011) Defect distributions in thin film solar cells deduced from admittance measurements under different bias voltages. *Journal of Applied Physics*, 110 (6):063722.

[93] Decock K, Khelifi S, Burgelman M (2011) An automated method to reduce noise of admittance spectra allowing assessment of a large number of complicated spectra. 4.

[94] Igalson M, Bodega M, Stolt L, Jasenek A (2003) The 'defected layer' and the mechanism of the interface-related metastable behavior in the ZnOyCdSyCu(In,Ga)Se_2 devices. *Thin Solid Films*, 5.

[95] Igalson M, Wimbor M, Wennerberg J (2002) The change of the electronic properties of CIGS devices induced by the 'damp heat' treatment. *Thin Solid Films*, 403–404:320–324.

[96] Igalson M, Zabierowski P (2003) Electron traps in Cu(In,Ga)Se_2 absorbers of thin film solar cells studied by junction capacitance techniques. *Opto-Electronics Review*, 11(4), 261–267.

[97] Igalson M, Platzer-Björkman C (2004) The influence of buffer layer on the transient behavior of thin film chalcopyrite devices. *Solar Energy Materials and Solar Cells*, 84(1–4):93–103.

[98] Igalson M, Urbaniak A, Edoff M (2009) Reinterpretation of defect levels derived from capacitance spectroscopy of CIGSe solar cells. *Thin Solid Films*, 517(7):2153–2157.

[99] Kask E, Grossberg M, Josepson R, Salu P, Timmo K, Krustok J (2013) Defect studies in $Cu_2ZnSnSe_4$ and $Cu_2ZnSn(Se_{0.75}S_{0.25})_4$ by admittance and photoluminescence spectroscopy. *Materials Science in Semiconductor Processing*, 16(3):992–996.

[100] Lauwaert J, Callens L, Khelifi S, *et al.* (2012) About RC-like contacts in deep level transient spectroscopy and Cu(In,Ga)Se_2 solar cells: about RC-like contacts in DLTS and Cu(In,Ga)Se_2 solar cells. *Progress in Photovoltaics: Research and Applications*, 20(5):588–594.

[101] Lauwaert J, Decock K, Khelifi S, Burgelman M (2010) A simple correction method for series resistance and inductance on solar cell admittance spectroscopy. *Solar Energy Materials and Solar Cells*, 94(6):966–970.

[102] Li JV, Crandall RS, Repins IL, Nardes AM, Levi DH (2011) Applications of admittance spectroscopy in photovoltaic devices beyond majority-carrier trapping defects. *2011 37th IEEE Photovoltaic Specialists Conference*, 000075–000078.

[103] Li JV, Ferarri, G. (Eds) (2018) *Capacitance Spectroscopy of Semiconductors*. Jenny Stanford Publishing, Singapore. ISBN 978-981-4774-54-3.

[104] Li JV, Levi DH (2011) Determining the defect density of states by temperature derivative admittance spectroscopy. *Journal of Applied Physics*, 109(8):083701.

[105] Losee DL (1972) Admittance spectroscopy of deep impurity levels: ZnTe Schottky barriers. *Applied Physics Letters*, 21(2):54–56.

[106] Maier D, Hug P, Fiederle M, Eiche C, Ebling D, Weese J (1995) High resolution method for the analysis of admittance spectroscopy data. *Journal of Applied Physics*, 77(8):3851–3857.

[107] Oueslati S, Brammertz G, Buffière M, *et al.* (2015) Physical and electrical characterization of high-performance $Cu_2ZnSnSe_4$ based thin film solar cells. *Thin Solid Films*, 582 (May):224–228.

[108] Pautrat JL, Katircioglu B, Magnea N, Bensahel D, Pfister JC, Revoil L (1980) Admittance spectroscopy: a powerful characterization technique for semiconductor crystals—Application to ZnTe. *Solid-State Electronics*, 23 (11):1159–1169.

[109] Singh DV, Rim K, Mitchell TO, Hoyt JL, Gibbons JF (1999) Admittance spectroscopy analysis of the conduction band offsets in $Si/Si_{1-x-y}Ge_xC_y$ and $Si/Si_{1-y}Cy$ heterostructures. *Journal of Applied Physics*, 85(2):985–993.

[110] Li JV, Johnston SW, Li X, Albin DS, Gessert TA, Levi DH (2010) Discussion of some "trap signatures" observed by admittance spectroscopy in CdTe thin-film solar cells. *Journal of Applied Physics*, 108(6):064501.

[111] Cohen JD, Zhong F (1995) Transient modulated photocurrent studies in hydrogenated amorphous silicon: a new look at defect relaxation dynamics. *Journal of Non-Crystalline Solids*, 190(1–2):123–132.

[112] Hamadani BH, Roller J, Kounavis P, Zhitenev NB, Gundlach DJ (2013) Modulated photocurrent spectroscopy of CdTe/CdS solar cells—equivalent circuit analysis. *Solar Energy Materials and Solar Cells*, 116 (September):126–134.

[113] Herberholz R, Walter T, Schock HW (1994) Density of states in $CuIn(SSe)_2$ thin films from modulated photocurrent measurements. *Journal of Applied Physics*, 76(5):2904–2911.

[114] Kleider JP, Longeaud C, Gueunier ME (2004) Investigation of bandgap states using the modulated photocurrent technique in both high and low frequency regimes. *Journal of Non-Crystalline Solids*, June, 338–340:390–399.

[115] Krysztopa A, Igalson M, Zabierowski P, et al. (2011) Defect levels in $CuGaSe_2$ by modulated photocurrent spectroscopy. *Thin Solid Films*, 519 (21):7308–7311.

[116] Krysztopa A, Igalson M, Aida Y, Larsen JK, Gütay L, Siebentritt S (2011) Defect levels in the epitaxial and polycrystalline $CuGaSe_2$ by photocurrent and capacitance methods. *Journal of Applied Physics*, 110(10):103711.

[117] Longeaud C, Kleider JP, Kaminski P, Kozlowski R, Pawlowski M, Cwirko J (1999) Investigation of defect levels in semi-insulating materials by modulated and transient photocurrent: comparison of methods. *Semiconductor Science and Technology*, 14(9):747–756.

[118] Schumm G, Bauer GH (1989) Spatially resolved and energy-resolved defect kinetics in a-Si:H: A comprehensive study by phase-shift analysis of modulated photocurrents. *Physical Review B*, 39(8):5311–5326.

[119] Li JV, Halverson AF, Sulima OV, *et al.* (2012) Theoretical analysis of effects of deep level, back contact, and absorber thickness on capacitance–voltage profiling of CdTe thin-film solar cells. *Solar Energy Materials and Solar Cells*, 100:126–131.

[120] Erslev PT, Lee J, Hanket GM, Shafarman WN, Cohen JD (2011) The electronic structure of $Cu(In_{1-x}Ga_x)Se_2$ alloyed with silver. *Thin Solid Films*, 519(21):7296–7299.

[121] Warren CW, Roe ET, Miller DW, Shafarman WN, Lonergan MC (2017) An improved method for determining carrier densities via drive level capacitance profiling. *Applied Physics Letters*, 110(20):203901.

[122] Mooney PM (1999) Defect Identification using capacitance spectroscopy. *Semiconductors and Semimetals*, (Eds) RK Willardson and E. Weber. Vol. 51b, pp. 93–152.

[123] Šantić B, Desnica UV (1990) Thermoelectric effect spectroscopy of deep levels—application to semi-insulating GaAs. *Applied Physics Letters*, 56 (26):2636–2638.

[124] Soundararajan R, Lynn KG, Awadallah S., Szeles C., Wei SH (2006) Study of defect levels in CdTe using thermoelectric effect spectroscopy. *Journal of Electronic Materials*, 35(6):1333–1340.

[125] Wessels BW (1976) Determination of deep levels in Cu-doped GaP using transient-current spectroscopy. *Journal of Applied Physics*, 47(3):1131–1133.

[126] Borsuk JA, Swanson RM (1980) Current transient spectroscopy: a high-sensitivity DLTS system. *IEEE Transactions on Electron Devices*, 27 (12):2217–2225.

[127] Farmer JW, Lamp CD, Meese JM (1982) Charge transient spectroscopy. *Applied Physics Letters*, 41(11):1063–1065.

[128] Gelatos AV, Mahavadi KK, Cohen JD, Harbison JP (1988) Transient photocapacitance and photocurrent studies of undoped hydrogenated amorphous silicon. *Applied Physics Letters*, 53(5):403–405.

[129] Hierro A, Kwon D, Ringel SA, *et al.* (2000) Optically and thermally detected deep levels in n-type Schottky and p+-n GaN diodes. *Applied Physics Letters*, 76(21):3064–3066.

[130] Armstrong A, Arehart AR, Moran B, *et al.* (2004) Impact of carbon on trap states in n-type GaN grown by metalorganic chemical vapor deposition. *Applied Physics Letters*, 84(3):374–376.

[131] Chantre A, Vincent G, Bois D (1981) Deep-level optical spectroscopy in GaAs. *Physical Review B*, 23(10):5335–5359.

[132] Dobaczewski L, Hawkins ID, Peaker AR (1994) Laplace transform deep-level transient s ectroscopic studies of defects in semiconductors. *Journal of Applied Physics*, 76(1):5.

[133] Dobaczewski L, Peaker AR, Bonde Nielsen K (2004) Laplace-transform deep-level spectroscopy: The technique and its applications to the study of point defects in semiconductors. *Journal of Applied Physics*, 96(9):4689–4728.

[134] Sozzi G, Lazzarini M, Menozzi R, *et al.* (2016) A numerical study of the use of C-V characteristics to extract the doping density of CIGS absorbers. *2016 IEEE Photovoltaic Specialists Conference*, 1–6

[135] Sozzi G, Troni F, Menozzi R (2010) Numerical analysis of the effect of grain size and defects on the performance of CIGS solar cells. *2010 International Conference on Compound Semiconductor Manufacturing Technology, CS MANTECH 2010.*

Chapter 10

Nanometer-scale characterization of thin-film solar cells by atomic force microscopy-based electrical probes

Chun-Sheng Jiang[1]

10.1 Introduction

The most prominent differences in the physical properties of polycrystalline thin-film materials from their single-crystalline counterpart are structural inhomogeneity and lattice periodic discontinuity, such as grain boundaries (GBs) between grains, individual grain orientations, and inhomogeneity among the grains. These structural nonuniformities can cause nonuniform electrical and optoelectrical distribution and affect solar cell power output [1]. Solar cell performance parameters and their most related optoelectronic and electrical properties—such as minority-carrier lifetime and majority-carrier doping concentration—are affected by the polycrystalline structure and are often discussed directly in terms of nanometer (nm)—and atomic-scale structures [1,2]. The nm-scale localized electronic properties—that is, direct results of local atomic structures—should be understood to better link the microstructure to macroscopic electrical properties of solar cells. However, such nm-scale properties are often overlooked [3].

A distinguishing feature of polycrystalline photovoltaic (PV) devices is the use of materials differing from the active absorber for electron/hole transport materials (ETM/HTM); thus, heterogeneous interfaces are formed on both sides of the electron and hole contacts [2]. All of the most popular polycrystalline thin-film devices have heterointerface structures. For example, the typical organic–inorganic perovskite has a TiO_2/perovskite interface on the ETM side and a Spiro/perovskite interface on the HTM side. CdTe has CdS/CdTe and ZnTe/CdTe interfaces, and Cu (In,Ga)Se_2 (CIGS) has CdS/CIGS and $MoSe_x$/CIGS interfaces [2]. The interface design is based on the band alignment, doping type and concentration, and defect tolerance and passivation, as well as the assumption of an abrupt interface [1,4]. However, because thin-film deposition and optimization often requires high-temperature and high-energy (e.g., sputter) processing, the interface chemistry and structures are not necessarily abrupt as designed; diffusion as well as intermixing is

[1]Materials Science Center, National Renewable Energy Laboratory, Golden, Colorado, USA

likely to occur [2]. Furthermore, the structure in the vicinity of interfaces on both sides can also be different from their inner material layers by the nature of the layer surface even before a top layer is deposited, or by the effect of deposition of adjacent layers [2]. These unexpected differences in the vicinity of an interface can significantly alter the junction quality because the junction usually depends very sensitively on some of the chemical/structural characteristics—for example, a trace element on the order of ppm for either *n*- or *p*-type doping, or a small conduction offset in either a "spike" or "cliff" shape [5–7]. Therefore, rather than making inferences from the chemical and structural data, direct characterization of these microelectronic/electrical properties at the nm-scale are highly sought to determine these crucial device factors and understand how the microchemistry and structure affect macroscopic PV performance.

Since the invention of atomic force microscopy (AFM) [8,9], the primary conventional use is to image surface morphology at a nm resolution laterally and sub-nm resolution vertically. Also, since the invention of AFM, electrical mapping simultaneously with morphology mapping has been developed greatly and applied successfully in many fields [10–14]. AFM detects a mechanical force in the sub-nN to μN range between the probe and a sample surface to image the surface morphology. In addition to the mechanical force, many electrical interactions occur between the probe and the local sample area beneath the probe that can be probed and used for mapping the local electrical properties in nm resolutions. The techniques closely related to thin-film PV include but are not limited to Kelvin probe force microscopy (KPFM) [10,15], scanning capacitance microscopy (SCM) [12,16], scanning spreading resistance microscopy (SSRM) [11,17–20], scanning microwave impedance microscopy (SMIM) [14,21], piezoresponse force microscopy (PFM) [13,22]. Respectively, these techniques probe the Coulomb force, capacitance, resistance, microwave reflection, and piezo resonance between the probe and sample—for mapping, respectively, the local electrostatic potential, charge-carrier type and concentration, local resistivity, local conductance and capacitance, and local piezo coefficient. These techniques provide unprecedented spatial resolutions—from a few to several tens of nanometers—for electrical mapping, depending on the nature of the electrical signal and sample and the probe sizes. The results fill the knowledge gap between nm- and atomic-scale structure and chemistry and cell-level electrical and optoelectrical behavior. Among these techniques, KPFM and SSRM, which relate to the following characterization reviews, will be introduced briefly in Section 10.2. Some recent nm-scale characterizations of perovskite, CIGS, and CdTe thin-film solar cell materials and devices will be reviewed in Sections 10.3–10.5 and closing remarks will be given in Section 10.6.

10.2 AFM-based nanoelectrical probes

10.2.1 Kelvin probe force microscopy

KPFM measures a surface electrical potential distribution, with a fine resolution of tens of nanometers, by probing the Coulomb force between the tip and sample [10,15]. A contact potential difference (CPD) occurs if the work function of the AFM tip is different from that of the sample, when the tip and sample are electrically connected or are both connected to the ground level. This CPD creates a charge transfer between the tip and sample that aligns the Fermi level and generates a Coulomb force between tip and sample. The charging energy E and Coulomb force F are

$$E = \frac{1}{2} C V_{CPD}^2 \tag{10.1}$$

$$F = \frac{\partial E}{\partial z}, \tag{10.2}$$

where C, V_{CPD}, and Z are, respectively, the capacitance, CPD, and separation distance between the tip and sample [10]. However, the Coulomb force is too small to be detected because of the small capacitance (~aF) due to the small AFM tip size (~10 nm). It is also difficult to separate the Coulomb force from the atomic force that dominates during topographic imaging. To enhance the Coulomb force and separate it from the atomic force, an AC voltage with an amplitude of 1–2 V is usually applied to the tip. In a one-dimensional approximation, the Coulomb force can be written as

$$F = F_0 + F_1 + F_2 \tag{10.3}$$

$$F_0 = \frac{1}{2} \frac{\partial C}{\partial z} \left(V_{CPD}^2 + \frac{1}{2} V_{ac}^2 \right) \tag{10.4}$$

$$F_1 = \frac{\partial C}{\partial z} V_{CPD} V_{ac} \sin(\omega t) \tag{10.5}$$

$$F_2 = -\frac{1}{4} \frac{\partial C}{\partial z} V_{ac}^2 \cos(2\omega t), \tag{10.6}$$

where V_{CPD} equals the work function difference ($W_{\mathrm{tip}} - W_s$) of the tip and sample in the thermal equilibrium state, and equals ($W_{\mathrm{tip}} - W_s - V_{\mathrm{tip}} + V_{\mathrm{s}}$) if the tip and sample were applied with bias voltages of V_{tip} and V_{s} [23]. The first term, F_0, is a constant with time and applies a constant bending on the cantilever. The second term, F_1, oscillates in the same frequency as the AC voltage, and its amplitude is proportional to V_{CPD}. The third term, F_2, oscillates in a double frequency of the AC voltage, and the amplitude is dominated by the capacitance characteristics and the amplitude of AC voltage. F_1 can be detected using a lock-in amplifier, and its amplitude is proportional to V_{CPD}. Therefore, the experimental setup for probing the term F_1 is called electrostatic force microscopy (EFM) [24]. EFM gives a two-dimensional mapping of the Coulomb force; it is a qualitative measurement of the sample surface potential.

Using $F_1 \propto V_{CPD}$ and $V_{CPD} = W_{tip} - W_s - V_{tip} + V_s$, if a proper voltage $V_{tip} = V_s + W_{tip} - W_s$ is applied to the tip, then V_{CPD} and F_1 are minimized, and the tip voltage V_{tip} in this case is the Kelvin probe signal, which is essentially similar to the classical Kelvin probe signal, with fine resolution concentrating on the local sample area [10,23]. In the experimental setup shown in Figure 10.1, AFM uses a noncontact mode for tracing the topography because probing the Coulomb force term F_1 requires that the tip oscillates. The cantilever oscillation signal is filtered to two components, with frequencies corresponding to topographic (AFM) and surface potential (KPFM) measurements. The electrical component is then detected by a lock-in amplifier, and the output of the amplifier is further sent to a negative feedback circuit. The output of the feedback circuit is the Kelvin probe V_{KP}. The qualitative description would be that if V_{tip} deviates from V_{KP}, then V_{CPD} increases and consequently F_1 increases, and then V_{tip} changes in the direction to reduce the V_{CPD} until V_{tip} reaches V_{CPD}.

A typical AFM tip used in KPFM has three local oscillation maxima. The first maximum in the 0–20 kHz range is the low frequency; the second in the ~50–90 kHz range is the first resonant oscillation; and the third in the ~300–500 kHz range is the second resonant oscillation. The first resonant oscillation is mostly used for noncontact-mode topographic measurement. Either a low frequency in the 5–20 kHz range or the second resonant oscillation is used for the KPFM measurement. For the majority of tips, the second resonant oscillation has a stronger oscillation amplitude; thus, it has a better signal sensitivity or energy resolution (~10 mV) than in the low-frequency range (~50 mV) [15]. However, the potential image obtained by using the second resonant frequency has been found to include more topographic effect than the low frequency, which is not suitable for imaging on a rough surface with corrugations larger than ~100 nm. During the tip scanning, topographic and electrical images were obtained simultaneously by using atomic force and Coulomb force at the corresponding frequencies. Thus, the oscillations cannot be completely separated by the filters and lock-in amplifiers due to the cross-talk atomic and Coulomb forces. The oscillations between first and second resonant frequencies are more closely related to each other than those between the first and the off-peak low frequency. So,

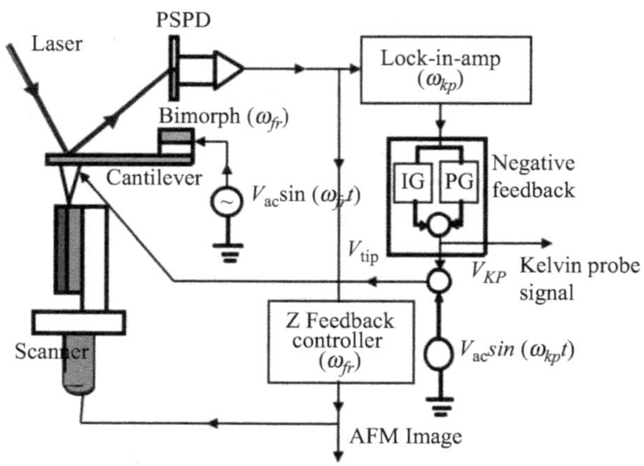

Figure 10.1 A schematic of KPFM setup

the potential image obtained by using the second resonant frequency includes more topographic effect than the low frequency when imaging on a rough surface.

10.2.2 Scanning spreading resistance microcopy

SSRM maps the local electrical conduction path and local resistivity of a semiconductor at nm resolution [20,25,26]. SSRM is based on the contact mode of AFM, where the probe is kept in contact with the sample surface at a constant contact force during the probe scanning. The resistance is measured by applying a DC bias voltage (V_s) between the probe and sample and measuring the current flow through the probe $R = V/I$, which involves all the resistances serially connected along the current route [17]. The principal contributors to this overall resistance are the probe/sample contact resistance R_c, the resistance through the sample and back contact R_b, and the spreading resistance R_s. The target R_s in this measurement is related only to the sample property, located beneath the probe and dominated by the nm-scale sampling volume. The local resistivity of a sample with lateral spatial resolution comparable to the probe size (~30 nm) is given by $\rho = 4 \cdot r \cdot R_s$, with r being the radius of the contact area [17]. The back-contact resistance R_b is generally much smaller than R_s, because the electronic conduction channel increases rapidly as it spreads out from a quasi-point contact, and the voltage drop is bound by the sample volume just beneath the probe (Figure 10.2(a)). Therefore, for R_s to dominate the overall R, R_c is the main factor to be reduced to a level sufficiently less than R_s; this is achieved by adequate probe contact forces (1–100 μN) and applying an adequately large V_s (0.5–10 V) [26–28]. An adequate contact force would deform the local area beneath the probe, create high-density strained and dangling bonds (Figure 10.2(c)), and make an ohmic contact with adequately small R_c [26]. However, if R_c has a significant non-ohmic component, the equivalent resistance can also be reduced by applying a relatively large V_s in the forward bias direction [27,28]. R_c can usually be reduced under R_s in thin-film PV materials and low-doped Si materials for PV applications. A decrease in the overall R with

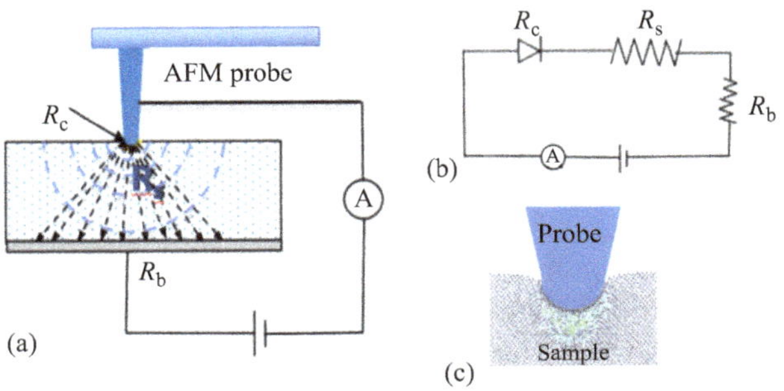

Figure 10.2 (a) A schematic illustration of SSRM setup; (b) an equivalent circuit along the measurement current loop; and (c) a schematic showing the local lattice distortion by the probe contact force that reduced the contact resistance R_c

increasing contact force and V_s results from the decrease in R_c because other contributors to the resistance in the current route should not change.

SSRM shares identical principles of two-probe resistance measurement with other versions of measurement, e.g., conductive AFM (C-AFM) and tunneling AFM (TUNA). But SSRM is equipped with a logarithmic current–voltage converter, which allows a wider measurement range of current from milli-ampere (10^{-3} A) to femto-ampere (10^{-15} A) but less accuracy than C-AFM, with a linear current amplifier in a current range from micro-ampere (10^{-6} A) to pico-ampere (10^{-12} A) and TUNA in 10 femto-ampere (10^{-14} A) to 100 pico-ampere (10^{-10} A). The wide range of the SSRM amplifier is necessary to measure a huge change of sample local resistivity and low-doped semiconductors, but it sacrifices bandwidth. The response of the current change is slow and current–voltage (I–V) measurement with changing V_s at a fixed probe position is usually not practical. Design of the current amplifier is due to specific detection of current range for the specific samples and research goals.

In SSRM instrumentation, bias voltage V_s is usually applied to the sample, and the probe is virtually grounded. A probe with stiff wear-resistance and high electrical conductance such as a highly doped diamond or diamond-coated tip is recommended for reliable resistance mapping and to last for many image scans [19] because a relatively large force is required for reducing R_c, and a highly conductive probe for less resistance contribution from the current loop. In addition to the two-dimensional (2D) resistance mapping, three-dimensional (3D) mapping has been developed by abrasive raster scans over the same planar location [29,30]. The sample material is scratched out controllably by the probe/sample contact force, simultaneously with the resistance mapping. The resistance maps at progressive depths are taken, and a 3D map of the resistance is constructed by the set of 2D resistance maps. To maximize reproducibility and data quality for the 3D mapping, it is preferable to perform measurements with the same probe on multiple samples using the same raster scan parameters. Changes to parameters (e.g., contact force, V_s, specific probe geometry) may influence quantitative resistance measurements. To monitor data reproducibility and probe condition, it is recommended that a reference sample is periodically measured before and between the measurement of multiple samples [30].

10.3 Potential profiling across perovskite devices

Inhomogeneous electrical properties in perovskite thin-film devices are exhibited in both planar and vertical directions. High-performance parameters and other optoelectronic characterizations of the perovskite device suggest relatively inert GBs and grain interiors for minority-carrier recombination [31–33]. However, significant electrical potential contrasts exist between GBs and grain interior and significant inhomogeneity among the grains [34–36]. These potential contrasts as revealed by the nanoelectrical probes may result from effects other than charged minority-carrier traps at GBs, such as phase segregation, which does not create deep levels (different from inorganic thin films) but passivates the GBs instead. The vertical inhomogeneous electrical properties depend on the ETMs and HTMs, interface passivation, and detailed band alignments [37–39]. Many nanoelectrical characterizations use AFM-based techniques, and in this section, potential profiling across perovskite devices using KPFM is reviewed [40–42].

The potential profile across a PV device under various conditions such as a bias voltage V_b and illumination is a direct assessment of the device operation such as the junction structure and interface quality [40,41]. KPFM images the electrical potential on a sample surface typically in spatial resolutions of ~30 nm and energy resolutions of ~10 mV. To image the potential across the device, the device is cleaved to expose cross-sections. However, the potential is affected strongly by charges trapped on the cross-sectional surface and in the near-surface region [43,44]. Any charges deeper than a screening length (in the range of hundreds of nanometers for most PV materials) have negligible effect on the surface potential. To avoid the surface charge effect and to profile the electrical potential in the device bulk, a V_b was applied to a working device, and the potential change was measured. This change in surface potential should be about the same as the potential change in the bulk provided that V_b is sufficiently small (such as 1–2 V) [43,44]. In other words, the V_b-induced potential change in the bulk can be evaluated by measuring the change in surface potential because the configuration of surface charges trapped at the localized surface states should not change significantly with such a small V_b. Figure 10.3(a) illustrates that the built-in potential of a *p-n* junction can be flattened by the surface band-bending (or surface charges), such that the bulk built-in potential (V_{bi}) of the junction does not appear properly on the surface. The junction location and electrical potential around it can be produced on the surface potential by applying a V_b to the junction (Figure 10.3(b)) [43,44].

This potential profiling across the device was applied to perovskite devices with various active and contact layer materials, and various device performances

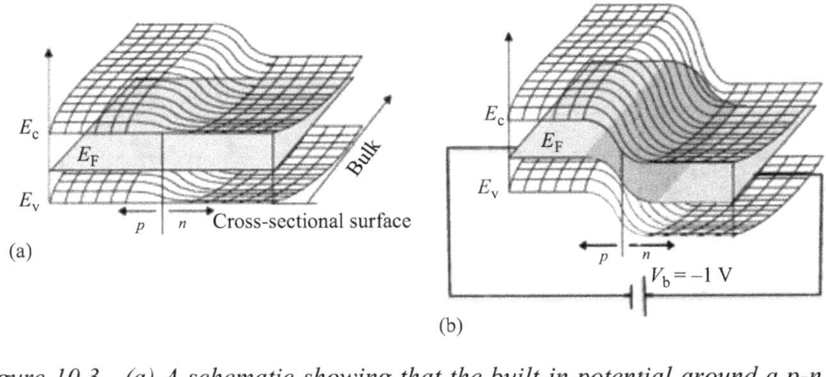

Figure 10.3 (a) A schematic showing that the built-in potential around a p-n junction is reduced on the cross-sectional surface and (b) shows the equivalent change of surface potential to the potential change in the bulk induced by a reverse-bias voltage

from different groups and at different times, which has gained knowledge about device fundamentals of junction configurations and carrier separation and transport [40–42]. In an earlier time of perovskite development, potential profiling showed a *p-n* junction-like ETM/perovskite interface on a planar device [40]. Figure 10.4(a) shows the layer structure of the device as illustrated by scanning electron microscopy (SEM); Figure 10.4(b) shows potential profiles with varying V_b from -1.5 V (reverse) to $+0.75$ V (forward); Figure 10.4(c) shows potential changes from that at $V_b = 0$, by simply subtracting the $V_b = 0$ potential from those obtained at various V_b values, to avoid the effect of static surface charge; Figure 10.4(d) shows the first derivative of the potential difference to obtain the V_b-induced electric-field distribution [40]. The potential drop of the electric field occurs around the interface between the transparent conductive oxide (TCO)/TiO_2 ETM and perovskite active layers, with the electric-field maximum being located at the TiO_2/perovskite interface. This potential/field picture demonstrates that the perovskite and TiO_2/TCO layers form a *p-n* junction-like interface, which is different from both an excitonic cell and an *n-i-p* junction of free carriers, because in both cases an electric field across the absorber layer would be expected [2].

The potential drop extends ~300 nm from the TiO_2/perovskite interface when $V_b = -1$ V, and no field is observed at deeper locations in the perovskite layer, which can be used to roughly estimate the depletion width of the *p-n* junction in the bulk [43,44], because the surface potential drop amplitude at $V_b = -1$ V roughly agrees with the bulk built-in potential $V_{bi} \sim 1.2$ V [45]. The carrier concentration in the perovskite layer is roughly estimated to be $\sim 7 \times 10^{16}/\text{cm}^3$, using the depletion width $W = 300$ nm, $V_{bi} = 1.2$ V, and dielectric constant $\varepsilon = 70$ [46]. This carrier concentration is in the range similar to high-performance CIGS, without intentional

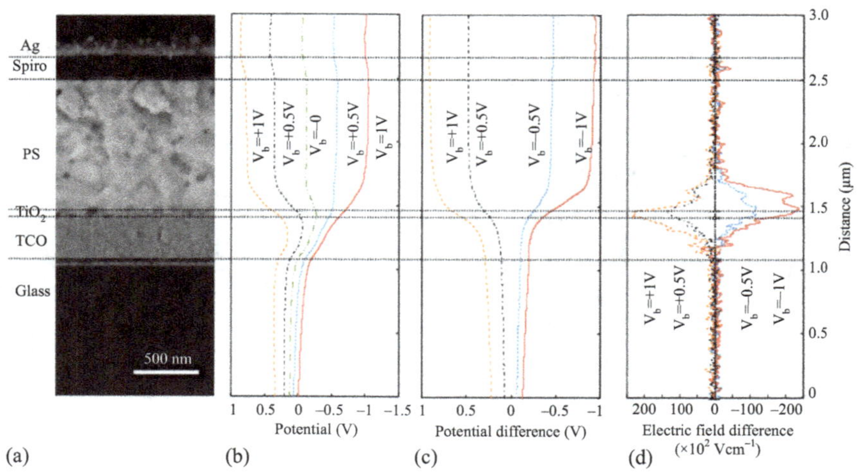

Figure 10.4 (a) An SEM image showing the layer structure of a perovskite device; (b) potential profiles across the device with the various V_b values; (c) potential changes by the V_b; and (d) electric-field changes

foreign doping during the film growth and other device fabrication steps. Referring to the intrinsic *p*-type doping of point defects in CIGS, which results from shallow acceptor levels, intrinsic defect doping could be responsible for the *p*-type carriers in perovskite film. This intrinsic doping was further supported by similar potential profiles taken on a device without the HTM layer, which excludes the foreign doping from the HTM [40].

Another interesting and important finding in this potential profiling is an additional electric field in the capping layer of a porous device structure, which acts as an unintentional back-surface field to reflect electrons and benefits the device performance [40].

This characterization work demonstrated that perovskite devices subjected in this work have a *p-n* junction structure with *p*-type perovskite in $\sim 10^{16}/cm^3$ doping concentration and the minority-carrier diffusion/drift operation, rather than the mechanisms of exciton separation and *p-i-n* structure of free carriers [40]. The same potential profiling on the later devices made by another group also found an electric-field peak at the HTM/perovskite interface, and this procedure was further expanded to assess and compare the qualities of different ETM/perovskite interfaces/junctions by measuring the ratio of potential drop at both the interfaces while keeping all the HTM/perovskite interfaces the same [41].

A potential profiling result across the device with an ETM-free cell is shown in Figure 10.5(a) [41]. In addition to the electric-field peak at the ETM/perovskite interface, another large electric field on a perovskite/Spiro interface was observed. If the perovskite active layers and the Spiro/perovskite interfaces at all the devices subjected are identical, then the large electric-field peak at the perovskite/HTM back interface suggests a poor main junction at the ETM/perovskite interface. A bias voltage is applied to the device because the electric current through the device or through the front junction and back barrier must be the same. If there is a significant voltage drop at the back side, then the voltage drop at the front junction must be reduced. This indicates a reduced equivalent shunt resistance in the front junction, resulting from poor junction quality or increased reverse saturation current J_0 and/or diode ideality factor. Therefore, the front junction can be assessed using the ratio of voltage drop between the front and back sides as identified from the electric-field profile. The larger field ratio of front/back junction indicates a better front-junction quality.

SAM is reported to improve the cell performance by promoting charge extraction of ETM, passivating the ETM/perovskite interface defect states, or making cleaner interface of ETM/perovskite [47–49]. The same profiling procedure was applied to two other devices with ETMs of intrinsic SnO_2 (15 nm) and intrinsic SnO_2 + SAM (self-assembled monolayer). The electric-field ratio between the front and back interfaces and all the PV performance parameters are listed in Table 10.1. The degree of voltage drops at the ETM/perovskite junction relative to that at the perovskite/Spiro interface is significantly larger in the SnO_2 device (Figure 10.5(b)) than in the ESL-free device. This result indicates a better diode quality factor of the ETM/perovskite junction and/or a smaller J_0 of the SnO_2 device than the ESL-free device, which is consistent with the open-circuit voltage

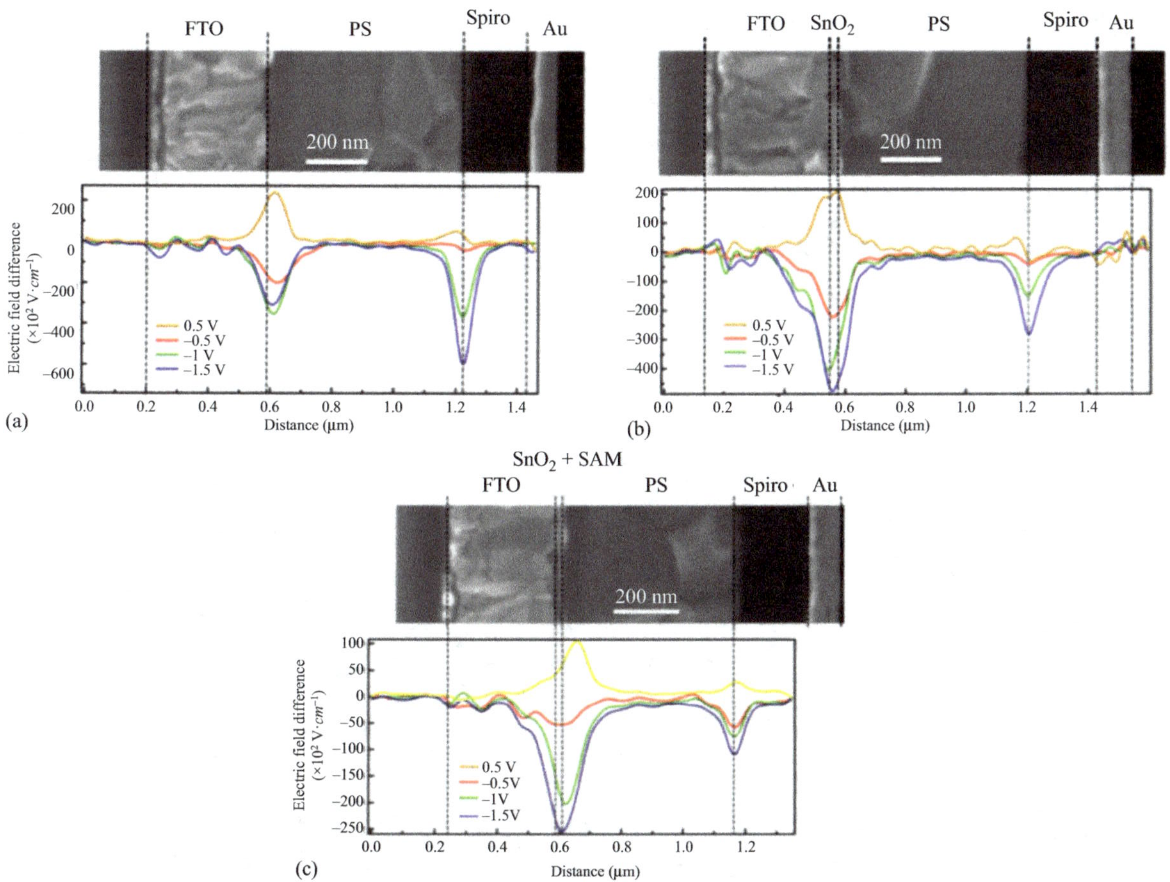

Figure 10.5 *SEM images and electric-field profiles taken on cross-sections of perovskite devices with ETMs of (a) ETM-free, (b) intrinsic SnO_2, and (c) SnO_2 + SAM*

Table 10.1 PV performance parameters and electric-field peak ratios of the three perovskite cells

Cell type	Scan direction	V_{OC}(V)	J_{SC}(mA/cm^2)	FF(%)	Eff.(%)	Peak ratio
ESL-free	Reverse	0.94	21.64	68.10	13.91	0.39
	Forward	0.78	21.62	58.08	9.85	
SnO_2 ESL	Reverse	1.07	22.40	73.85	17.78	1.65
	Forward	1.03	22.40	70.68	16.33	
SnO_2 + SAM	Reverse	1.09	23.20	76.39	19.28	3.25
	Forward	1.09	23.20	76.35	19.25	

(V_{oc}) difference of the two devices. In other words, if the front junction is better formed and the reverse current flowing through the junction is reduced, then the potential/voltage drop at the backside interface will decrease. In this case, the two junctions will compete less, which leads to better performance. This result highlights that SnO_2 can work as an effective hole-blocking layer because it prevents photo-generated hole recombination at the FTO/perovskite interface. On the other hand, the smaller voltage drop at the perovskite/Spiro interface of the SnO_2 device is consistent with the pronounced fill factor (FF) and V_{oc} gain. FF should be a significantly affected parameter by this "back-contact" voltage drop. This is because voltage loss at the back side of the device greatly affects the voltage at the maximum power output point of a current density–voltage curve, and hence, greatly affects FF.

The reduction in defect states at the interface may lead to a lower nonradiative recombination rate and thus improve the junction quality. As expected, the KPFM result on the SnO_2 + SAM ETL device shows the largest peak ratio of ETM/perovskite versus perovskite/HTM among the three types of SnO_2-based cells. The main potential drop is at the *p-n* junction formed by SnO_2/SAM/perovskite; the peak at perovskite/Spiro (relative scale) is smaller than the SnO_2 cell without SAM. Although the back-contact materials are the same among the three devices, the voltage drop at the backside is different because the equivalent shunt resistance of the front junction—and thus, the total equivalent shunt resistance—is different among the three devices. The smaller peak at the perovskite/Spiro side indicates a larger shunt resistance, and thus, smaller J_0 and a better interface quality of SnO_2/SAM/perovskite than SnO_2/perovskite.

These KPFM results are consistent with cell performances (Table 10.1); the larger the electric-field peak ratio, then the better the front-junction quality, the lower the voltage loss at the back side, and the better the FF and V_{oc}. The improvement by adding the SAM layer is likely caused by defect-state passivation at the SnO_2/perovskite interface. The SAM layer may also promote charge extraction [49–51]. However, its effect may be minor because the layer is very thin, and carriers can tunnel through it. The consistency of electric field and PV performance also showed up with the *I–V* hysteresis: the larger the peak ratio, the smaller the hysteresis [41]. However, this voltage drop measured by potential profiling cannot deduce detailed mechanisms of the junction improvement by changing the SnO_2-based ETM processing, such as interface passivation, band alignment, or prevention of interface reaction.

The potential profiling was also applied to measure potential changes at the ETM/perovskite interface by Li-ion migration [42]. Li ions migrated from an external doping source in HTM to the ETM/perovskite interface and improved the device performance by enhancing electron transport. With a pre-bias voltage polling, the ions move under the electric field and accumulate further at or are swept away from the ETM/perovskite interface. After the pre-polling, Li ions moved back to their thermal equilibrium states after minutes, which was observed by the potential profiling (Figure 10.6)

10.4 Determination of junction location in Cu(In,Ga)Se_2 devices

As minority-carrier lifetime of absorber materials in the state-of-the-art thin- film devices has greatly improved, interfaces can be the limiting factor that should be improved to significantly advance thin-film technologies [52]. In addition to the junction optoelectronic properties such as band alignment and recombination velocity, junction location can be one of the critical characteristics that changes the carrier transport and mitigates recombination losses around the junction [53,54]. Junction location in heterointerface thin-film solar cells can be much more complicated than device designs because of interdiffusion of elements during and after deposition of the multiple thin layers [53–55]. The diffusion can be either extensive at μm-scale or shallow at nm-scale, but it may all significantly affect device functions [56]. Therefore, determining the junction location in nm-resolution is greatly needed to further engineer the interface. With the most prominent capability

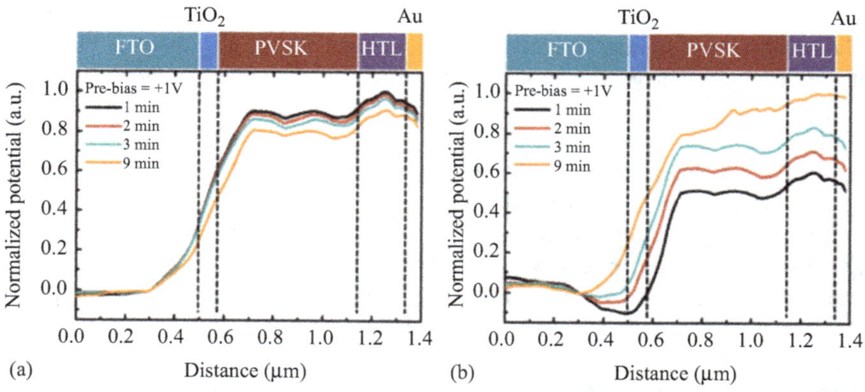

Figure 10.6 Potential profiles across devices with HTM with Li-salt doping after pre-bias voltage polling under (a) forward bias $V_b = +1$ V and (b) reverse bias $V_b = -1$ V

of nm-resolution potential imaging, KPFM was applied to determine the junction location in Cu(In,Ga)Se_2 (CIGS) devices, and it found critical differences in devices with different window layer materials. A buried homojunction and a heterointerface junction were determined in the CdS/CIGS and Zn(S,O) devices, respectively, which is one of the most important mechanisms leading to the difference in device performance [5].

KPFM and AFM measurements taken simultaneously can accurately align the electrical potential and surface morphology. But they cannot align them with the CdS/CIGS interface because of the lack of chemical information in the KPFM/AFM technique. On another hand, SEM can align surface morphology and different layer materials because of different second electron yield in the materials. One example of the alignments near a metal-contact edge of the CdS/CIGS device is shown in Figure 10.7, where a dashed oval (likely a nonuniform TCO feature) and the shape of the metal contact, as seen in all three images, were used for the alignment. The dashed line across the images indicates the center of the CdS layer, as identified from the SEM image. Because of the inadequate SEM resolution, the CdS/CIGS interface is not clearly shown on the SEM image; however, the center of the CdS layer was located instead. The dashed rectangle in Figure 10.7(b) shows the area where the potential data were averaged along the horizontal direction for improving the signal/noise ratio, and the potential profile is shown in Figure 10.8(a).

As described in Section 10.3, the potential changes in the device bulk (Figure 10.8(b)) by V_b applied to the device can be deduced by subtracting the profile at $V_b = 0$ V from those at the various V_b values to avoid the effect of surface charge [40,44]. The electric-field peaks in Figure 10.8(c) as deduced by taking the first directive of the potential difference in Figure 10.8(b) correspond to the location of the *p-n* junction because the electric field as well as the V_b-induced field change should be at their maxima at the location of the *p–n* junction. Based on the

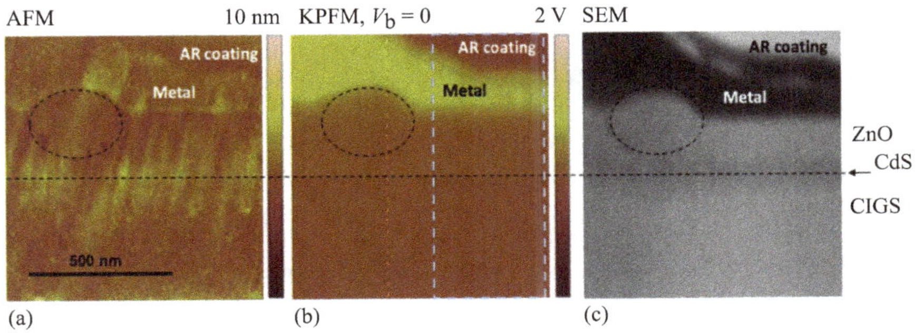

Figure 10.7 *(a) AFM, (b) the corresponding KPFM, and (c) SEM images taken on the same cross-section area of the CdS/CIGS device. Dashed circles and the metal layer were used to align these images. The dashed line indicates the center of the CdS layer*

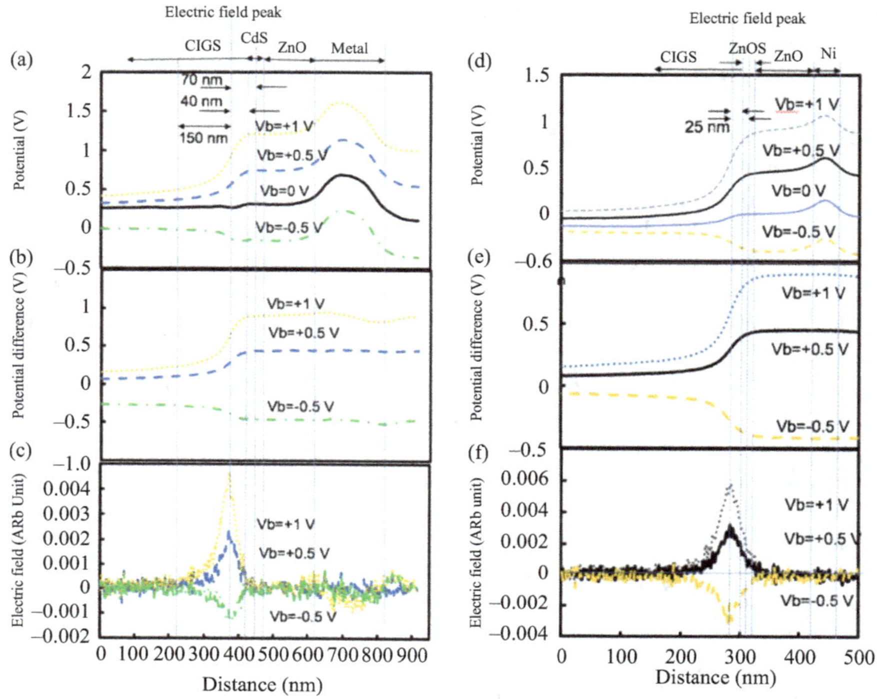

Figure 10.8 Potential profiling and junction location determination in (a)–(c) CdS/CIGS device, and (d)–(f) Zn(S,O)/CIGS device; (a) and (d) potential profiles at various V_b values; (b) and (e) potential changes by applying the V_b values; (c) and (f) Changes in electric field. The vertical lines indicate the locations of electric-field peak, positions of the multiple layers, the center of window layers, and the edge of the depletion region

alignment with the SEM image, the junction location is ~70 nm away from the center of the CdS layer. Because the CdS thickness (~60 nm) is well determined during the deposition of the layer, the junction should be ~40 nm away from the CdS/CIGS interface (Figure 10.8). This result reveals a buried homojunction at the CIGS subsurface region.

Cd diffusion during and after the deposition of the CdS layer might be a mechanism for forming a shallow buried homojunction [53–55,57]. The CIGS subsurface region is heavily off-stoichiometry because it is Cu-poor. Cd_{Cu} defects can form donor states replacing the V_{Cu} acceptor defects. Because of the surface charge effect, the potential profile around the junction at $V_b = 0$ V is flat (Figure 10.8(a)). However, the surface potential/field at reverse $V_b = +1$ V may present a similar profile to the bulk at $V_b = 0$ V. The electric-field profile at $V_b = +1$ V (Figure 10.8(c)) shows a ~150-nm field expansion at the CIGS side, which is consistent with the depletion width with a carrier concentration on the order of $10^{16}/cm^3$. The field at $V_b = +1$ V extends ~50 nm at the CdS side, indicating that the depletion at this side ends at about the location of the CdS/CIGS interface. Therefore, the junction and depletion region are all in the CIGS side, which should be beneficial to the device by suppressing recombination loss for minority-carrier transport over the junction. In other words, the effect of heterointerface recombination should be mitigated by the minority-carrier diffusion over the homojunction if structural defect concentration in the homojunction region is adequately low. Therefore, the high performance of the CdS/CIGS device might be contributed, in part, by the homojunction formation.

Figure 10.8(d)–(f) shows, respectively, the potential, potential change, and electric-field profiles taken at the different V_b values on the device with a different window layer of Zn(S,O). The electric-field peak or the junction location is ~25 nm away from the center of the ZnOS layer, which is significantly different from the CdS/CIGS device (~70 nm). Considering the well-known ZnOS thickness (~20 nm), the junction location is ~15 nm from the ZnOS interface (Figure 10.8(f)). This distance is smaller than the KPFM resolution (~30 nm) and alignment uncertainty. Therefore, it cannot be well determined whether the junction is right at the heterointerface or at an extremely shallow location in the CIGS subsurface region. Because the CIGS films are identical in the two devices, this junction formation may imply a smaller diffusivity of Zn than Cd in CIGS. The center of the ZnOS layer—and thus, the ZnOS/CIGS interface—is well within the depletion region (Figure 10.8(f)). The recombination at the interface cannot be avoided for carrier transport, which can negatively affect the device performance.

10.5 Nm-scale photocurrent and resistance mapping on CdTe solar cells

CdTe thin-film PV technology has progressed significantly in recent years and has reached a conversion efficiency beyond 22% [58]. The improvements are mainly in short-circuit current density (J_{sc}) due to improvements of the TCO layer, the buffer layer, and absorber minority-carrier lifetime by alloying Se in the CdTe thin-film material [58,59]. However, open-circuit voltage (V_{oc}) of the high-performance cell remains relatively low compared with the CdTe bandgap of 1.45 eV [52]. It is believed that the major challenges for improving the V_{oc} are defects in the CdTe thin film, junction quality, and low concentration of Cu doping [52]. Therefore, defect physics in CdTe has been studied actively [60], with most characterizations being macroscopic investigations that involve a large amount of point and extended defects. Two microscopic electrical characterizations using AFM-based nanoelectrical probes will be reviewed below: one is three-dimensional (3D) photocurrent mapping in the absorber layer [29] and the other is resistance mapping across the device junction [28].

By using a conductive AFM, Luria *et al.* reported 3D photocurrent transport pathways with illumination [29]. A conducting diamond probe simultaneously records topography and photocurrent between the grounded probe and TCO. The 3D tomography was obtained by applying forces from the AFM probe that continuously scratched out and removed sample materials. Electrical current for a given bias voltage at a position (in *X*, *Y*, and *Z*) was computationally reconstructed for 3D tomography in a resolution comparable to the probe size of 20 nm [29,30,61]. Figure 10.9(a) shows a 3D photocurrent throughout a 2.2-μm thickness of a CdTe/CdS/FTO cell based on 147 consecutive commutated tomography (CT-AFM) images [29]. The images were collected from top to bottom of the film during 15-suns illumination and at zero-bias voltage; so, bright contrast correlates to a locally enhanced short-circuit current (I_{sc}) for the solar cell.

Figures 10.9(b) and (c) show two regions from Figure 10.9(a); each has a series of five evenly sliced *XZ* cross-sections with 10-nm inter-distance between the slices [29]. In Figure 10.9(b), two GBs are highlighted with dashed lines. Planar defects in Figure 10.9(b) appear and converge as indicated by the solid circles, and they sometimes approach obliquely or bifurcating, as indicated by the dotted circles. These periodic planar defects are stacking faults with a spacing of 100 nm, as further evidence by using scanning transmission electron microscopy (STEM) [29].

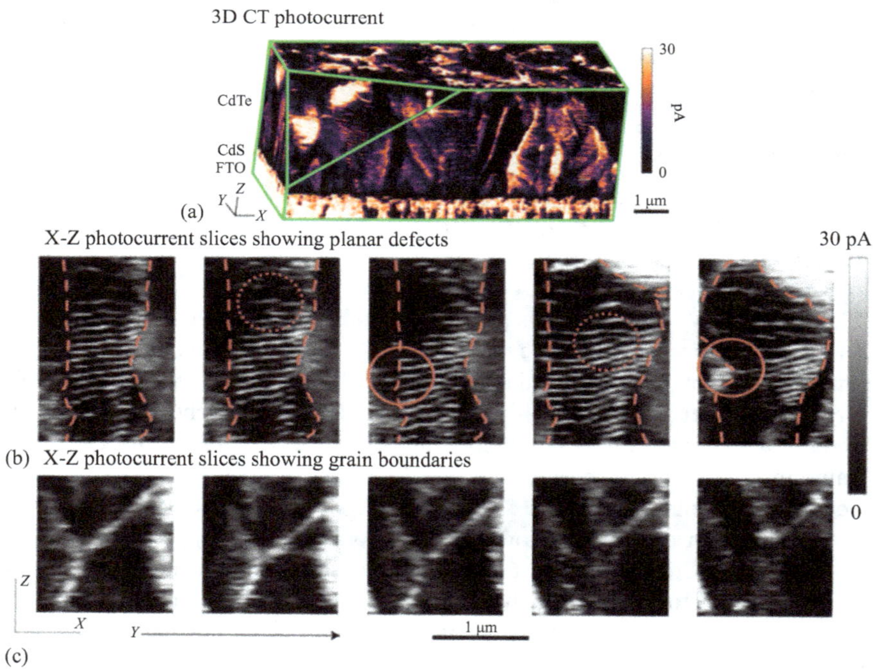

Figure 10.9 (a) A 3D photocurrent CT-AFM taken on a CdTe device with 15-suns illumination and at short-circuit condition; (b) and (c) example 2D consecutive photocurrent images taken from Figure 10.9(a) in the X–Z planes in regions containing (b) planar defects and (c) GBs

Figure 10.9(c) shows three GBs converging at high angles, probably dependent on the orientation of the adjacent grains. These results illustrate photocurrent pathways through the GB and intragrain planar defects in CdTe.

In another work, SSRM was used to map active charge-carrier distributions with and without illumination and under various V_s to investigate the junction electrical property and inhomogeneity [28]. Figure 10.10(a) and (b) shows the SSRM resistance images taken on a cross section of the CdTe device with a sample bias voltage $V_s = 4$ V in the dark and with the AFM laser being turned ON, respectively [28]. Figure 10.10(c) shows the corresponding AFM topography image. The middle values of the resistance scale bars are averages of the resistance over the images. The TCO, CdTe, and back-contact layers can be distinguished from both the resistance and topography images. The CdS layer (~50 nm) is too thin to be identified in this 10-μm-scale image. Figure 10.10(a) and (b) illustrates that the resistance of the CdTe layer is highly nonuniform along both the lateral and

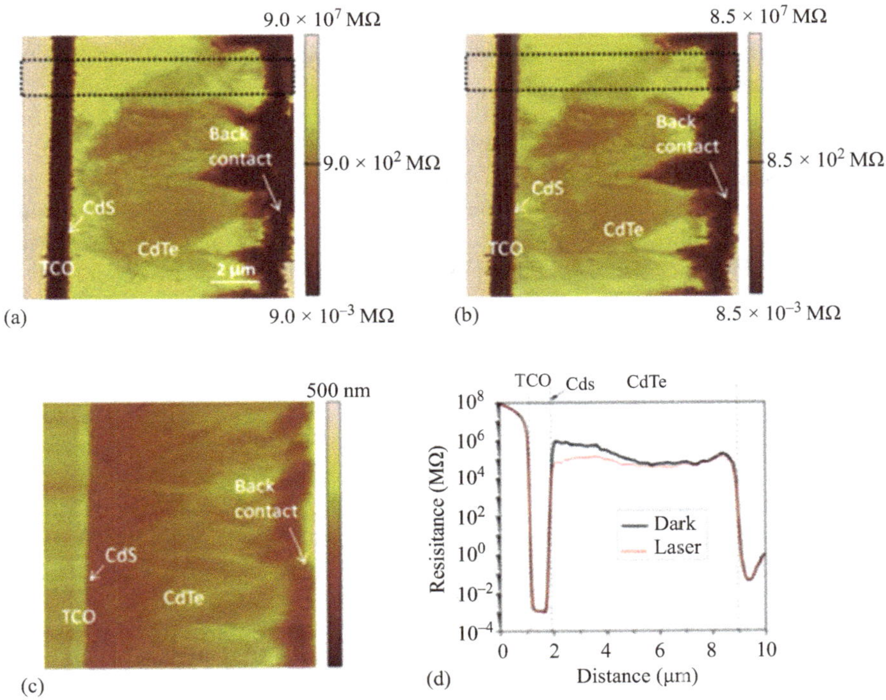

Figure 10.10 SSRM resistance images taken on the cross-section of a CdTe device under short-circuit condition and (a) in the dark and (b) with the AFM laser illumination; (c) shows the corresponding AFM image to (a) and (b); (d) shows the resistance profiles averaged from the regions indicated in the dashed rectangular in Figure 10.10(a) and (b)

vertical directions. The nonuniformity in the device lateral direction illustrates carrier or doping nonuniformity of this polycrystalline material. The nonuniformity along the device vertical direction can originate from both the doping nonuniformity and the carrier depletion around the device junction. Figure 10.10(d) shows resistance profiles in the dark and under laser illumination, averaged from 20 lines in an area as marked by the dashed rectangles in Figure 10.10(a) and (b). From the resistance profiles, the resistance of the CdTe layer in the region near the CdTe/CdS interface is significantly different between the dark and illumination. Under the dark, the resistance increases significantly approaching the junction (the CdTe/CdS interface). The increase of this resistance occurs over a width (~3 μm) that is consistent with the expected depletion width of the device. Under the AFM laser illumination, the high resistance in the depletion region decreases significantly; the resistance should be dominated by photocarriers generated by the AFM laser.

The power of the AFM laser is 1 mW, and the area of the laser spot size is about 10^{-4} cm^2. The cantilever of the probe prevents the AFM laser from directly illuminating the cross-sectional sample beneath the probe. However, the laser beam is larger than the width of the cantilever. Scattering of the laser also indirectly illuminates the sample area beneath the probe. Assuming that 1% of the laser was received by the local sample area (0.1 W/cm^2) and that the quantum efficiency is 100%, then the increase of the carrier concentration induced by the AFM laser is $2I\tau/lh\upsilon = 3.8 \times 10^{12}/\text{cm}^3$, using a carrier lifetime of $\tau \sim 0.5$ ns [62] and a penetration depth of $l \sim 1$ μm [2], which is consistent with the measurement of the photoconductance in the junction area. The changes in resistance by applying a bias voltage (V_b) to the working device were further investigated. The results show that the resistance decreased significantly by applying a forward V_b that injects majority carriers.

10.6 Closing remarks

The thin-film PV technologies have shown great progress in recent years in all the mainstream materials of CdTe, CIGS, and perovskite—with efficiencies now comparable with state-of-the-art crystalline Si PV [58]. In addition, thin-film technologies have been applied to a wide range of uses such as flexible PV, building-integrated PV, and portable PV [2]. The CdTe technology is currently focusing on group-V (P, As, Sb) doping to increase carrier concentration, which has become the main limiting factor to further advancing the conversion efficiency [63,64]. The nanoelectrical probe can directly image the active carrier distribution at nm resolution and is expected to be a powerful characterization tool for this subject. In recent years, CIGS has progressed by use of alkaline post-treatment [65–67]. Detailed electrical effects of the treatment pose demands on nanoelectrical characterization [68]. Perovskite thin-film PV has demonstrated >24% efficiency in small cell areas of <0.1 cm^2 [69].

This technology uses low-cost processing such as spin-coating, blading, and casting, which pose problems related to device and module nonuniformity.

Nanoelectrical probes fit this characterization need well, in addition to the advantages of non-modification and non-damage that is critical for characterizing perovskite.

On another hand, nanoelectrical probes are constantly being developed and improved. For example, a current sensor for electrical current or resistance mapping has been improved in both its sensitivity and bandwidth, which allows imaging of low-doped materials and use of instant-contact imaging for imaging soft and fragile materials. Another mode of KPFM, "frequency modulation," has been developed to enhance spatial resolution. SMIM can image both the capacitance and conductance simultaneously and mitigate the strict requirement of a superior thin insulating layer (10–30 nm) on the top of the sample [14,21] that is required by SCM [70]. With these developments, nanoelectrical probes can be more suitable and powerful for the characterization needs of thin-film PV.

AFM-based electrical techniques do have some drawbacks: for example, artifacts, reproducibility issues induced by the probe change, and convolution of surface roughness into the small electrical signals. Significant efforts have been made to improve the probe materials and to control the probe size and contamination during scanning.

Developments in thin-film PV and nanoelectrical characterization are expected to result in further significant impacts on thin-film PV science and technologies.

References

[1] J. Nelson, *The Physics of Solar Cells*. Imperial College Press, London UK, World Scientific Publishing Company, 2003.

[2] A. Luque and S. Hegedus, *Handbook of Photovoltaic Science and Engineering*. The AtriumSouthern Gate, Chichester, UK, John Wiley & Sons, 2011.

[3] C.-S. Jiang, *Chapter 24* in *Scanning Probe Microscopy in Nanoscience and Nanotechnology Volume 2, Edited by* B. Bhushan. Tiergartenstraße, Heidelberg, Germany, Springer Science & Business Media, 2010.

[4] M.A. Green, *Solar Cells: Operating Principles, Technology and System Applications*. University of New South Wales, 1986.

[5] C.-S. Jiang, L. Mansfield, S. Glynn *et al.*, "Effect of window-layer materials on p-n junction location in Cu(In,Ga)Se_2 solar cells," *IEEE J. Photovolt.*, vol. 9, no. 1, pp. 308–312, 2019.

[6] C. Xiao, C.-S. Jiang, H. Moutinho *et al.*, "Locating the electrical junctions in Cu(In,Ga)Se_2 and $Cu_2ZnSnSe_4$ solar cells by scanning capacitance spectroscopy," *Prog. Photovolt. Res. Appl.,* vol. 25, no. 1, pp. 33–40, 2017.

[7] T. Song, A. Kanevce, and J. R. Sites, "Emitter/absorber interface of CdTe solar cells," *J. Appl. Phys.*, vol. 119, no. 23, p. 233104, 2016.

[8] G. Binnig, H. Rohrer, Ch. Gerber, and E. Weibel, "Surface studies by scanning tunneling microscopy," *Phys. Rev. Lett.*, vol. 49, no. 1, pp. 57–61, 1982.

[9] G. Binnig, C. F. Quate, and Ch. Gerber, "Atomic force microscope," *Phys. Rev. Lett.*, vol. 56, no. 9, pp. 930–933, 1986.

[10] M. Nonnenmacher, M.P. O'Boyle, and H.K. Wickramasinghe, "Kelvin probe force microscopy," *Appl. Phys. Lett.*, vol. 58, no. 1, pp. 2921–2923, 1991.

[11] P. De Wolf, M. Geva, T. Hantschel, W. Vandervorst, and R.B. Bylsma, "Two-dimensional carrier profiling of InP structures using scanning spreading resistance microscopy," *Appl. Phys. Lett.*, vol. 73, no. 15, pp. 2155–2157, 1998.

[12] C.C. Williams, W.P. Hough, and S.A. Rishton, "Scanning capacitance microscopy on a 25 nm scale," *Appl. Phys. Lett.*, vol. 55, no. 2, pp. 203–205, 1989.

[13] T. Hidaka, T. Maruyama, M. Saitoh, and N. Mikoshiba, "Formation and observation of 50 nm polarized domains in $PbZr_{1-x}Ti_xO_3$ thin film using scanning probe microscope," *Appl. Phys. Lett.*, vol. 68, no. 17, pp. 2358–2359, 1996.

[14] C. Gao, T. Wei, F. Duewer, Y. Lu, and X.-D. Xiang, "High spatial resolution quantitative microwave impedance microscopy by a scanning tip microwave near-field microscope," *Appl. Phys. Lett.*, vol. 71, no. 13, pp. 1872–1874, 1997.

[15] A. Kikukawa, S. Hosaka, and R. Imura, "Silicon pn junction imaging and characterizations using sensitivity enhanced Kelvin probe force microscopy," *Appl. Phys. Lett.*, vol. 66, no. 25, pp. 3510–3512, 1995.

[16] C.C. Williams, J. Slinkman, W.P. Hough, and H.K. Wickramasinghe, "Lateral dopant profiling with 200 nm resolution by scanning capacitance microscopy," *Appl. Phys. Lett.*, vol. 55, no. 16, pp. 1662–1664, 1989.

[17] P. Eyben, M. Xu, N. Duhayon, T. Clarysse, S. Callewaert, and W. Vandervorst, "Scanning spreading resistance microscopy and spectroscopy for routine and quantitative two-dimensional carrier profiling," *J. Vac. Sci. Technol. B Microelectron. Nanometer Struct. Process. Meas. Phenom.*, vol. 20, no. 1, pp. 471–478, 2002.

[18] P. Eyben, T. Janssens, and W. Vandervorst, "Scanning spreading resistance microscopy (SSRM) 2d carrier profiling for ultra-shallow junction characterization in deep-submicron technologies," *Mater. Sci. Eng. B Solid-State Mater. Adv. Technol.*, vol. 124–125, pp. 45–53, 2005.

[19] T. Hantschel, P. Niedermann, T. Trenkler, and W. Vandervorst, "Highly conductive diamond probes for scanning spreading resistance microscopy," *Appl. Phys. Lett.*, vol. 76, no. 12, pp. 1603–1605, 2000.

[20] L. Zhang, K. Ohuchi, K. Adachi, K. Ishimaru, M. Takayanagi, and A. Nishiyama, "High-resolution characterization of ultrashallow junctions by measuring in vacuum with scanning spreading resistance microscopy," *Appl. Phys. Lett.*, vol. 90, no. 19, p. 192103, 2007.

[21] K. Lai, M. B. Ji, N. Leindecker, M. A. Kelly, and Z. X. Shen, "Atomic-force-microscope-compatible near-field scanning microwave microscope with separated excitation and sensing probes," *Rev. Sci. Instrum.*, vol. 78, no. 6, p. 063702, 2007.

[22] J.A. Christman, R.R. Woolcott, A.I. Kingon, and R.J. Nemanich, "Piezoelectric measurements with atomic force microscopy," *Appl. Phys. Lett.*, vol. 73, no. 26, pp. 3851–3853, 1998.

[23] J.M.R. Weaver and D.W. Abraham, "High resolution atomic force microscopy potentiometry," *J. Vac. Sci. Technol. B Microelectron. Nanometer Struct. Process. Meas. Phenom.*, vol. 9, no. 3, pp. 1559–1561, 1991.

[24] B.D. Terris, J.E. Stern, D. Rugar, and H.J. Mamin, "Contact electrification using force microscopy," *Phys. Rev. Lett.*, vol. 63, no. 24, pp. 2669–2672, 1989.

[25] P. Eyben, S. Denis, T. Clarysse, and W. Vandervorst, "Progress towards a physical contact model for scanning spreading resistance microscopy," *Mater. Sci. Eng. B*, vol. 102, no. 1, pp. 132–137, 2003.

[26] P. Eyben, F. Clemente, K. Vanstreels *et al.*, "Analysis and modeling of the high vacuum scanning spreading resistance microscopy nanocontact on silicon," *J. Vac. Sci. Technol. B*, vol. 28, no. 2, pp. 401–406, 2010.

[27] C.-S. Jiang, H.R. Moutinho, R.G. Dhere, and M.M. Al-Jassim, "The nanometer-resolution local electrical potential and resistance mapping of CdTe thin films," *IEEE J. Photovolt.*, vol. 3, no. 4, pp. 1383–1388, 2013.

[28] H. Li, C.-S. Jiang, W.K. Metzger, C.-K. Shih, and M. Al-Jassim, "Microscopic Real-Space Resistance Mapping Across CdTe solar cell junctions by scanning spreading resistance microscopy," *IEEE J. Photovolt.*, vol. 5, no. 1, pp. 395–400, 2015.

[29] J. Luria, Y. Kutes, A. Moore, L. Zhang, E. A. Stach, and B. D. Huey, "Charge transport in CdTe solar cells revealed by conductive tomographic atomic force microscopy," *Nat. Energy*, vol. 1, no. 11, pp. 1–6, 2016.

[30] C. Stetson, T. Yoon, J. Coyle *et al.*, "Three-dimensional electronic resistivity mapping of solid electrolyte interphase on Si anode materials," *Nano Energy*, vol. 55, pp. 477–485, 2019.

[31] Y. Yang, M. Yang, D.T. Moore *et al.*, "Top and bottom surfaces limit carrier lifetime in lead iodide perovskite films," *Nat. Energy*, vol. 2, no. 2, pp. 1–7, 2017.

[32] Y. Chen, H.T. Yi, X. Wu *et al.*, "Extended carrier lifetimes and diffusion in hybrid perovskites revealed by Hall effect and photoconductivity measurements," *Nat. Commun.*, vol. 7, no. 1, pp. 1–9, 2016.

[33] D. Zhao, Y. Yu, C. Wang *et al.*, "Low-bandgap mixed tin–lead iodide perovskite absorbers with long carrier lifetimes for all-perovskite tandem solar cells," *Nat. Energy*, vol. 2, no. 4, p. 17018, 2017.

[34] M. Yang, Y. Zeng, Z. Li *et al.*, "Do grain boundaries dominate non-radiative recombination in $CH_3NH_3PbI_3$ perovskite thin films?" *Phys. Chem. Chem. Phys.*, vol. 19, no. 7, pp. 5043–5050, 2017.

[35] C. Chen, Z. Song, C. Xiao *et al.*, "Achieving a high open-circuit voltage in inverted wide-bandgap perovskite solar cells with a graded perovskite homojunction," *Nano Energy*, vol. 61, pp. 141–147, 2019.

[36] C. Li, Z. Song, D. Zhao *et al.*, "Reducing saturation-current density to realize high-efficiency low-bandgap mixed tin–lead halide perovskite solar cells," *Adv. Energy Mater.*, vol. 9, no. 3, p. 1803135, 2019.

[37] C.M. Wolff, P. Caprioglio, M. Stolterfoht, and D. Neher, "Nonradiative recombination in perovskite solar cells: The role of interfaces," *Adv. Mater.*, 2019. doi: 10.1002/adma.201902762.

[38] K. K. Wong, A. Fakharuddin, P. Ehrenreich *et al.*, "Interface-dependent radiative and nonradiative recombination in perovskite solar cells," *J. Phys. Chem. C*, vol. 122, no. 20, pp. 10691–10698, 2018.

[39] I. Gelmetti, N.F. Montcada, A. Pérez-Rodríguez *et al.*, "Energy alignment and recombination in perovskite solar cells: weighted influence on the open circuit voltage," *Energy Environ. Sci.*, vol. 12, no. 4, pp. 1309–1316, 2019.

[40] C.-S. Jiang, M. Yang, Y. Zhou *et al.*, "Carrier separation and transport in perovskite solar cells studied by nanometre-scale profiling of electrical potential," *Nat. Commun.*, vol. 6, no. 1, pp. 1–10, 2015.

[41] C. Xiao, C. Wang, W. Ke *et al.*, "Junction quality of SnO_2-based perovskite solar cells investigated by nanometer-scale electrical potential profiling," *ACS Appl. Mater. Interfaces*, vol. 9, no. 44, pp. 38373–38380, 2017.

[42] Z. Li, C. Xiao, Y. Yang *et al.*, "Extrinsic ion migration in perovskite solar cells," *Energy Environ. Sci.*, vol. 10, no. 5, pp. 1234–1242, 2017.

[43] C.-S. Jiang, F.S. Hasoon, H.R. Moutinho, H.A. Al-Thani, M.J. Romero, and M.M. Al-Jassim, "Direct evidence of a buried homojunction in $Cu(In,Ga)Se_2$ solar cells," *Appl. Phys. Lett.*, vol. 82, no. 1, pp. 127–129, 2002.

[44] C.-S. Jiang, H.R. Moutinho, D.J. Friedman, J.F. Geisz, and M.M. Al-Jassim, "Measurement of built-in electrical potential in III–V solar cells by scanning Kelvin probe microscopy," *J. Appl. Phys.*, vol. 93, no. 12, pp. 10035–10040, 2003.

[45] A. Guerrero, E.J. Juarez-Perez, J. Bisquert, I. Mora-Sero, and G. Garcia-Belmonte, "Electrical field profile and doping in planar lead halide perovskite solar cells," *Appl. Phys. Lett.*, vol. 105, no. 13, p. 133902, 2014.

[46] Q. Lin, A. Armin, R.C.R. Nagiri, P. L. Burn, and P. Meredith, "Electro-optics of perovskite solar cells," *Nat. Photonics*, vol. 9, no. 2, pp. 106–112, 2015.

[47] K. Wojciechowski, T. Leijtens, S. Siprova *et al.*, "C_{60} as an efficient n-type compact layer in perovskite solar cells," *J. Phys. Chem. Lett.*, vol. 6, no. 12, pp. 2399–2405, 2015.

[48] W. Ke, D. Zhao, C. Xiao *et al.*, "Cooperative tin oxide fullerene electron selective layers for high-performance planar perovskite solar cells," *J. Mater. Chem. A*, vol. 4, no. 37, pp. 14276–14283, 2016.

[49] C.-C. Chueh, C.-Z. Li, and A.K.-Y. Jen, "Recent progress and perspective in solution-processed Interfacial materials for efficient and stable polymer and organometal perovskite solar cells," *Energy Environ. Sci.*, vol. 8, no. 4, pp. 1160–1189, 2015.

[50] Y. Shao, Y. Yuan, and J. Huang, "Correlation of energy disorder and open-circuit voltage in hybrid perovskite solar cells," *Nat. Energy*, vol. 1, no. 1, pp. 1–6, 2016.

[51] C. Wang, C. Xiao, Y. Yu *et al.*, "Understanding and eliminating hysteresis for highly efficient planar perovskite solar cells," *Adv. Energy Mater.*, vol. 7, no. 17, p. 1700414, 2017.

[52] A. Kanevce, M.O. Reese, T.M. Barnes, S.A. Jensen, and W.K. Metzger, "The roles of carrier concentration and interface, bulk, and grain-boundary

recombination for 25% efficient CdTe solar cells," *J. Appl. Phys.*, vol. 121, no. 21, p. 214506, 2017.

[53] K. Ramanathan, R. Noufi, J. Granata, J. Webb, and J. Keane, "Prospects for in situ junction formation in $CuInSe_2$ based solar cells," *Sol. Energy Mater. Sol. Cells*, vol. 55, no. 1, pp. 15–22, 1998.

[54] T. Sugiyama, S. Chaisitsak, A. Yamada *et al.*, "Formation of pn homo-junction in $Cu(InGa)Se_2$ thin film solar cells by Zn doping," *Jpn. J. Appl. Phys.*, vol. 39, no. 8R, p. 4816, 2000.

[55] D. Abou-Ras, G. Kostorz, A. Romeo, D. Rudmann, and A.N. Tiwari, "Structural and chemical investigations of CBD- and PVD-CdS buffer layers and interfaces in $Cu(In,Ga)Se_2$-based thin film solar cells," *Thin Sol. Films*, vol. 480–481, pp. 118–123, 2005.

[56] X. Mathew, J. Santos, D.R. Coronado *et al.*, "CdS thin film post-annealing and Te–S interdiffusion in a CdTe/CdS solar cell," *Sol. Energy*, vol. 86, no. 4, pp. 1023–1028, 2012.

[57] A. Rockett, D. Liao, J.T. Heath *et al.*, "Near-surface defect distributions in $Cu(In,Ga)Se_2$," *Thin Sol. Films*, vol. 431–432, pp. 301–306, 2003.

[58] M.A. Green, E.D. Dunlop, D.H. Levi, J. Hohl-Ebinger, M. Yoshita, and A. W.Y. Ho-Baillie, "Solar cell efficiency tables (version 54)," *Prog. Photovolt. Res. Appl.*, vol. 27, no. 7, pp. 565–575, 2019.

[59] M. Gloeckler, I. Sankin, and Z. Zhao, "CdTe solar cells at the threshold to 20% efficiency," *IEEE J. Photovolt.*, vol. 3, no. 4, pp. 1389–1393, 2013.

[60] "CdTe and Related Compounds; Physics, Defects, Hetero- and Nano-structures, Crystal Growth, Surfaces and Applications – 1st Edition." [Online]. Available: https://www.elsevier.com/books/cdte-and-related-compounds-physics-defects-hetero-and-nano-structures-crystal-growth-surfaces-and-applications/triboulet/978-0-08-046409-1. [Accessed: 22-Oct-2019].

[61] D. Pires, J.L. Hedrick, A. De Silva *et al.*, "Nanoscale three-dimensional patterning of molecular resists by scanning probes," *Science*, vol. 328, no. 5979, pp. 732–735, 2010.

[62] W.K. Metzger, D. Albin, D. Levi *et al.*, "Time-resolved photoluminescence studies of CdTe solar cells," *J. Appl. Phys.*, vol. 94, no. 5, pp. 3549–3555, 2003.

[63] W.K. Metzger, S. Grover, D. Lu, "Exceeding 20% efficiency with in-situ group V doping in polycrystalline CdTe solar cells," *Nat. Energy*, vol. 4, no. 10, pp. 837–845, 2019.

[64] B.E. McCandless, W.A. Buchanan, C.P. Thompson *et al.*, "Overcoming carrier concentration limits in polycrystalline CdTe thin films with in-situ doping," *Sci. Rep.*, vol. 8, no. 1, pp. 1–13, 2018.

[65] A. Chirilă, P. Reinhard, F. Pianezzi *et al.*, "Potassium-induced surface modification of $Cu(In,Ga)Se_2$ thin films for high-efficiency solar cells," *Nat. Mater.*, vol. 12, no. 12, pp. 1107–1111, 2013.

[66] P. Jackson, R. Wuerz, D. Hariskos, E. Lotter, W. Witte, and M. Powalla, "Effects of heavy alkali elements in $Cu(In,Ga)Se_2$ solar cells with efficiencies up to 22.6%," *Phys. Status Solidi RRL – Rapid Res. Lett.*, vol. 10, no. 8, pp. 583–586, 2016.

[67] T.P. Weiss, B. Bissig, T. Feurer, R. Carron, S. Buecheler, and A.N. Tiwari, "Bulk and surface recombination properties in thin film semiconductors with different surface treatments from time-resolved photoluminescence measurements," *Sci. Rep.*, vol. 9, no. 1, pp. 1–13, 2019.

[68] Y. Sun, S. Lin, W. Li *et al.*, "Review on alkali element doping in Cu(In,Ga) Se_2 thin films and solar cells," *Engineering*, vol. 3, no. 4, pp. 452–459, 2017.

[69] E.H. Jung, N.J. Jeon, E.Y. Park *et al.*, "Efficient, stable and scalable perovskite solar cells using poly(3-hexylthiophene)," *Nature*, vol. 567, no. 7749, pp. 511–515, 2019.

[70] C.C. Williams, "Two-dimensional dopant profiling by scanning capacitance microscopy," *Annu. Rev. Mater. Sci.*, vol. 29, no. 1, pp. 471–504, 1999.

Chapter 11
STEM characterization of solar cells

Jinglong Guo[1], Tadas Paulauskas[1,2] and Robert F. Klie[1]

11.1 Introduction

Photovoltaic (PV) devices consisting of a thin-film CdTe absorber layer on glass support have long been considered one of the leading technologies for reaching the Department of Energy's 2030 cost target of less than \$0.03/kWh. The currently used device architectures include a thin transparent conductive oxide (TCO) on top of the glass superstrate following by very thin (<100 nm) polycrystalline emitter layer, typically CdS or more recently MgZnO (MZO) [1]. The CdTe absorber layer is usually up to 4 μm thick and polycrystalline, followed by the back contact (Figure 11.1). Typical CdTe thin-film layers are formed from rapid deposition techniques, such as close-spaced sublimation (CSS) or high-rate vapor-transport deposition. Both processes result in a polycrystalline CdTe photo-absorbing film [2], and such devices can be manufactured cheaply and efficiently at large scale. Record cell efficiencies as high as 22.1% have been demonstrated [3]. However, such cell performances are still far from the theoretical limits (~30% for CdTe) [4], and improvements have mostly been due to better light collection, rather than improvements in absorber conversion efficiency. The remaining obstacles to further improving these devices toward their theoretical limits concern improvements in the fill factor (FF) and, more importantly, open-circuit voltage (V_{oc}), which is still far from its limit [5]. In fact, it is widely accepted that V_{oc} is the limiting factor since it reaches only about 60% of the CdTe bandgap. This is in contrast to GaAs solar cells, which have reached a V_{oc} of 77% of the bandgap. The reasons for this discrepancy have been identified as the low *p*-type doping concentrations of CdTe due to the formation of compensating defect states, a poor *p–n* junction, and the need for better backside ohmic contacts. High interfacial defect density at grain boundaries (GBs) and intragrain dislocations is also one of the main concerns because they contribute to the dark current [6]. This is supported by the study of single-crystal-based CdTe solar cells that were able to reach V_{oc} in excess of 1 V [7,8].

[1]Department of Physics, University of Illinois, Chicago, IL, USA
[2]Department of Optoelectronics, Center for Physical Sciences and Technology, Vilnius, Lithuania

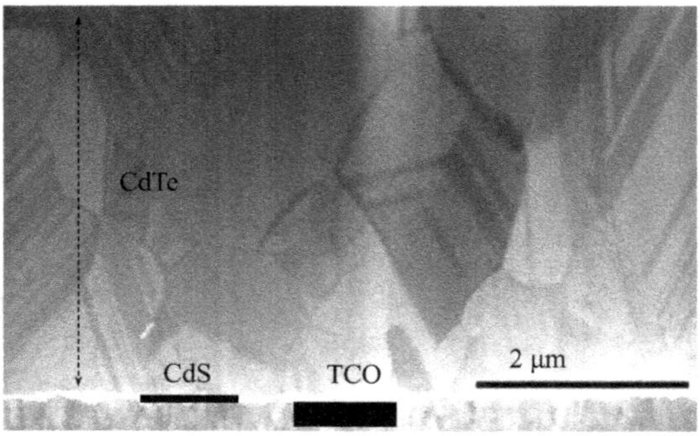

Figure 11.1 Cross-sectional ABF image of a polycrystalline CdTe solar cell. CdTe, CdS, and TCO layers are marked

Minority-carrier recombination therefore severely limits the V_{oc} and can play a powerful role in determining the FF. The carrier lifetime is primarily limited by defect-mediated, nonradiative, Shockley–Read–Hall (SRH) processes [9,10]. Such nonradiative recombination within the CdTe grain-boundary network limits V_{oc} and to a lesser extent the FF. Thus, improvements in the FF and V_{oc}—and ultimately, efficiency—can be accomplished through defect mitigation to prevent the recombination of carriers as they move through the device. Over the years, many approaches have been developed to mitigate these defects, most notably the $CdCl_2$ treatment, which has numerous beneficial effects [11,12]. Yet, the work-function mismatch between available contact materials and the CdTe absorber causes significant forward resistance and a reduction in the conversion efficiency by further reducing the FF [13]. Therefore, a better understanding of the electronic behaviors of defects and of the atomic structure–property relationships is needed.

Due to the structural complexity of the GB network and the bulk defect structures found within the grains, transmission electron microscopy (TEM) has been used extensively to determine the structure–property relationships of CdTe thin-film devices. In 1989, Lu and Smith studied the atomic structure of dislocation cores in CdTe using TEM [14]. The authors found two types of partial dislocations, 30° and 90°, resulting from the dissociation of 60° dislocations, and the dislocations exhibited glide and shuffle sets. Yet, the chemical composition of such partial dislocations could not be analyzed directly until the introduction nearly 25 years later of aberration correctors and large-area X-ray detectors [15,16].

The structure of twin boundaries, where the stacking sequence of the (111) planes in the twin region is changed with respect to the host lattice, was first characterized by Durose and Russell using TEM [17]. The authors identified two types of twin-boundary structures: a Σ3 (first-order) and Σ9 (second-order) twin lying along different crystallographic planes. Such twin-boundary structures are prevalent in most grains within thin-film CdTe PV devices and have been studied extensively since 1990. The atomic structures have been resolved using aberration-corrected scanning transmission electron microscopy (STEM) imaging [18], and the chemical composition of twin boundaries in as-grown and $CdCl_2$-treated grains was determined using electron energy-loss spectroscopy (EELS) and X-ray energy-dispersive spectroscopy (XEDS) [19]. It is widely accepted that twin boundaries in $CdCl_2$-treated CdTe forming stoichiometric $\Sigma 3$ interfaces that penetrate the entire grain are electrically inactive and do not introduce mid gap states. However, Li *et al.* [20] found that when twin boundaries terminate within a grain, a Cl-doped dislocation core at either end of the twin boundary establishes an electro-static potential that helps separate charge carriers which is beneficial for carrier-collection efficiency [20]. Meanwhile, in 2003, Yan *et al.* [21] imaged the atomic structure of edge dislocations in CdTe, finding that the dangling bonds at the dislocation core create midgap states in CdTe, which are responsible for the high SRH recombination rates found in CdTe GBs. The presence of Cl at CdTe GBs after the $CdCl_2$ annealing, as reported by Li *et al.* [20], passivated the dislocation cores and reduced the density of midgap states.

In this chapter, we will summarize some of the more recent results of atomic-resolution transmission electron microscopy of defects and GBs in CdTe. Aberration-corrected STEM has developed into a powerful tool to characterize the local atomic and electronic structures of materials. Due to the incoherent nature of the imaging process, the atomic-resolution high-angle annular dark-field (HAADF) images can be used directly to determine the atomic configurations at GBs or dislocation cores. Moreover, recent developments in spectrometer technology now enable atomic-resolution chemical quantification and measurements of the local density of unoccupied states above the Fermi level. When combined with first-principles modeling, electron-beam-induced current (EBIC) measurements, or device-level modeling, STEM characterization can provide powerful information about the role that GBs and individual defect complexes play on V_{oc}, FF, and thus, the overall cell conversion efficiency.

The remainder of this chapter is organized as follows: We will first describe the working principle of analytical STEM, followed by several examples where atomic-resolution STEM imaging was combined with EELS and XEDS to study GBs in CdTe. We will then briefly discuss how STEM imaging can be combined with first-principles modeling to explore new passivants. We conclude by providing an outlook of possible new developments in STEM approaches for CdTe devices that can potentially address the outstanding issues limiting the performance of polycrystalline CdTe thin-films devices.

11.2 Scanning transmission electron microscopy

Imaging and chemical analysis in an aberration-corrected STEM has become one of the preeminent tools for studying the local atomic and electronic structure of materials. The state-of-the-art aberration-corrected STEM allows for atomic-resolution incoherent imaging using HAADF imaging [22], but it can also be used to perform high spatial-resolution microanalysis using EELS and XEDS [23]. The atomic-resolution HAADF detector, which forms images by registering electrons that have scattered to high angles, provides contrast where the image intensity depends directly on the atomic number of the imaged atomic columns.

While aberration correction has made crucial contributions toward improving the spatial resolution and beam current in conventional STEM setups, the integration of cold field-emission gun (CFEG) electron sources now allows for sub-angstrom (sub-Å) spatial resolution with sub-eV energy resolution to determine the atomic structure as well as atomic-level chemical composition and local bonding. When HAADF imaging is coupled with simultaneous XEDS or EELS acquisitions, the atomic structures of the crystal lattice can be resolved directly (if the sample is orientated along a zone axis) and the chemical composition and local bonding can be measured at similar spatial resolutions. This makes STEM an indispensable tool for identifying the defect structures in CdTe, GaAs, and other thin-film materials.

Figure 11.2 shows a schematic drawing of the STEM optical configuration. A series of magnetic lenses is required to provide the desired demagnification of the CFEG electrons onto the sample surface to form an atomic-scale electron probe. The condenser lenses provide initial demagnification for the objective lens (OL), which provides the final and largest demagnification. Aberrations introduced by this strongest focusing lens dominate the optical system, which can only be corrected by using a series of multipoles lenses that provide negative spherical aberrations to compensate positive spherical aberrations introduced by OL.

The aberration-corrected electron probe is scanned across the surface of the sample using a set of scan coils. Electrons scattered by the specimen in the forward direction are detected by scintillator detectors in the far-field diffraction plane and

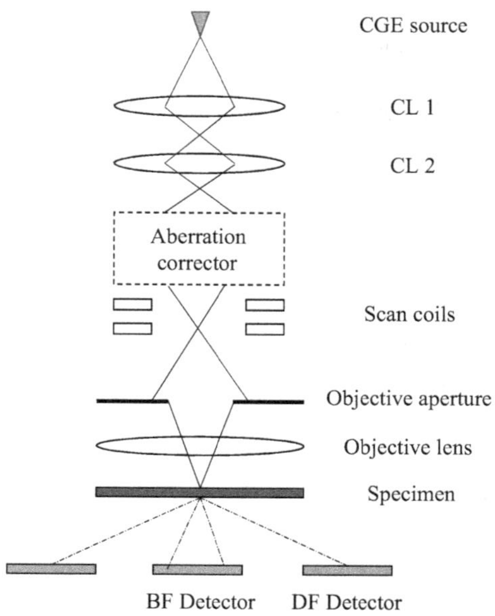

Figure 11.2 Schematic of the aberration–corrected STEM

plotted as functions of the probe position to form STEM images. STEM imaging modes are defined based on which angular ranges are used for signal integration within annular detectors centered on the optic axis. These are typically distinguished whether the bright-field (BF) or dark-field (DF) regions are used. The former is defined by the convergence semi-angle of the electron probe and for atomic-resolution imaging is typically 15–30 mrad or more. Signal in the BF region exhibits pronounced interference between elastically scattered and the unscattered electrons. However, placing an annular detector in the outer half of the BF disk results in the so-called annular bright-field (ABF) imaging mode that still shows little-to-none image contrast reversal in typical imaging conditions [24,25]. ABF is widely used for atomic-resolution imaging of crystals that may contain light element atomic-columns, such as oxygen, lithium, or even hydrogen [26–28]. HAADF is by far the most popular imaging mode within the DF region that utilizes quasi-elastically and thermally scattered electrons [22,29–34]. The image contrast in HAADF scales directly with atomic number of the atomic-column species ($\sim Z^{1.8}$) and so images can be directly interpreted in terms of the lattice structure. Low-angle DF region (LAADF) shows some coherence effects and is also useful since LAADF enhances contrast from locally strained lattice. The above-mentioned images can be acquired simultaneously within a STEM instrument equipped with several annular detectors.

STEM image simulations can be used to refine an atomic lattice structure or to interpret images within highly strained regions, such as dislocation cores. Two main approaches exist to simulate dynamical electron propagation in STEM: (1) The Bloch wave method is based on solving the three-dimensional time-independent Schrödinger equation within the crystal structure. The electron wavefunctions at the exit surface of the specimen are calculated after solving the eigenvalues and eigenvectors of the scattering matrix with appropriate boundary conditions. The Bloch wave method provides an intuitive picture of electrons propagating within the crystal in terms of Bloch waves. However, in complicated structures with defects and interfaces, the Bloch wave matrix solution is impractical [35]; (2) the multislice method divides the specimen into thin two-dimensional slices along the incident-beam direction and implements wavefunction propagation within the crystal. The defect structures and 3-D crystal shape can easily be accounted for in this method [36,37].

Whereas the high-angle quasi-elastic scattering is used for HAADF images, the small-angle inelastically scattered electrons can be coupled into an electron energy-loss to form an EEL spectrum. The inelastically scattered electrons are separated according to their kinetic energy by a magnetic prism, and the signal intensity is plotted as a function of energy channels. EEL spectrum contains three major parts: the (1) zero-loss peak, (2) low-loss, and (3) core-loss parts. The zero-loss peak contains spectroscopic information about the energy resolution of the electron microscopy system, but also acts as the zero-energy reference point for the rest of the spectrum. The specimen thickness information in terms of inelastic-scattering mean free path can also be extracted from the low-loss regime [23]. Low-loss spectra are characteristic of the outer-shell electron excitations, including the optical properties of the specimen, reflecting the dielectric information of the material. Core-loss spectra reflect the density of unoccupied states above the Fermi level, and the near-edge fine-structure contains valuable information about the local bonding and long-range order. EELS excels at the chemical analysis of light elements such as C, N, O, or Cl, which have well-behaved core-loss edges within the 1.5keV range. In characterizing CdTe solar cells, EELS is used to show Cl segregation along GBs and other interfaces.

XEDS is another approach to characterize the chemical composition of materials in STEM. In XEDS, heavy and light elements can be analyzed using characteristic X-rays that are generated when an electron transitions from an excited state to the ground state. The characteristic X-ray peaks are little affected by the chemical environment, which makes XEDS highly reliable in quantifying chemical composition of a specimen. The development of aberration correctors and cold field emission sources makes it possible to acquire atomic-resolution XEDS in STEM. The modern silicon drift detector (SDD), such as the Oxford XMAX 100TLS, considerably increases the X-ray-collection solid angle, which improves the counts rate and acquisition speed. Atomic-resolution XEDS mapping enables identification of chemical composition within defect cores appearing in CdTe solar cell thin films [38].

Specimen preparation is essential for the successful atomic-resolution STEM characterization of materials. To achieve atomic-resolution STEM imaging, the specimen must be electron transparent and representative of the material. For STEM characterization of CdTe thin film, focused ion beam (FIB) is a commonly used technique to make cross-sectional TEM specimen. The constant thickness of the lamella makes FIB an ideal technique for quantitative analysis of chemical composition. For the dual-beam system with a scanning electron microscope (SEM), precise areas can easily be selected under secondary-electron imaging. However, high-voltage ion bombardment may cause artificial structure change and gallium ion implantation. Particularly for CdTe-based specimens, low-voltage gallium or argon ionic cleaning is necessary. There are some indications that the Cl concentration within CdTe thin films might be affected by the FIB process. In this case, conventional polishing with Ar milling at liquid nitrogen might be needed.

11.3 Structural and chemical characterization of CdTe solar cell devices

Polycrystalline CdTe-based solar cells have undergone improvements and optimization in their growth conditions and post-processing steps over the years to reach the present-day cell efficiencies of more than 22%. High-resolution transmission electron microscopy has played a pivotal role in this process by providing insights into the device morphology and chemical composition, thus helping to optimize the synthesis methods. Here, we highlight how aberration-corrected STEM characterization approaches have been applied to CdTe solar cells, resulting in advances in today's fundamental understanding of this technology. This sets the stage for the atomic-resolution analysis of defect structures, as well as the need for correlative experimental and first-principles computational methods that assume increasing importance in developing CdTe-based PV.

A cross-sectional image of a polycrystalline CdTe solar cell is shown in Figure 11.1. The image was acquired with aberration-corrected STEM using ABF mode by integrating electrons scattered within a cone from 11 to 22 mrad with respect to the optic axis [25]. This imaging mode can provide superior contrast for polycrystalline materials at low magnification, as well as at atomic resolution of low atomic-number (Z)-containing materials [39,40]. Here, a typical CdTe solar cell grown in a superstrate configuration after a post-deposition $CdCl_2$ heat treatment is shown with marked layers: TCO composed of fluorine-doped tin oxide (F: SnO_2), CdS emitter/window, and *p*-type CdTe absorber layer, the latter forming the heterojunction. Numerous micrometer-sized CdTe grains are visible, with smaller grains often found near the interface with CdS. Low interfacial energy of stacking faults in CdTe and typical growth conditions favor formation of numerous twin boundaries within the interior of randomly oriented grains. These can be clearly seen in Figure 11.1 as brighter and darker stripes extending through an entire grain.

The performance of the CdS–CdTe heterojunction is crucial to high conversion efficiencies. The CdS layer is preferably kept as thin as possible, typically 30–50 nm, to reduce its absorption in the blue portion of the visible spectrum, which can otherwise reduce the short-circuit current. However, studies showed that CdS synthesized via phase-vapor deposition (PVD) in the absence of oxygen undergoes large interdiffusion with CdTe in the subsequent growth and high-temperature $CdCl_2$ processing [6,41–44]. This leads to the formation of CdS_yTe_{1-y} alloy nanocrystals with reduced optical bandgap and possibly produces graded-like lattice matching of CdS and CdTe. However, in some cases, this interdiffusion can consume parts of CdS entirely, creating local shunting pathways. Introducing O_2 into the PVD environment, as well as its presence in CSS and other sputtering deposition approaches, has been credited with suppressing such interdiffusion. Furthermore, it has been suggested that CdS with higher oxygen content exhibits higher bandgap, which would be beneficial to reduce the aforementioned absorption [41].

It is crucial to be able to map the distribution of light elements with nanometer spatial resolution to elucidate how the mixing occurs. Such characterization has been performed using STEM in combination with EELS or XEDS [23,45]. EELS generally provides a better signal-to-noise ratio over XEDS because most of the inelastically scattered electrons are collected for analysis. However, this depends on whether the element of interest has a strong core-electron ionization fingerprint. Furthermore, present-day dual-EELS systems allow the selection of two rather narrow energy windows for acquisition; so, a complete elemental sample mapping can rarely be performed. The introduction of large solid-angle, windowless, SDDs in recent years now allows increased sensitivity to X-rays from lighter elements. In addition, semi-automated quantification based on *k*-factors and ability to simultaneously map emissions from all elements present in a sample has made XEDS a highly attractive technique [16,45].

An illustration of CdTe solar cell XEDS elemental mapping is shown in Figure 11.3. In this study, the authors examined the role of oxygen on the properties of the *p–n* junction [41]. CdS layer thickness and its composition can be seen to vary considerably. Although oxygen appears to be retained in some parts of CdS, other parts show oxygen-rich clusters in a form of $CdSO_x$ phase. Intermixing of CdS–CdTe at the interface as well as small parts of nearly consumed CdS layers

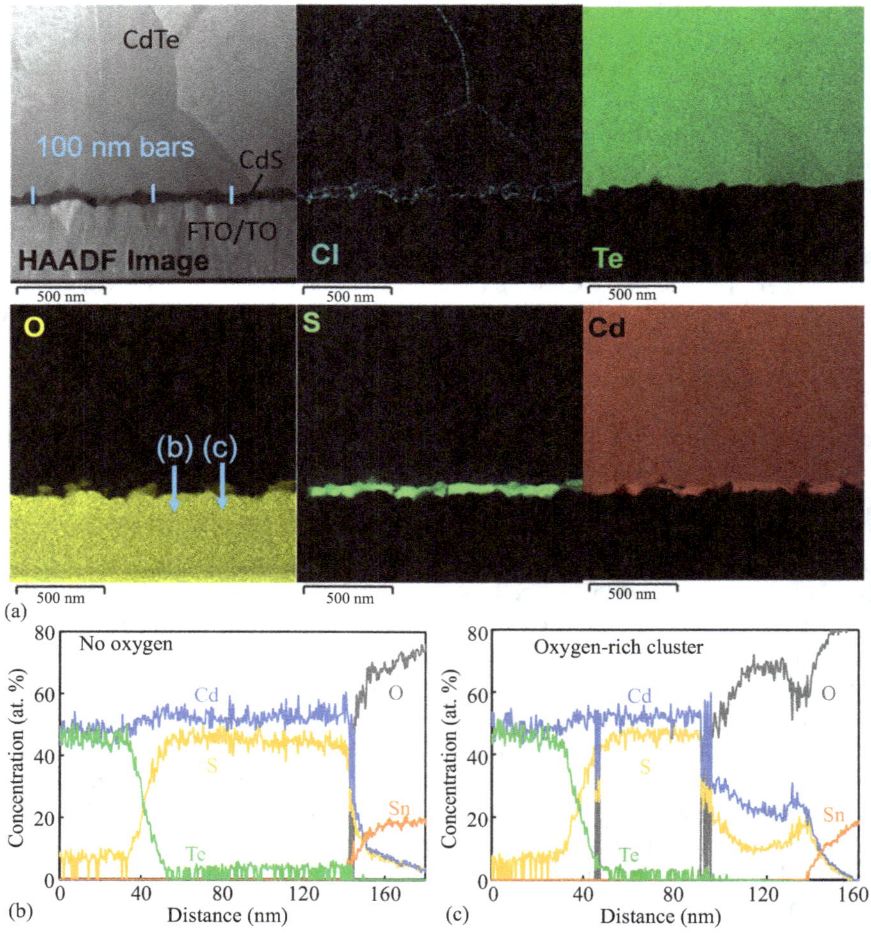

Figure 11.3 HAADF image (a) and X-ray energy dispersive spectroscopic mapping of the top layers of a CdTe solar cell. Cl enrichment of the p–n junction and grain boundaries is visible. Presence of oxygen-rich CdS clusters is indicated by the line profile in (c). Taken with permission from Ref [41]

can be seen in the images and extracted line profiles shown in Figure 11.3(b) and (c).

The post-deposition annealing of CdTe solar cells in $CdCl_2$ is another standard post-processing step that is crucial for high performance of the solar cells. Several changes occur during this so-called activation of CdTe cells, which are reflected in all the key PV parameters, increasing the conversion efficiency by as much as 10 percentage points [6,41]. First, CdTe recrystallizes, leading to coalescence and the growth of grains. This morphological rearrangement likely reduces the number of interfacial defects associated with GBs and bulk defects that are detrimental to device performance. Furthermore, Cl is found to diffuse strongly throughout CdTe and is expected to occur much faster along GBs than in the bulk [46]. Enrichment of the CdS–CdTe interface with Cl as well as of CdTe GBs is widely reported in the literature [6,19]. This can also be seen in the XEDS elemental maps shown in Figure 11.3. Combined EBIC and STEM-density functional theory (DFT) studies suggested that Cl helps to passivate the dangling bonds associated with interfacial and bulk defects [18,47–49]. This is reflected in enhanced carrier-collection efficiency and reduced interfacial recombination.

Support for this view also comes from atomistic simulations of defect structures in CdTe. Segregation of Cl at GBs in CdTe is often found to correlate with reduction of Te at these interfaces. Figure 11.4 shows GB EELS chemical mapping at an intersection of three randomly oriented grains [49]. Enrichment of Cl is apparent and is coupled with reduced Te signal with no significant change in Cd concentration. This type of Cl–Te substitution was reported to be energetically favorable for the studied dislocation cores. The authors found that the exchange can lead to removal of midgap states introduced by defects, which, in turn, reduces recombination centers that impact charge-carrier lifetimes [50,51]. Another benefit

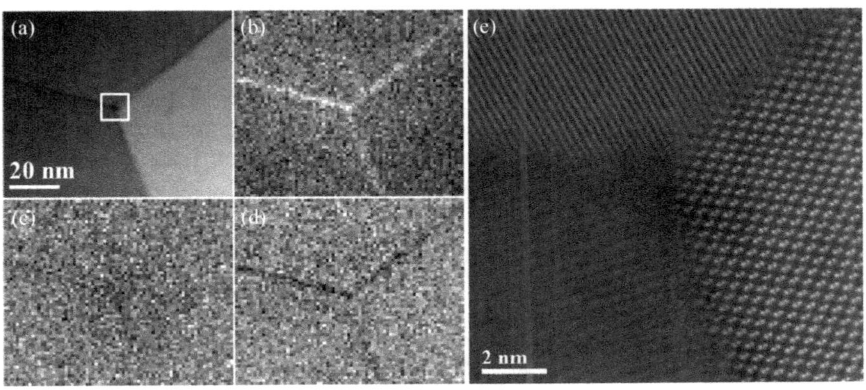

Figure 11.4 Electron energy-loss spectroscopic (b)–(d) and HAADF (c) and (e) imaging of random CdTe grain boundaries after $CdCl_2$ annealing treatment. Enrichment with Cl and the associated reduction of Te is visible without significant changes in Cd concentration. Taken with permission from Ref [49]

of GB enrichment with Cl is suggested to arise if local *p–n–p* junctions are established across these interfaces [20,52]. The local electric field would promote carrier separation and enhance their collection efficiency.

Post-deposition $CdCl_2$ annealing of CdTe solar cells is typically followed by Cu diffusion treatment [53,54]. Although the interplay between these treatments is still an active research topic, Cu is thought to serve in several ways. As-grown CdTe is slightly *p*-type due to Cd vacancies and is hole-doped with diffused Cu to the typically reported levels on the order of 10^{15} at/cm^3 [6,55,56]. No significant segregation of Cu at GBs in CdTe and the *p–n* junction has been reported in the literature. However, such Cu segregation would be a concern because it could lead to detrimental effects by compensating the emitter or creating shunts. Another important role played by Cu is formation of the back contact. A more heavily doped *p*-type surface layer combined with Te alloying is usually employed to promote formation of ohmic contacts.

In the next section, we will show how advanced atomic-resolution STEM studies can be used to characterize the defect structures in CdTe. These studies have paved the way to fundamental understanding of atomic-scale effects, and combined with atomistic simulations, they have provided new insights on how to advance CdTe PV technology.

11.4 High-resolution imaging and spectroscopy

Recent developments in aberration-corrected STEM instrumentation have allowed for investigations of defect structures with sub-Ångstrom spatial resolution. Incoherent HAADF imaging mode is typically employed, which provides images where the contrast can be directly correlated to the sample thickness and the atomic number (*Z*) of the constituent atomic columns [39]. Readily available and fast quantum-mechanical image simulations can be further employed to refine the structure and identify, for example, positions of light or heavy dopant atoms [37,57]. Sample imaging can be coupled with simultaneous acquisition of XEDS or EELS signals to obtain information about the chemical composition of the atomic columns. This has led to rapid cataloguing of a plethora of different types of dislocations and their atomic-core structures that naturally form within polycrystalline CdTe films [15,18,20,52,58].

Figure 11.5 shows several atomic-resolution HAADF images of CdTe along its <110> crystallographic direction [49]. In this projection, CdTe appears as a collection of closely spaced atomic columns (dumbbells) that are each natively composed of either Cd or Te atoms. Due to the difference in the atomic number of Cd and Te ($Z_{Cd} = 48$, $Z_{Te} = 52$), inspection of the HAADF image contrast allows one to determine their polarity in the bulk. However, atomic-column-resolved XEDS or EELS is still preferred for unusual dislocation-core configurations that may also contain point defects or be strained, where the contrast in HAADF images is less reliable in gauging chemical composition. Although atomic-resolution information can be obtained along several CdTe zone axes, such as [112], mixed-column [111],

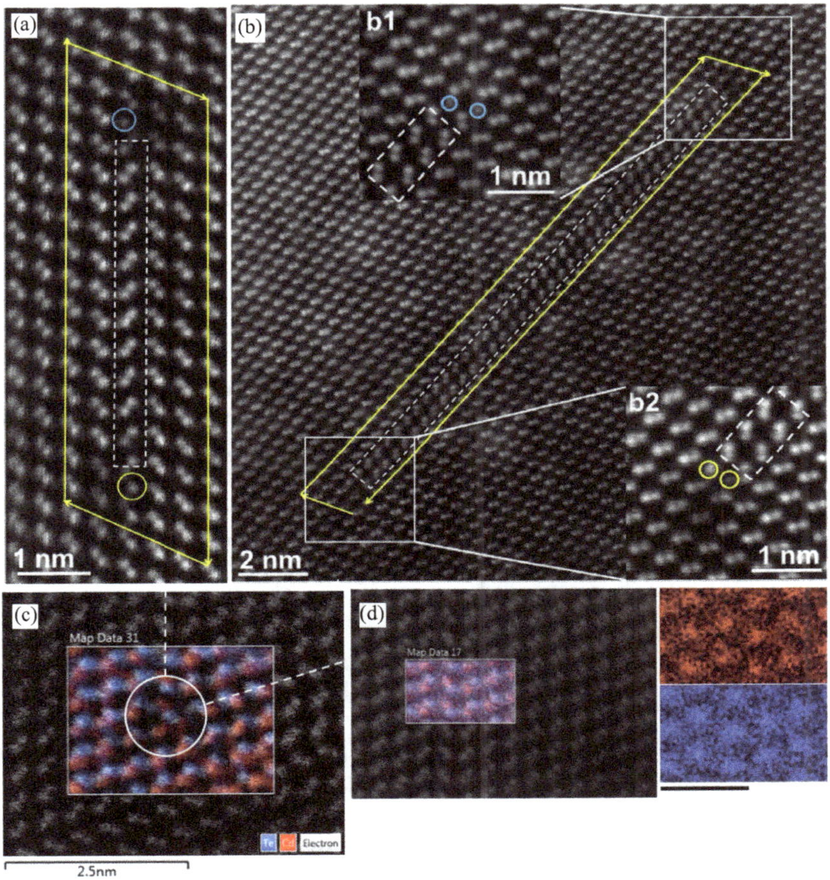

Figure 11.5 *(a), (b) HAADF images of dissociated dislocations in CdTe. Circles in dislocation cores indicate Te (yellow) and Cd (blue) atomic-columns. Taken with permission from Ref [49]; (c) XEDS mapping of Lomer–Cottrell dislocation junction (circled), overlaid on HAADF image; and (d) XEDS image of a [111] stacking fault with raw Cd (red) and Te (blue) Lα1 emissions*

and others, the <110>-type zone axis is by far the most preferred. Burgers vector of an elemental glide-set dislocation system (111) [110] in zinc-blende materials makes a 60° angle with respect to the dislocation line, thus having both edge and screw components. Dislocation lines are oriented parallel to the [110] direction, so that their core structures as well as their Burgers vector edge-component can be determined by imaging along this zone axis. Although there is ambiguity regarding the screw component, knowledge of the edge part is often sufficient for determining the type of dislocation in an HAADF image [15].

Due to small interfacial stacking fault energy, dislocations in bulk CdTe are usually found dissociated into a pair of Shockley partial dislocations joined by an intrinsic or extrinsic stacking fault. Two such dissociated dislocations are shown in Figure 11.5(a) and (b), respectively. The Burgers circuits drawn around the dislocations show enclosures, or lack thereof, and define the edge component of the dislocation. Similar circuits can be drawn around each partial dislocation core crossing through the stacking fault. For example, it was found that the two partials in Figure 11.5(a) are both 30° Shockley partial dislocations but with opposite edge components, concluding that this is a dissociated screw dislocation. The cores are terminated with single atomic columns that are Cd and Te, respectively. These core configurations indicate an atomic under-coordination with respect to the natural tetrahedral bonding in CdTe, suggesting the presence of dangling bonds. Similar analysis can be performed for the dissociated elemental dislocation in Figure 11.5 (b). More complex interacting dislocations were also reported in CdTe crystals. Figure 11.5(c) shows an atomic-column-resolved XEDS image of a Lomer–Cottrell dislocation junction-core overlaid on top of the HAADF image [15]. This type of dislocation junction can be formed by the intersection of two 30° Shockley partial dislocations from different sets of (111) planes making the acute angle, or a dislocation cross-slip onto another (111) plane. The dislocation core is seen to be Cd_3Te type. Due to the crystallographic polarity of CdTe, a $CdTe_3$ variant of Lomer–Cottrell junction also exists and was reported in the literature. First-principles DFT calculations show electrical activity on 30° and 90° dislocation cores as well as the structural unit Cd_3Te, and it will be discussed in more detail in the following sections [18,20,52,59].

Before we go on to discuss atomic-resolution characterization of general CdTe GBs, we briefly come back to twin boundaries seen in Figure 11.1. Stacking faults and lamellar coincidence-site lattice (CSL) $\Sigma 3$ (111) twin boundaries are the simplest and most abundant interfacial defects in CdTe. Atomic-column-resolved XEDS maps of a typical stacking fault on a (111) plane are shown in Figure 11.5(d) [16]. Unlike the defects discussed above, this interface has no dangling bonds and its atoms satisfy the tetrahedral coordination. Computational studies suggest that this configuration is indeed electrically benign. Interaction of twin boundaries with native point defects was also reported in the literature using combined STEM and DFT approaches [19].

Five macroscopic degrees of freedom are required to define a most general GB, in addition to microscopic ones related to atomic relaxations. Electron backscattering diffraction analysis of grain distributions in polycrystalline CdTe indicates no special preference for their orientations [60]. This excludes the abovementioned intragrain lamellar twins, which may be viewed as a GB formed by a 180° twist around the interface normal [111]. Although the grains in CdTe are randomly oriented, a great majority of them reported in the literature nevertheless share the tendency for Cl segregation and the accompanying Te depletion along their boundaries. This provides motivation to investigate GB structures—or, more generally, CdTe interfaces—in more detail to find the underlying reasons for this tendency and possibly shed new light on how to introduce more potent passivating agents. Due to random GB orientations, the practical limitation in this approach is locating two or more grains sharing boundaries that can be simultaneously imaged at atomic resolution to reveal the interface structure. There are only a handful of crystallographic zone axes suitable for STEM imaging, at least with present-day instrumentation, and this limits the types of GBs that are possible to observe.

Various tilt, as well as CSL GBs, were studied in polycrystalline CdTe [15,19,20,58]. A tilt GB imaged along [110] is shown in HAADF Figure 11.6(a). Here, the interface is found to be composed of repeating structural units of the Te_3Cd-type, similar to the one shown in Figure 11.5(c). Chlorine segregation and

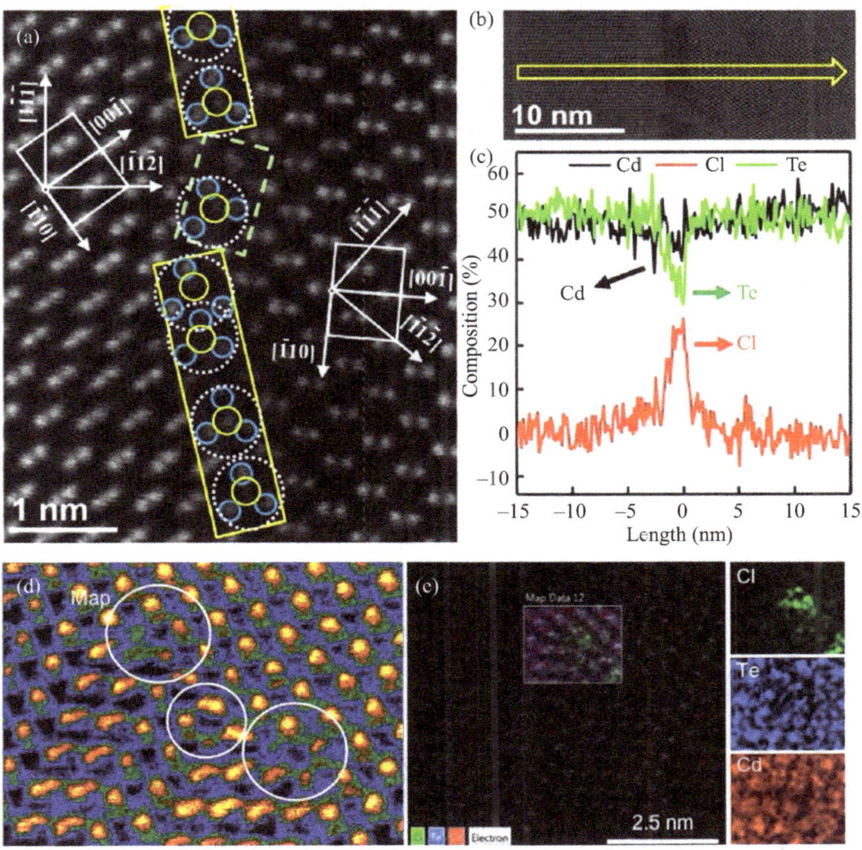

Figure 11.6 *(a) HAADF image of a CdTe tilt grain boundary composed of structural units. Crystallographic relationships are marked. (b) and (c) EELS mapping of Cl segregation and Te depletion in grain boundary shown in (a). Taken with permission from Ref [20]. (d) and (e) Heavily twinned CSL Σ3 [111] grain boundary with XEDS imaging of the structural units*

Te depletion is observed to occur in this interface as well, as seen in EELS line profiles in Figure 11.6(b) and (c). An HAADF image in Figure 11.6(d) shows the same structural units in the CSL Σ3 (112) GB with dense twinning. The high-resolution XEDS mapping in Figure 11.6(e) indicates a localized Cl segregation within the structural units and Te depletion. Thus, it appears that certain atomic structural units may be more prevalent in CdTe.

Another approach, as opposed to the heuristic scan of polycrystalline CdTe samples for suitably oriented grains, is to produce artificial GBs with pre-determined crystallographic orientations. This entails bonding single-crystal CdTe wafers with chosen interfacial planes, twist, and tilt angles such that the zone axes of the crystallites allow for reliable atomic-resolution STEM characterization. An added benefit of this approach is that auxiliary electro-optical measurements can be performed and that processing steps can be tested on such structures, for example, $CdCl_2$ treatment, or of other passivating agents. One concern of this approach is how reliably these structures represent GBs that naturally occur within poly-crystalline CdTe. Some very high interfacial-energy interfaces would likely have to be ruled out. However, this could still allow for a nontrivial part of the GB configuration to be investigated.

In the following, we look at a correlative STEM–DFT study of a low-angle tilt artificial GB created by wafer bonding [61,62]. Two single-crystal CdTe wafers with (110) surfaces and a 2.4° symmetrical off-cut were used, as shown in Figure 11.7, producing a 4.8°-tilt GB. Tilt GBs in simple cubic materials are predicted to be composed of discrete edge dislocations along their interface, with the spacing increasing toward small tilt angles. However, it is not easy to predict what resulting dislocations and their arrangements—or some other structural units altogether—are going to be found for any given interfacial boundary plane [63]. In the GBs studied here, dislocations were mostly pure-edge Lomer dislocations, which can be seen as resulting from two merged 60° elemental dislocations with opposite screw components. Atomic-resolution HAADF imaging of the dominant Lomer dislocation core is shown in Figure 11.8. It is seen to be composed of 5–7 and 6–8 atomic rings. A seed structure based on the HAADF image was used for DFT structural relaxations. The relaxed model is shown overlaid on the experimental image, giving a nearly perfect match. The electronic structure analysis of these defect cores and the behavior of such a GB are discussed in the following paragraphs.

Two other examples of wafer-bonded GBs are shown in Figure 11.9. A 8°-tilt GB with (111) interface plane is seen in Figure 11.9(a). An extra (111) plane emerges at each dislocation core every ~5 nm. This GB is composed solely of 60° elemental dislocations that carry large screw components. A possible explanation is that the dislocations' screw components alternate between parallel and anti-parallel orientations along the interface to accommodate the pure tilt. An inset shows a geometrical phase-analysis (GPA) strain-field mapping of the interface [64,65]. In particular, local lattice compression (green) and tension (red) in the horizontal image direction (e_{xx}) is mapped, generating dipole-like strain fields by the edge

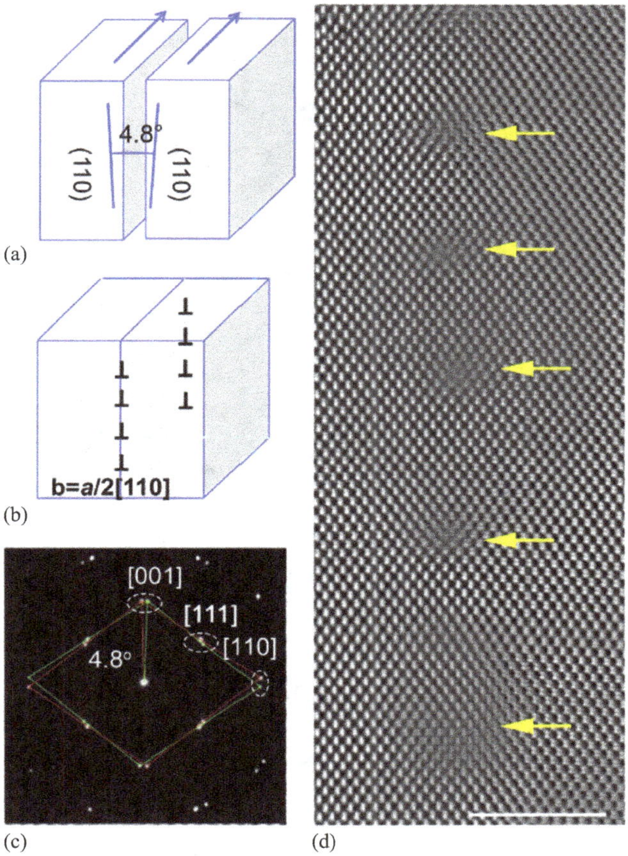

Figure 11.7 *(a), (b) CdTe wafer bonding geometry in creation of a low angle (110) tilt grain boundary; (c) electron diffraction image of the interface; and (d) HAADF image of the interface showing an arrangement of discrete dislocations. Taken with permission from Ref [62]*

components. The strain fields of adjacent dislocations overlap negligibly so that the interface electronic properties could still be approximated by considering individual defects.

A more exotic GB formed by wafer bonding is shown in Figure 11.9(b). Polar Te-terminated (001) crystal surfaces for both crystallites were chosen here as the interface plane, thus defining a 180°-tilt GB. An atomic-column-resolved XEDS map overlaid on the simultaneously acquired HAADF image shows chemical composition of the interface marked with a dashed yellow line. The interface is Te-rich and demonstrates long-range order. A closer look also reveals formation of triple-column structures on the left-hand side of the interface.

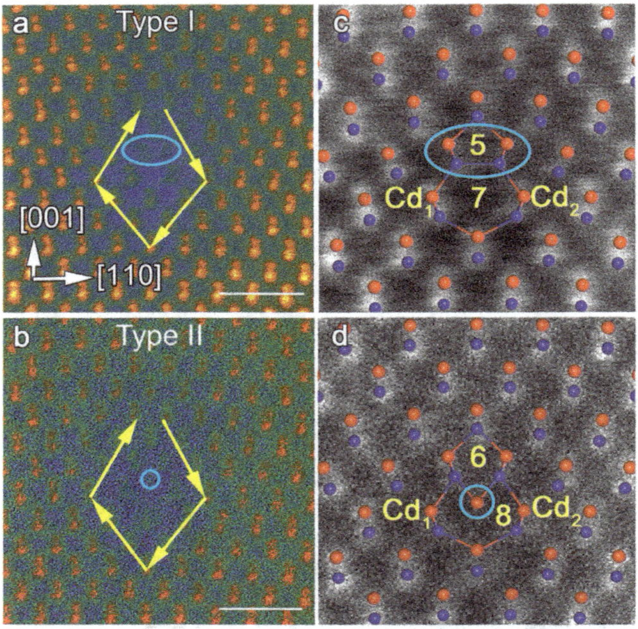

Figure 11.8 *(a), (b) HAADF image of two Lomer dislocation cores types that dominate along the tilt grain boundary in Figure 11.6, with drawn Burgers vectors; and (c), (d) relaxed atomic model with DFT of the dislocation cores overlaid on the HAADF images. Taken with permission from Ref [62]*

11.5 Correlating STEM imaging with DFT modeling

Atomic-resolution imaging and EELS as well as XEDS spectroscopies can be used to determine the local composition and crystal structure. First-principles modeling is needed to correlate the observed feature with the local density of states (DOS), the transport properties, and carrier lifetime properties. Here, we will show how a combination of STEM imaging and DFT modeling can be used to investigate the roles of dislocation cores and GBs in CdTe polycrystalline thin-film devices.

First, the HAADF images are used to determine the dislocation core structure by extracting the atomic-column locations directly from the atomic-resolution images. These two-dimensional projections are then extended in the third dimension, and the positions of the atomic columns are refined using first-principles DFT modeling, as shown in Figure 11.10(a) and (b). Next, the DOS and local charge density are calculated using Bader charge analysis for edge dislocations as well as $\Sigma 3$ (111) (or twin) GBs. It was found that dislocation cores show significant charge transfer compared to bulk CdTe. This charge transfer results from the changes in coordination at the dislocation core center, which can introduce additional states in

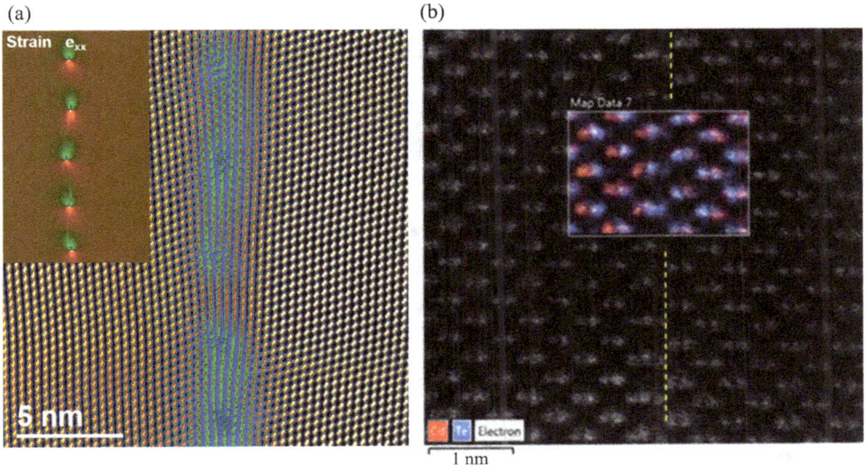

Figure 11.9 *(a) HAADF image of wafer bonded CdTe (111) tilt grain boundary composed of elemental dislocations. Inset shows local lattice strain-field in the horizontal image direction (e_{xx}) and (b) XEDS mapping overlaid on HAADF image of a wafer bonded high-angle twist grain boundary with polar Te-terminated (100) interface*

the bandgap. This result for the dislocation is in contrast to the Σ3 (111) GBs, which exhibit features very similar to that of bulk CdTe.

The electronic structures, or DOS, calculated for the dislocations core structures as seen in Figure 11.10(a) and (b) are shown in Figure 11.10(c) and (d), which clearly demonstrates that the dislocation cores give rise to additional states within bulk bandgap. Those extra midgap states act as SRH recombination centers, which will lower the minority-carrier lifetime considerably, and thus, they affect the overall efficiency of the CdTe solar cell. Σ3 (111) twin boundaries do not exhibit any midgap states compared to bulk CdTe because there are no dangling bonds found at the twin boundaries. This finding is in good agreement with the experimental results that found Σ3 (111) twin GBs to be electrically inactive and having no effect on the electron–hole recombination rates.

Next, we will show how the effects of Cl segregation along GBs in polycrystalline CdTe thin films can be addressed using DFT calculations. Figure 11.11 (a) shows the relaxed structure of an edge dislocation, where Te atomic columns have been replaced by Cl. The DOS of such Cl-passivated dislocation cores is shown in Figure 11.11(b). It can clearly be seen that the Cl-passivated dislocation cores exhibit a substantial reduction in midgap states compared to pristine cores. This effect has been attributed to the presence/passivation of Cl at the dislocation core center, which can result in significantly reducing recombination sites and increasing minority-carrier lifetimes in CdTe thin films that have been treated with a $CdCl_2$ annealing step. Therefore, the correlation of experimental STEM with theoretical DFT calculations explains the role of dislocation as recombination

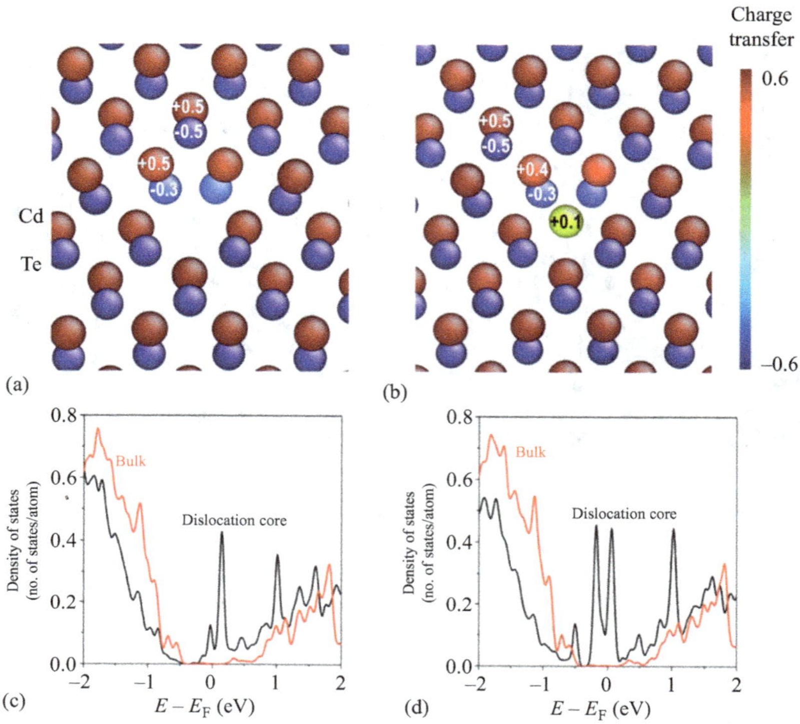

Figure 11.10 Electronic structure of type I and type II dislocation cores. Charge transfer with dislocation core structures are shown in (a) and (b). (c) and (d) show DOS compared with bulk area for type I and type II, respectively. Taken with permission from Ref [62]

centers in CdTe thin films and demonstrates the important role of Cl segregation in passivating midgap states as well as improving CdTe solar cell efficiency.

In addition to the GB passivation effect, Cl segregation was also considered to form a local *p–n–p* structure along GBs. Using aberration-corrected HAADF imaging and electron spectroscopy, Li *et al.* studied the atomic structure of Σ9 GBs in CdTe thin films. The structure of the dislocation cores is shown in Figure 11.12(a). First-principles DFT calculations were used to determine the electronic structures of the Σ9 GB dislocation cores with Cl passivation, as shown in Figure 11.12(b). These electron-structure calculations show that at sufficient Cl concentrations, the GB inverts from *p*-type to *n*-type. A local *p–n* junction is created by the *n*-type GB and *p*-type grain interior. The thus-formed built-in field now attracts electrons toward the GB and repels holes, which enhances the carrier separation. Figure 11.12(c) shows the proposed band diagram of a *p–n–p* junction at the GB. This Cl-induced band bending at the GBs can also contribute to the observed

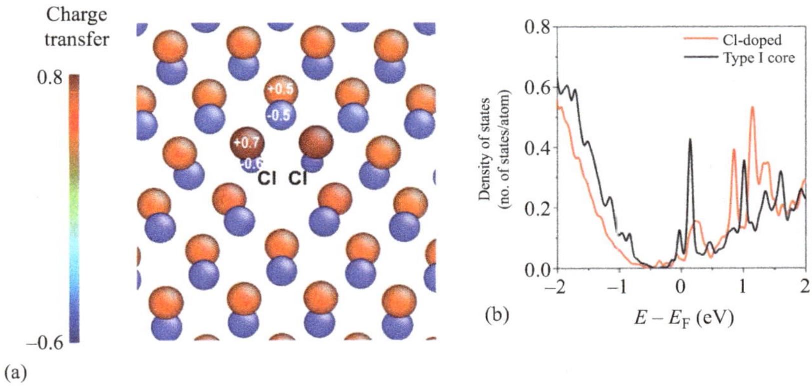

Figure 11.11 *Electronic structure of Cl-doped type I dislocation core. Charge transfer around dislocation with Cl dopant is shown is (a). DOS of Cl-doped dislocation core compared with original dislocation core is shown in (b). Taken with permission from Ref [62]*

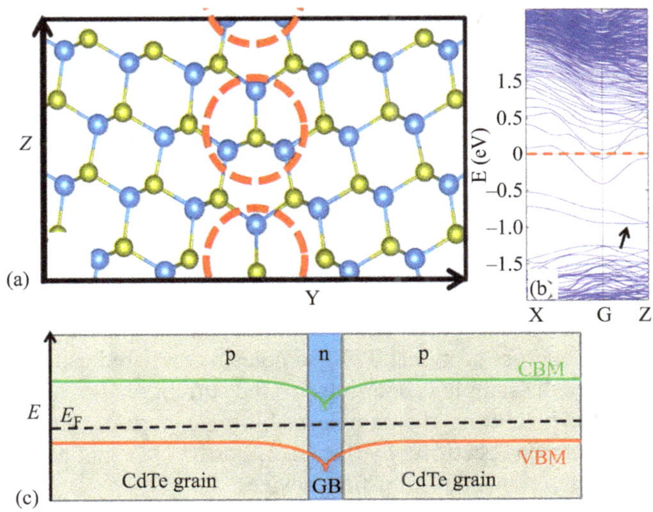

Figure 11.12 *(a) Center part of the Cd-core Σ9 GB model for DFT calculation in red dashed circles. (b) Band structure of dislocation core containing five Cl atoms. Te(Cl)-Te(Cl) antibonding states are indicated by black arrow. (c) Diagram of local p–n–p junction at the grain boundary. Taken with permission from Ref [20]*

improved conversion efficiency of polycrystalline CdTe-based solar cell after the $CdCl_2$ annealing step.

In addition to GBs, stacking faults and partial dislocation cores were also characterized using the combination of aberration-corrected HAADF imaging and first-principles DFT calculations. Stacking-fault structures, which terminate at two partial dislocations within a grain, were identified and their DOS was determined. It was found that the charge transfer caused by partial dislocation cores introduces significant band bending, which encourages the separation of carriers. The authors suggested that an increased density of intragrain stacking faults and partial dislocation cores may further improve the performance of CdTe solar cells [52].

11.6 Conclusions and outlook

We have shown how atomic-resolution imaging, EELS, XEDS spectroscopy, and DFT can be combined to determine the fundamental structure–property relationships of defects and GBs in polycrystalline CdTe thin-film solar cell devices. Although these approaches have been used successfully to explain the effects of Cl passivation, numerous questions related to device performance still remain unanswered, including the role of heterointerfaces on carrier lifetimes and the role of grain-boundary potential on carrier separation and carrier mobility.

STEM imaging has developed into a critical tool for atomic-resolution structural characterization. However, an "information gap" still exists between the atomic-level resolution that can be achieved with these imaging approaches and the approaches used to measure electrical activity of a device, such as EBIC measurements. EBIC measurements are often performed in an SEM with nanometer resolution at best [66]. Poplawski *et al.* acquired cross-sectional EBIC maps of polycrystalline CdTe devices combined with secondary-electron SEM imaging and reported that GBs and twinning defects appear bright whereas the bulk of grains appears dark [43].

Future developments of EBIC measurement capabilities are therefore needed to fill this information gap by developing a novel experimental approach to measure the local EBIC signal during atomic-resolution HAADF image acquisitions. This will require that cross-sectional TEM samples be prepared using conventional FIB lamellas, where the sample is connected to the current read-outs using e-beam-deposited Pt, without limiting the sample stability to perform atomic-resolution STEM imaging, electron spectroscopy (i.e., EELS or XEDS), and potentially EBIC analysis of single-junction CdTe thin-film devices.

Another exciting development recently is 4D STEM [67] and differential-phase contrast imaging (DPC) [68]. These STEM imaging methods acquire a convergent-beam diffraction pattern at each electron-probe position and will enable us to determine the GB potential and local electric field. If used in polycrystalline CdTe thin-film devices, the quantification of the role of Cl or other elements, such as Se, As, or Cu segregation, on the interfacial potential as a function of treatment parameters could be possible. Such additional analytical capabilities will certainly enable the acquisition of more complete datasets that will facilitate device modeling and the discovery of novel dopants, passivants, or device architectures—thus allowing polycrystalline CdTe thin-film devices to finally approach the Shockley–Queisser limit.

Acknowledgments

The authors acknowledge contributions from Maria Chan and Fatih Sen at the Argonne National Laboratory, Moon Kim at the University of Texas at Dallas, Angus Rockett at the Colorado School of Mines, as well as Sampath Walajabad and Jim Sites at Colorado State University. The work by J.G., T.P., and R.F.K. was supported by the U.S. Department of Energy's Office of Energy Efficiency and Renewable Energy (EERE) under Solar Energy Technology Office (SETO) Award Number DE-EE00007545 and DE-EE00005956. The acquisition of the UIC JEOL JEM-ARM200CF was supported by NSF MRI-R^2 Grant (no. DMR-0959470), and the acquisition of the new Gatan Continuum spectrometer was supported by a grant from the National Science Foundation MRI program (DMR-1626065). Support from the UIC Research Resources Center (RRC), in particular, A.W. Nicholls and F. Shi, is acknowledged.

References

[1] Munshi, A.H., J. Kephart, A. Abbas, *et al.*, *Polycrystalline CdSeTe/CdTe absorber cells with 28 mA/cm^2 short-circuit current*. IEEE Journal of Photovoltaics, 2018. **8**(1): p. 310–314.

[2] X. Wu, R. G. Dhere, D. S. Albin, *et al.*, *High-Efficiency CTO / ZTO / CdS / CdTe Polycrystalline Thin-Film Solar Cells*. NCPV Program Review Meeting, 2001.

[3] NREL, *Best Research Cell Efficiencies*. 2019. https://www.nrel.gov/pv/cell-efficiency.html.

[4] Shockley, W. and H.J. Queisser, *Detailed balance limit of efficiency of p-n junction solar cells*. Journal of Applied Physics, 1961. **32**(3): p. 510–519.

[5] Gloeckler, M., I. Sankin, and Z. Zhao, *CdTe solar cells at the threshold to 20% efficiency*. IEEE J. Photovoltaics, 2013. **3**(4): p. 1389–1393.

[6] McCandless, B.E. and J.R. Sites, *Cadmium Telluride Solar Cells*. Handbook of Photovoltaic Science and Engineering 2003: The Atrium, Southern Gate, Chicester, England, John Wiley & Sons, Ltd.

[7] Zhao, Y., M. Boccard, S. Liu, *et al.*, *Monocrystalline CdTe solar cells with open-circuit voltage over 1 V and efficiency of 17%*. Nature Energy, 2016. **1**(6): p. 16067.

[8] Burst, J.M., J.N. Duenow, D.S. Albin, *et al.*, *CdTe solar cells with open-circuit voltage breaking the 1V barrier*. Nature Energy, 2016. **1**.

[9] Read, W.T. and W. Shockley, *Dislocation models of crystal grain boundaries*. Physical Review, 1950. **78**(3): p. 275–289.

[10] Hall, R., *Electron–hole recombination in germanium*. Physical Review, 1952. **87**(2): p. 387.

[11] Dharmadasa, I., *Review of the $CdCl_2$ treatment used in CdS/CdTe thin film solar cell development and new evidence towards improved understanding.* Coatings, 2014. **4**(2): p. 282–307.

[12] Moutinho, H.R., M.M. Al-Jassim, D.H. Levi, P.C. Dippo, and L.L. Kazmerski, *Effects of CdCl2 treatment on the recrystallization and electro-optical properties of CdTe thin films*. Journal of Vacuum Science & Technology A – Vacuum Surfaces and Films, 1998. **16**(3): p. 1251–1257.

[13] Shen, K., R. Yang, D. Wang, *et al.*, *Stable CdTe solar cell with V_2O_5 as a back contact buffer layer*. Solar Energy Materials and Solar Cells, 2016. **144**: p. 500–508. DOI: 10.1016/j.solmat.2015.09.036.

[14] Lu, P. and D.J. Smith, *Dissociated 60° dislocations in CdTe studied by high-resolution electron microscopy*. Philosophical Magazine B, 1990. **62**(4): p. 435–450.

[15] Paulauskas, T., C. Buurma, E. Colegrove, *et al.*, *Atomic scale study of polar Lomer–Cottrell and Hirth lock dislocation cores in CdTe*. Acta Crystallographica A – Foundation and Advances, 2014. **70**(6): p. 524–531.

[16] Phillips, P.J., T. Paulauskas, N. Rowlands, *et al.*, *A new silicon drift detector for high spatial resolution STEM-XEDS: performance and applications*. Microscopy and Microanalysis, 2014. **20**(4): p. 1046–1052.

[17] Durose, K. and G.J. Russell, *Twinning in CdTe*. Journal of Crystal Growth, 1990. **101**(1): p. 246–250.

[18] Li, C., J. Poplawsky, Y. Yan, and S.J. Pennycook, *Understanding individual defects in CdTe thin-film solar cells via STEM: from atomic structure to electrical activity*. Materials Science in Semiconductor Processing, 2017. **65**: p. 64–76. DOI: 10.1016/j.mssp.2016.06.017.

[19] Paulauskas, T., C. Buurma, E. Colegrove, *et al.*, *Atomic-resolution characterization of the effects of $CdCl_2$ treatment on poly-crystalline CdTe thin films*. Applied Physics Letters, 2014. **105**(7): p. 071910.

[20] Li, C., Y. Wu, J. Poplawsky, *et al.*, *Grain-boundary-enhanced carrier collection in CdTe solar cells*. Physical Review Letters, 2014. **112**(15): p. 156103.

[21] Yan, Y., M.M. Al-Jassim, and K.M. Jones, *Structure and effects of double-positioning twin boundaries in CdTe*. Journal of Applied Physics, 2003. **94** (5): p. 2976–2979.

[22] Browning, N.D., M.F. Chisholm, and S.J. Pennycook, *Atomic-resolution chemical-analysis using a scanning-transmission electron-microscope*. Nature, 1993. **366**(6451): p. 143–146.

[23] Egerton, R.F., *Electron Energy Loss Spectroscopy in the Electron Microscope.* 2nd edition. 1986, New York: Plenum Press.

[24] Findlay, S.D., N. Shibata, H. Sawada, *et al.*, *Robust atomic resolution imaging of light elements using scanning transmission electron microscopy*. Applied Physics Letters, 2009. **95**(19): p. 191913–3.

[25] Findlay, S.D., N. Shibata, H. Sawada, E. Okunishi, Y. Kondo, and Y. Ikuhara, Dynamics of annular bright field imaging in scanning transmission electron microscopy Ultramicroscopy, 2010. **110**(7): p. 903–923.

[26] Kim, Y.J., R.Z. Tao, R.F. Klie, and D.N. Seidman, *Direct atomic-scale imaging of hydrogen and oxygen interstitials in pure niobium using atom-probe tomography and aberration-corrected scanning transmission electron microscopy*. ACS Nano, 2013. **7**(1): p. 732–739.

[27] Phillips, P.J. and R.F. Klie, *On the visibility of very thin specimens in annular bright field scanning transmission electron microscopy*. Applied Physics Letters, 2013. **103**(3). DOI: 10.1063/1.4816081.

[28] Phillips, P.J., H. Iddir, D.P. Abraham, and R.F. Klie, *Direct observation of the structural and electronic changes of Li_2MnO_3 during electron irradiation*. Applied Physics Letters, 2014. **105**(11): p. 113905.

[29] Pennycook, S.J. and D.E. Jesson, *High-resolution incoherent imaging of crystals*. Physical Review Letters, 1990. **64**(8): p. 938–941.

[30] McGibbon, M.M., N.D. Browning, M.F. Chisholm, *et al.*, *Direct determination of grain-boundary atomic-structure in $SrTiO_3$*. Science, 1994. **266** (5182): p. 102–104.

[31] Nellist, P.D. and S.J. Pennycook, *Direct imaging of the atomic configuration of ultradispersed catalysts*. Science, 1996. **274**(5286): p. 413–415.

[32] Chisholm M. F., A.R. Lupini, S.J. Pennycook, *et al.*, *Simultaneous Z-contrast and phase contrast imaging of oxygen in ceramic interfaces*. Microscopy and Microanalysis, 2004. **10**: p. 256–257. DOI: 10.1063/1.4816081.

[33] van Benthem, K., A.R. Lupini, M. Kim, *et al.*, *Three-dimensional imaging of individual hafnium atoms inside a semiconductor device*. Applied Physics Letters, 2005. **87**(3): p. 034104.

[34] Krivanek, O.L., N. Dellby, M.F. Murfitt, *et al.*, *Gentle STEM: ADF imaging and EELS at low primary energies*. Ultramicroscopy, 2012. **110**(8): p. 935–945.

[35] Peng, Y., P.D. Nellist, and S.J. Pennycook, *HAADF-STEM imaging with sub-angstrom probes: a full Bloch wave analysis*. Microscopy, 2004. **53**(3): p. 257–266.

[36] Kirkland, E.J., R.F. Loane, and J. Silcox, *Multislice simulation of ADF-STEM and CBED images*. Journal Of Metals, 1988. **40**(11): p. 84–84.

[37] Kirkland, E.J., *Advanced Computing in Electron Microscopy*. 2nd edition. 2010, New York: Plenum.

[38] Klie, R.F., A. Gulec, Z. Guo, *et al.*, *The new JEOL JEM-ARM200CF at the University of Illinois at Chicago*. Crystal Research and Technology, 2014. **49**(9): p. 653–662.

[39] Loane, R.F., P. Xu, and J. Silcox, *Incoherent imaging of zone axis crystals with ADF STEM*. Ultramicroscopy, 1992. **40**(2): p. 121–138.

[40] Paulauskas, T. and R.F. Klie, Decay of high-energy electron bound states in crystals. Ultramicroscopy, 2019. **196**: p. 99–110. DOI: 10.1016/j.ultramic.2018.10.001.

[41] Meysing, D.M., M.O. Reese, C.W. Warren, *et al.*, *Evolution of oxygenated cadmium sulfide (CdS:O) during high-temperature CdTe solar cell fabrication*. Solar Energy Materials and Solar Cells, 2016. **157**: p. 276–285. DOI: 10.1016/j.solmat.2016.05.038.

[42] Kumar, S.G. and K.S.R.K. Rao, *Physics and chemistry of CdTe/CdS thin film heterojunction photovoltaic devices: fundamental and critical aspects*. Energy Environ. Sci., 2014. **7**(1): p. 45–102.

[43] Poplawsky, J.D., W. Guo, N. Paudel, *et al.*, *Structural and compositional dependence of the $CdTe_xSe_{1-x}$ alloy layer photoactivity in CdTe-based solar cells*. Nature Communications, 2016. **7**: p. 12537. DOI: 10.1038/ncomms12537.

[44] Albin, D.S., Y. Yan, and M.M. Al-Jassim, *The effect of oxygen on interface microstructure evolution in the CdS/CdTe solar cells*. Progress in Photovoltaics, 2002. **10**(5): p. 309–322.

[45] Allen, L.J., A.J. D'Alfonso, B. Freitag, and D.O. Klenov, *Chemical mapping at atomic resolution using energy-dispersive X-ray spectroscopy*. MRS Bulletin, 2012. **37**(1): p. 47–52.

[46] Colegrove, E., S.P. Harvey, J.-H. Yang, *et al.*, *Phosphorus diffusion mechanisms and deep incorporation in polycrystalline and single-crystalline CdTe*. Physical Review Applied, 2016. **5**(5): p. 054014.

[47] Poplawsky, J.D., C. Li, N.R. Paudel, W. Guo, Y. Yan, and S.J. Pennycook, *Nanoscale doping profiles within CdTe grain boundaries and at the CdS/CdTe interface revealed by atom probe tomography and STEM EBIC*. Solar Energy Materials and Solar Cells, 2016. **150**: p. 95–101. DOI: 10.1016/j.solmat.2016.02.004.

[48] Poplawsky, J.D., N.R. Paudel, C. Li, *et al.*, *Direct imaging of Cl- and Cu-induced short-circuit efficiency changes in cdte solar cells*. Advanced Energy Materials, 2014. **4**(15): p. 1400454.

[49] Li, C., J. Poplawsky, Y. Wu, *et al.*, *From atomic structure to photovoltaic properties in CdTe solar cells*. Ultramicroscopy, 2013. **134**: p. 113–125. DOI: 10.1016/j.ultramic.2013.06.010.

[50] Kanevce, A., M.O. Reese, T.M. Barnes, S.A. Jensen, and W.K. Metzger, *The roles of carrier concentration and interface, bulk, and grain-boundary recombination for 25% efficient CdTe solar cells*. Journal of Applied Physics, 2017. **121**(21): p. 214506.

[51] Barnard, E.S., E.T. Hoke, S.T. Connor, *et al.*, *Probing carrier lifetimes in photovoltaic materials using subsurface two-photon microscopy*. Scientific Reports, 2013. **3**: p. 2098. DOI: 10.1038/srep02098.

[52] Li, C., Y. Wu, T.J. Pennycook, *et al.*, *Carrier separation at dislocation pairs in CdTe*. Physical Review Letters, 2013. **111**(9): p. 096403.

[53] Yang, J.-H., W.-J. Yin, J.-S. Park, W. Metzger, and S.-H. Wei, *First-principles study of roles of Cu and Cl in polycrystalline CdTe*. Journal of Applied Physics, 2016. **119**(4): p. 045104.

[54] Ma, J., S.-H. Wei, T.A. Gessert, and K.K. Chin, *Carrier density and compensation in semiconductors with multiple dopants and multiple transition energy levels: Case of Cu impurities in CdTe*. Physical Review B, 2011. **83**(24): p. 245207.

[55] Ablekim, T., S.K. Swain, W.-J. Yin, *et al.*, *Self-compensation in arsenic doping of CdTe*. Scientific Reports, 2017. **7**(1): p. 4563.

[56] McCandless, B.E., W.A. Buchanan, C.P. Thompson, *et al.*, *Overcoming carrier concentration limits in polycrystalline CdTe thin films with in situ doping*. Scientific Reports, 2018. **8**(1): p. 14519.

[57] Ophus, C., *A fast image simulation algorithm for scanning transmission electron microscopy*. Advanced Structural and Chemical Imaging, 2017. **3** (1): p. 13.

[58] Yan, Y., K.M. Jones, M.M. Al-Jassim, R. Dhere, and X. Wu, *Transmission electron microscopy study of dislocations and interfaces in CdTe solar cells*. Thin Solid Films, 2011. **519**(21): p. 7168–7172.

[59] Feng, C., W.-J. Yin, J. Nie, *et al.*, *Possible effects of oxygen in Te-rich Σ3 (112) grain boundaries in CdTe*. Solid State Communications, 2012. **152** (18): p. 1744–1747.

[60] Burst, J.M., S.B. Farrell, D.S. Albin, *et al.*, *Carrier density and lifetime for different dopants in single-crystal and polycrystalline CdTe*. APL Materials, 2016. **4**(11): p. 116102.

[61] Sun, C., N. Lu, J. Wang, *et al.*, *Creating a single twin boundary between two CdTe (111) wafers with controlled rotation angle by wafer bonding*. Applied Physics Letters, 2013. **103**(25): p. 252104.

[62] Sun, C., T. Paulauskas, F.G. Sen, *et al.*, *Atomic and electronic structure of lomer dislocations at CdTe bicrystal interface*. Scientific Reports, 2016. **6**: p. 27009. DOI: 10.1038/srep27009.

[63] Tong, W., H. Yang, P. Moeck, M.I. Nandasiri, and N.D. Browning, *General schema for [001] tilt grain boundaries in dense packing cubic crystals*. Acta Materialia, 2013. **61**(9): p. 3392–3398.

[64] Hytch, M.J., E. Snoeck, and R. Kilaas, *Quantitative measurement of displacement and strain fields from HREM micrographs*. Ultramicroscopy, 1998. **74**(3): p. 131–146.

[65] Hÿtch, M.J. and T. Plamann, Imaging conditions for reliable measurement of displacement and strain in high-resolution electron microscopy. Ultramicroscopy, 2001. **87**(4): p. 199–212.

[66] Leamy, H.J., *Charge collection scanning electron microscopy*. Journal of Applied Physics, 1982. **53**(6): p. R51–R80.

[67] Jiang, Y., Z. Chen, Y. Han, *et al.*, *Electron ptychography of 2D materials to deep sub-ångström resolution*. Nature, 2018. **559**(7714): p. 343–349.

[68] Shibata, N., S.D. Findlay, Y. Kohno, H. Sawada, Y. Kondo, and Y. Ikuhara, *Differential phase-contrast microscopy at atomic resolution*. Nature Physics, 2012. **8**: p. 611. DOI:10.1038/nphys2337.

Chapter 12
Photoelectron spectroscopy methods in solar cell research

Glenn Teeter[1] *and Philip Schulz*[2]

12.1 Introduction

Photoelectron spectroscopy (PES), also referred to as photoemission spectroscopy, is a direct experimental method for assessing the chemical and electronic properties of materials. The technique is becoming increasingly important in the research of photovoltaic (PV) devices—where, more specifically, X-ray photoelectron spectroscopy (XPS) is used primarily to measure the chemical properties such as composition and contamination of solar cell materials, whereas ultraviolet photoelectron spectroscopy (UPS) reveals key electronic properties such as work function and electronic energy-level positions. PES is a surface-sensitive technique ideally suited for the analysis of thin films and interfaces, either completed ones or during their formation process. Because the new generation of PV devices comprise a multitude of complex interfaces—each of which plays a critical role for performance and functionality—PES analysis of functional cell components has gained even more relevance.

To assess the scope of PES characterization methods in the context of research on PV technologies, it is useful to first identify key solar cell material properties and interface functionalities that require iterative co-optimization. On its face, the solar cell is a conceptually simple semiconductor device that functions primarily as a photodiode with optoelectronic properties tailored to efficiently extract energy from the solar spectrum. Thus, at the core of photovoltaics is the understanding and control of the principles of semiconductor physics, specifically with respect to charge-carrier generation, recombination, and transport processes. The most fundamental requirements for bulk properties of a PV absorber include a suitable bandgap (BG), long minority-carrier lifetime, and the ability to control doping. Once these basic requirements have been satisfied, building an efficient solar cell requires integrating the absorber into a device structure that enables efficient charge-carrier collection. Consequently, the conceptual simplicity of a

[1]National Center for Photovoltaics, National Renewable Energy Laboratory, Golden, Colorado, USA
[2]CNRS, Institut Photovoltaïque d'Ile de France (IPVF), Palaiseau, France

solar cell translates into a complex set of inherently coupled challenges relating to control of bulk and interfacial properties—all of which must be simultaneously optimized for device performance and stability.

In particular, the interfaces between the absorber and the charge-transport-layer materials that extract charge carriers to the solar cell terminals play a crucial role in performance. Interestingly, in most relevant PV technologies, the classical p–n homojunction has been largely abandoned in favor of heterojunction architectures. Indeed, heterojunctions offer greater degrees of freedom in device design and, ultimately, higher performance. However, heterojunction architectures also impose stringent requirements on the characterization and understanding of interfacial properties. This argument becomes even more salient considering the trend in photovoltaics toward multijunction devices, which offer higher power conversion efficiencies (PCEs), but also exhibit a higher degree of complexity and a larger number of interfaces.

In general, the dominant loss process in a solar cell is nonradiative electron-hole recombination via defect states, for example, those associated with bulk defects, grain boundaries, or interfaces. Bulk or grain-boundary recombination processes must be controlled through optimized growth and processing tailored to a particular material. Heterointerfaces represent important extended defects for the solid-state system, and at the same time, they constitute primary contributors to the mode of operation and functionality of the device. To suppress interfacial recombination, the interfacial energy-level alignment and defect types and densities must be characterized, understood, and optimized, particularly for new absorber materials (see Figure 12.1) [1].

Thus, emerging challenges in the field of photovoltaics demonstrate a trend toward a focus on the need to characterize interfaces and interfacial materials. For several PV technologies (including Si, III–Vs, and halide perovskites), the quality of light-absorber materials has reached such a high level that once an incident photon has been absorbed, the carrier lifetime (e.g., on the order of milliseconds in monocrystalline silicon) vastly exceeds typical drift or diffusion times to interfaces with charge extraction and transport layers (e.g., on the order of microseconds in monocrystalline Si). As bulk absorber properties become increasingly excellent, bulk recombination becomes negligible, and interfacial recombination and charge-extraction processes begin to dominate observed loss mechanisms and performance limitations. For a comprehensive picture of approaches for determining the energy-level offsets and elucidating interfacial recombination mechanisms, we refer the reader to a recent discussion on contact selectivity in PV applications by Rau *et al.* [2]

Finally, chemical interactions at solar cell interfaces are often implicated in degradation processes and are hence a limiting factor for technological advances. This statement holds true particularly for novel and metastable absorber and transport-layer materials, which puts the analysis of chemical and electronic properties of the involved interfaces, and related isolated surfaces, into a new spotlight.

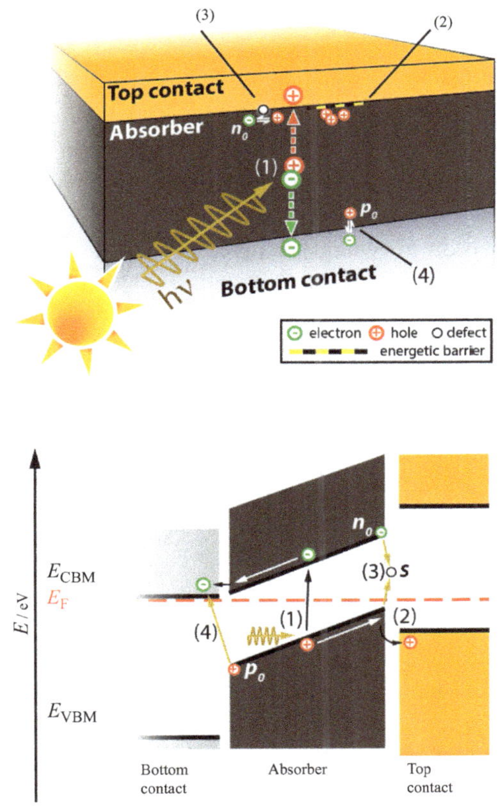

Figure 12.1 Interfaces in a generic n-i-p solar cell (top) and a corresponding energy-level diagram under short-circuit conditions (bottom). After photoexcitation (1), primary performance-loss channels include interface barrier formation (2), interface recombination S (3), and back recombination (4). Key electronic properties of the individual materials in the cell stack are the position of the conduction-band minimum (E_{CBM}), the valence-band maximum (E_{VBM}), and the Fermi level (E_F). $n_0(x)$ and $p_0(x)$ denote equilibrium carrier-concentration profiles. From [1]

As introduced in the beginning of this section, PES is a powerful tool for experimentally accessing interfacial properties. In this spirit, this chapter incorporates four sections relating to the principles and practice of PES for PV research applications. First, we summarize the fundamental working principles of PES and the technical realization within a typical XPS instrument in Section 12.2. The details of PES analysis of new absorber materials along with their interface formation to transport layers are contained in Section 12.3. A few selected examples for the more rarely applied yet powerful advanced analysis methods for structural and morphological aspects by PES are discussed in Section 12.4, and Section 12.5 gives an outlook on opportunities, best practices, and the future development of PES methods for PV research applications.

12.2 Basic principles of photoelectron spectroscopy

PES analysis is based on the energy-resolved measurements of electrons that are emitted due to the interaction of photons with a material. The first high-energy-resolution PES spectrum was reported in 1954 by Kai Siegbahn and his group in Uppsala, Sweden, and in the decades since, it has been well established as a standard surface characterization technique both in industry and academia. The fundamental working principle of PES techniques is the photoelectric effect, first described in 1905 by Albert Einstein, who established the relation between incident photon energy, $h\nu$, and the kinetic energy of the emitted photoelectron, E_{kin}. Photoemission is initiated by the absorption of a photon by a core-level (CL) or valence electron in a material. The photon is annihilated, and in the process, the total energy of the electron is increased by $h\nu$. Absent additional inelastic scattering effects, the photoelectron carries information about its initial binding energy, E_{B}, according to the Einstein relation,

$$E_{\mathrm{B}} = h\nu - E_{\mathrm{kin}} - \phi, \tag{12.1}$$

where ϕ is the work function, defined as the work needed to lift an electron in the solid from the Fermi level (E_{F}) to the vacuum level (E_{vac}). A basic consequence of (12.1) is that photoemission is only possible when $h\nu > \phi$.

At its core, the Einstein relation is an expression of energy conservation, and the inclusion of ϕ equates zero on the binding-energy scale with E_{F} of the sample. E_{B} therefore represents the energy separation between the electron initial-state and E_{F}. PES spectra are traditionally plotted vs. E_{B}, in part because this practice eliminates peak shifts due only to changes in work function, such as those that arise from surface dipoles associated with particular surface terminations or adsorbed contaminants. On the other hand, Fermi-level referencing of PES spectra leads to peak shifts stemming from changes in doping levels in semiconductors. This fact has additional important consequences that will be discussed in more detail in Section 12.5.

The photoemission process that connects the density of electrons in the solid $N(E)$ to the intensity of emitted photoelectrons $I(E)$ is depicted schematically in Figure 12.2. The radiation used to excite photoelectrons is produced by ultraviolet (UV) or X-ray sources, often with the use of monochromators to enhance the energy resolution and suppress measurement artifacts stemming from X-ray satellite peaks. For laboratory-scale measurements, soft X-ray and UV sources have fixed excitation energies (commonly 1,253.6 eV and 1,486.6 eV for the Mg- and Al-Kα lines, respectively, and 21.22 eV for He-I), whereas synchrotron radiation can be tuned continuously over a broad range. The energy resolution in the

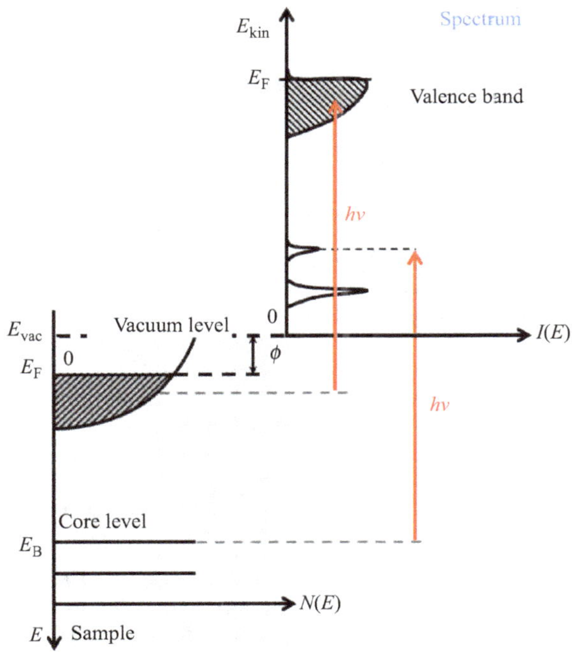

Figure 12.2 Schematic of the relation between the electronic energy levels in the sample on the scale of the binding energy E_B, referenced to the Fermi level E_F, and the corresponding energy distribution $I(E)$ of the photoelectrons emitted upon excitation with photons of energy hv referenced to the vacuum level E_{vac} [4]

acquired spectra is determined in part by the line width of the excitation source and can vary from about 0.3 to 0.7 eV in laboratory X-ray-based PES, to 100 meV for PES based on laboratory UV sources, down to the sub-meV range in synchrotron-based experiments performed at low temperatures and precisely calibrated analyzer optics [3].

The most commonly used spectrometer for measuring photoelectron spectra is the hemispherical electrostatic electron energy analyzer, illustrated in Figure 12.3. These analyzers are designed to create a radial electrostatic field with $1/r^2$ dependence between the inner and outer hemispheres, where r is the distance from the center of the hemispheres. Electrons in the analyzer therefore follow circular or elliptical orbits that are familiar from Newtonian gravitational mechanics. The $1/r^2$ field is achieved by applying voltages to the inner and outer hemispheres that obey a $1/r$ dependence:

$$V(r) = -V_0\left[\frac{r_0}{r} - 1\right] + V_{sw}. \tag{12.2}$$

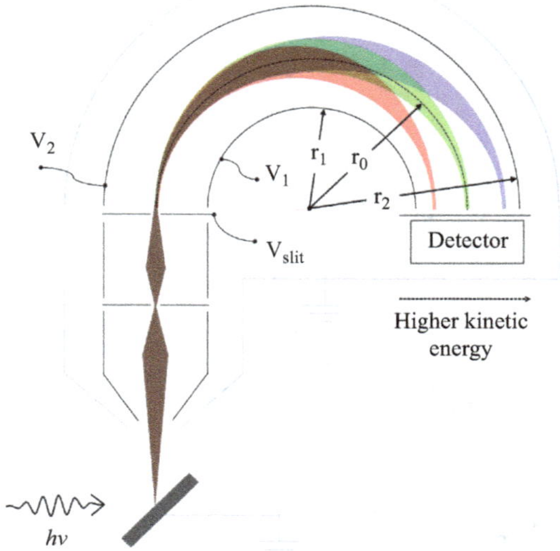

Figure 12.3 Simplified schematic of a hemispherical electrostatic electron energy analyzer. Key elements include the lens column, the inner and outer hemispheres, and the entrance slit. The lens column typically includes one or more apertures that define analysis area and the acceptance angle. Electrons are accelerated or decelerated as they pass from the sample (at or near ground potential) to the entrance slit. Only electrons that pass through the entrance slit with kinetic energy equal to the pass energy will traverse the analyzer to the center of the detector

The parameter V_{sw} represents the sweep voltage that scans the analyzer voltages to measure a spectrum over the required range of kinetic energies. The parameter r_0 is the midpoint between the hemispheres (radii r_1 and r_2):

$$r_0 = (r_1 + r_2)/2. \tag{12.3}$$

From (12.2), the inner and outer hemisphere voltages are

$$V_1 = -V_0\left[\frac{r_0}{r_1} - 1\right] + V_{sw}, \tag{12.4}$$

and

$$V_2 = -V_0\left[\frac{r_0}{r_2} - 1\right] + V_{sw}. \tag{12.5}$$

The parameter V_0 is related to the analyzer pass energy, E_p, according to

$$E_p = qV_0, \tag{12.6}$$

where q is the magnitude of the electron charge. The pass energy is defined as the kinetic energy required for an electron within the analyzer to travel from the entrance slit to the center of the detector. In the above formulation, the analyzer entrance slit voltage is simply given by

$$V_{slit} = V_{sw}. \tag{12.7}$$

In the typical mode of operation, E_p is set to a fixed value, and analyzer voltages are swept relative to ground potential to collect a spectrum over the desired range of kinetic energies:

$$E_{kin} = E_p - qV_{sw}. \tag{12.8}$$

Counting rates are roughly proportional to E_p, whereas energy resolution increases with decreasing E_p. Thus, there is an inherent tradeoff between signal-to-noise (S/N) ratios and peak widths.

For the sake of simplicity, the above analysis assumes that the spectrometer and sample work functions (ϕ_{sp} and ϕ, respectively) are the same. This is not the case in general, and the difference in work functions can lead to a slight offset in measured kinetic energies as photoelectrons travel from the sample to the entrance slit of the spectrometer. This issue is discussed in more detail in Section 12.2.2.

12.2.1 Compositional and chemical-state analysis

12.2.1.1 Electron inelastic scattering effects

Photoelectron inelastic scattering significantly impacts the information depth in PES measurements. Photoelectrons that escape the sample without energy loss are called primary electrons, whereas inelastically scattered electrons that nevertheless manage to escape are labeled secondary electrons. Typical X-ray attenuation lengths in solids for Mg- and Al-Kα sources are on the order of several hundred nanometers, and therefore, photoelectrons are generated in the sample over similar depth scales. Electron inelastic mean-free path (IMFP) values, on the other hand, are much shorter. IMFP values depend on both material composition and E_{kin}, but typically, they range over ~0.2–30 nm for E_{kin} values in the range from 10–1,500 eV [5]. A database of IMFP values is maintained by the National Institute of Standards and Technology (NIST) that is available for download free of cost [6]. Due to short IMFP values, only primary photoelectrons that originate near a free surface can escape the sample and be collected. Hence, the effective escape length of primary electrons is on the order of 2–10 nm. This high degree of surface sensitivity is a key feature and strength of PES measurements in general that must nevertheless be carefully considered when interpreting data. Higher excitation energies result in higher primary photoelectron kinetic energies that can escape from greater depths, as will be discussed in Section 12.4.1.

12.2.1.2 XPS compositional analysis

We now turn to a specific application space for PES—the chemical analysis of sample surfaces—that yields information about the composition of the sample and, to some degree, insight into the chemical bonds of the elemental constituents of the material investigated. XPS is the most commonly used PES method for chemical quantification. In standard XPS measurements, photoelectrons are excited by X-ray sources typically operating at photon energies between 150 and 2,000 eV, i.e., in the range of soft X-rays. Excitation energies in this range give access to the CL spectra of the sample, so that recorded spectra exhibit peaks at specific binding energies for each element present in the investigated material. With the exception of hydrogen and helium, XPS enables CL peak measurements from all elements in the periodic table. XPS peaks thus serve as signatures or "fingerprints" for the presence of chemical elements in the analysis region on the sample surface.

12.2.1.3 XPS survey spectrum

A relatively coarse-grained XPS analysis is achieved by recording a so-called survey spectrum at low energy resolution over the energy range accessible by the employed excitation energy, which facilitates adequate S/N ratios even for short acquisition times. A survey spectrum of a monocrystalline silicon surface in (100) orientation and native oxide on top is shown in Figure 12.4. Several peaks are indexed in this plot, corresponding to the atomic CLs that are detected at known E_B. The notation of the peaks is first the element symbol (Li, Be, ...), then the principal quantum number n (1,2,3, ...), followed by the orbital quantum number l (s, p, d, ...). Here, Si 2p thus stands for an electron with angular momentum 1 emitted from the second shell of a Si atom. In addition, CL peaks with angular

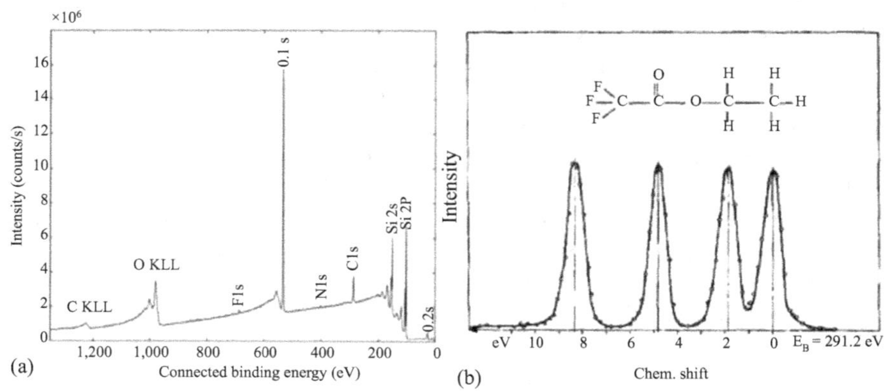

Figure 12.4 Examples of acquired XPS data: (a) XPS survey spectrum of a Si (100)/SiO$_x$ surface. Principal elemental components are Si and O with C surface contaminations and traces of F and N [7]. (b) XPS chemical shifts observed in the C 1s signal from a gas-phase fluorinated organic molecule [8]

momentum >1 come as doublets in the XPS spectra due to the spin-orbit energy splitting. In the standard notation, the total angular momentum $j = l + s$, where $s = \pm½$ is the spin, is added as a subscript after the orbital quantum number. Also seen in the spectrum are Auger electron peaks indexed by the electron shells involved in the Auger process, for example, O_{KLL} for an Auger electron emitted from an oxygen atom involving transitions from the K and L electronic shells.

12.2.1.4 Elemental relative sensitivity factors

The straightforward observation of CLs allows for qualitative assessments of the elemental constituents of materials as well as contamination and trace elements. Quantitative determinations are based on the comparison of integrated CL peak intensities, which are proportional to the atomic concentrations weighted by the energy-level-dependent photo-ionization cross-section σ for each element detected. Other factors that affect peak intensities include IMFP values, which depend on both material and photoelectron kinetic energy, and the transmission function of the electron spectrometer. The combined effects of all factors that affect intensities are captured by empirically derived relative sensitivity factor (RSF) values. Over the decades of XPS research, RSF values have been determined and tabulated for many CLs investigated under standard XPS conditions [9–11]. The higher the RSF, the more intense is the CL signal of a specific element in the XPS spectrum. In very favorable cases, XPS detection limits can range down to ~0.01 atomic % for elements that exhibit CLs with a high RSF in the accessible binding-energy range, but detection limits of ~0.5 atomic % are more typical. RSF values derived from calibration measurements on dedicated, specially prepared reference samples for a particular spectrometer generally provide the highest accuracy, especially for analysis of novel materials.

12.2.1.5 XPS compositional analysis

Atomic percent compositions (C_x) for each element x in a sample are calculated from background-subtracted, integrated peak intensities (I_x) weighted by RSF values (S_x) according to

$$C_x = \frac{I_x/S_x}{\sum I_i/S_i}, \tag{12.9}$$

where the summation extends over all i elements detected in the sample. This approach is straightforward to apply and generally provides useful information. On the other hand, the basic analysis implicitly assumes that atomic species are uniformly distributed within the XPS information volume. In many cases, this assumption breaks down, resulting in misleading results (e.g., in cases where significant concentration gradients or lateral inhomogeneities exist in a sample within the information depth or across the analysis region). In these situations, information from complementary measurement techniques can supplement XPS results to provide a more accurate understanding of a sample's composition, structure, and morphology. Also, variations of XPS measurements including sputter depth profiling and angle-resolved measurements can provide additional information on the depth distribution of constituents in a sample.

12.2.1.6 XPS background subtraction and related issues

A couple of additional points must be considered for the accurate quantification of the sample composition via XPS. First, a careful background subtraction is required to account for the primary electrons only. To remove the secondary-electron background, several models exist, of which the Shirley background has found the most widespread practical application for CL peak fitting [12]. The Tougaard background [13] is another well-known example. A simple linear background is also commonly used, and many XPS curve-fitting packages also facilitate the use of combined linear plus Shirley backgrounds. Secondly, inelastic scattering processes and "ghost peaks" from excitation sources that are not perfectly monochromatic can be observed in the spectra, and these need to be filtered and properly accounted for in the primary-electron contribution [9].

12.2.1.7 XPS energy resolution considerations

At this point in the quantification process, the energy resolution of XPS spectra can become critical for the precise determination of electron binding energies, particularly for distinct but closely spaced CL peaks. As laid out in the previous section, the resolution depends on the electron energy analyzer as well as the X-ray UV photon source. The energy width from X-ray monochromators is generally about 0.3 eV [14]. At the same time, the electron energy analyzer has a finite energy resolution that is tied directly to the chosen pass energy. Loosely speaking, the physical width of the entrance slit to the electron energy analyzer defines the minimum energy spread that can be measured for a given pass energy. The angular acceptance determined by the electron optics that comprise the lens system also plays a role. Depending on the specific electron spectrometer and chosen pass energy, the energy broadening imparted by the spectrometer might take characteristic values in the range of 0.25–1.5 eV. Ultimately, the width of measured peaks is a convolution of all of the aforementioned effects:

$$I(E) = \alpha * \beta * g(E). \tag{12.10}$$

As previously defined, $I(E)$ represents the measured spectrum vs. kinetic energy, and $g(E)$ is the distribution of photoelectron kinetic energies within the spectrometer acceptance angle. Here, α represents the energy distribution of the photon source, and β captures broadening due to finite energy resolution of the spectrometer. Accounting for these factors, it is typically found that high-resolution CL spectra are characterized by widths roughly in the range of 0.5–2 eV.

In practice, it is difficult and generally not necessary to individually characterize both α and β in (12.10). On the other hand, a relatively straightforward method exists to determine the overall instrument response function of the PES measurement system, which corresponds to the convolution $\alpha * \beta$. This procedure is based on the effect that the instrument response function has on the Fermi edge of a metallic sample. The Fermi-Dirac distribution has the form

$$f(E) = \frac{1}{e^{(E-E_F)/K_B T} + 1}, \tag{12.11}$$

where K_B is Boltzmann's constant and T is temperature. The distribution approaches a step function at $E = E_F$ as $T \rightarrow 0K$ and broadens around $E = E_F$ as temperature increases. Therefore, at a given sample temperature, the true width of the Fermi edge can be directly calculated, providing a means for assessing the effect of instrument broadening on the observed Fermi edge. This is illustrated in

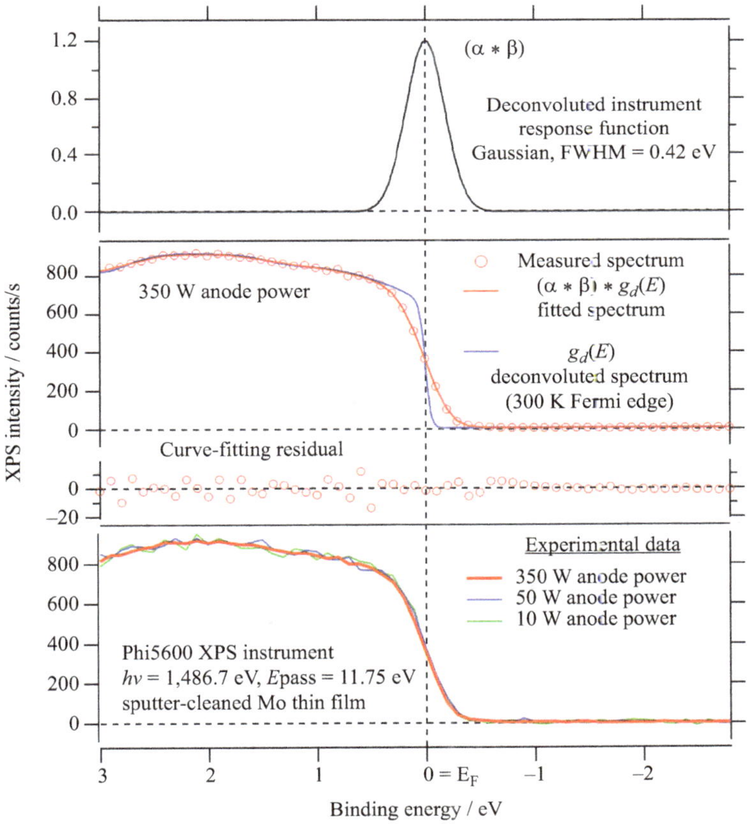

Figure 12.5 Demonstration of a method for extracting the instrument response function $(\alpha * \beta)$ from Fermi-edge broadening in the valence-band spectrum from a metallic sample. The method also provides the deconvoluted valence-band spectrum, $g_d(E)$. The lower panel shows that measurement resolution is not noticeably impacted by changing the X-ray anode power (Adapted from SI in: [15])

Figure 12.5 for an X-ray-excited valence-band measurement on a sputter-cleaned Mo thin-film sample. In the curve-fitting procedure, a deconvoluted valence-band spectrum $g_d(E)$ is synthesized by summing an arbitrary set of Gaussian constituents modulated by $f(E)$ at $T = 300K$ to introduce the Fermi edge. A Gaussian is also used to represent the instrument response function $(\alpha * \beta)$, and the combined set of fitting parameters are varied until the best fit is achieved for the convolution $(\alpha * \beta) * g_d(E)$. The resulting fit thereby provides both the deconvoluted spectrum as well as an empirical fit for the instrument response function for a particular pass energy. In the example, the response function is well described by a Gaussian with a full-width at half-maximum (FWHM) equal to 0.42 eV. A final note of caution is that, in principle, the instrument response function might depend to some extent on initial electron kinetic energy—for example, due to variations in acceptance angle at different values of V_{slit}. If deemed necessary, the procedure outlined here can be carried out at two or more photon energies to characterize the magnitude of variations in instrumental broadening vs. electron kinetic energy.

12.2.1.8 Accuracy vs. precision of binding-energy determinations

It should be noted that the energy resolution of the PES measurement process is identical to neither the precision nor the accuracy with which the binding energies of specific peaks can be determined. CL binding energies are generally determined via curve fitting with Gaussian, Voigt (Gaussian–Lorentzian), or Doniach–Sunjic [16] line shapes. Accuracy in peak-energy determinations depends principally on careful calibration of the binding-energy scale, typically accomplished by using sputter-cleaned, high-purity metal foils. For example, Cu and Au foils provide access to the Cu $2p_{3/2}$ and Au $4f_{7/2}$ CLs, and the metallic Fermi level E_F at 932.62 eV, 83.96 eV, and 0.00 eV binding energies, respectively [17]. Precision, on the other hand, is largely a function of measurement conditions including acquisition times and choice of pass energy, both of which affect S/N ratios of measured spectra, and hence, standard deviations in curve-fitting parameters including E_B values. Figure 12.6 illustrates the variability over time of XPS CL spectra measured on Au foil. In this experiment, the Au $4f_{7/2}$ was measured repeatedly, with a 5-min integration time for each spectrum. The time series in Figure 12.6(a) shows that the variability in binding energies between successive spectra is quite low: curve-fitting standard deviations from these spectra are typically on the order of ± 0.002 eV or less. The observed variability presumably caused by spectrometer temperature changes that occur over the course of the day-night cycle is substantially larger. The statistical analysis of these spectra summarized in Figure 12.6(b) shows that an empirical uncertainty on the order of ± 0.02 eV adequately accounts for these effects.

The example in Figure 12.6 illustrates that the precision of XPS binding-energy determination, based on curve-fitting standard deviations, can be on the order of ± 0.001 eV, even for modest integration times on the order of several minutes. On the other hand, subtle effects such as apparent changes in binding energy stemming from laboratory temperature fluctuations can affect the accuracy of measurements over days or even hours, resulting in systematic errors that are

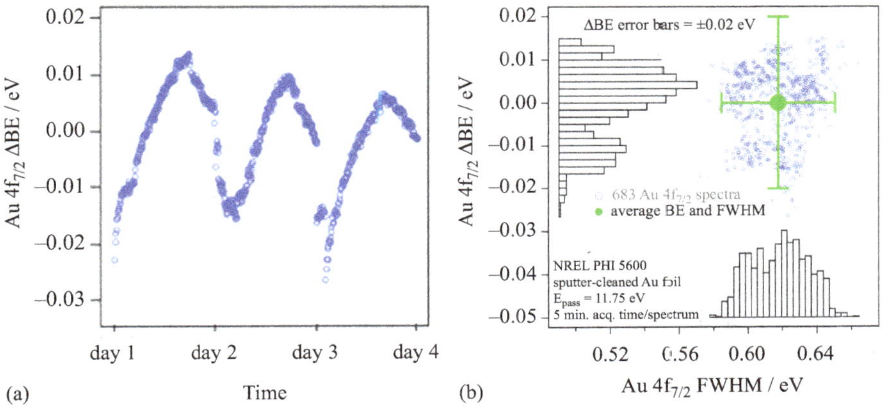

Figure 12.6 *(a) Relative Au 4f$_{7/2}$ peak positions measured repeatedly over the course of several days, showing the effects of ambient temperature variations on measured CL binding energies and (b) statistical analysis of these spectra showing variability of binding energies and peak FWHM values (Figure adapted from SI in: [15])*

significantly larger than the precision of a single measurement. Ultimately, achieving an optimal combination of energy resolution, accuracy, and precision for a particular application requires the performance of a particular spectrometer to be characterized under specific measurement conditions. At the same, a balance must be struck between competing factors including measurement integration time and spectrometer settings such as pass energy, both of which directly impact S/N ratios of measured spectra.

12.2.1.9 XPS chemical-state analysis

High-resolution measurements also provide an interpretation of the chemical environments of the probed elemental species through chemical-state analysis, historically referred to as electron spectroscopy for chemical analysis (ESCA). The approximate binding energy for a particular CL is determined by the identity of the probed element, but more subtle shifts are related to the chemical state of the species. These chemical-state shifts can be extremely useful for identifying particular bonding configurations and phase compositions in samples. For instance, the signal from the metal CL of an oxidized metal is usually found at a higher, i.e., deeper, binding energy than the reduced counterpart.

In chemical-state analysis, CL peak shifts can be traced back to localization of valence electrons in chemical bonds, which affects the ionic character of the atoms. To first order, the chemical shift observed in XPS correlates with the oxidation state of the atom, which, depending on the element and its bonding environment, can be only a few meV or amount to several eV. Chemical shifts are determined not only by oxidation states, but more generally by the partial charge on the probed atomic species. Atoms bonded to strongly oxidizing (electron-withdrawing) groups become positively charged, yielding XPS peak shifts to higher binding energies. A prime example for this effect is carbon in organic molecules, where the C 1s peak is found between 284.5 eV and 285 eV binding energy in C–C bonds but shifts to 289 eV or even higher binding energies for carbon atoms in C–O or C–F bonds (see Figure 12.4(b)).

12.2.1.10 Charging effects during PES measurements on insulators

It is worth noting a specific technical aspect that can become relevant for determining CL binding energies on poorly conducting or insulating samples. In such cases, one can find CLs at erroneously high binding energies. This charging effect is related to photoholes that accumulate at the sample surface, leading to a net positive electrostatic potential, ΔV. The positive surface potential alters the electric-field distribution between the sample and the spectrometer such that photoelectrons' kinetic energies are lowered by $-q\Delta V$ by the time they reach the entrance slit. In turn, this leads to apparent peak shifts to higher binding energies by $+q\Delta V$. For this type of electrostatic charging, charge compensation can become necessary. Usually, charge compensation can be achieved by "flooding" the sample surface with an excess of low-energy electrons to counterbalance the positive charges [11]. Although elemental quantification is still feasible even when charge compensation is used for the measurement, identification of chemical environments becomes more difficult. Furthermore, the determination of band alignments as discussed in the following section is no longer possible because the electron current on the sample surface translates into an arbitrary and indeterminable shift of E_F.

The chemical quantification discussed in this section has become an essential approach to access the composition of new absorber materials and functional films in PV devices. Moreover, and with particular relevance for multicomponent systems such as halide perovskite (HaP) light absorbers, the evolution of the chemical environment of the observed atomic species can be determined by tracking chemical shifts in the photoelectron binding energy, which enables identification of the compositional space and effects, such as alloying and phase segregation. Further comprehensive descriptions and guidelines for the performance of PES measurements can be found in the literature for a broad range of material systems, in general, and for PV-relevant materials, in particular [10,11].

12.2.2 Determining surface and interface energetics

PES measurements directly assess the valence region (VR) and CL region of metallic and semiconducting materials. These values, along with measurements of the sample work function, enable determination of the key energetic parameters—including the position of the valence-band maximum (VBM) and the ionization energy (IE), which together determine the energy-level alignments at PV interfaces.

Although in principle following (12.1), the actual measurement and interpretation of photoemission spectra to extract the energetic parameters of the investigated surface requires some additional analytical steps. The relation between E_B and E_{kin} as established in (12.1) holds only for primary electrons, i.e., photoelectrons that were not subject to inelastic scattering in the emission process. As discussed previously, these primary electrons retain information of their initial state's energy and momentum in the solid.

12.2.2.1 Extracting the valence-band maximum

For any given excitation energy, the primary electrons originating from the highest-occupied states in the valence band have the maximum kinetic energy, E_{kin}^{max}, corresponding to the lowest binding energy detected in the PES measurement. This valence-band onset will further be used to determine the VBM for semiconductor samples. The commonly used procedure for extracting VBM from a valence-band spectrum is to identify and curve-fit linear regions of the spectrum that represent the (i) valence-band onset and (ii) background signal at binding energies below the onset. The intersection of these straight-line sections is then defined to be the VBM. In some cases, the region of the valence-band spectrum that corresponds to the valence-band onset is relatively easy to identify; but in other cases, it can be much less clear. Therefore, a potential source of error or inconsistency between measurements performed at different times or by different researchers is how the valence-band onset is identified and fit. Consequently, a best practice is to clearly illustrate the fitting procedure used to derive a particular VBM value in a figure whenever possible, especially in cases where there is room for interpretation.

12.2.2.2 Extracting the work function (ϕ)

On the high-binding-energy (low-kinetic-energy) side of the PE spectrum, the measured intensity drops abruptly to zero at the so-called secondary-electron cutoff (SEC). The SEC corresponds to zero on the kinetic-energy scale, i.e., these are electrons that lost essentially all of their kinetic energy due to inelastic scattering events. The SEC therefore corresponds to the secondary electrons with the minimum kinetic energy to overcome the sample work function to escape the solid and be detected. The work function ϕ can be determined from the SEC of the photoemission spectra according to

$$\phi = h\nu - E_{SEC}, \tag{12.12}$$

where E_{SEC} is the SEC on the binding-energy scale. The value of E_{SEC} is determined in a manner directly analogous to that described above for extracting the VBM, except in this case it is the onset of the SEC that is fitted with a straight line and subsequently intersected with the background signal.

The concrete experiment underlying the PES measurement—from photoelectron excitation to kinetic-energy determination—is laid out in detail in the schematic representation shown in Figure 12.7(a). The schematic captures two regimes in which the experiment can be performed. In UPS, the sample is irradiated by UV light (5–150 eV) to measure the VR, whereas in XPS, the sample is irradiated by X-rays (>150 eV) to measure CLs. It should be noted that X-ray excitation can also be used to measure the VR. In the following, the photoelectron emitted upon

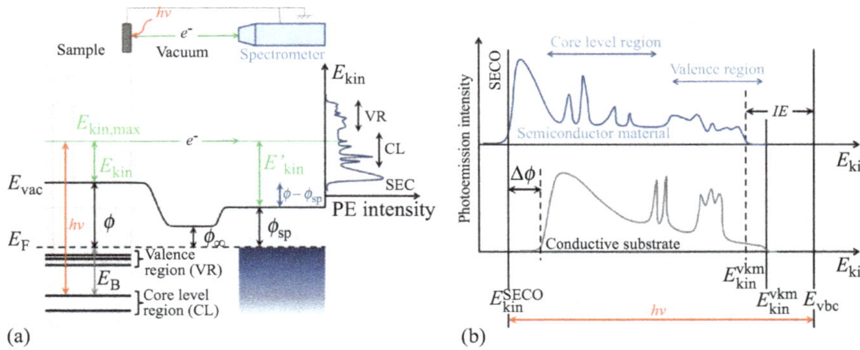

Figure 12.7 (a) Schematic representation of the PES process including the energy-level diagram of the sample and the electron spectrometer in equilibrium. ϕ and ϕ_{SP} correspond to the work function of the surface and of the spectrometer, respectively. E_B is the binding energy with respect to the Fermi level E_F, E_{kin} the kinetic energy of the photoemitted electron, and E_{kin}^{max} is the maximum kinetic energy of the photoelectrons originating from the VBM of a semiconductor. In practice, E'_{kin} is the measured kinetic energy and depends only on the potential difference $\phi - \phi_{SP}$. The PE intensity corresponds to $I(E)$ on the right and is further broken down in (b) the PE spectra, plotting the intensity of the emitted photoelectrons over their kinetic energy. Key energetic parameters, such as E_{kin}^{SEC}, can be determined from these spectra as further explained in the text [4]

photoexcitation from an energy level at a specific binding energy E_B takes on the kinetic energy E_{kin} as described in (12.1). Now, the particular referencing of the energy levels becomes the crucial parameter to generate an accurate read-out from the PES experiment.

12.2.2.3 Spectrometer work function (ϕ_{SP})

In the typical configuration, the sample and the electron spectrometer are in electronic equilibrium, i.e., their Fermi levels are aligned, $E_F = E_{F,SP}$. In the general case, the work function of the sample ϕ and the work function of the spectrometer ϕ_{SP} are not the same. Effectively, this constitutes a variation of the vacuum level E_{vac} along the electron path through the analyzer from the sample to the spectrometer through vacuum. However, to reach the detector, the kinetic energy of the photoelectrons needs to be higher than the difference in the work functions, $E_{kin} > \phi - \phi_{SP}$. Thus, to measure the true SEC from the sample, the photoelectrons need to overcome this possible potential barrier. In practice, this can be achieved by applying a negative bias V_{bias} to the sample such that $\phi + qV_{bias} > \phi_{SP}$; but it should be noted that this situation is only likely to occur when measuring at very low analyzer pass energies, or samples with very low work functions, or both. In practice, it is standard to apply a bias voltage (typically ~ -5 to -10 V relative to ground) to enable efficient collection of very low-kinetic-energy photoelectrons, which otherwise are quite susceptible to stray electric or magnetic fields within the analysis chamber.

This apparent change in the vacuum level means that the measured kinetic energy E'_{kin} is also determined by this work function difference, such that E'_{kin} only depends on ϕ_{SP}, leading to:

$$E'_{kin} = h\nu - E_B - \phi_{SP}, \tag{12.13}$$

where E_B is the binding energy with respect to the Fermi level in the sample.

Now, ϕ_{SP} can be determined by measuring the Fermi edge of a metal sample $E^{E_F}_{kin}$, which is in electronic equilibrium with the spectrometer, and then calculated according to

$$\phi_{SP} = h\nu - E^{E_F}_{kin}. \tag{12.14}$$

Typical acquired spectra corresponding to $I(E)$, sometimes referred to as an energy distribution curve, are depicted in Figure 12.7(b). Note that $I(E)$ is now referred to as photoemission intensity, and the x-axis scale denotes the kinetic energy, which can be converted into E_B according to (12.2) once $E^{E_F}_{kin}$ and hence ϕ_{SP} have been determined.

The main features of interest, indicated in these PES spectra, are the SEC energy (E^{SEC}_{kin}), the CL region with the signal from electrons that are strongly bound to the atomic nucleus, as well as the VR. The key parameters for establishing interfacial energy-level alignment such as the sample work function ϕ, the VBM (E^{VBM}_{kin}), and the IE (E_i) can be extracted from the spectra following simple arithmetic operations:

$$\phi = h\nu - \left(E^{E_F}_{kin} - E^{SEC}_{kin}\right) - qV_{bias} = E^{SEC}_{kin} - qV_{bias} + \phi_{SP} \tag{12.15}$$

$$E_{VBM} = E^{E_F}_{kin} - E^{VBM}_{kin} \tag{12.16}$$

$$E_i = E_{VBM} + \phi. \tag{12.17}$$

These parameters are highly sensitive to changes in key sample properties such as structure and composition, as well as to surface modifications including adsorbates. So, PES is ideally suited to complement investigations of structure-property relationships of materials used in PV applications.

12.2.2.4 Practical considerations relating to binding-energy calibrations

The detailed description above on the role of spectrometer work function (ϕ_{SP}) is provided for completeness; but nevertheless, it is worthwhile to emphasize the practical day-to-day procedures for accurately and repeatably calibrating the binding-energy scale. For example, for UPS measurements, the commonly used single-point binding-energy calibration consists in measuring a valence-band spectrum from a well-characterized metal reference sample, then finding the binding-energy offset that places the center of the metallic Fermi edge at $E_B = E_F = 0$. Critically, this calibration must be performed using identical measurement conditions as will be used for subsequent measurements, including, in particular, the bias voltage V_{bias} discussed above. If this procedure is followed, then work functions extracted from subsequent measurements can be determined using

$$\phi = h\nu - E^{SEC}_{BE,bias}, \tag{12.18}$$

where $E^{SEC}_{BE,bias}$ is the binding energy of the SEC when measured with V_{bias} applied to the sample. In this simplified (but nevertheless correct) approach, there is no need to directly characterize ϕ_{SP}.

For XPS measurements, it is possible to perform calibrations using one or more tabulated binding energies from well-known reference materials. Typical examples are sputter-cleaned Cu, Ag, and Au foils [17], which together provide access to CLs across a broad range of binding energies as well as the metallic Fermi level. A two-point calibration might use, for example, the Fermi edge and the Cu $2p_{3/2}$ CL, so that a linear (two-parameter) correction can be applied to the binding-energy scale. In general, the measurement of N distinct calibration points on the binding-energy scale will enable (or require, depending on one's perspective) the use of a polynomial of order $(N - 1)$ to correct the binding-energy scale. In practice, there is rarely much to be gained by going beyond a first-order polynomial (linear) correction. The considerations described above for UPS binding-energy calibrations also apply to XPS binding-energy calibrations, i.e., to achieve the highest accuracy, one must always perform calibration measurements using measurement conditions identical to those to be used in subsequent experiments. For example, binding-energy calibrations can change slightly with different analyzer pass energies, or even photoelectron exit angles.

12.2.2.5 Momentum-resolved PES measurements

The PES technique can offer an even more detailed view into the electronic structure of the materials investigated. As mentioned above, the primary electrons also carry information about the momentum of the electron in the solid state. By pursuing angle-resolved photoemission spectroscopy (ARPES), the information on the momentum distribution can be assessed. Thus, one can resolve the electronic dispersion $E(\boldsymbol{k})$, where $\boldsymbol{k}$ is the crystal momentum. In principle, changing the angle of emission and detection allows one to track the parallel component of $\boldsymbol{k}$, whereas the observed range in the Brillouin zone as well as the orthogonal component of $\boldsymbol{k}$ can be sampled by changing the excitation energy. The intrigued reader can find additional in-depth discussion on the possibilities of ARPES analysis in further literature sources, also delving deep into the discussion of fundamental physics and strongly correlated systems [18].

12.2.3 Complementary electron and X-ray spectroscopy techniques

In light of the preceding applications, PES has proven to be a powerful probe of the chemical composition of samples as well as determining key energetic and electronic surface properties. Nevertheless, XPS and UPS analyses are restricted to $N(E)$ of the occupied states at the sample surface. A wide variety of complementary PES techniques and closely related methods has been established to further access the broad parameter space of optoelectronic and chemical properties.

Moving beyond the classical combination of UPS/XPS experiments, we find Auger electron spectroscopy (AES), X-ray excited Auger electron spectroscopy (XAES), and inverse photoemission spectroscopy (IPES) as laboratory-based methods that can yield additional chemical information (AES/XAES) and map the unoccupied electronic states in the conduction band (IPES). If one includes synchrotron-based techniques, the approach can be expanded by soft X-ray absorption spectroscopy (XAS), soft X-ray emission spectroscopy (XES), and the resonantly excited combination of the two, i.e., resonant inelastic soft X-ray scattering (RIXS). An overview of these electron and X-ray spectroscopic techniques and their principal insights into the electronic structure of a sample has been given by Weinhardt *et al.* in the scope of applying these techniques for solar electric materials [19]. Similar to the schematic on the working principle of PES presented in Figure 12.2, an overview of the mentioned spectroscopy tools is visualized in Figure 12.8. Note that although the schematic electronic structure of the sample—and hence, the density of states (DOS) (here: $D(E)$)—remains the same for all cases depicted in the figure, the type of detected signal—and hence, the spectroscopic information—changes, as explained in the following.

The already discussed cases of XPS, usually probing the CLs of the sample, and UPS, to assess the valence-band information and work function, are depicted in Figure 12.8(a) and (b), respectively. The XAES process is added in Figure 12.8(a). Therein, a core hole is created by an X-ray absorption process and filled by an electron from an outer shell. By moving from the outer shell to this core hole, the electron will transfer a portion of its binding energy to an electron on its initial shell, which is emitted from the sample and collected in the electron analyzer. The surplus in kinetic energy of this Auger electron corresponds to the difference in binding energy between the atomic levels; hence, it carries information of the element and chemical environment. This process can also be induced by bombardment of the sample with electrons, in which case the technique refers to regular Auger electron spectroscopy.

Concerning the analysis of the electronic structure, IPES is the complement to UPS. The sample is irradiated with low-kinetic-energy electrons that relax into unoccupied electronic states, emitting UV photons. In the so-called isochromat mode, the kinetic energy of the electrons is varied, and the UV photons are detected as a function of this kinetic-energy variation, as depicted in Figure 12.8(c). In this way, the conduction-band region is probed and the onset energy of the spectra corresponding to the conduction-band minimum (E_{CBM}) with respect to E_F can be

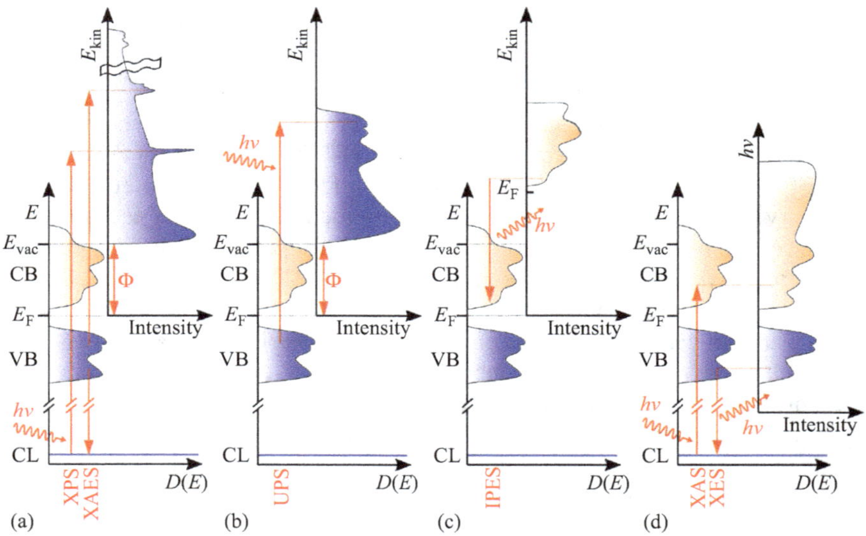

*Figure 12.8 Schematic diagram for the measurement principle of (a) XPS and XAES, (b) UPS, (c) IPES, and (d) XAS and XES, projecting the ground-state electronic structure (left) onto the acquired spectrum (right). D(E) denotes the energy-dependent DOS; E_{vac} is the vacuum level, and E_F is the Fermi level. CL denotes a representative core level, VB is the valence band, CB is the conduction band, and **Φ** is the work function. Blue and beige coloring indicate occupied and unoccupied electronic states, respectively, as well as their corresponding spectra. Vertical red arrows represent excitations and de-excitations of electrons into unoccupied electronic states/holes [19]*

obtained. Together with a successively run UPS measurement to determine the sample work function, the electron affinity (E_{EA}) can be determined by

$$E_{EA} = E_{CBM} + \phi \tag{12.19}$$

While the above techniques can be employed using monoenergetic laboratory sources, XAS, XES, and RIXS need to be performed at a synchrotron light source that provides X-rays with tunable photon energies. In addition, the high flux and small energy band width of synchrotron radiation improves the resolution and S/N ratio, enabling a more accurate assessment of the electronic structure parameters and accelerated data acquisition.

In XAS (see Figure 12.8(d)), the sample is irradiated with X-rays, as in XPS. The main difference between the two techniques is that in XAS, the X-ray energy is scanned over a range that corresponds to the binding energy of a CL of the probed sample, i.e., across the absorption edge. In a one-electron picture (as sketched in

Figure 12.8(d)), an electron from a CL is then excited into an unoccupied electronic state. The technique thus probes details of the conduction band of semiconductors and is sensitive to the chemical environment of the probed atoms. Furthermore, the technique probes conduction-band features with elemental specificity by measuring across the corresponding absorption edges [19]. The acquired data and information depth are strongly dependent on the employed experimental setup and signal-detection mode. In Figure 12.8(d), the fluorescence mode is depicted, which is limited by the attenuation length of X-rays, following the Lambert-Beer law. Alternatively, one can also track the X-ray absorption by detecting the emitted photoelectrons, for example, via the previously introduced PES analyzer optics, which results in a high surface sensitivity of the method. Or one can record the sample current corresponding to the current of emitted photoelectrons, i.e., the total electron yield, which makes the technique sensitive to the subsurface region of the sample down to several tens of nanometers.

On a general note, the analysis of the acquired spectra involves a peak assignment in the peak structures around the X-ray absorption edge, and it is less straightforward compared to the previously presented XPS analysis. The analysis is complicated by the fact that the conduction-band features strongly depend on the molecular bonds, chemical environment of the individual atoms, and also, scattering from neighboring atoms—and thus, local correlation as well as symmetry effects and lattice periodicity. These effects, along with comprehensive introductions and further details on the technical aspects, can be found in a broad range of literature reports and textbooks [20].

In this scope, XES spectra are acquired in a similar experimental configuration, with the key difference being the detection of the fluorescence with a high-resolution X-ray spectrometer [21]. The working principle of the technique is based on the process that a core hole can also be filled with a valence electron while emitting an X-ray photon instead of being filled in an Auger process (Figure 12.8(d)). In XES, the emitted X-ray photons are detected with a high-resolution X-ray spectrometer giving a spectrum of the occupied valence states modulated by the interaction with the probed core hole. Similar to XAS, the technique gives insight into the local chemical environment of the core hole and core-hole relaxation processes. The quality of this information is contingent on the spectrometer resolution, which is thus the focus of current efforts on instrument development [22].

12.3 PES analysis of new absorber materials

The previously described PES and X-ray based spectroscopies continue to occupy a critical role in the development of thin-film solar cells, as outlined in the progress report by Weinhardt *et al.* [20]. Beyond established thin-film PV technologies such as CIGS and CdTe, new photoabsorber materials define a separate class of emergent solar cells [23]. Notably, this includes nonconventional semiconductors such as organic materials, i.e., aromatic hydrocarbons, or hybrid organic–inorganic compounds. The chemical and electronic properties of these materials can differ strongly from those of their purely inorganic counterparts. Here, PES methods have proven invaluable to determine key energetic parameters and the alignment of electronic energy levels at the interface between new absorber material and adjacent transport layers.

12.3.1 Organic photovoltaics

Organic solar cells (OSCs), also referred to as organic photovoltaic devices, aim to combine the mechanical properties and processability of plastics (lightweight, flexibility, solubility) with optoelectronic properties that can be tuned and controlled on the molecular level, further differentiating them from standard inorganic compounds.

The crystal structure of conventional inorganic semiconductors, for example, silicon or III–V compound semiconductors, consists of atoms that are tightly bound by predominantly covalent bonds. The overlap of atomic wave functions between neighboring atoms results in a quasi-continuum of occupied and unoccupied states that form bands in momentum space separated by energy gaps, with this band structure dictating the transport of the delocalized charge carriers [24]. In this regard, we already laid out the importance of the VBM and conduction-band minimum to describe the transport of charge carriers in the solid, which can be assessed by PES methods as discussed in the previous section.

Organic semiconductors show marked differences in the properties described above, as well as their (opto)electronic properties, and thus, the features observed in PES spectra. The organic molecule exhibits strong covalent intramolecular bonds, but the bonds between individual molecules or polymer chains are facilitated by van der Waals interactions with a bond energy that is roughly one order-of-magnitude lower than in a covalent bond. Hence, the electronic structure of organic solids can be described better in terms of discrete molecular energy levels, as opposed to band structure, for which these levels would constitute very narrow bands. In addition, organic molecules exhibit a non-negligible polarization energy upon excitation, which is accounted for in the picture of a single-particle gap, when elevating one electron from the highest occupied state to the lowest unoccupied state.

In consequence, the feature at the highest kinetic energy in the UPS spectra of organic semiconductors represents the DOS of electrons excited from the highest-occupied molecular orbital (HOMO), which is analogous to the VBM in an inorganic semiconductor. Similarly, the feature at the lowest kinetic energy in the IPES spectra represents the lowest-unoccupied molecular orbital (LUMO), analogous to the conduction-band minimum in an inorganic semiconductor.

The alignment of these energy levels becomes of paramount importance for the OSC, which operates as a heterojunction device, i.e., the open-circuit voltage is constrained by the difference between the position of the HOMO of the donor material and the LUMO of the acceptor material. Combined UPS and IPES spectra of separate films of the pure donor polymer poly(3-hexylthiophene-2,5-diyl) (P3HT) and the acceptor 1′,1″,4′,4″-tetrahydro-di[1,4]methanonaphthaleno[5,6] fullerene-C_{60} ICBA are shown in Figure 12.9(a) [25]. Measured IE and EA of

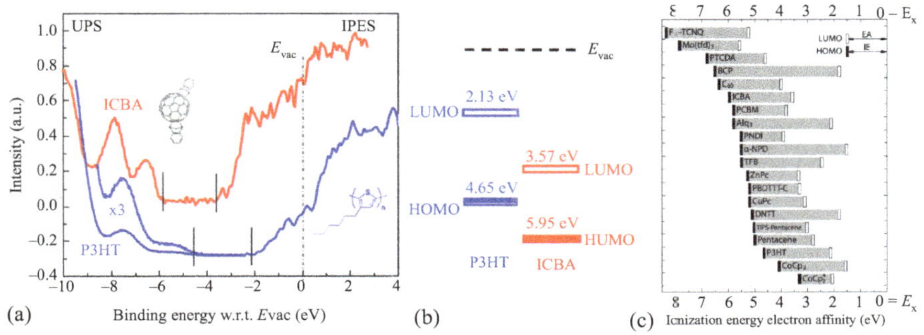

Figure 12.9 Energy-level determination by PES and IPES for the functional layer in OSCs: (a) UPS and IPES spectra of the acceptor molecule ICBA and the donor polymer P3HT. The vacuum level (E_{vac}) has been determined from the secondary-electron cutoff in the UPS spectra. (b) Corresponding energy diagram with IE and EA determined from (a) [25]. (c) Tabulated IE and EA values from PES/IPES measurements for a set of organic molecules typically employed in OSCs [27]

the pristine P3HT are 4.65 eV and 2.13 eV, respectively, resulting in a 2.52-eV single-particle gap, whereas for the ICBA, IE and EA are 5.95 eV and 3.57 eV, respectively, with a single-particle gap of 2.38 eV. Given these values, a simple vacuum-level alignment scheme for the P3HT:ICBA junction would lead to a gap of 1.08 eV between the LUMO of the ICBA and the HOMO of P3HT.

However, the real case of energy-level alignment at the donor/acceptor interface can deviate significantly from the scenario of vacuum-level alignment implied in Figure 12.9(b). Guan *et al.* performed additional PES/IPES measurements of the blend systems (not shown here) and demonstrated that the occurrence of interface dipoles can lead to a departure from the simple vacuum-level alignment scheme. They find the HOMO/LUMO gap of the blend to be 1.68 eV, which is 0.6 eV larger than the value expected from an assumed vacuum-level alignment at the donor/acceptor interface [25].

This result underlines that although PES/IPES measurements of the individual components are already helpful to build libraries of materials energetic parameters, dedicated measurements of the joint system are often required to pinpoint the actual energy-level alignment process.

One of the most promising attributes of organic semiconductors is the large number of materials that enable layer combinations with tailored optoelectronic and chemical properties. Figure 12.9(c) illustrates the expansive range of IE and EA values as measured by UPS and IPES methods, for a small selection of commonly used organic donor and acceptor molecules [27]. This assessment enables the precise engineering of complex multi-heterojunction devices that would be impossible to achieve with the limited choice of thermally, structurally, and electronically compatible inorganic semiconductors [26].

12.3.2 Perovskite solar cells

12.3.2.1 Materials properties and interfaces in perovskite solar cells

Perovskite solar cells (PSCs) are defined by their namesake absorber material, hybrid organic-inorganic metal HaPs. These compounds crystallize in cubic, tetragonal, or orthorhombic ABX_3 structures with an organic cation (e.g., methylammonium $CH_3NH_3^+$ (MA) or formamidinium $CH_5N_2^+$ (FA)) or an alkali atom (e.g., Cs^+) on the A-site; divalent metal cation (e.g., Pb^{2+} or Sn^{2+}) on the B-site; and a halide anion (e.g., I^-) on the X-site; see Figure 12.10(a). They show a unique combination of remarkable semiconductor properties, for example, long-lived charge carriers, room-temperature optical Stark effect [28–31], with "soft" mechanical properties and dynamic disorder [32]. HaPs exhibit a widely tunable optical BG and can be grown in thin films by evaporation or solution-based processes. The BG tunability and facile processability offer the opportunity to implement PSC subcells in multijunction solar cell architectures to increase PCE in a scalable and cost-effective manner. However, key challenges remain with regard to engineering the interface between the HaP absorber and adjacent buffer and transport layers because the models describing the detailed nature, density, and origin of defect states, as well as interfacial chemistry, are still evolving.

At present, PSCs have achieved record PCEs beyond 25% [23], which is attributed to the strong optical absorption coupled with low bulk recombination rates. Despite the high degree of disorder and large density of defects in the bulk and at internal interfaces, this leads to long diffusion lengths for both types of charge carriers. Nevertheless, a precise characterization of these properties in thin films must account for perturbations at grain boundaries in the material as well as electronic coupling at interfaces to the charge-transport layers (see Figure 12.10(b)). These interdependencies between bulk and interface properties are particularly hard to decouple in HaPs. For example, changing the interface between perovskite absorber and charge-transport layer impacts multiple properties simultaneously. First and foremost, recombination of charge carriers at the interface is affected. At the same time, a change in the chemical potential can drive ion migration in the perovskite absorber, and thus, alter the bulk optoelectronic properties. These limitations in understanding and differentiation have been identified as the major obstacle to advancing the rapidly evolving PSC technology [1].

12.3.2.2 Energy-level alignment in PSCs

PES methods marked a decisive step toward a better assessment of the energetic parameters at the interface between the perovskite film and typical transport-layer materials such as organic semiconductors or conductive oxides. The measured energy-level positions allow for the construction of the energy diagram of the layer stack (Figure 12.10(c)), and they add to identifying physical mechanisms such as interface dipoles (ΔE_{vac}), interface barrier heights (ΔE_{VB}), and the degree of band bending that are either beneficial or detrimental for the device operation.

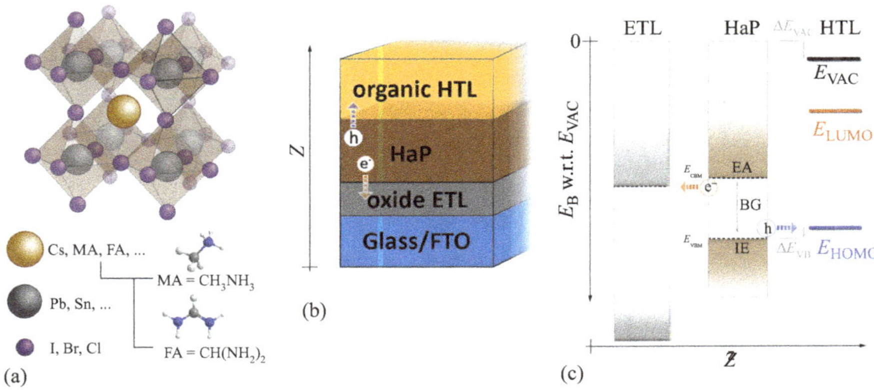

Figure 12.10 *(a) ABX_3 crystal structure and most commonly employed elemental and molecular components of HaPs. (b) Schematic representation of layer stack as realized in a prototypical HaP-based solar cell with organic hole-transport layer (HTL) and oxide electron-transport layer. (c) Schematic energy-level diagram for the device pictured in (b). Characteristic energies, such as the IE and EA, as well as the relative positions of the band edges (E_{CBM}, E_{VBM}) of the HaP absorber of a given BG with respect to the energy levels of the adjacent transport levels determine charge-carrier extraction at the interfaces. ΔE_{VB} indicates the band offset between the valence-band maxima of two adjacent layers (here, the HaP and HOMO level in the hole transporter), and ΔE_{VAC} is the difference between the vacuum-level positions of the HaP and HTL. Reproduced from [41] with permission from The Royal Society of Chemistry*

In a key experiment, the formation of the interface between a set of various HaP films ($MAPbI_3$, $MAPbI_{3-x}Cl_x$, and $MAPbBr_3$) and the typically employed organic hole-transport material 2,2,7,7-tetrakis[*N*,*N*-di(4-methoxyphenyl)amino]-9,9-spirobifluorene (spiro-OMeTAD) was tracked by evaporating thin spiro-OMeTAD films on top of the three different HaP layers [33]. The UPS/IPES data of this experiment of the incrementally deposited (undoped) spiro-OMeTAD layer grown on top of the $MAPbI_3/TiO_2$/glass-layer stack is presented in Figure 12.11(a). Note that although in principle the valence- and conduction-band onsets were determined according to the methodology presented in Section 12.2, the actual fitting procedure of the leading edge in the spectra can be quite intricate. The most technologically relevant perovskite materials are lead-halide-based, which exhibit a low DOS at the band edges, and therefore, a very weak signal in the PES and IPES measurements. A specific requirement was thus to refine the fitting accordingly to capture the best approximation to the true VBM and CBM energy-level positions, which could be visualized by plotting the spectra on a logarithmic scale [34,35].

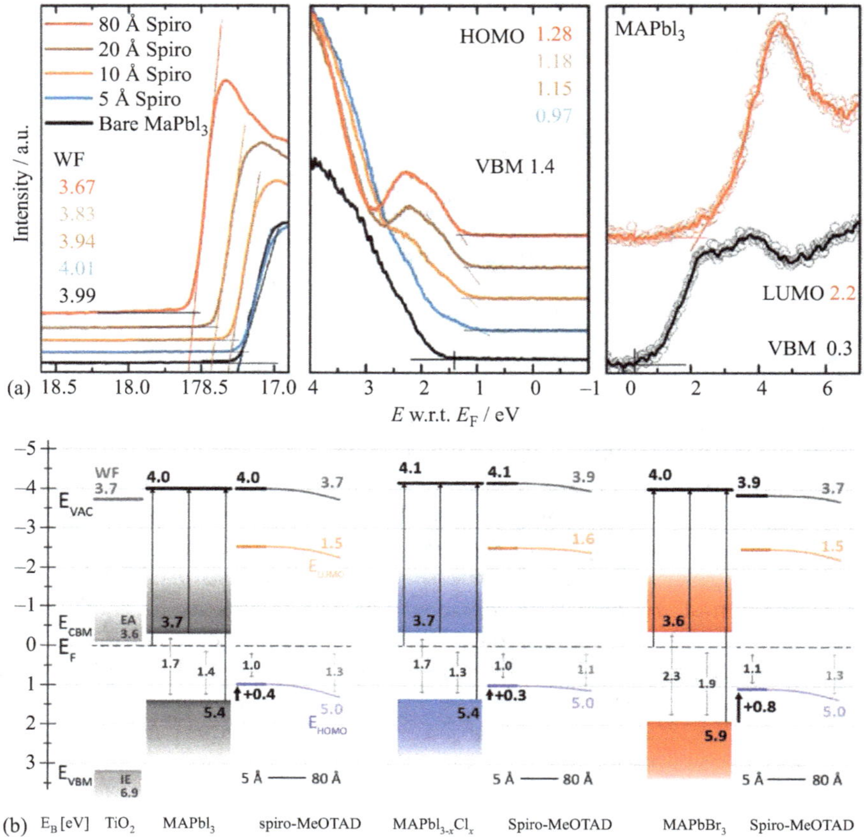

Figure 12.11 Energy-level alignment between the HTM spiro-OMeTAD and HaP thin films. (a) UPS (left and middle panel) and IPES (right panel) of incrementally evaporated spiro-OMeTAD films on top of a $MAPbI_3$/TiO_2/glass-layer stack. (b) Energy-level diagram derived from UPS/IPES measurements of incrementally grown spiro-OMeTAD films on top of $MAPbI_3$, $MAPbI_{3\text{-}x}Cl_x$, and $MAPbBr_3$. Reproduced from [33] with permission from The Royal Society of Chemistry

Three results concerning the energetics at the interface can be summarized as follows. First, no work-function change is observed when the first thin layer of spiro-OMeTAD is deposited, which indicates the absence of any interface dipole between the HaP and spiro-OMeTAD films. Second, the offset between the VBM of the $MAPbI_3$ film and the spiro-OMeTAD HOMO level amounts to 0.4 eV, with the spiro-OMeTAD HOMO being closer to E_F. This corresponds to the case that no barrier for hole extraction from the perovskite film onto the HTL is expected, whereas hole injection in the other direction would be impeded. Third, focusing on the spiro-OMeTAD film, the HOMO-level onset shifts toward higher binding energies (i.e., away from E_F) with increasing thickness. The work function changes accordingly, as the IE of the organic layer remains equal to the equilibrium value of 5.0 eV.

The absence of any interface dipole equates to vacuum alignment, as is analogously observed for the interface between spiro-OMeTAD and the $MAPbI_{3-x}Cl_x$ and $MAPbBr_3$, respectively (see Figure 12.11(b)). XPS measurements of the HaP/spiro-OMeTAD interface systems (not shown here) found the Pb and I CL positions to be constant, which indicates that no band bending occurs in the HaP film as a result of the deposition of the spiro-OMeTAD layer.

These results must be regarded with care because the energy diagram—and particularly, the energy-level misalignment between the charge-transport levels of the organic semiconductor and the HaP—could denote limitations for the attainable solar cell device parameters. For instance, the achievable photovoltage (open-circuit voltage) could be affected because the quasi-Fermi-level splitting (QFLS) in the HaP could be limited by pinning of the electron Fermi level ($E_{F,n}$) close to the TiO_2 CBM at the bottom interface and pinning the hole Fermi level ($E_{F,p}$) close to the organic HTL HOMO on the other side of the absorber film. However, testing devices with different organic hole-transport materials that span a wide range of IEs—and hence, presumably increased/decreased offset between the HaP VBM and the organic transport layer's HOMO level—showed no changes in performance outside of the margin of error [36,37]. Observed exceptions included cases where the IE of the hole-transport materials exceeded that of the HaP, leading to a decrease in device performance, presumably due to the formation of an extraction barrier for holes.

These results indicate that drawing strong correlations between PES data and device metrics can be limited. However, several important findings could be extracted for the energy-level alignment at the HaP/organic transport-layer interface: (i) If the *IE* of the (undoped) organic HTM (IE_{HTM}) is smaller than the IE of the MAPb-based HaP (IE_{HaP}), then vacuum-level alignment occurs. (ii) In a PSC, for the case that $IE_{HTM} < IE_{HaP}$, the quasi-Fermi-level splitting, and hence, attainable photovoltage, does **not** seem to be limited by pinning $E_{F,p}$ at the HTL HOMO level. (iii) If $IE_{HTM} > IE_{HaP}$, then the formation of a hole-extraction barrier can limit charge transport across the HaP/transport-layer interface, which ultimately leads to a decline in short-circuit current and fill factor of PSCs that constitute such an interface.

12.3.2.3 Challenges and limits to the PES characterization of HaPs

In the scope of determining the energy-level alignment between HaP and transport film, we underline the difficulty to access basic properties such as the proper band-edge onsets, which requires a detailed analysis and complex complementary calculations to capture low DOS [34]. This leads to only partially conclusive statements as shown above, while eventually performing PES analysis of HaP-based semiconductors remains quite uncharted territory.

Another important consideration when performing PES experiments to assess the energetic parameters of HaPs is the notorious difficulty in measuring their absolute energy-level positions, which leads to large fluctuations of literature values for any single HaP composition. In materials with low background carrier concentration and few mid-gap defect states, as is the case for $MAPbI_3$ and other lead-based HaPs, the Fermi level is strongly influenced by the Fermi-level position of the substrate. For instance, $MAPbI_3$ exhibits a Fermi-level position that would indicate it to be n-type on TiO_2, but p-type on NiO_x (Figure 12.12) [38]. These results have been in line with studies on $MAPbI_3$ on a variety of substrate materials, which show the perovskite to be n-type on F-doped SnO_2, Al_2O_3, ZnO, TiO_2, and ZrO_2. On p-type substrates, such as PEDOT:PSS, NiO, and Cu_2O, the Fermi level of $MAPbI_3$ shifts closer to mid-gap [39]. These examples show the importance of

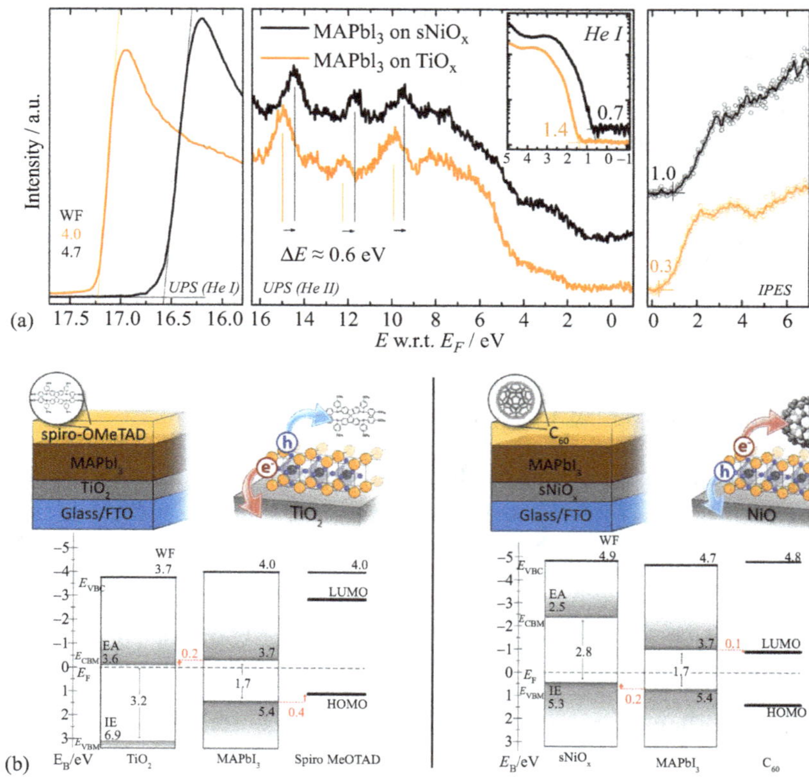

Figure 12.12 Comparison of the Fermi-level position in $MAPbI_3$ thin films on different substrates. (a) UPS and IPES spectra used for band-structure determination. (b) The energy-level diagram of $MAPbI_3$ adjacent to different charge-transport layers. Reproduced with permission from [38]

characterizing the absorbers on the charge-transport layers that they would be paired with and to report the substrate with the PES measurements.

Finally, a specific challenge for the PES analysis of HaP films is the high reactivity and metastability of the material, which shows in transient changes of the signal during the experiment. First and foremost, this behavior can be observed by studying the degree of degradation during the measurements when characterizing these soft materials. The CLs of the elemental components in the HaP layer can shift substantially with prolonged X-ray exposure during XPS measurements, as has been the case for the $MAPbI_3$ layer on top of the NiO_x substrate [38]. After exposure to X-rays in the XPS measurements, the Fermi level moved by more than 0.7 eV, such that initially p-type $MAPbI_3$ on NiO_x became n-type. X-ray-induced degradation of $MAPbI_3$ on TiO_2 in ultrahigh-vacuum conditions was further explored by Steirer *et al.*, who found that under a constant X-ray photon flux density of 1.5×10^{11} photons cm^{-2} s^{-1}, the methylammonium cation deprotonated [40]. At the same time, the lead-to-iodide ratio reduced exponentially over time, indicating that the $MAPbI_3$ converted back to PbI_2. Considering degradation of HaP is particularly important for studies with prolonged exposure to the X-ray beam. However, even for studies involving only short exposure times of well below one hour, it is strongly advised to carefully track changes in the XPS signal to rule out degradation-related measurement artifacts. This can be realized by sequential and alternating acquisition of UPS and XPS spectra to better evaluate the influence of radiation exposure and beam damage from each technique.

Ultimately, this apparent drawback of XPS measurements for HaP films can be turned into an analytical tool for assessing the chemistry at HaP interfaces. This is of high relevance for the many compositional variants of HaP films that find application in solar cells, many of which comprise volatile chemical species and are subject to ion migration, for example, of the dynamically disordered halide species. In addition, the compounds are prone to react with extrinsic species, i.e., atmospheric gases, dopant molecules from adjacent transport layers, or metal atoms from the cell terminals. All of these effects can impact the final device, either in terms of accelerated degradation or inferior functionality such as large hysteresis in the current-voltage characteristic.

Although there are detailed reviews on the use of XPS for unraveling the chemistry at the interface between HaP and buffer and transport layers [41], we point to Section 12.5 for the emerging use of *operando* XPS techniques. In this specific case, XPS could be used to simultaneously track the chemistry at the interface and correlate chemical changes to changes in the device operation.

12.4 Advanced analysis for structural and morphological aspects

The most stringent limitations of PES methods are, as laid out in Section 12.2, the limited probe depth. In addition, typical electron analyzer optics collect electrons from a fixed surface area. This confines the spatial region of analysis to the beam spot and analyzer focus (several tens to hundreds of micrometers) and the sample surface layer of only a few nm thickness without further discrimination.

Information about spatial inhomogeneity on the micro- and nanoscales are therefore usually not resolved in conventional PES experiments. Yet, details on phase separation, material segregation, and species aggregation at grain boundaries are of high interest for many new absorber materials and interface systems in solar cell applications. Buried interfaces carry a plethora of hidden information, critical for device performance. Access to these layers beyond the sample surface can be realized by depth-profiling such as sputter cleaning, ion milling, or reactive etching. However, all these techniques can be regarded as highly invasive. This means that, although a general trend in the concentration profile—and thus, vertical gradient—of species in the sample can be retrieved, preferential sputtering (i.e., higher sputtering rates for more volatile species) usually perturbs any information about the chemical environment of the probed atoms and correlated electronic properties in the revealed surface layer. Nonetheless, XPS depth-profiling is regularly employed for solar-cell-relevant layers to unravel compositional changes over the layer stack and layer thickness. The results are comparable to, but often times less precise than, those of an analysis by time-of-flight secondary-ion mass spectrometry (ToF-SIMS).

Alternatively, hard X-ray photoemission spectroscopy (HAXPES) measurements, using high-energy photon sources, lead to longer photoelectron escape depths, and hence, an assessment of films in the subsurface region of the sample.

On the other end of the toolkit spectrum lies photoemission electron microscopy (PEEM), which combines the energy-dispersive analytics of PES with the lateral resolution of electron microscopy. However, the technique usually requires smooth samples to reach high-resolution images, which can limit the applicability for standard films used in PV applications.

12.4.1 Hard X-ray photoelectron spectroscopy

12.4.1.1 General considerations and technical parameters for HAXPES experiments

As already described in Section 12.2.1, the probe depth in a PES experiment is limited by the inelastic mean-free path (λ_e) of photoelectrons leaving the sample surface. Although this surface selectivity is a fundamental characteristic of PES techniques and can be leveraged as a strength to measure sequentially constructed interface systems (see Section 12.3), it restricts the applicability of PES for the characterization of bulk material properties.

The values for λ_e as a function of the kinetic energy of the electrons compiled by Seah and Dench can be plotted in a "universal curve" (Figure 12.13) to estimate the attenuation length of photoelectrons in various PES (and IPES) techniques depending on the corresponding range of typical electron energies. The actual probe depth equals about three times λ_e. Therefore, typically, the probe depth—from which information can be obtained in classical UPS, XPS, and IPES experiments—will vary between 1 nm and 10 nm.

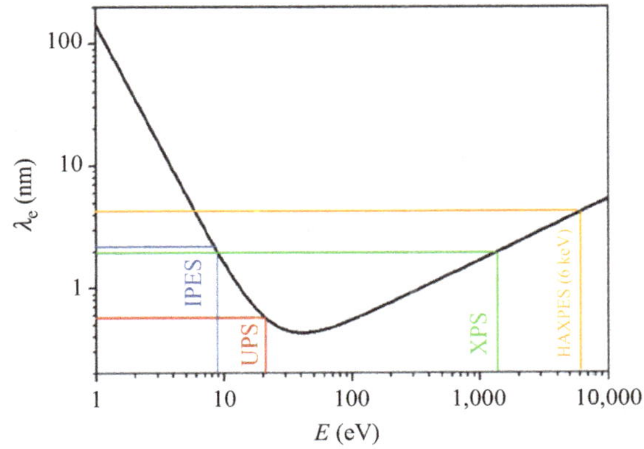

Figure 12.13 *"Universal curve" of the inelastic mean free path (λ_e) of electrons in pure elemental materials as a function of the electron kinetic energy after Seah and Dench [6]. The range of electron kinetic energies for several PES techniques is indicated by colored lines for an estimation of the typical values of λ_e, adapted from Seah and Dench [6]*

Typical laboratory spectrometers for standard XPS measurements are operated with Al or Mg X-ray sources from which the Kα radiation (1,486.7 and 1,253.6 eV, respectively) is used, leading to a maximum probe depth of 10 nm for electrons at low binding energies (and thus, high kinetic energy following (12.1)). An expansion of the probe depth is achieved by the advancement of XPS methods that focus on excitation sources with high photon energies as in HAXPES-compatible experimental configurations. Increasing the photon energy of the source thus translates to an increase of the kinetic energy of the emitted photoelectrons from a given energy level compared to excitation from the same energy level with soft X-ray excitation, i.e., low-energy photons. On the universal curve, depicted in Figure 12.13, the regime of HAXPES measurements is indicated in yellow with typical electron kinetic energies of 6 keV and beyond leading to λ_e-values of several tens of nanometers. For higher excitation energies, i.e., beyond 10 keV, HAXPES can be employed to directly assess the electronic properties and elemental composition hundreds of nm below the surface, and thus, even down to the interface with the transport layer or substrate underneath.

With this preface, the initial motivation for pursuing HAXPES measurements has been to reduce the effect of surfaces (contamination for instance) and instead, to acquire more signal from electrons from the bulk of the probed materials [42]. By systematically varying the excitation energy, one can tune the probe depth and gradually assess the composition at the surface and in the subsurface region. The approach is particularly valuable for new absorber materials such as HaPs films that exhibit strong deviations from the nominal stoichiometry due to the segregation of chemical species and corresponding compositional gradients.

The most common way to generate hard X-rays for HAXPES measurements is at synchrotron beam lines, which allow a tunable energy range for the employed X-ray photons. Alternatively, a set of X-ray anodes exist for laboratory-scale experiments. For example, by using the Ag Lα line (2,984.2 eV) instead of the Al Kα (1,486.6 eV), the probe depth is extended to beyond 15 nm from the sample surface. Today, sources that emit X-rays at even higher photon energies have become available such as X-ray anodes producing Ga Kα (9,252 eV) radiation.

12.4.1.2 HAXPES measurements for PSCs

The mode of data acquisition in a HAXPES experiment is the same as in XPS measurements using lower-energy X-rays in the conventional experimental setup. A key disadvantage of HAXPES over conventional XPS measurements is the lower photoionization cross-section, which leads to a lower signal-to-noise ratio and hence longer acquisition times. In addition, accurate data for photoionization cross-sections for excitation with high-energy photons are not as routinely available as for Al or Mg X-ray anodes. Therefore, the quantitative assessment of CL intensities and determination of elemental ratios and sample compositions becomes more ambiguous. In addition, the high-energy photons can potentially be more damaging to the probed specimen and hence lead to accelerated degradation, which has, for instance, been observed in the case of HaP samples [43].

Nonetheless, the method yields valuable chemical information of new absorber systems. To reveal inhomogeneities, HAXPES measurements have been performed on various HaP compositions [43,44]. Therein, spectra taken at various excitation energies gave a first assessment of chemical and electronic structure gradients of the investigated perovskite. For instance, in an approach to probe the effect of non-stoichiometric ionic distributions, i.e., deviating from the ABX_3 perovskite composition, Jacobsson *et al.* measured CL spectra of $FA_{0.85}MA_{0.15}PbBr_{0.45}I_{2.55}$ layers made from precursors that were either in a stoichiometric ratio or had either 10% excess or deficiency in the PbI_2 content [44]. Spectra of the Br 3d, I 4d, and Pb 5d CL regions and the valence band were acquired at 758-eV, 2,100-eV, and 4,000-eV excitation energies corresponding to probe depths of t_p ~5, ~11, and ~18 nm, respectively. The data from the measurements performed at 758-eV and 4,000-eV excitation energy are depicted in Figure 12.14. The results indicate that the iodine-to-lead ratio (a) increases with PbI_2 deficiency, as the ratio is 2:1 for PbI_2 and 3:1 for the perovskite, and (b) decreases at higher excitation energies, i.e., for photoelectrons originating from further down in the film. The compositional analysis also indicated a lower iodine-to-bromine ratio at the surface and enabled Jacobsson *et al.* to propose a model in which the perovskite layer is terminated with an overlayer of unreacted formamidinium iodide, followed by a bromine-rich perovskite phase as shown in Figure 12.14(c).

In further examples, the technique was also applied successfully on perovskite samples to explore the distribution of trace species such as chlorine in methylammonium lead iodide chloride films or to track the presence of metallic lead throughout a perovskite layer [45,46]. HAXPES has thus become an invaluable

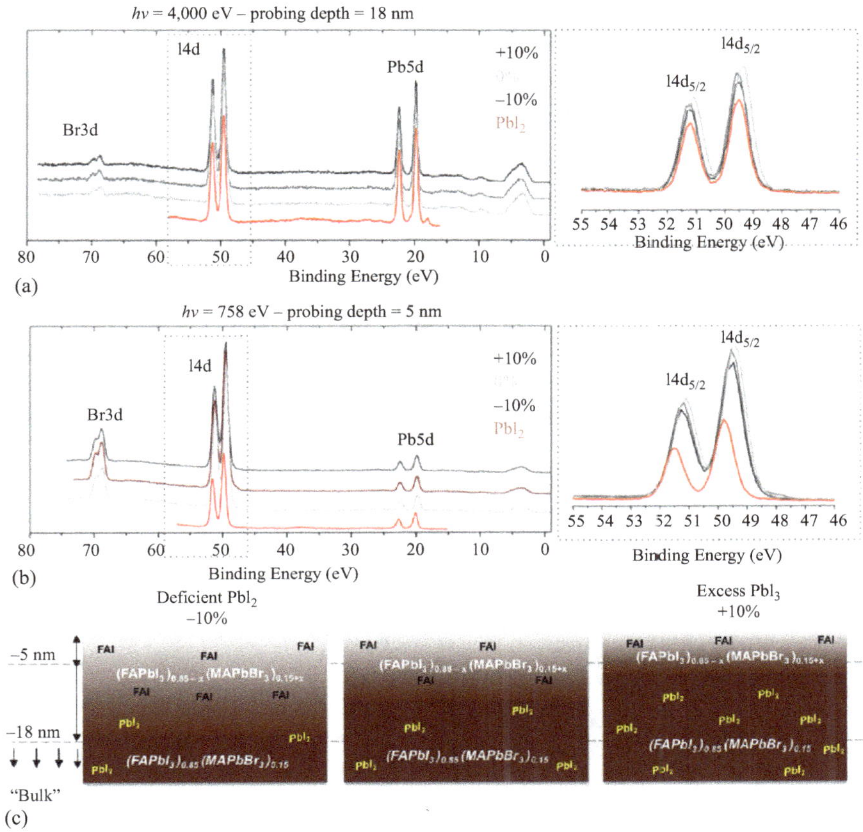

Figure 12.14 HAXPES measurements of $FA_{0.85}MA_{0.15}PbBr_{0.45}I_{2.55}$ at excitation energies of (a) 4,000 eV and (b) 758 eV. The spectra show Br 3d, I 4d, and Pb 5d CL regions for films that were either stoichiometric or prepared with excess (+10%) or deficiency (−10%) in PbI_2. Reference spectra of a PbI_2 film are plotted for comparison. (c) Coarse model of the component distribution in the surface and subsurface region derived from the HAXPES measurements indicating a PbI_2-rich bulk film (dark brown) with a surface FAI layer (light brown). Reprinted with the permission from [44]. Copyright 2016 American Chemical Society

technique to gain insight into the chemical make-up of novel solar electric materials and their interfaces.

12.4.1.3 Advanced electronic characterization

Aside from this compositional analysis, systematic HAXPES measurements can be employed to unravel critical electronic properties of a given absorber film, heterojunction, or interface. This is exemplified quite readily in the case of a conventional absorber material such as microcrystalline silicon (μc-Si).

This system is relevant for the current generation of silicon heterojunction solar cells, whose functionality is based on thin doped layers of alloyed silicon on either side of an embedded Si wafer. The heterojunction usually consists of crystalline and hydrogenated, amorphous silicon (a-Si:H), while the front contact is realized by a transparent conductive oxide layer.

Wippler *et al.* investigated such a system by first depositing a 38.5-nm-thick hydrogenated and heavily boron-doped μc-Si (μc-Si:H:B) layer on top of an Al-doped ZnO film on top of a glass substrate [47]. In the dedicated HAXPES study, the information depth was varied by measuring the Si 1s and Si 2s CLs at different excitation energies, thus giving access to a large range of electron kinetic energies. The results indicate the existence of a Si–O$_x$ surface layer, whose influence decreases rapidly with increasing probe depth. More importantly, depth-dependent

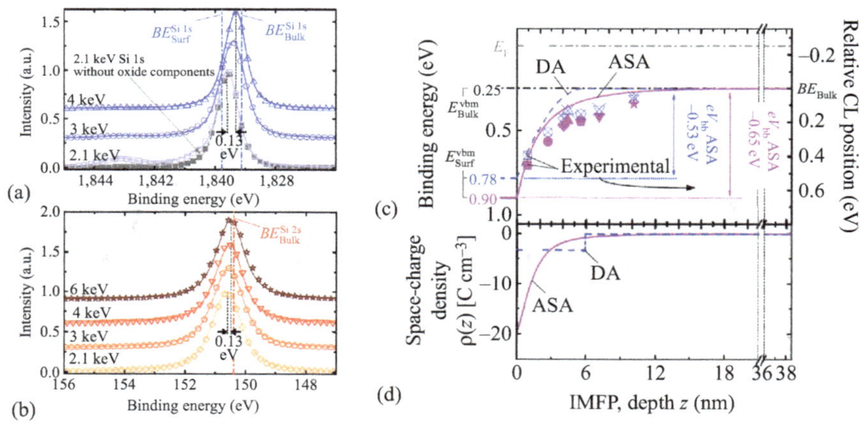

Figure 12.15 HAXPES measurements of a μc-Si:H:B film showing the Si 1s (a) and Si 2s (b) CL regions. Measured CL shifts and model-derived CL positions for bulk and surface are indicated by vertical lines. (c) Measured Si 1s and Si 2s CL peak position vs. corresponding λ_e (here: IMFP)and the associated depth-resolved band profile/CL position vs. depth z (lines) for the model best fit concerning ASA (magenta solid line, filled symbols) and the depletion approximation DA (blue dashed line, open-crossed symbols). The symbols represent Si 1s [Si 2s] with increasing excitation energies hν ∈ {2.1, 3, 4, [6]} keV: square [diamond], circle [pentagon], upward-pointing [downward-pointing] triangle, [asterisk]; the left and right axes illustrate binding energy (VBM) relative to E_F and CL position relative to the Si 1s or Si 2s bulk CL, respectively. (d) Related space charge density ρ(z) for the ASA and the DA model. From [47]

shifts of the Si-Si feature in the Si 1s and Si 2s CL spectra were observed. These shifts are more pronounced for low kinetic energy of the photoelectrons, i.e., lower excitation energy and/or higher binding energy as in the case of the Si 1s CLs.

The fitted data and extracted CL positions with respect to the bulk Si binding energy are depicted in Figure 12.15. The spectra show the (deep) Si 1s CL at high binding energy (Figure 12.15(a)) and Si 2s CL at lower binding energy (Figure 12.15(b)) for excitation energies varied between 2 and 6 keV. The shift is apparent by the guide to the eyes, but a more accurate and quantitative assessment of the band bending requires a sophisticated modeling approach. In the current example, the modeling was done by advanced semiconductor analysis (ASA) to generate a band profile. ASA simulates space-charge effects associated with a partially occupied quasi-continuum DOS within the BG, thereby effectively simulating band-tail states and dangling bonds. For further comparison, the profile was also modeled by a first-order depletion approximation, which emulates the electrical properties of crystalline semiconductors without defect states in the BG. In this model, the surface charge that causes the band bending is compensated by ionized donors or acceptors in the space-charge region [47]. As shown in Figure 12.15(c) and (d), the data obtained from the HAXPES experiment yield a good representation of the band bending at the Si surface, in line with the model approximations.

12.4.2 Imaging and photoemission electron microscopy (PEEM)

The PEEM technique is characterized by recording the two-dimensional intensity distribution of photoelectrons emitted from the sample area, and thus, it results in an enlarged image of the sample surface with chemical and electronic specificity.

Analogously to conventional PES experiments, in PEEM, electrons are emitted from the sample surfaces through the photoelectric effect. However, in contrast to PES, the primarily measured quantity is not the number of electrons of a kinetic energy selected by the analyzer, but rather, the intensity distribution of the photoelectrons from a two-dimensional area of the sample. In this mode, the emitted photoelectrons are collected through a strong electrostatic field between the sample and the imaging column of the instrument, and the electron image is subsequently enlarged by a series of coaxial electron lenses. The electrons are deflected either onto a luminescent screen to produce an image that can be captured via a CCD camera or more directly into a microchannel plate detector, i.e., a large number of two-dimensionally arranged channel electron multipliers.

So far, there are only limited examples of studies that correlate PEEM results of PV materials with the functionality in solar cells. However, moving beyond the static images obtained in the standard PEEM experiments, the technique becomes particularly interesting in time-resolved mode of operation to track dynamic events across semiconductor heterostructures. For instance, the combination of high spatial and temporal resolution in PEEM imaging allowed researchers to track the motion of photoexcited electrons from high-energy to low-energy states at an InSe/GaAs heterostructure [48]. In that specific case, the PEEM setup is realized in a pump-probe experiment, in which carriers were excited into the conduction band by a 800-nm pump probe and emitted by a time-delayed 266-nm probe pulse for collection via the electron analyzer optics.

The observed electron dynamics were described in a simple model with the creation of uncorrelated electrons (in the GaAs) and holes (in the InSe) and subsequent to lower-energy conduction- and valence-band states, respectively. The resulting charge accumulation and depletion regions were suggested to form internal electric fields, which could impede carrier transport and correspond to photovoltage and band-bending effects in the semiconductor heterostructure.

The experiment thus gives insight into the operation of the solar-cell-relevant interface and underlines the importance of turning toward dynamic measurement conditions for PES methods to gain insight into the parameters that determine device operation. In the following section, the basic principles of such *operando*-style measurements for a generalized application across various PES methods will be laid out in more detail.

12.5 Surface and junction photovoltage, transient behavior, and beam damage: opportunities, best practices, and *Operando* studies

12.5.1 Consequences of Fermi-level energy referencing

As described in Sections 12.1 and 12.2, the concept of Fermi-level alignment plays a central role in determining surface energetics, interfacial energy offsets, and chemical-state assignments. Figure 12.16 illustrates Fermi-level positions in a metal and a generic semiconductor with different doping types. Under dark equilibrium conditions, the spectrometer and sample Fermi levels are aligned. A direct consequence of the Fermi-level-referenced binding-energy scale is that changes in the equilibrium Fermi-level position produce an equivalent rigid shift of the PES spectrum, so that both the valence bands and CLs of the semiconductor shift to lower binding energy as doping type/level shifts from n- to p-type, such that $E_{CL} - E_{\mathrm{VBM}}$ is constant. This fact is leveraged in band-alignment measurements on junctions, where both materials forming the junction are within the XPS information depth. In these measurement, CL shifts can serve as proxies for valence-band shifts in cases where individual valence-band spectral contributions are not easily separated. A further consequence of Fermi-level referencing is that CL binding energies for a particular semiconductor can vary over the range defined by the BG. At the same time, a potential pitfall of XPS measurements on optoelectronic materials is that light bias can create unanticipated shifts in the measurement reference level. Failure to properly account for these shifts can lead to errors when measurements are performed on materials or devices with non-negligible photovoltages.

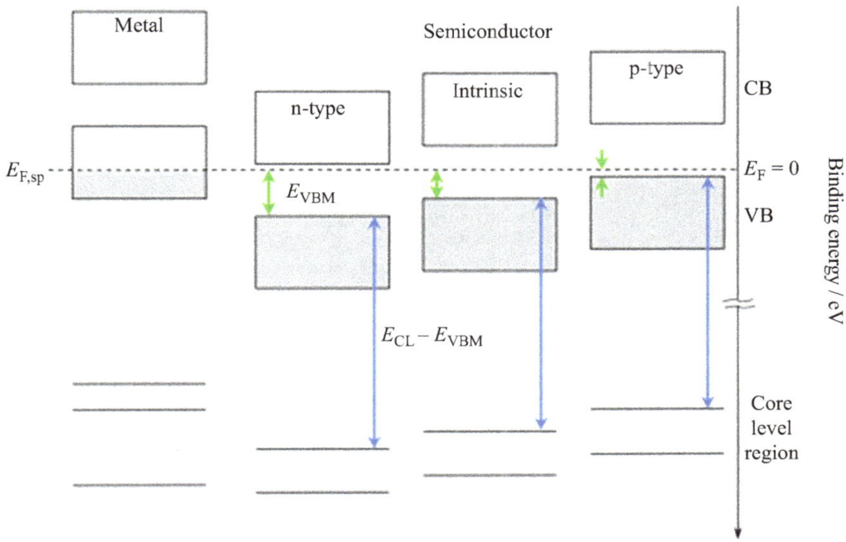

Figure 12.16 Schematic illustration of native bulk Fermi-level alignment in a metal and in a semiconductor doped n-type, intrinsic, and p-type. For the semiconductor, CLs and valence-band features shift by equivalent amounts on the binding-energy scale as doping type and degree vary

12.5.2 Surface photovoltage effects

Figure 12.17 illustrates how surface photovoltage (SPV) associated with light bias during PES measurements can alter measured valence-band and CL binding energies. Figure 12.17(a) represents equilibrium and light-biased surface band diagrams for a p-type semiconductor. In the absence of light bias, there is no quasi-Fermi-level splitting, and charged defect states at the free surface pin the Fermi level near mid-gap and create downward band bending, characterized as the free-surface built-in voltage V_{bi}. For a p-type bulk semiconductor, the resulting surface band bending shifts the PES spectrum to higher binding energy. The converse situation is illustrated in Figure 12.17(b), where surface band bending for the n-type material shifts the PES spectrum to lower binding energy. The effect of light bias in both cases is to generate electron-hole pairs and create quasi-Fermi-level splitting (μ),

$$\mu = E_{Fp} - E_{Fn}, \tag{12.20}$$

that tends to remove surface band bending. Both cases illustrated in Figure 12.17 tacitly assume that no junction exists between the semiconductor and the spectrometer Fermi level $E_{F,sp}$, so that the majority-carrier quasi-Fermi level remains aligned with $E_{F,sp}$ under light-biased conditions.

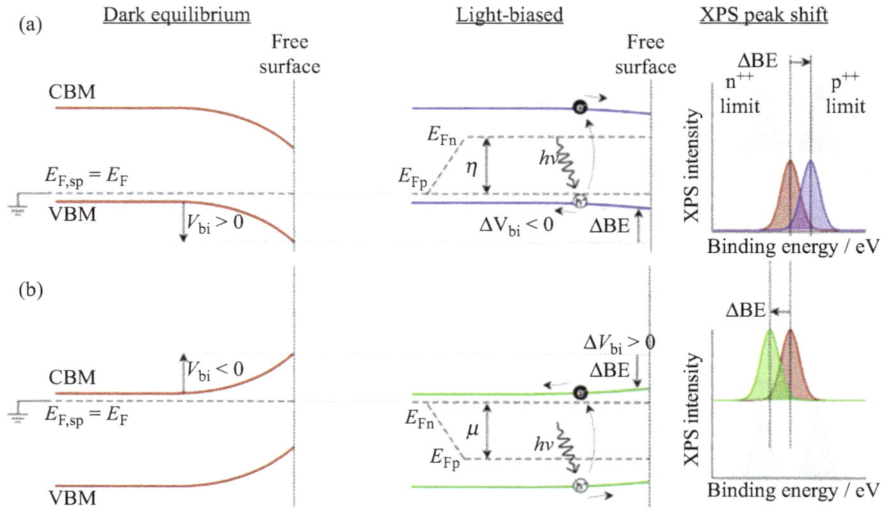

Figure 12.17 Schematic band diagrams comparing surface photovoltage (SPV) effects on free-surface band alignments for unpassivated (a) p-type vs. (b) n-type semiconductors. Also shown are schematic XPS CL shifts that occur due to quasi-Fermi-level splitting that accompanies applied light bias. The gray spectra represent the upper and lower limits for CL binding energies for degenerate n-type vs. p-type carrier densities at the semiconductor surface

12.5.3 Junction photovoltage effects

Figure 12.18 shows schematic band diagrams for p-n and n-p homojunctions with and without applied light bias. For simplicity, in these examples, there is no surface band bending, and hence, no SPV effects. On the other hand, the existence of the junction means that there will be a junction photovoltage (JPV) when light bias is applied. The magnitude of the JPV is given by

$$V_{OC} = \mu / q, \tag{12.21}$$

where V_{OC} denotes open-circuit voltage, recognizing that the JPV magnitude in these examples is identical to the standard solar-cell performance metric.

In Figure 12.18(a), the p-type side of the homojunction is in direct contact with $E_{F,sp}$, whereas the free surface is n-type. Consequently, in the absence of light bias, the measured CL approaches the n^{++} limit, and light bias results in a shift toward lower binding energy. The situation is reversed for the n-p homojunction in Figure 12.18(b), where light bias creates a peak shift to higher binding energy.

12.5.4 Combined surface and junction photovoltage effects

The final example illustrated in Figure 12.19 shows the combined effects of SPV and JPV. This example depicts an n–p homojunction, where the surface of the

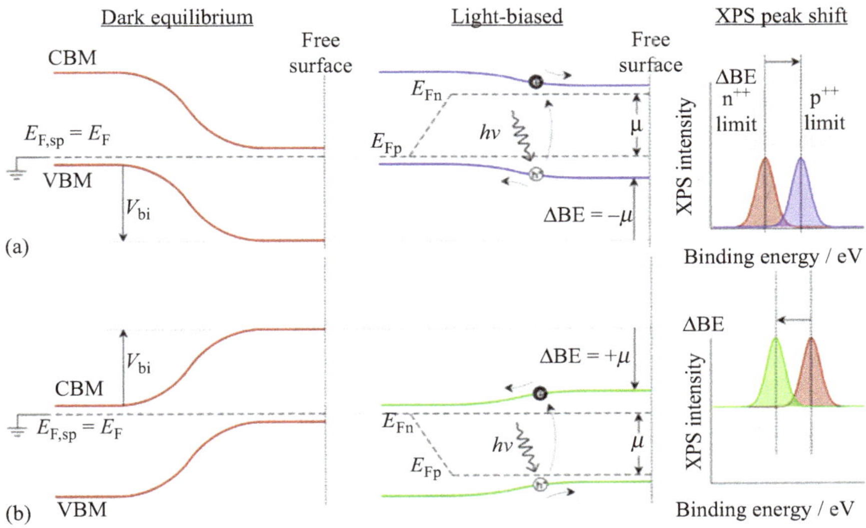

Figure 12.18 *Schematic band diagrams comparing the effects of light bias on free-surface band alignment for (a) p–n vs. (b) n–p semiconductor homojunctions. Also shown are schematic XPS CL shifts that occur with light bias. For the p-n junction, light bias shifts spectral features upward to lower binding energies by an amount equal to quasi-Fermi-level splitting, μ. The opposite effect is seen in the n-p junction*

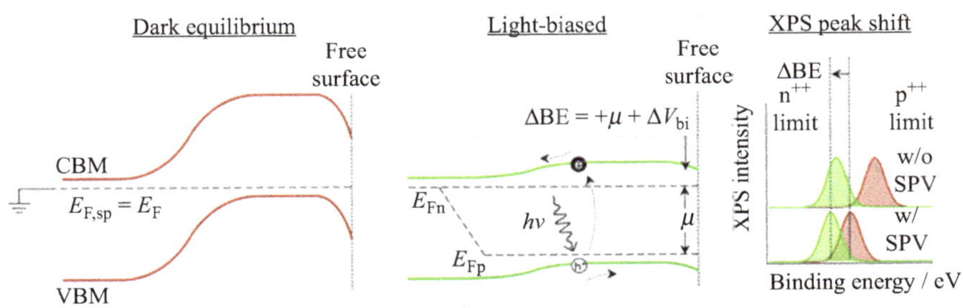

Figure 12.19 *Schematic illustration of combined effects of JPV and SPV on measured XPS binding-energy shifts. In comparison to the JPV-only case (cf. Figure 12.18(b)), the coexistence of SPV results in different measured binding energies for both the dark equilibrium and light-biased cases, as well as a lower overall binding-energy shift*

p-type junction partner also exhibits downward band bending. In the absence of light bias, the observed Fermi level is pinned near mid-gap, as was the case for Figure 12.17(a). But light bias now has two distinct effects, JPV and SPV, which act in opposite directions. The JPV effect tends to shift XPS CLs to higher binding energies due to quasi-Fermi-level splitting in the junction, whereas the SPV effect tends to shift spectral features to lower binding energies. As illustrated in Figure 12.19, the net effect of these countervailing effects is that there is a relatively small overall shift relative to the JPV-only case in Figure 12.18(b). In practice, SPV vs. JPV effects can be separated by performing additional measurements on suitable reference samples. These effects were studied in detail in a study of CIGS/CdS and CZTSe/CdS junctions, where SPV for the CdS layers were characterized by measuring SPV on a Mo/CdS reference sample [16].

Acknowledgments

P.S. thanks the French Agence Nationale de la Recherche for funding under contract no. ANR-17-MPGA-0012.

This work was authored in part by the National Renewable Energy Laboratory, operated by Alliance for Sustainable Energy, LLC, for the U.S. Department of Energy (DOE) under Contract No. DE-AC36-08GO28308. Funding provided by U. S. Department of Energy Office of Energy Efficiency and Solar Energy Technologies Office. The views expressed in the contribution do not necessarily represent the views of the DOE or the U.S. Government. The U.S. Government retains and the publisher, by accepting the contribution for publication, acknowledges that the U.S. Government retains a nonexclusive, paid-up, irrevocable, worldwide license to publish or reproduce the published form of this work, or allow others to do so, for U.S. Government purposes.

References

[1] Schulz, P.; Interface design for metal halide perovskite solar cells. ACS Energy Lett. 2018, 3, 1287–1293.

[2] Rau. U.; Kirchartz T. Charge carrier collection and contact selectivity in solar cells. Adv. Mater. Interfaces 2019, 6, 1900252.

[3] Doh, W. H.; Papaefthimiou, V.; Zafeiratos, S. "Applications of synchrotron-based X-ray photoelectron spectroscopy in the characterization of nanomaterials." In Surface Science Tools for Nanomaterials Characterization, pp. 317–366. Springer, Berlin, Heidelberg, 2015.

[4] Ralaiarisoa, M. Electronic properties of hybrid organic-inorganic perovskite films: Effects of composition and environment. 2019, DOI: 10.18452/20194

[5] Seah, M. P; Dench, W. A. Quantitative electron spectroscopy of surfaces: a standard data base for electron inelastic mean free paths in solids. Surf. Interface Anal. 1979, 1, 2–11.

[6] https://www.nist.gov/srd/nist-standard-reference-database-71 (as of January 10th, 2020)

[7] Jensen, D. S.; Kanyal, S. S.; Madaan N. Silicon (100)/SiO_2 by XPS. Surf. Science Spectra 2013, 20, 36.

[8] Gelius, U.; Basilier, E.; Svensson, S.; Bergmark, T.; Siegbahn, K. A high resolution ESCA instrument with X-ray monochromator for gases and solids. J. Electron Spectrosc. Relat. Phenom. 1973, 2, 405–434.

[9] Hüfner, S. Photoelectron Spectroscopy, Springer Berlin Heidelberg, Berlin, Heidelberg, 2003.

[10] Van Der Heide, P. X-Ray Photoelectron Spectroscopy: An Introduction to Principles and Practices, Wiley, Hoboken, NJ, 2011.

[11] Watts, J. F.; Wolstenholme, J. An Introduction to Surface Analysis by XPS and AES, John Wiley & Sons, Ltd, Chichester, UK, 2003.

[12] Shirley, D. A. High-resolution X-ray photoemission spectrum of the valence bands of gold Phys. Rev. B 1972, 5, 4709–4714.

[13] Tougaard, S. Quantitative analysis of the inelastic background in surface electron spectroscopy. Surf. Interface Anal. 1988, 11, 453–472.

[14] Morton, S. in Handb. Spectrosc., Wiley-VCH Verlag GmbH & Co. KGaA, Weinheim, Germany, 2014, pp. 779–820.

[15] Teeter, G.; Harvey, S. P.; Perkins, C. L.; Ramanathan, K.; Repins I. L. Comparative operando XPS studies of quasi-Fermi level splitting and open-circuit voltage in CZTSe/CdS and CIGS/CdS junctions and device structures. Journal of Vacuum Science & Technology A 2019, 37, 031202.

[16] Doniach, S.; Sunjic, M. Many-electron singularity in X-ray photoemission and X-ray line spectra from metals. J. Phys. C 1970, 3, 285–291.

[17] Seah, M. P.; Gilmore, L. S.; Beamson, G. XPS: binding energy calibration of electron spectrometers 5—Re-evaluation of the reference energies. Surf. Interface Anal. 1998, 26, 642–649.

[18] Comin, R.; Damascelli. A. "ARPES: A Probe of Electronic Correlations" in Strongly Correlated Systems, Springer, Berlin, Heidelberg, 2015.

[19] Weinhardt, L.; Hauschild, D.; Heske, C. Surface and interface properties in thin-film solar cells: using soft X-rays and electrons to unravel the electronic and chemical structure. Adv. Mater. 2019, 31, 1806660.

[20] Van Bokhoven, J. A.; Lamberti, C. X-Ray Absorption and X-Ray Emission Spectroscopy: Theory and Applications, John Wiley & Sons, Ltd, 2016

[21] Fuchs, O.; Weinhardt, L.; Blum, M.; *et al.* High-resolution, high-transmission soft X-ray spectrometer for the study of biological samples. Rev. Sci. Instrum. 2009, 80, 063103.

[22] Qiao, R.; Li, Q.; Zhuo, Z.; Sallis, S.; *et al.* High-efficiency in situ resonant inelastic x-ray scattering (iRIXS) endstation at the advanced light source. Rev. Sci. Instrum. 2017, 88, 033106.

[23] https://www.nrel.gov/pv/cell-efficiency.html (as of January 10th, 2020)

[24] Ashcroft N.; Mermin, N. Solid State Physics. (Brooks/Cole Cengage Learning, 1976).

[25] Guan, Z.; Kim, J. B.; Wang, H.; *et al.* Electronic structure of the poly(3-hexylthiophene):indene-C60 bisadduct bulk heterojunction, J. Appl. Phys.
[26] Hwang, J.; Wan A.; Kahn, A. Energetics of metal–organic interfaces: new experiments and assessment of the field, Mater. Sci. Eng. R 2009, 64, 1–31.
[27] Man, G.; Endres, J.; Lin, X.; Kahn A.; Experimental characterization of interfaces of relevance to organic electronics in The WSPC Reference on Organic Electronics: Organic Semiconductors, World Scientific Publishing Co Pte Ltd, 2016.
[28] Yang, Y.; Yang, M.; Zhu, K.; *et al.* Large polarization-dependent exciton optical Stark effect in lead iodide perovskites. Nat. Commun. 2016, 7, 12613.
[29] Azarhoosh, P.; McKechnie, S.; Frost, J. M.; Walsh, A.; van Schilfgaarde, M. Research update: relativistic origin of slow electron-hole recombination in hybrid halide perovskite solar cells. APL Mater. 2016, 4, 091501.
[30] Yang, Y.; Ostrowski, D. P.; France, R. M.; *et al.* Observation of a hot-phonon bottleneck in lead-iodide perovskites. Nat. Photonics 2016, 10, 53–59.
[31] Semonin, O. E.; Elbaz, G. A.; Straus, D. B.; *et al.* Limits of carrier diffusion in n type and p type $CH_3NH_3PbI_3$ perovskite single crystals. J. Phys. Chem. Lett. 2016, 7, 3510–3518.
[32] Egger, D. A.; Bera, A.; Cahen, D.; *et al.* What remains unexplained about the properties of halide perovskites? Adv. Mater. 2018, 30, 1800691.
[33] Schulz, P.; Edri, E.; Kirmayer, S.; Hodes, G.; Cahen, D.; Kahn, A. Interface energetics in organo-metal halide perovskite-based photovoltaic cells. Energy Environ. Sci. 2014, 7, 1377–1381.
[34] Endres, J.; Egger, D. A.; Kulbak, M.; *et al.* Valence and conduction band densities of states of metal halide perovskites: a combined experimental–theoretical study. J. Phys. Chem. Lett. 2016, 7, 2722–2729.
[35] Zu, F.; Amsalem, P.; Egger, D. A.; *et al.* Constructing the Electronic Structure of $CH_3NH_3PbI_3$ and $CH_3NH_3PbBr_3$ Perovskite Thin Films from Single-Crystal Band Structure Measurements. J. Phys. Chem. Lett. 2019, 10, 601.
[36] Belisle, R. A.; Jain, P.; Prasanna, R.; Leijtens, T.; McGehee, M. D. Minimal effect of the hole-transport material ionization potential on the open-circuit voltage of perovskite solar cells. ACS Energy Lett. 2016, 1, 556–560.
[37] Park, S. M.; Mazza, S. M.; Liang, Z.; *et al.* Processing dependent influence of the hole transport layer ionization energy on methylammonium lead iodide perovskite photovoltaics. ACS Appl. Mater. Interfaces 2018, 10, 15548–15557.
[38] Schulz, P.; Whittaker-Brooks, L. L.; MacLeod, B. A.; Olson, D. C.; Loo, Y.-L.; Kahn, A. Electronic level alignment in inverted organometal perovskite solar cells. Adv. Mater. Interfaces 2015, 2, 1400532.
[39] Miller, E. M.; Zhao, Y.; Mercado, C. C.; *et al.* Substrate-controlled band positions in $CH_3NH_3PbI_3$ perovskite films. Phys Chem Phys 2014, 16, 22122–22130.

[40] Steirer, K. X.; Schulz, P.; Teeter, G.; *et al.* Defect tolerance in methylammonium lead triiodide perovskite. ACS Energy Lett. 2016, 1, 360–366.

[41] Philippe, B.; Man, G. J.; Rensmo, H. "Chapter 5 – Photoelectron spectroscopy investigations of halide perovskite materials used in solar cells" In Characterization Techniques for Perovskite Solar Cell Materials, pp. 109–137. Elsevier, Amsterdam, Oxford, Cambridge 2020; Béchu, S.; Ralaiarisoa, M.; Etcheberry, A.; Schulz, P. Photoemission Spectroscopy Characterization of Halide Perovskites. Adv. Energy Mater. 2020, 1904007.

[42] Fadley, C. S. in Hard X-Ray Photoelectron Spectrosc. (Ed.: J. Woicik), Springer International Publishing, Cham, 2016.

[43] Jacobsson, T. J.; Svanström, S.; Andrei, V.; Rivett, J. P. H.; Kornienko, N.; Philippe, B.; Cappel, U. B.; Rensmo, H.; Deschler, F.; Boschloo, G., Extending the Compositional Space of Mixed Lead Halide Perovskites by Cs, Rb, K, and Na Doping. J. Phys. Chem. C 2018, 122, 13548.

[44] Jacobsson, T. J.; Correa-Baena, J.-P.; Halvani Anaraki, E.; *et al.* Unreacted PbI_2 as a double-edged sword for enhancing the performance of perovskite solar cells. J. Am. Chem. Soc. 2016, 138, 10331–10343

[45] Starr, D. E.; Sadoughi, G.; Handick, E.; *et al.* Direct observation of an inhomogeneous chlorine distribution in $CH_3NH_3PbI_{3-x}Cl_x$ layers: surface depletion and interface enrichment. Energy Environ. Sci. 2015, 8, 1609–1615.

[46] Sadoughi, G.; Starr, D. E.; Handick, E.; *et al.* Observation and mediation of the presence of metallic lead in organic–inorganic perovskite films. ACS Appl. Mater. Interfaces 2015, 7, 24, 13440–13444.

[47] Wippler, D.; Wilks, R. G.; Pieters, B. E.; *et al.* Pronounced surface band bending of thin-film silicon revealed by modeling core levels probed with hard X-rays. ACS Appl. Mater. Interfaces 2016, 8, 17685–17693.

[48] Man, M. K. L.; Margiolakis, A.; Deckoff-Jones, S.; *et al.* Imaging the motion of electrons across semiconductor heterojunctions. Nature Photonics 2017, 12, 36–40.

Chapter 13

Time-of-flight secondary-ion mass spectrometry and atom probe tomography

Steven P. Harvey[1] *and Oana Cojocaru-Mirédin*[2]

13.1 Introduction

Time-of-flight secondary-ion mass spectrometry (TOF-SIMS) and atom probe tomography (APT) have many similarities. Both detect the chemical makeup of a solid material at ppm or better sensitivity, while retaining the spatial location information from the signal. Both use a similar principle to measure a signal associated with a given charged secondary ion—the flight time it takes a secondary ion to reach the detector after it is generated. Both are destructive techniques—albeit destructive of very small volumes—because the signal measured is generated by the removal of material from the sample itself. Importantly, both techniques require ultra-high vacuum to ensure that the secondary ions, once generated, can reach the detector without collision with other atoms in the gas phase. In the case of TOF-SIMS, the samples can vary from several mm to several inches in size, whereas for APT, the samples are cones several hundred microns in size that are prepared in a special manner (covered in detail later in the chapter). We will discuss the basic fundamentals of each technique and cover several examples of the technique applied to photovoltaic materials, which should give a good idea of the types of information one can gain from TOF-SIMS and APT and how they are complementary as they both operate at different length scales: at the hundreds of microns to hundreds of nanometer scale for TOF-SIMS and at the nanometer to a few angstroms scale for APT.

13.2 Time-of-flight secondary-ion mass spectrometry

13.2.1 Introduction and comparison to dynamic SIMS

Any secondary-ion mass spectrometry measurement is based on a sputtering process due to an incident high-energy ion beam that causes a collision cascade when

[1]Materials and Chemical Science and Technology Directorate, National Renewable Energy Laboratory, Golden Colorado, USA
[2]Institute of Physics, RWTH Aachen University, Düsseldorf, Germany

it hits the sample, as illustrated in Figure 13.1. The collision cascade is the result of an elastic transfer of energy from the primary ion to atoms in the sample matrix through a series of elastic collisions. This causes some damage to the sample surface within the collision cascade region, with the volume of this region changing depending on the energy of the incident ion (note that the cascade depth is greatly exaggerated in Figure 13.1 for illustration of the effect only). The elastic collisions can result in the ejection of sample material from the first few monolayers of the sample surface to vacuum. Charged secondary ions are a small fraction of the ejected material—typically less than 1%—and they can be collected for SIMS analysis.

SIMS is a versatile and powerful technique, but it is not without its limitations. One significant limitation is the complex relationship between intensity and concentration, which makes quantification difficult. The SIMS intensity equation is shown in (13.1):

$$i_A^S = I^p Y \alpha_A \eta_A \theta_A X_A, \tag{13.1}$$

where the measured SIMS intensity $\left(i_A^S\right)$ depends on the primary-ion intensity (I^p), sputter yield (Y) of the ion, ionization probability (α_A) of that ion, transmission efficiency (η_A) of the detection system, species isotopic abundance (θ_A), as well as its fractional concentration in the material (X_A). The sputter yield and ionization

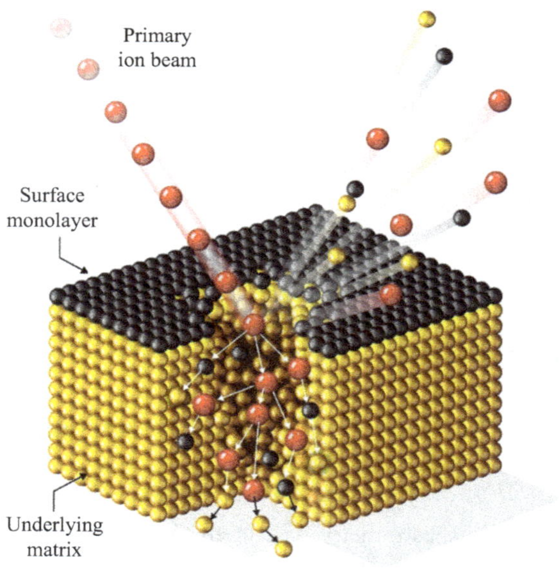

Figure 13.1 The SIMS primary-ion collision cascade results in sample material being ejected from the first few monolayers of the sample surface. A small amount of the ejected material is charged, and the mass of these charged secondary ions can then be analyzed. Taken from [8]

probability are both affected by the matrix, which means that the intensity of any secondary ion measured could be different when present in the same concentration in two different matrixes.

The complicated relationship between intensity and concentration manifests itself in SIMS data in many ways. One common artifact, often referred to as the "matrix effect," is that the ionization probability for a species depends strongly on the elements or molecules that surround that species in the solid material being probed. A common example of this matrix effect is that oxygen tends to enhance secondary-ion signals. Thus, for example, when profiling through poly-silicon on a silicon wafer, a small spike in all signals is observed when the profile reaches the buried native oxide between the polysilicon and the silicon wafer. The same matrix effect also means that if two samples contain the same amount of a dopant—e.g., phosphorous at 1×10^{17} atoms/cm^3—but one matrix is silicon and the other is germanium, then different intensity levels would be seen for phosphorous in each sample, even though the actual phosphorous concentration is the same. Indeed, quantification of SIMS signals is at best cumbersome, and ion-implantation of an element of interest at a known fluence into the matrix of interest is the most common method of quantification [1].

Changing the sputtering ion can also increase the secondary-ion yield because implantation of the sputter ions into the matrix can change the ionization probability of certain secondary ions. The two most common sputter ions (and primary ions in the case of dynamic-SIMS, or D-SIMS) are oxygen and cesium. Oxygen is very electronegative, so it increases the probability of positive secondary-ion formation and enhances positive secondary-ion signals. Cesium is very electropositive, and it enhances the yield of negative secondary ions. For a much more thorough background on SIMS principles and measurement details, see the excellent book by Prof. Fred Stevie [1].

SIMS excels at detecting small quantities of an element or species in a matrix of differing composition. But quantification of SIMS data at matrix-level compositions is difficult because at high concentration (generally $>1\%$) the matrix effect changes the ion yield as the material alloys with the dopant. In such cases, other standardless characterization techniques such as X-ray photoelectron spectroscopy or Auger depth profiling may be more appropriate for matrix-level quantification of a material.

When the global photovoltaic industry was in its infancy, D-SIMS was the technique of choice for measuring dopant concentrations in semiconductors, because the difference in capabilities (specifically detection limits) between D-SIMS and TOF-SIMS was stark at that time. Because, as shown in (13.1), the signal measured (and thus the detection limits) is proportional to the primary ion intensity (I^p), D-SIMS is operated with primary-ion currents several orders of magnitude higher than TOF-SIMS. In D-SIMS, the primary-ion currents are large enough that there is no secondary sputter beam like with TOF-SIMS; the primary beam itself is doing bulk sputtering. This is not the case for TOF-SIMS, where the analysis beam is typically operated under the static limit of 1×10^{12} primary ions/cm^2. D-SIMS still boasts lower detection limits than TOF-SIMS in many cases; however, due to continued improvement and development of TOF-SIMS hardware, the difference between the two techniques is now much smaller than it once was. In addition, TOF-SIMS boasts unique capabilities such as parallel detection combined with high-resolution chemical imaging and tomography (<60 nm lateral resolution is possible).

A real strength of TOF-SIMS lies in the extensive post-acquisition data analysis options. The entire mass spectrum is saved for each and every data point in a depth profile, so no data are thrown away (which is not the case for D-SIMS). There are many ways to extract additional information from a dataset after it is collected, as will be shown in several examples in the chapter. In addition, the primary beam is used at low currents, so a second ion gun is used for sputtering, unlike D-SIMS. In this case, both ion guns can be tuned for optimum performance to get the best data for the desired experiment. These technical advances in TOF-SIMS hardware, in addition to these inherent strengths of the technique listed above, give TOF-SIMS the ability to provide unique insight for photovoltaic development, and the technique has gained in popularity over the last decade.

13.2.2 High-resolution imaging, tomography, and diffusion in CdTe photovoltaics

SIMS excels at detecting small concentrations of impurities or dopants in a matrix of different composition. Combined with high lateral-resolution imaging (sub 60 nm is achievable, but ~100 nm is more common without extensive setup time spent on focusing), TOF-SIMS can elucidate the spatial distribution of dopants and/or impurities and greatly assist the development of new and next-generation photovoltaic technologies. For example, a $CdCl_2$ vapor annealing process has been used extensively by the CdTe community since the mid-1980s. But until recently, the exact lateral distribution of chlorine in polycrystalline CdTe grown via industrially relevant methods subjected to an anneal similar to industrial processes had not been observed. Figure 13.2 shows a TOF SIMS image taken from Harvey *et al.* [2] that definitively reveals that chlorine is concentrated at grain boundaries in these samples. In this work, 3-D tomography was also used and showed the chlorine content enhanced at grain boundaries well over a micron into the sample [2].

Using the strength of SIMS in detecting small concentrations of impurities or dopants in a matrix of different composition makes it an excellent tool to measure dopant concentrations. And with the right set of samples, it can be used to extract diffusion profiles that can be used to determine the diffusion solution for a given dopant. In their works on next-generation group-V dopants in CdTe, Colegrove and Harvey studied diffusion of phosphorous [3], antimony [4], and arsenic [5] dopants in polycrystalline CdTe. The work used standard depth profiling to obtain profiles with a high data density to allow for accurate fitting of the diffusion profiles, in addition to 3-D tomography and imaging to reveal the spatial location of additives within the polycrystalline CdTe material [3–5]. Figure 13.3(a) and (b) shows an electron backscatter diffraction image (showing grain-boundary locations) and a

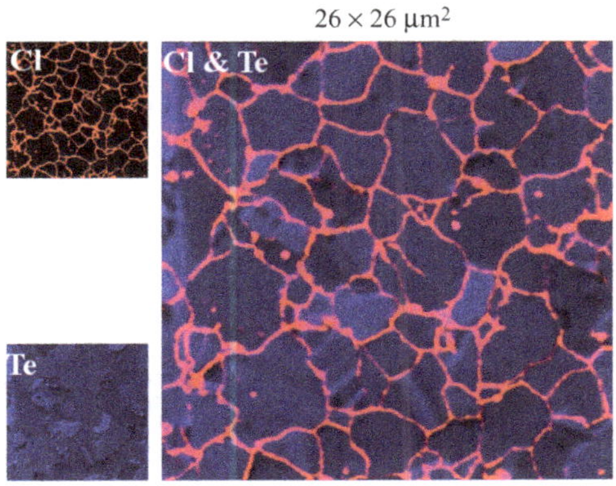

Figure 13.2 *TOF-SIMS 2-D imaging results for chlorine and tellurium (26 × 26 μm²). The large image is a red-blue overlay, with the chlorine signal in red and the tellurium signal in blue. The small images at the left show each signal individually in the same color scheme. Taken from [2]*

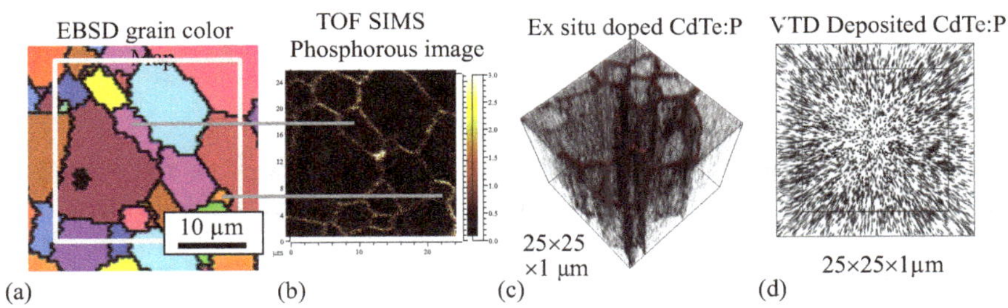

Figure 13.3 *(a) EBSD grain orientation map and (b) corresponding TOF-SIMS phosphorous image (25 × 25 μm) from the same area, showing enhanced phosphorous at the grain boundaries. (c) 3-D rendering of the phosphorous signal from the same area as in B (25 × 25 × 1 μm), where the dopant was incorporated ex situ to growth. (d) A TOF-SIMS 3-D rendering of phosphorous (25 × 25 × 1 μm), where the dopant was incorporated in situ to material growth by vapor-transport deposition (VTD). The grain structure is not visible in the phosphorous rendering, even though the samples in C and D have similar grain sizes. Taken from [7]*

TOF-SIMS image of the same area, the phosphorous dopant, which was incorporated by an *ex-situ* post-growth anneal. Figure 13.3(b) shows that phosphorous is accumulated at grain boundaries. This is further revealed by the TOF-SIMS 3-D tomography shown in Figure 13.3(c), where the enhancement along boundaries is observed for the entire depth of the dataset (over one micron). These works revealed the full diffusion solution for bulk and grain-boundary diffusion for each dopant, which allows one to predict the incorporated dopant concentration given a certain annealing time and temperature. It was shown that when the dopant is incorporated via an *ex-situ* anneal, the incorporation proceeds via a combination of fast grain-boundary diffusion and slow bulk (grain-core) diffusion. Interestingly, for phosphorous, a second fast bulk-diffusion mechanism was noted for phosphorus in single-crystal CdTe, and theory suggested that this was due to a phosphorous interstitial diffusion mechanism. This fast bulk-diffusion mechanism allowed phosphorous to incorporate at higher concentrations more uniformly given similar annealing conditions when compared to arsenic and antimony dopants [3,5].

TOF-SIMS 3-D tomography has also shown that the dopant location is influenced by the incorporation method. The phosphorous incorporation for the sample in Figure 13.3(c) was incorporated via a post-growth anneal in cadmium phosphate vapor, which leads to preferential diffusion and thus higher concentration of the dopant along grain boundaries. Conversely, when group-V dopants are incorporated into the material during growth, the dopant is uniformly distributed in the CdTe material [6,7], as shown in Figure 13.3(d).

TOF-SIMS has been shown to be a powerful tool to understand dopant concentration, spatial distribution, incorporation mechanisms, and diffusion in CdTe materials, and the methods and examples shown here would be applicable to other absorber materials, as well.

13.2.3 Insight into perovskite photovoltaics

The current state of development of perovskite photovoltaics was covered in depth in Chapter 4. Understanding chemical gradients and inhomogeneities in perovskite absorbers and devices appears to be a key to continued advancement of the technology, both in terms of performance and reliability. TOF-SIMS is one of the few analytical techniques that can obtain chemical information from both the organic and inorganic components of halide organic-inorganic perovskite photovoltaics in one dimension (standard depth profiling), two dimensions (high-resolution 100-nm imaging), as well as three dimensions (tomography combines high-resolution imaging with depth profiling). In their paper titled "Probing Perovskite Inhomogeneity Beyond the Surface: TOF-SIMS Analysis of Halide Perovskite Photovoltaic Devices," Harvey *et al.* [8] illustrate through examples the breadth of insight that can be gained from the technique including: cation uniformity (depth and lateral), changes in chemistry upon alternate processing, changes in chemistry upon degradation (including at interfaces), and lateral distribution of passivating additives. Known measurement artifacts and best practices related to perovskites were also discussed, with a few such examples highlighted here.

TOF-SIMS can sample organic species due to the low primary-beam dosage, which does not break up the bonds of organics as occurs with D-SIMS, and secondary-ion signals from large molecular-weight ions can be observed. Additionally, the strength of the two-beam operation allows for the sputter gun to be tuned separately from the primary-ion beam, allowing for collection of data on halide perovskite solar cell (HPSC) materials with minimal artifacts. The importance of the information that TOF-SIMS can provide to HPSC development is illustrated in Figure 13.4.

The most common method of making HPSC materials is via spin coating from a solution and subsequent annealing around 80–100 °C. It is common practice in the field to quote the bath chemistry as the film composition, even though the film has gone through both a drying and heating step after deposition from the precursor solution. This practice—even if the underlying assumption holds that the bath chemistry transfers 100% to the processed film—is inadequate in capturing any information about the cation gradients within the film. Figure 13.4 illustrates this point, where 1-D depth profiles through a HPSC film are shown for two films processed similarly with different Cs to formamidinium (FA) ratios. Only the profile for the organic species FA is shown. There is a vastly different gradient for FA in these two films: in the 50:50 film, there is more FA at the back, and in the 25:75 film, there is more FA at the front. These different organic cation gradients could have strong implications on the performance or reliability of the material, and without the insight provided by TOF-SIMS, one could erroneously attribute a performance metric to composition that is actually the result of cation gradients.

SIMS excels at detecting small concentrations of impurities in a matrix of differing composition. Thus, TOF-SIMS is an excellent tool to investigate ion

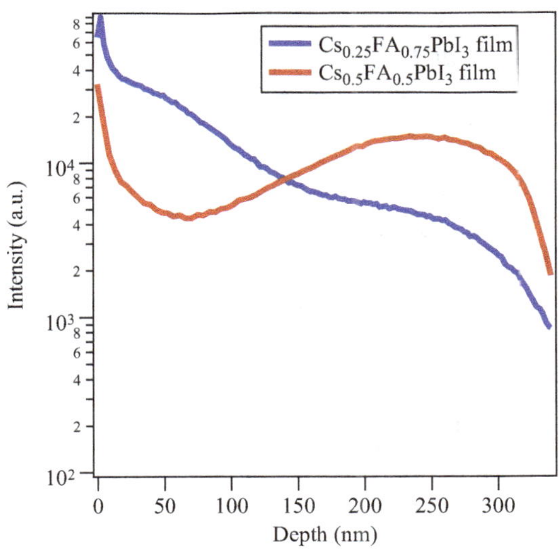

Figure 13.4 TOF-SIMS depth profile of the organic cation formamidinium for two films of slightly different composition, subjected to identical post-deposition processing. This illustrates how simply using the precursor bath chemistry as a proxy for film composition fails to capture actual film properties that would relate to the cation distribution in the film. Taken from [8]

migration and degradation in perovskite devices. In the work by Li *et al.* [9], it was shown using TOF-SIMS that lithium migrates from a lithium-doped organic hole-transport layer through the entire device stack upon operation of the device. Ion migration in a device was also investigated in real time by using a light bias inside the TOF-SIMS [10]. The lack of ion migration over time was shown in the work by Christians *et al.* [11], detailing more than 1,000 h of operational device stability by tailoring the interfaces of the perovskite device stack. Lin *et al.* [12] used deuterium tracer water to investigate the degradation of a perovskite active layer *in-situ* with TOF-SIMS.

As the HPSC field is moving to incorporate more additives at small concentrations to improve performance and/or reliability, TOF-SIMS can step in and reveal the spatial location of such additives, or other nonuniformities of the material [13]. Figure 13.5 shows the location of cyanide at apparent grain boundaries from a lead thiocyanate post-deposition treatment [8]. Although this example is from a very large-grained material, with better than 100-nm lateral resolution possible, such images are possible in perovskite material with several-hundred nanometer grain sizes [14].

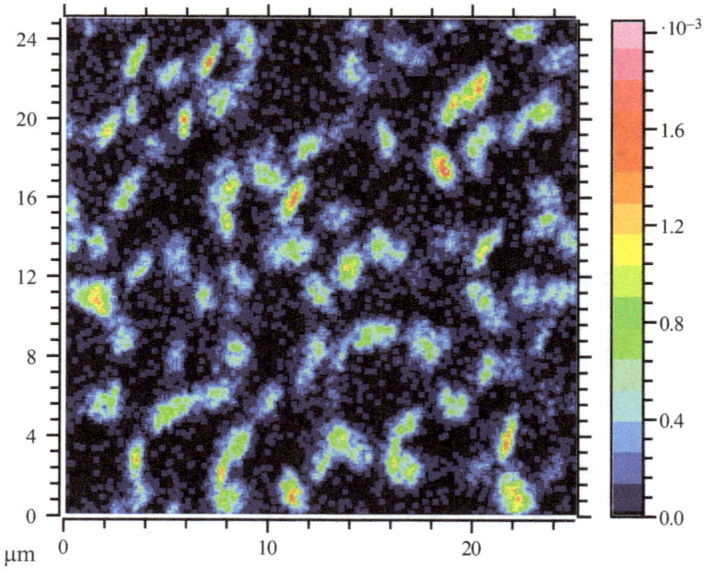

Figure 13.5 TOF-SIMS 2-D image of a passivating additive ($Pb(SCN)_2$) at the grain boundaries after a high-temperature process step (SCN signal shown). An enrichment of sulfur and cyanide at the grain boundaries due to the ($Pb(SCN)_2$) post-deposition treatment is seen. The color-based intensity scale at right is counts/pixel normalized to the total counts image, which minimizes measurement artifacts as outlined in [8]. Taken from [8]

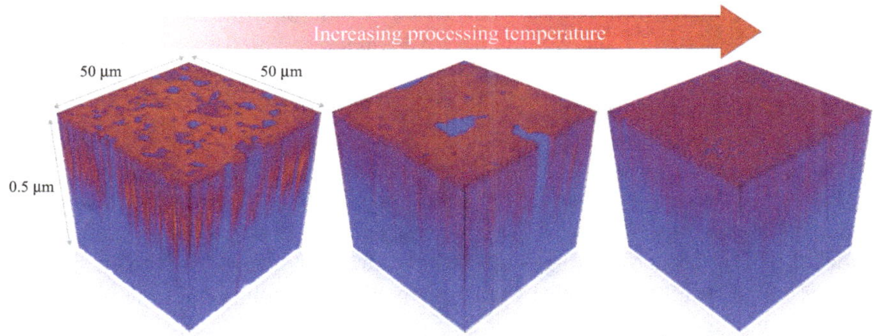

Figure 13.6 TOF-SIMS 3-D tomography results for a two-cation perovskite, where one cation is an organic species and the other is inorganic, processed at different temperatures. Each 3-D reconstruction is 50 × 50 × 0.5 µm. The organic cation is red and inorganic cation is blue. The material processing temperature increased from left to right, and the lateral segregation of the inorganic cation is shown to be mitigated with the higher processing temperatures. With a lateral resolution of 100 nm, such TOF-SIMS tomography data can be used to investigate phase segregation and cation distributions in detail through the depth of the film. Taken from [8]

3-D tomography combines imaging with depth profiling, and a 3-D representation of the material can be measured. This can yield valuable insight into dopant distribution, ion migration, cation gradients, and how those properties change with changes in processing, operation, or degradation. Indeed, tomography of perovskites has already seen a number of applications in the field [8,12]. Figure 13.6 shows both the organic (red) and inorganic (blue) cation distributions in three samples of the same composition, subjected to different processing temperatures. One can clearly see a lateral nonuniformity of the organic cation at low processing temperatures, which becomes more uniform at higher processing temperatures.

Several examples have been covered that illustrate only a fraction of the wealth of information that TOF-SIMS can provide to assist in developing HPSC materials. This field is moving at an incredibly fast pace, but as the field matures, TOF-SIMS will continue to play an important role in advancing the performance and reliability of HPSC materials and devices.

13.2.4 Investigating photovoltaic module degradation fundamentals with TOF-SIMS

Potential-induced degradation (PID) is a challenging reliability issue for silicon solar cells, where sodium migrates from the front glass due to an electric-field gradient present across the cell into defects in the silicon. Sodium can accumulate at structural defects and shunt the solar cell junction. This can cause significant module power loss [15]. TOF-SIMS profiling and tomography can yield unique insight into the symptoms of PID shunting. Combining TOF-SIMS with other characterization techniques can provide a clear picture of the root-cause mechanisms for PID shunting [16].

Figure 13.7 outlines results from a multi-technique, multi-scale characterization effort that investigated PID shunting defects by using laser and focused ion beam (FIB) marks to enable TOF-SIMS characterization of a single defect on the nanometer length scale that was causing PID shunting issues at the module length scale. Figure 13.7(a) shows thermography image results for an entire module, where a cell showing a hot-spot due to localized PID shunting is highlighted. Figure 13.7(b) and (c) are higher-resolution thermography images, where in Figure 13.7(c) individual shunts are observed. At this point, laser marks are placed around these few defects of interest to allow location of the areas for further characterization. Figure 13.7(d) shows the defect highlighted in the thermography image of Figure 13.7(c), and the electron-beam-induced current (EBIC) image shows the defect in the center of the image, which is dark, as well as four square FIB marks that were placed around the defect of interest and that also appear dark due to beam damage from the FIB. Figure 13.7(e) shows a 2-D TOF-SIMS image of the same area, where the FIB marks are shown in red from the residual gallium from the FIB, and the shunt shows up as a high concentration of sodium. TOF-SIMS 3-D tomography results from the same area are shown in Figure 13.7(e); the rendering is 200 $\times$ 200 microns by 500 nm in depth. The high concentration of sodium in the defect extends through the entire depth of the measurement. One can use a strength of TOF-SIMS in post-data-acquisition data processing to gain further insight from the defect region. Figure 13.7(f) shows two dashed circles, and 1-D depth profiles can be created from each of these areas, masking all other data in the dataset. These selected-area depth profiles are shown in Figure 13.7(g); the shunted region has more than one order of magnitude more sodium than the non-shunted area, and sodium peaks at the Si/SiN interface in the shunted region at around 0.1% sodium. This has been consistently observed in tens of defects with TOF-SIMS, and it is consistent with the sodium amount in defects seen with APT and STEM-EDS line profiling of defects completed using similar methods as outlined in Figure 13.7 [16].

The recovery mechanisms for PID shunting has also been investigated with TOF-SIMS. Module performance is shown to be able to recover from PID shunting by subjecting the modules to a reversed electrical bias for an extended period [15]. A recovery of single shunts was noted subsequent to high-resolution imaging in the scanning electron microscope, presumably due to a localized heating of the defect due to the electron beam. Figure 13.8 shows the results of an experiment meant to elucidate the recovery mechanism, where the EIBC image in Figure 13.8(a) shows both FIB marks for locating the area, as well as several annealed and recovered shunts, and one non-annealed shunt. The TOF SIMS results in Figure 13.8(b) show a significant bloom of sodium on the surface of the

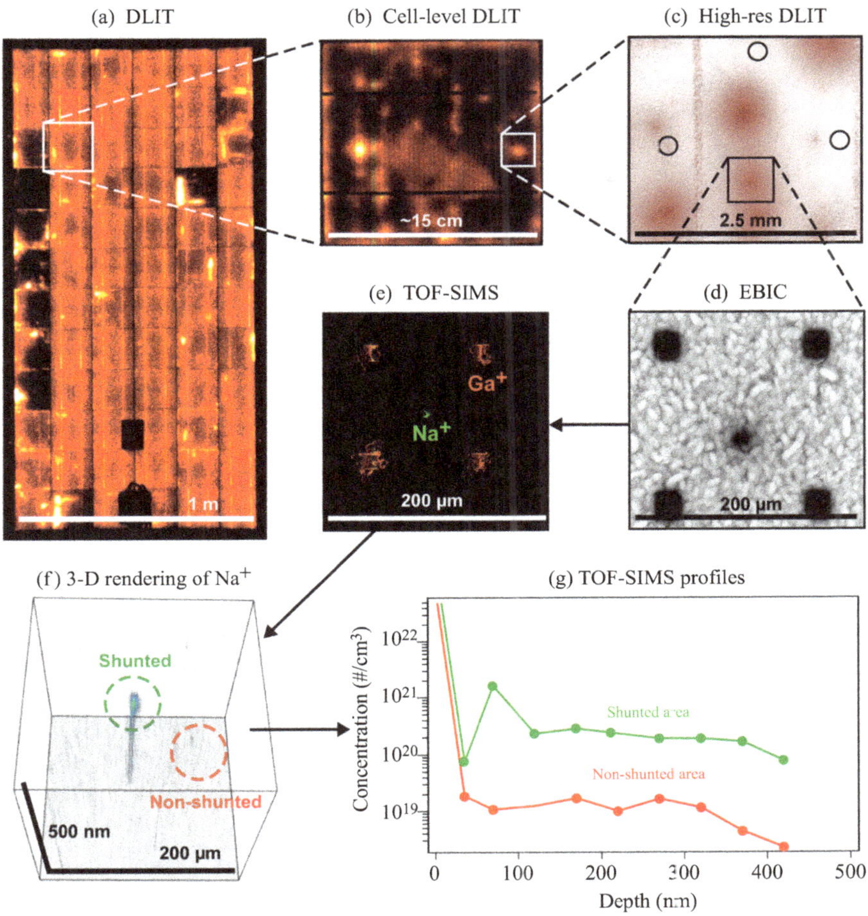

Figure 13.7 *(a) Dark lock-in thermography (DLIT) image of entire module (1–2 m); (b) DLIT image of a single cell from the module; (c) High-resolution DLIT of a shunted area; the black circles show the locations of four laser marks around shunted area; (d) FIB marks were made centered around single shunt identified with EBIC; (e) TOF-SIMS image (200–200 μm), showing the Ga signal in red (corresponding to the FIB marks), and sodium in green. The sodium spot matches the location of the shunt seen in EBIC; (f) TOF-SIMS 3-D rendering of the shunt area. A single shunt is seen to persist through the depth of the measurement; and (g) Selected-area depth profiles from the shunted and an unshunted region, marked with circles in D. The sodium concentration peaks at ~1% at the SiN/Si interface in the shunted region. Taken from [16]*

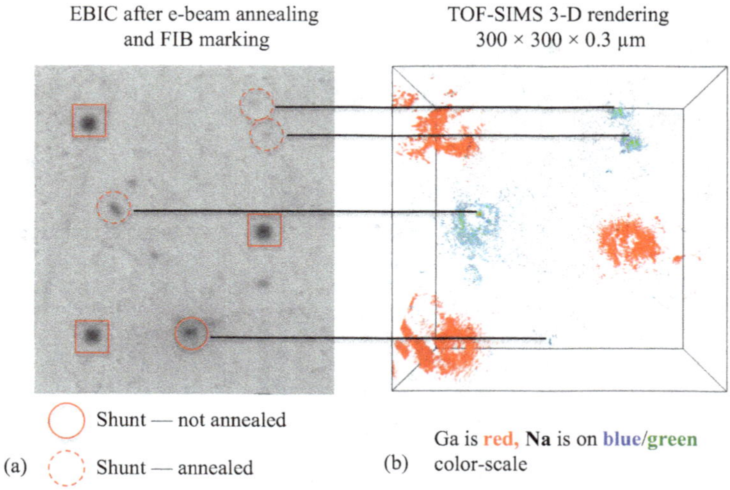

Figure 13.8 Elucidates a mechanism for PID recovery after selected annealing of several PID shunts: (a) EBIC image after FIB marking (outlines with boxes). The image shows several shunts that were subjected to e-beam annealing (circled—dashed), and one that was not (circled). (b) TOF-SIMS 3-D rendering of same area, showing gallium near the FIB marks; comparison to the one non-annealed shunt reveals significant sodium out-diffusion to the surface from the annealed shunts. Taken from [16]

device for the annealed shunts, which contrasts with the non-annealed shunt where the sodium is still mostly localized in the defect region. This illustrates that the recover mechanism in this case is an out-diffusion of sodium from the defect to the sample surface, and the sodium diffuses only slightly away from the annealed defect on the surface due to the localized nature of the annealing source—the electron beam in this case. In an actual module, this would mean that upon recovery, sodium would be accumulated at the SiN/encapsulant interface [16].

TOF-SIMS has also proven useful in investigating the PID degradation mechanism in CIGS mini-modules [17]. 1-D profiling showed that sodium migration through the CIGS absorber is a major contributor to module degradation, contrasted to the larger alkali ion potassium, which does not appear to be as mobile as sodium under the stress conditions investigated. 3-D tomography revealed that fast grain-boundary diffusion of sodium in CIGS is the reason for large-scale diffusion of sodium from the back to the front of the device, which contributed to the degradation of performance. Figure 13.9 shows the 3-D tomography renderings for sodium in three CIGS samples. Figure 13.9(a) is the as-prepared absorber, where the sodium is uniformly distributed. Figure 13.9(b) is a sample subjected to only a

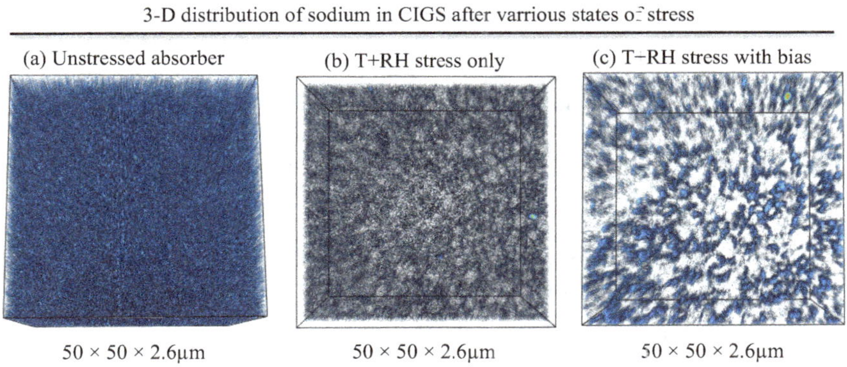

Figure 13.9 *TOF-SIMS 3-D renderings of sodium for three CIGS devices. Each rendering is 50 × 50 × 2.6 μm in total; however, for the stressed samples, some of the data from the back of the CIGS have been cropped for clarity to better illustrate the grain boundaries, which may give the false impression that the unstressed absorber appears to have more sodium. (a) Unstressed absorber; sodium distribution is uniform. (b) Temperature and relative humidity stress only; some slight segregation of sodium to apparent grain boundaries is noted. (c) Temperature and relative humidity and voltage-biased sample; strong segregation of sodium to apparent grain boundaries is noted. Taken from [17]*

heat and humidity stress, and a slight increase of sodium at grain boundaries is noted. Figure 13.9(c) shows that when an electrical bias is added to the heat and humidity stress, strong accumulation of sodium at grain boundaries is noted. Selected-area images of the dataset (not shown) reveal that sodium is more concentrated at grain boundaries near the front of the device furthest from the molybdenum contact, the source for sodium ingress.

Several examples were highlighted to illustrate ways in which TOF-SIMS can provide valuable information related to photovoltaic reliability. TOF-SIMS will continue to make further contributions in this area on new reliability issues as they become known to this increasingly more important sector of the photovoltaic development landscape.

13.3 Atom probe tomography and correlated microscopies

13.3.1 Introduction

Atom probe tomography (APT) has demonstrated over the years to be a powerful technique to access the structure and composition of materials down to the atomic level in some cases. It also allows exploration of the elemental distributions within the material in three-dimensional (3D) view, with typical dimensions of the analyzed volume of about 50 × 50 × 300 nm^3 for state-of-the-art instruments. And when combined with other microscopy techniques—in a correlative microcopy approach—it has the potential to help understand the interconnected relationship between structure, composition, and performance of many semiconductor devices such as solar cells [18,19], transistors [20], thermoelectrics [21–23], and shape-memory materials [24].

13.3.2 Theoretical principles of atom probe tomography

The unique characteristic of APT is the atom-by-atom analysis of materials prepared into the form of a tip. This controlled *field-evaporation* process is initiated and coordinated through a high electric field applied on the tip (about a few tens of V/nm). To remove an atom from the tip surface, two processes occur: the desorption of an atom from the surface and the ionization of the removed atom [25,26]. Desorption involves breaking the atomic bonds of a specific atom to remove it from the surface, whereas ionization involves pulling away this atom from the surface as a charged ion.

The electric field generated at the apex of the tip with a maximum radius of curvature of 50 nm is defined as:

$$F = \frac{V}{kR}, \tag{13.2}$$

where k is the field factor, V is the voltage, and R is the radius of curvature. The constant k can range from 3 to 8, as calculated for a tungsten tip, and its value depends primarily on the tip shape [27].

A typical APT experiment is as follows: First, the tip-shaped specimen is placed in an ultra-high-vacuum chamber with a base pressure of ~10^{-9} Pa (see Figure 13.10). Second, the tip is cooled to between 20 and 80 K to preserve the position of the surface atoms subjected to the high electric field and to reduce the thermal agitation energy, k_BT, of the atoms at the instant of their evaporation, thus providing high spatial resolution. Third, the tip is positioned in front of a local electrode or microelectrode (e.g., for a LEAP 4000X Si^{TM}) to enhance the field at the apex of the tip without further increasing the applied voltage. This ensures a longer lifetime of the tip during the APT analysis. Finally, depending on the material's nature (i.e., metal, semiconductor, or insulator), voltage or laser pulses are applied on the tip. In this chapter, we are focusing on laser-assisted APT because photovoltaic materials are mostly semiconductor materials.

Applying laser pulses to a positively biased APT tip induces controlled field evaporation of surface atoms. The field-evaporated atoms are accelerated toward a position-sensitive detector that records their time-of-flight and impact positions. A 3D elemental map can be reconstructed from the collected data using an inverse projection of the detector coordinates (x_D, y_D) to the tip surface (x, y) [27].

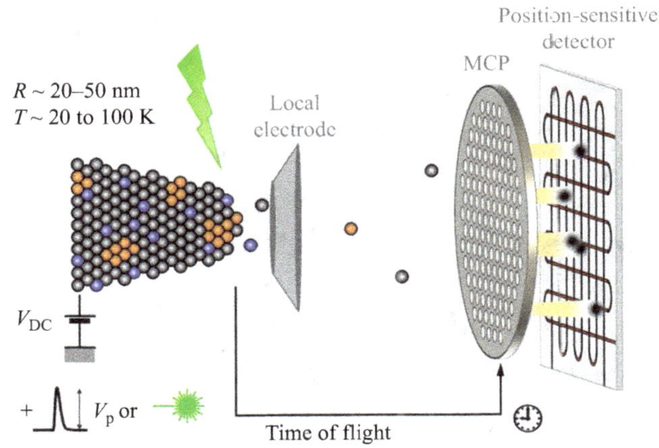

Figure 13.10 Schematic of the experimental setup used for atom probe tomography, and more specifically, of the local electrode atom probe, LEAP. Taken from [54]

13.3.3 Sample preparation

The importance of shaping the APT sample like a *nanoscale needle* is to induce a high electric field at its apex, as highlighted in the previous section. Obtaining such small and atypical samples was established long ago by using either electropolishing (for metallic samples) [28] or FIB (for semiconductor and insulator samples) [29]. Hence, in the present section, we will focus on nonstandard sample preparation methods developed for specific tasks, such as investigating crystallographic defects or heterojunctions. Several nonconventional, site-specific sample preparation methods are already published [30–32], but here, we describe a recently developed method by the Atomprobe group of Dr. Oana Cojocaru-Mirédin at the Institute of Physics, RWTH Aachen University in Germany. This preparation method is specifically adapted to provide unique insight into photovoltaic materials.

Figure 13.11 describes the main steps in the procedure developed for investigating *grain boundaries* [33]. The goal of this procedure is to investigate the composition of a specific grain boundary (GB) in a $Cu(In,Ga)Se_2$ (CIGSe) absorber, but also, its structural and electrical properties prior to the APT experiment. For that, the standard lift-out method described in [29] was combined with electron backscatter diffraction (EBSD) and EBIC measurements. The main steps are as follows:

1. The surface of the absorber layer is polished and cleaned using the FIB system. A smooth surface is mandatory for EBSD investigations.
2. EBSD measurement is performed on the smoothed CIGSe surface (Figure 13.11(a)). By doing so, the character of each GB (based on its disorientation angle and axis) is revealed, as well as the grain size.
3. EBIC measurement is then performed on the same polished area, where the EBSD investigation was done (Figure 13.11(b)). This specific configuration—an

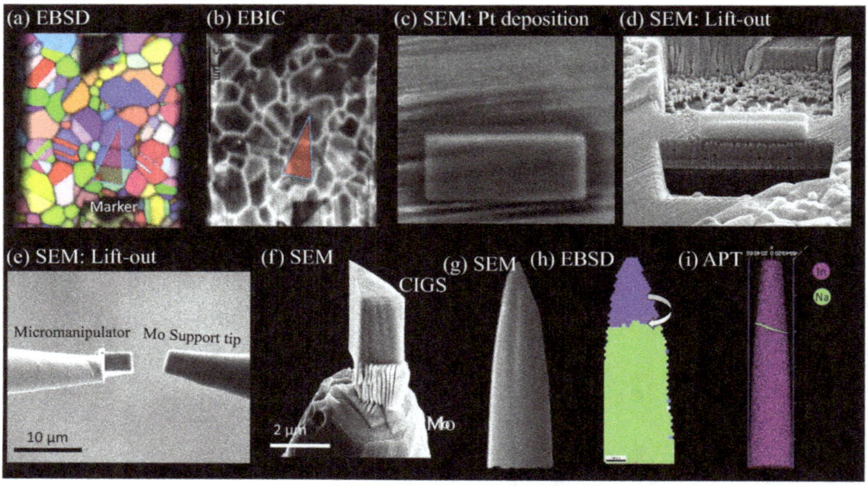

Figure 13.11 Novel site-specific sample preparation with its main steps: (a) EBSD and (b) EBIC measurements performed on the smoothed CIGSe surface; (c) Pt deposition to mark the position of the selected grain boundary; (d)–(g)) standard lift-out technique necessary to fabricate the APT tip containing the selected grain boundary; (h) transmission EBSD measurement performed directly on the APT tip, revealing the exact position and nature of the boundary; (i) APT measurement clearly revealing the Na segregation at the boundary position

"edge scan configuration"—is unconventional, but it helps to definitively identify whether or not the GBs in the CIGSe absorber are electronically active.

4. After analysis of the EBSD and EBIC data, specific GBs are chosen to be investigated by APT. For that, the position of each GB is marked by depositing a thin Pt layer using a gas-injection system (Figure 13.11(c)).
5. Lamellae containing the selected GB are prepared using the standard lift-out procedure [29] (Figure 13.2(d)–(f)). APT tips are then prepared using the annular milling procedure [34] (Figure 13.11(g)).
6. To confirm the position and the character of each selected grain boundary, the APT tips are pre-investigated by EBSD in transmission mode (t-EBSD [35], Figure 13.11(h)). This clearly shows the disorientation angle and axis of each selected GB down to 10-nm resolution.
7. Lastly, the APT experiment and analysis is completed (Figure 13.11(i)).

13.3.4 Possible artifacts when analyzing semiconductor materials in APT

Two types of artifacts are encountered when a semiconductor material is investigated by APT. The first type is an *experimental artifact*, such as DC evaporation and preferential evaporation or retention. It often appears when experimental

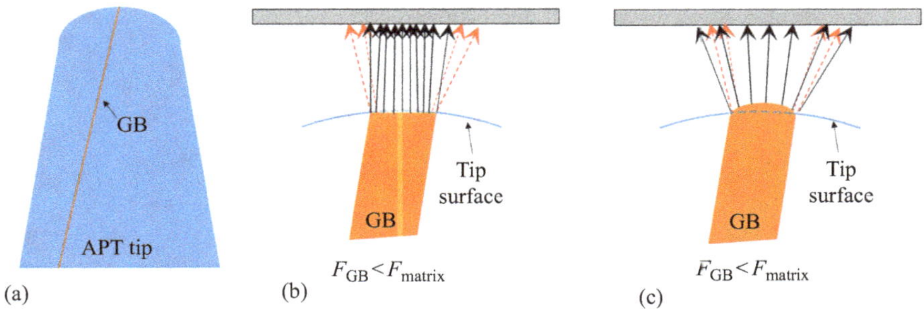

Figure 13.12 (a) Schematics of an APT tip containing a grain boundary (GB). The local magnification effect in the case of (b) low- and (c) high-field GB. Taken from [54]

parameters are not properly considered during the APT investigations, as covered in [34]. The second type, an *inherent artifact,* appears independently of the experimental conditions used. Only this latter type of artifact will be discussed here because it is the one that most often limits the APT analysis. Possible solutions to overcome such artifacts are also given.

A typical inherent artifact that occurs when analyzing a semiconductor material with APT is the *local magnification effect* [36]. This effect occurs due to local variation in the evaporation field when nanostructures such as a grain or phase boundaries, precipitates, and nanovoids are present in the matrix. These nanostructures often lead to a local protrusion on the tip surface, which results in a trajectory deviation in flight path. Figure 13.12 explicitly shows the trajectory aberrations and the density variations due to the presence of a GB, which might have either a lower or a higher evaporation field than the matrix. An effective way to reduce the local magnification effect is to place the GB perpendicular to the tip axis (*z*-axis), because it has been demonstrated that the local magnification effect increases continuously as the GB approaches an orientation parallel to the tip axis [37].

Another inherent artifact is the *peak* and/or *tail overlap* observed in the mass spectrum. An example of peak overlap between Zn and S_2 is given in Figure 13.13. This artifact inevitably appears when investigating materials containing S and Zn, such as ZnS, and it can lead to an erroneous material composition.

Recent efforts have made great strides toward correcting the deviation in elemental concentration due to peak overlap. For example, a new algorithm has been developed that can correct the elemental concentration directly on the one-dimensional concentration profile. This novel algorithm is implemented in a new user-friendly software called EPOSA [38].

13.3.5 Application to multicrystalline silicon solar cells

Atom probe tomography and correlated microscopy has been applied successfully to study GBs in multicrystalline silicon solar cells [39–42]. Figure 13.14 shows the

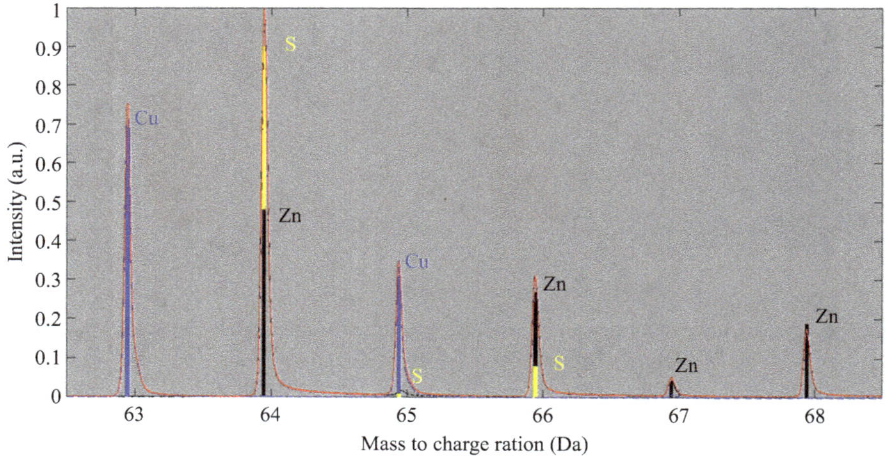

Figure 13.13 Mass spectrum of the ZnS/Cu(In,Ga)Se$_2$ heterointerface showing the peak overlap between $_{64}Zn^+$ (in black, second peak) and $_{64}S_2^{2+}$ (in yellow, second peak) or between $_{66}Zn^+$ (in black, fourth peak) and $_{66}S_2^{2+}$ (in yellow, fourth peak). Data courtesy of Jens Keutgen

procedure for determining the electrical, structural, and chemical properties of a selected Σ27 twin boundary in a multicrystalline silicon absorber. First, an EBIC scan on the sample surface is performed, as shown in Figure 13.14(a). With this method, a change in contrast is typically observed at the three detected twin boundaries: Σ3, Σ9, and Σ27. Yet, the strongest change in contrast was observed for the Σ27 twin boundary. Then, the EBSD investigations allowed for the determination of the structural characteristics of the GBs such as disorientation angle and its deviation from the ideal misorientation. Furthermore, for low coincidence site lattice twin boundaries, it is even possible to determine the most probable boundary plane and its deviation from ideal boundary-plane positions—and thus, the degree of boundary coherence. The region marked by the yellow arrow in Figure 13.14(b) represents the portion of the GB chosen to be analyzed by APT. This specific region was lifted out with a micromanipulator and attached to Mo support tips that were positioned on a half-grid, as shown in Figure 13.14(c) and (d). A specially designed and fabricated sample retainer allows the transfer of the Mo grid between FIB/scanning electron microscopy (SEM), EBSD, transmission electron microscopy (TEM), and APT, as described in [31,43]. Then, an APT tip with an apex diameter of ~50 nm is obtained by annular milling using 16-kV Ga$^+$, as shown in Figure 13.14(d). Final Ga$^+$ milling is performed at 5 kV to minimize the Ga contamination and damage of the sample. Additionally, TEM is performed on the final APT samples, as shown in Figure 13.14(e). This final TEM imaging ensures that the selected boundary lies within a maximum distance of 200 nm from the tip apex (the typical depth of APT analysis for a sample tip). Furthermore, the knowledge of the exact tip shape—e.g., tip radius, shank angle, and GB position provided by the

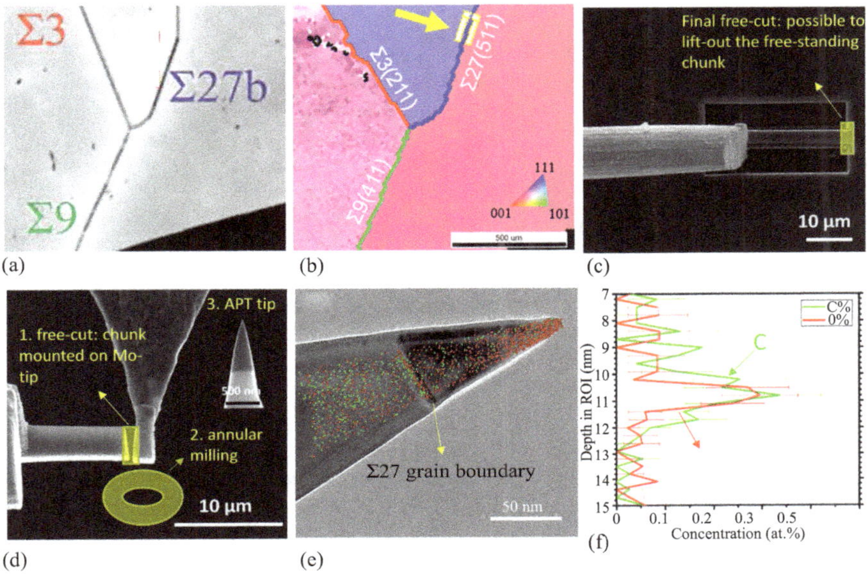

Figure 13.14 (a) EBIC map and (b) EBSD inverse-pole figure map with the lift-out position indicated by the yellow marker. (c) Lift-out of a wedge-shaped lamella containing the boundary to be investigated. (d) SEM image of the lifted-out material placed on the top of a Mo post and glued with Pt weld. The APT specimen prepared by annular milling is also shown. (e) TEM image of the final APT tip overlaid with the reconstructed APT data. (f) One-dimensional concentration profile for C and O across the GB plane

TEM analysis—enables a more accurate 3D reconstruction of the analyzed volume by APT. The TEM image in Figure 13.14(e) is superimposed by the 3D map of the detected impurities. Carbon and oxygen enrichment at the position of the GB is clearly detected (Figure 13.14(f)). This GB decoration is influenced by the GB type and can lead to increased recombination activity, as discussed in [44]. Note that the asymmetric and incoherent GB portions are especially prone to impurity segregation.

Another example of correlative microscopy investigation of an electrically active and structurally curved Σ9 twin boundary is shown in Figure 13.15. In this case, the impurities are not segregated homogeneously at the boundary position; rather, they are accumulated on the lower kink position of the boundary, as highlighted in Figure 13.15(a). This unconventional segregation phenomenon has been called asymmetric line segregation [42] and was observed for the first time thanks to APT-TEM correlative high-resolution investigations (Figure 13.15(b) and (c)), which identified the lattice planes around the twin boundary.

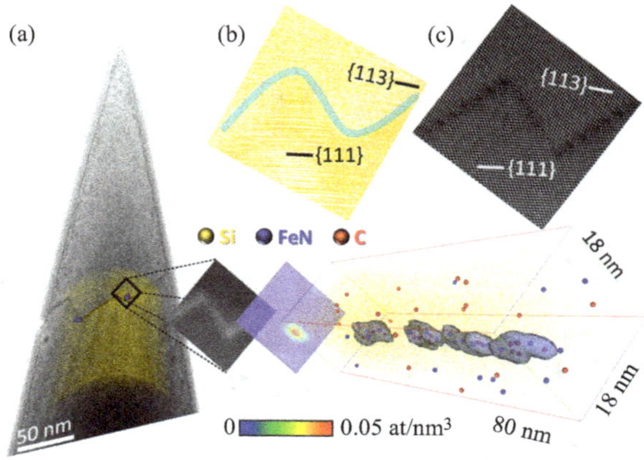

Figure 13.15 Asymmetric line segregation observed at Σ9 twin boundary using 1:1 APT–high-resolution STEM correlative studies. (a) Overview annular bright-field (ABF) STEM image of APT/TEM specimen resolving the faceted structure of the Σ9 twin boundary superimposed by APT 3D map. Extended magnification of the highlighted area showing a strain-sensitive low-angle annular dark field–STEM micrograph, a two-dimensional density map of impurities and a 3D atom map. (b) Corresponding projection of 3D atom map with reconstructed lattice planes and twin-boundary topology. (c) High-resolution high-angle annular dark field–STEM image resolving the atomic structure in the vicinity of the right-hand twin-boundary step. Taken from [42]

13.3.6 Application of correlative microscopy to thin-film solar cells

13.3.6.1 Cu(In,Ga)Se_2 thin-film solar cells

Grain boundaries are one of the main limiting factors for cell efficiency in Cu(In, Ga)Se_2 thin-film solar cells due charged defects that lead to increased recombination of minority carriers before they can be collected from the cell. By inserting alkali impurities in the absorber layer, most of the GBs become either electrically inactive or even highways for electron accumulation and diffusion. Yet, only a few techniques can quantify the impurity segregation at the GB down to the nanometer level. One of these techniques is APT [45–48]. Moreover, by combining the APT analysis with those of TEM, EBSD, and EBIC—through the correlative microscopy approach described earlier—the nature of the relationship between composition, structure, and performance can be determined definitively.

Figure 13.16 shows an example of correlative TEM-APT studies performed on a Σ3 twin boundary. Taking into account the boundary-plane orientation, i.e., the trace of the boundary observed in the TEM image, one can readily determine the boundary-plane normal. We note here that the dotted line inserted in Figure 13.16 marks the

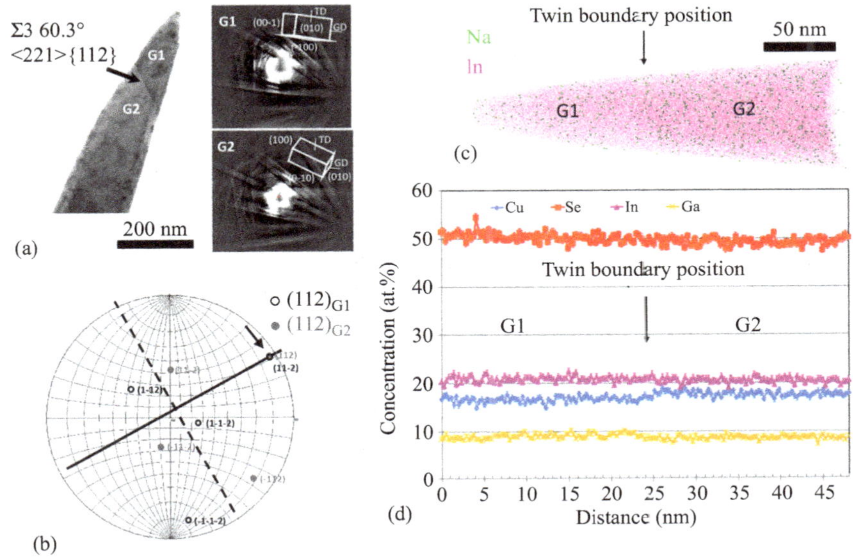

Figure 13.16 *Correlative TEM-APT investigations of a Σ3 twin boundary. (a) Bright-field TEM picture of the APT sample showing a Σ3 twin-boundary plane and the two neighboring grains, referred here to as G1 and G2. Kikuchi patterns taken from both grains G1 and G2 are shown. By indexing the transmission Kikuchi patterns of regions G1 and G2, the calculated disorientation angle was 60.3° with a deviation angle of 0.1°. (b) Stereographic projection displaying all possible symmetrically equivalent planes from G1 and G2, the boundary trace (dashed line), and the boundary-plane normal (thick continuous line), proving the (112) misorientation plane for this specific Σ3 twin boundary. (c) 3D APT map of a Σ3 twin boundary and (d) the corresponding elemental distribution profiles showing neither a change in elemental concentration, nor Na or O segregation at the Σ3 twin boundary. Taken from [49]*

boundary trace that is perpendicular to the GB plane normal shown by a thick continuous line. Thus, it is clearly demonstrated that the Σ3 twin boundary lies exactly on a (112) plane. Moreover, this Σ3 twin boundary should be coherent because no deviation between boundary-plane orientation and (112) plane is observed.

The APT investigations on this specific Σ3 twin boundary, as shown in Figure 13.16, demonstrate clearly that no change in elemental concentration is seen at the boundary position. These results are in line with EBIC measurements, where no change in contrast was observed at the twin boundary position. Hence, we can conclude that this specific twin boundary behaves like the bulk material, where the recombination processes are drastically reduced [49].

Two more examples of correlative TEM-APT studies are shown in Figure 13.17. Two random high-angle GBs with disorientation angles of 28° and 85° were investigated in a similar way as described previously for the twin boundary in Figure 13.16. For both GBs, changes in elemental concentration are observed, but so is Na, K, and O impurity co-segregation. A few differences are also observed. For the first random high-angle GB, Figure 13.17(a) shows that Cu and Ga are indeed depleted, whereas In is enriched (In on Cu anti-site defects). This is in agreement to what was postulated by Persson *et al.* [50] on the basis of *ab-initio* density functional theory (DFT) calculations—namely, that Cu vacancy rows are preferentially formed at the CIGSe grain boundaries. The In enrichment and the Cu depletion could be explained by the low formation energy of the charge-neutral charge defect pair $2V_{Cu}^{-} + In_{Cu}^{2+}$ [51]. For the second random high-angle GB, Figure 13.17(b) shows the opposite behavior: the Cu is not depleted anymore, but is enriched. The same trend was observed for In, which in this case is depleted, maintaining an anti-correlative behavior with respect to Cu (Cu on In anti-site defects) [52].

Another important aspect is the Na, K, and O impurity segregation that was noted at the boundaries. These impurities are believed to diffuse out of the soda-lime-glass substrate into the absorber layer during the deposition of the CIGSe layer at ~600 °C, because no barrier diffusion layer between soda-lime glass and Mo was present in this case. Moreover, random high-angle GBs containing no enrichment in oxygen have also been observed, which confirms that the oxygen at the boundaries seen in Figure 13.17 comes from the material itself and not from the APT chamber. (H, H_2, and H_2O are common contaminants to vacuum and are often found inside the APT chamber.) The Na, K, and O concentrations at the first high-angle GB are 1.7±0.16 at. %, 0.045 ± 0.02 at. %, and 0.5 ± 0.08 at. %, respectively. For the second high-angle GB, higher Na, K, and O concentration values were calculated (3.5 at. % Na, 0.04 at. % K, and 1.5 at. % O, respectively) [49].

Based on the results from the correlative STEM-APT approach shown here, several scenarios could be determined for the random high-angle GBs as follows [49]:

- Se-depletion at random high-angle GB: strong O segregation and Cu-on-In anti-site defects;
- Se-enrichment at random high-angle GB: no O segregation and In-on-Cu behavior;
- No change in Se at random high-angle GB: slight O segregation and "In on Cu" anti-site defects;
- Na is always present at random high-angle GBs. K is also believed to be found at most of the GBs, but it could not be detected for some GBs because of its very low concentration (limit of detection for K in APT under these conditions is 15 ppm).

13.3.6.2 Perovskite solar cells

Understanding chemical gradients and instabilities is a main factor that limits the performance and/or reliability of halide perovskite solar cells. Therefore, APT has

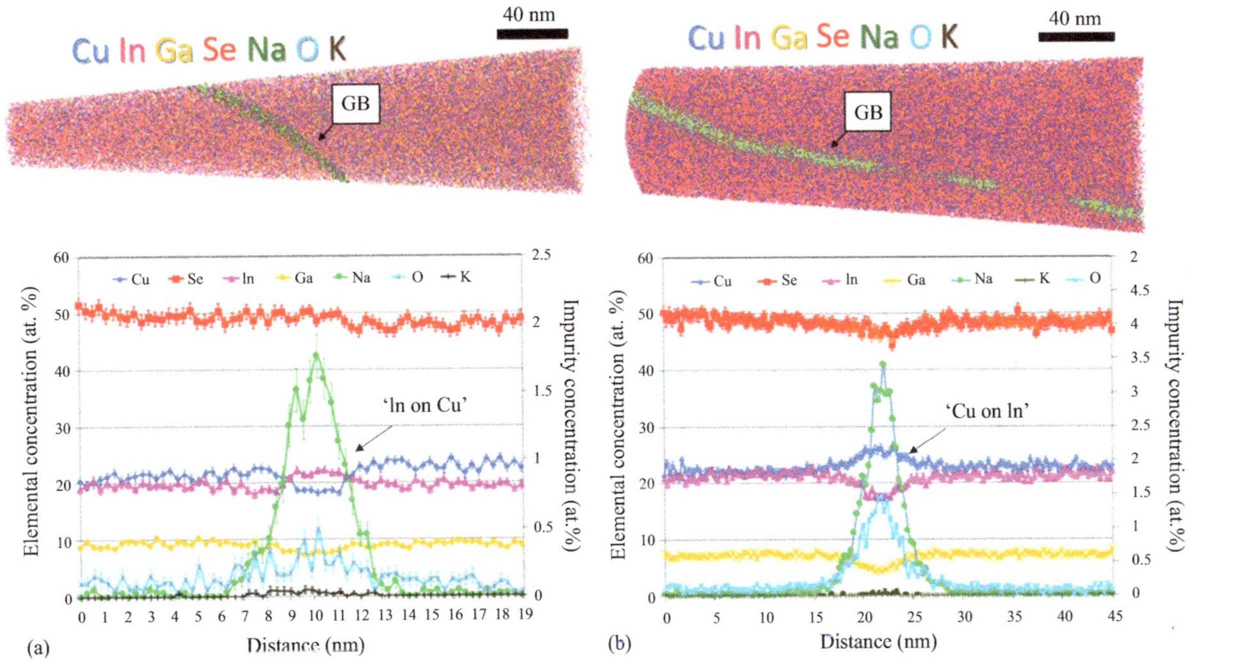

Figure 13.17 *APT investigations of random high-angle grain boundaries in CIGSe absorber. (a) and (b) 3D reconstruction showing two of the investigated grain boundaries by APT showing the distribution of Cu (blue), In (pink), Ga (yellow), Se (red), Na (green), O (light blue), and K (khaki). Linear concentration profiles of the elements and impurities showing both In-on-Cu and Cu-on-In behaviors at the boundary are presented below the APT reconstruction for each sample. Na and O impurity segregation was also detected. Taken from [49]*

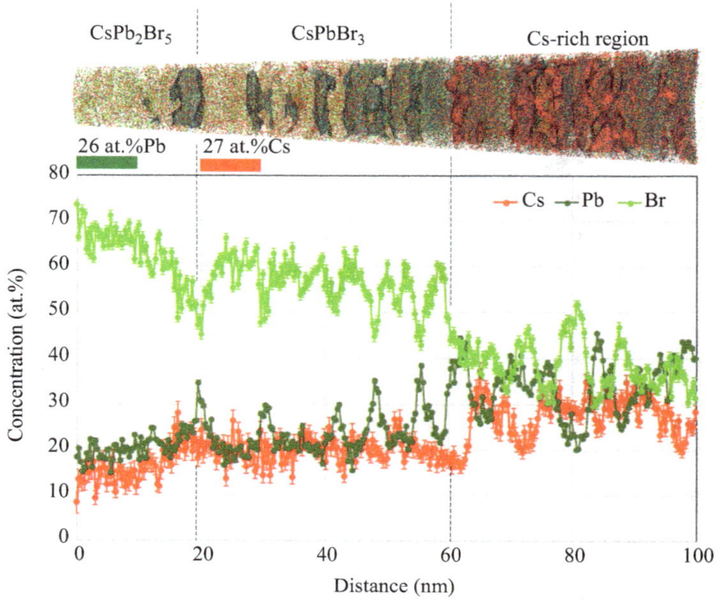

Figure 13.18 APT 3D elemental map together with the one-dimensional concentration profiles for Br (light green), Pb (dark green), and Cs (red) of the sample containing a co-evaporated perovskite thin film on a Si substrate. In addition to the expected $CsPbBr_3$ phase, this APT measurement reveals the presence of a Br-rich surface that approaches the composition of $CsPb_2Br_5$ phase and of a Cs-rich region deep inside the sample. The latter one could not be correlated to any existing phases on the Cs-Pb-Br ternary diagram. Taken from [53]

been applied to investigate the phase formation in $CsPbBr_3$ perovskite. Figure 13.18 reveals three distinctive regions: (1) a Br-rich surface with the average elemental concentrations, (2) the expected $CsPbBr_3$ phase, and (3) a Cs-rich region. The Cs-rich region does not correspond to any known phases in the existing Cs-Pb-Br ternary diagram, which suggests that it might be a metastable phase. These APT results demonstrate that in addition to the existing perovskite $CsPbBr_3$, other metastable phases are present in the material [53].

13.4 Summary

The similarities and differences between TOF-SIMS and APT can be inferred from the examples that have been presented. Both tools are powerful analytical techniques with low (ppm or better) detection limits, well suited to investigating dopants and impurities in solar cell materials and devices. In each technique, imaging and

3D tomography can provide powerful insight into different performance or reliability issues. TOF-SIMS operates on a length scale of micrometers to nanometers, with a relatively high throughput possible, whereas APT operates on a length scale of nanometers to Angstroms, with a comparatively slower throughput due to more specialized sample preparation needed. Examples were given of each technique applied to conventional silicon solar cells as well as to next-generation thin-film materials (both organic and inorganic).

What is also shown is that whereas each technique is powerful on its own, it can enable some next-level insight when combined with other correlative characterization methods—thus bridging the gaps between inferring information from just TOF-SIMS or APT analysis alone, to understanding correlated degradation or performance mechanisms in full. The combined approach has used TOF-SIMS and other techniques to investigate PID in silicon solar cells [16], and the correlative microscopy approach has helped to understand the performance limitations of grain boundaries in CIGS with APT [49].

Acknowledgments

The work on TOF SIMS was funded by the U.S. Department of Energy (DOE) at the National Renewable Energy Laboratory (NREL) under contract No. DE-AC36-08GO28308, with funding provided by the Solar Energy Technology Office. The work on APT was funded by the German Federal Ministry of Education and Research (Contract 03X5522A). Special thanks to Pyuck-Pa Choi, Torsten Schwarz, and Purvesh Soni for help in APT data acquisition, analysis, and understanding.

References

[1] 'Stevie, F., *Secondary Ion Mass Spectrometry: Applications for Depth Profiling and Surface Characterization*. 2015: New York, NY: Momentum Press.

[2] Harvey, S.P., G. Teeter, H. Moutinho, and M.M. Al-Jassim, 'Direct evidence of enhanced chlorine segregation at grain boundaries in polycrystalline CdTe thin films via three-dimensional TOF-SIMS imaging'. *Progress in Photovoltaics*, 2015. **23**(7): p. 838–846.

[3] Colegrove, E., S.P. Harvey, J.-H. Yang, *et al.*, 'Phosphorus diffusion mechanisms and deep incorporation in polycrystalline and single-crystalline CdTe'. *Physical Review Applied*, 2016. **5**(5): p. 054014.

[4] Colegrove, E., S.P. Harvey, J.H. Yang, *et al.*, 'Antimony diffusion in CdTe'. *IEEE Journal of Photovoltaics*, 2017. **7**(3): p. 870–873.

[5] Colegrove, E., J.H. Yang, S.P. Harvey, *et al.*, 'Experimental and theoretical comparison of Sb, As, and P diffusion mechanisms and doping in CdTe'. *Journal of Physics D: Applied Physics*, 2018. **51**(7): 075102.

[6] Harvey, S.P., E. Colegrove, B. McCandless, D. Albin, M. Al-Jassim, and W. K. Metzger. 'Using time-of-flight SIMS to investigate group V dopant

distribution in CdTe'. in *2017 IEEE 44th Photovoltaic Specialist Conference (PVSC)*. 2017. p. 3319–3322.
[7] McCandless, B.E., W.A. Buchanan, C.P. Thompson, *et al.*, 'Overcoming carrier concentration limits in polycrystalline CdTe thin films with in situ doping'. *Scientific Reports*, 2018. **8**(1): p. 14519.
[8] Harvey, S.P., Z. Li, J.A. Christians, K. Zhu, J.M. Luther, and J.J. Berry, 'Probing perovskite inhomogeneity beyond the surface: TOF-SIMS analysis of halide perovskite photovoltaic devices'. *ACS Applied Materials & Interfaces*, 2018. **10**(34): p. 28541–28552.
[9] Li, Z., C. Xiao, Y. Yang, S.P. Harvey, *et al.*, 'Extrinsic ion migration in perovskite solar cells'. *Energy & Environmental Science*, 2017. **10**(5): p. 1234–1242.
[10] Xu, D., X. Hua, S.-C. Liu, *et al.*, 'In situ and real-time ToF-SIMS analysis of light-induced chemical changes in perovskite $CH_3NH_3PbI_3$'. *Chemical Communications*, 2018. **54**(43): p. 5434–5437.
[11] Christians, J.A., P. Schulz, J.S. Tinkham, T.H. Schloemer, *et al.*, 'Tailored interfaces of unencapsulated perovskite solar cells for >1,000 hour operational stability'. *Nature Energy*, 2018. **3**(1): p. 68–74.
[12] Lin, W.-C., H.-Y. Chang, K. Abbasi, J.-J. Shyue, and C. Burda, '3D in situ ToF-SIMS imaging of perovskite films under controlled humidity environmental conditions'. *Advanced Materials Interfaces*, 2017. **4**(2): p. 1600673.
[13] Schelhas, L.T., Z. Li, J.A. Christians, *et al.*, 'Insights into operational stability and processing of halide perovskite active layers'. *Energy & Environmental Science*, 2019. **12**(4): p. 1341–1348
[14] Tong, J., Z. Song, D.H. Kim, *et al.*, 'Carrier lifetimes of >1 μs in Sn-Pb perovskites enable efficient all-perovskite tandem solar cells'. *Science*, 2019: p. eaav7911.
[15] Luo, W., Y.S. Khoo, P. Hacke, *et al.*, 'Potential-induced degradation in photovoltaic modules: a critical review'. *Energy & Environmental Science*, 2017. **10**(1): p. 43–68.
[16] Harvey, S.P., J. Moseley, A. Norman, *et al.*, 'Investigating PID shunting in polycrystalline silicon modules via multiscale, multitechnique characterization'. *Progress in Photovoltaics: Research and Applications*, 2018. **26**(6): p. 377–384.
[17] Harvey, S.P., H. Guthrey, C.P. Muzzillo, *et al.*, 'Investigating PID shunting in polycrystalline CIGS devices via multi-scale, multi-technique characterization'. *IEEE Journal of Photovoltaics*, 2019. **9**(2): p. 559–564.
[18] Choi, P.-P., O. Cojocaru-Mirédin, D. Abou-Ras, *et al.*, 'Atom probe tomography of compound semiconductors for photovoltaic and light-emitting device applications'. *Microscopy Today*, 2012. **20**(03): p. 18–24.
[19] Schwarz, T., O. Cojocaru-Mirédin, P. Choi, *et al.*, 'Atom probe tomography study of internal interfaces in $Cu_2ZnSnSe_4$ thin-films'. *Journal of Applied Physics*, 2015. **118**(9): p. 095302.
[20] Mangelinck, D., K. Hoummada, O. Cojocaru-Mirédin, E. Cadel, C. Perrin-Pellegrino, and D. Blavette, 'Atom probe tomography of Ni silicides: First

stages of reaction and redistribution of Pt'. *Microelectronic Engineering*, 2008. **85**(10): p. 1995–1999.

[21] Yu, Y., S. Zhang, A.M. Mio, *et al.*, 'Ag-segregation to dislocations in PbTe-based thermoelectric materials'. *ACS Appl Mater Interfaces*, 2018. **10**(4): p. 3609–3615.

[22] Cojocaru-Miredin, O., L. Abdellaoui, M. Nagli, *et al.*, 'Role of nanostructuring and microstructuring in silver antimony telluride compounds for thermoelectric applications'. *ACS Appl Mater Interfaces*, 2017. **9**(17): p. 14779–14790.

[23] Yu, Y., D.-S. He, S. Zhang, *et al.*, 'Simultaneous optimization of electrical and thermal transport properties of $Bi_{0.5}Sb_{1.5}Te_3$ thermoelectric alloy by twin boundary engineering'. *Nano Energy*, 2017. **37**(July): p. 203–213.

[24] Zhu, M., O. Cojocaru-Mirédin, A.M. Mio, *et al.*, 'Unique bond breaking in crystalline phase change materials and the quest for metavalent bonding'. *Advanced Materials*, 2018. **30**(18): p. 1706735.

[25] Lefebvre, W., F. Vurpillot, and X. Sauvage, *Atom Probe Tomography: Put Theory Into Practice*. 2016: Paris, Elsevier.

[26] Gault, B., M.P. Moody, J.M. Cairney, and S.P. Ringer, *Atom Probe Microscopy*. Springer Series in Material Science, ed. R. Hull, *et al.* 2012, New York, Heidelberg Dordrecht, London. 396.

[27] Kelly, T.F. and M.K. Miller, 'Atom probe tomography'. *Review of Scientific Instruments*, 2007. **78**: p. 031101-1–031101-20.

[28] M.K. Miller, A. Cerezo, M.G. Hetherington, and G.W. Smith, *Atom Probe Field Ion Microscopy*, ed. Oxford, UK, Oxford science publication. 1996.

[29] Kelly, T.F., D.J. Larson, K. Thompson, *et al.*, 'Atom probe tomography of electronic materials'. *Annual Review of Materials Research*, 2007. **37**(1): p. 681–727.

[30] Felfer, P.J., T. Alam, S.P. Ringer, and J.M. Cairney, 'A reproducible method for damage-free site-specific preparation of atom probe tips from interfaces'. *Microscopy Research and Techniques*, 2012. **75**(4): p. 484–491.

[31] Cojocaru-Miredin, O., T. Schwarz, P.P. Choi, M. Herbig, R. Wuerz, and D. Raabe, 'Atom probe tomography studies on the Cu(In,Ga)Se_2 grain boundaries'. *Journal of Visualized Experiments*, 2013. **74**: p. e50376.

[32] Prosa, T.J., D. Lawrence, D. Olson, D.J. Larson, and E.A. Marquis, 'Backside lift-out specimen preparation: reversing the analysis direction in atom probe tomography'. *Microscopy and Microanalysis*, 2009. **15**(S2): p. 298–299.

[33] Raghuwanshi, M., B. Thöner, P. Soni, M. Wuttig, R. Wuerz, and O. Cojocaru-Mirédin, 'Enhanced carrier collection at grain boundaries in Cu(In, Ga)Se_2 absorbers: correlation with microstructure'. *ACS: Applied Materials and Interfaces*, 2018. **10**(17): p. 7.

[34] Larson, D.J., T.J. Prosa, R.M. Ulfig, B.P. Geiser, and T.F. Kelly, *Local Electrode Atom Probe Tomography*. Springer. Vol. 1. 2013: Springer.

[35] Schwarz, T., O. Cojocaru-Mirédin, R. Wuerz, and D. Raabe, 'Correlative transmission Kikuchi diffraction and atom probe tomography study of Cu(In,

Ga)Se_2 grain boundaries'. *Progress in Photovoltaics: Research and Applications*, 2018. **26**(3): p. 196–204.

[36] Vurpillot, F., A. Bostel, and D. Blavette, 'Trajectory overlaps and local magnification in three-dimensional atom probe'. *Applied Physics Letters*, 2000. **76**(0): p. 3127.

[37] Blavette, D., P. Duval, L. Letellier, and M. Guttmann, 'Atomic-scale APFIM and TEM investigation of grain boundary microchemistry in astroloy Ni base superalloys'. *Acta Materialia*, 1996. **44**(12): p. 4995–5005.

[38] Keutgen, J., M. Wuttig, and O. Cojocaru-Mirédin, 'EPOSA: user friendly software for compositional corrections in linear profiles'. *in preparation*, 2019.

[39] Stoffers, A., O. Cojocaru-Miredin, O. Breitenstein, W. Seifert, S. Zaefferer, and D. Raabe, 'Grain boundary characterization in multicrystalline silicon using joint EBSD, EBIC, and atom probe tomography'. *Photovoltaic Specialist Conference (PVSC), 2014 IEEE 40th*, 2014: p. 0042–0046.

[40] Stoffers, A., B. Ziebarth, J. Barthel, O. Cojocaru-Miredin, C. Elsasser, and D. Raabe, 'Complex nanotwin substructure of an asymmetric sigma9 tilt grain boundary in a silicon polycrystal'. *Physical Review Letters*, 2015. **115** (23): p. 235502.

[41] Stoffers, A., J. Barthel, C.H. Liebscher, *et al.*, 'Correlating atom probe tomography with atomic-resolved scanning transmission electron microscopy: example of segregation at silicon grain boundaries'. *Microsc Microanal*, 2017. **23**(2): p. 291–299.

[42] C. H. Liebscher, A. Stoffers, M. Alam, *et al.*, 'Strain-induced asymmetric line segregation at faceted Si grain boundaries'. *Physical Review Letter*, 2018. **121**: p. 015702.

[43] Herbig, M., P. Choi, and D. Raabe, 'Combining structural and chemical information at the nanometer scale by correlative transmission electron microscopy and atom probe tomography'. *Ultramicroscopy*, 2015. **153**(June): p. 32–39.

[44] Stoffers, A., O. Cojocaru-Mirédin, W. Seifert, S. Zaefferer, S. Riepe, and D. Raabe, 'Grain boundary segregation in multicrystalline silicon: correlative characterization by EBSD, EBIC, and atom probe tomography'. *Progress in Photovoltaics: Research and Applications*, 2015. **23** (-): p. 1742–1753.

[45] Cojocaru-Mirédin, O., P. Choi, D. Abou-Ras, S. Siebentritt, R. Caballero, and D. Raabe, 'Characterization of Gbs in CIGS films using atom-probe tomography'. *IEEE Journal of Photovoltaics*, 2011. **1**(2): p. 207–212.

[46] Cojocaru-Miredin, O., P. Choi, R. Wuerz, and D. Raabe, 'Atomic-scale distribution of impurities in $CuInSe_2$-based thin-film solar cells'. *Ultramicroscopy*, 2011. **111**(6): p. 552–556.

[47] Cadel, E., N. Barreau, J. Kessler, and P. Pareige, 'Atom probe study of sodium distribution in polycrystalline Cu(In,Ga)Se_2 thin film'. *Acta Materialia*, 2010. **58**(7): p. 2634–2637.

[48] Schlesiger, R., C. Oberdorfer, R. Wurz, *et al.*, 'Design of a laser-assisted tomographic atom probe at Munster University'. *Review of Scientific Instruments*, 2010. **81**(4): p. 043703.

[49] Cojocaru-Mirédin, O., T. Schwarz, and D. Abou-Ras, 'Assessment of elemental distributions at line and planar defects in Cu(In,Ga)Se_2 thin films by atom probe tomography'. *Scripta Materialia*, 2018. **148**(April): p. 106–114.
[50] Persson, C., Y.-J. Zhao, S. Lany, and A. Zunger, 'n-type doping of $CuInSe_2$ and $CuGaSe_2$'. *Physical Review B*, 2005. **72**(3):035211.
[51] Zhang, S.B., S.H. Wei, A. Zunger, and H. Katayama-Yoshida, 'Defect physics of the $CuInSe_2$ chalcopyrite semiconductor'. *Physical Review B*, 1998. **57**(16): p. 9642–9656.
[52] Abou-Ras, D., S.S. Schmidt, R. Caballero, *et al.*, 'Confined and chemically flexible grain boundaries in polycrystalline compound semiconductors'. *Advanced Energy Materials*, 2012. **2**(8): p. 992–998.
[53] Caicedo-Davila, S., H. Funk, R. Lovrinčić, *et al.*, 'Spatial phase distributions in solution-based and evaporated Cs-Pb-Br thin films'. *Journal of Physics C*, 2019. **123**(29), p. 17666–17677
[54] Abou-Ras, D., T. Kirchartz, and U. Rau, *Advanced Characterization Techniques for Thin Film Solar Cells*. 2nd edn, Wiley-VCH. Vol. 2. 2016, Germany. 760.

Chapter 14

Solid-state NMR characterization for PV applications

Elizabeth Pogue[1]

14.1 Introduction

Polycrystalline photovoltaic (PV) materials are sensitive to defects, which may be difficult to identify using traditional spectroscopy-based technologies. Although scientists can model a variety of defects present in absorber materials, spectroscopy-based techniques are not chemically specific. Although X-ray diffraction is the standard technique for structural characterization, it is less sensitive to local motifs that average out over the broader structure. Furthermore, its use is limited when trying to distinguish disorder between atoms with similar scattering factors (for example, Cu/Zn disorder in Cu_2ZnSnS_4). Solid-state nuclear magnetic resonance (ssNMR) is both chemically specific (it senses up to 2–3 atomic distances away from the atom being studied) and sensitive to local symmetry and coordination. In a sense, one probes the structure of a material by starting at the atom and moving out rather than the more conventional approach of looking in at the overall structure as a whole. Consequently, ssNMR is sensitive to defects and defect complexes, provided the concentrations of these defects are large enough. Nonetheless, there are some important limitations that have, to date, limited the application of ssNMR to PV materials. More recent PV materials such as CdTe, Cu_2ZnSnS_4, and halide perovskites are more amenable to ssNMR characterization compared to Si because of the nature and abundance of the nuclei being probed. There have been a variety of recent of advances in the ssNMR field that have made it much easier to use for solar PV material characterization and will likely make it even more relevant in the future.

The purpose of this chapter is to show the capabilities of ssNMR in the context of PV materials, to give enough background to understand research that uses the technique, and to help the reader evaluate whether ssNMR would be helpful to their research endeavors. This is not a comprehensive review of the ssNMR field as a whole. This chapter introduces the basics of ssNMR on a practical level. First, the terminology and basics surrounding ssNMR are introduced. Data analysis and

[1]Department of Chemistry, The Johns Hopkins University, Baltimore, MD, USA

sample requirements are explained. Although ssNMR is highly capable, it does have many limitations that are also highlighted, specifically in the context of studying polycrystalline PV materials. To demonstrate the utility of this technique, some recent applications of ssNMR for studying PV materials are discussed. Finally, this chapter highlights recent developments in ssNMR and nuclear magnetic resonance (NMR) crystallography that may broaden the applicability of ssNMR in the future.

14.2 Solid-state NMR basics and terminology

This section provides a basic description of physics of ssNMR and introduces the terminology necessary to understand work in the field. This is not a comprehensive discussion of the physics of ssNMR. A classic short NMR reference book that focuses heavily on the fundamental physics and concepts is Farrar and Becker's 1971 book *Pulse and Fourier Transform NMR Introduction to Theory and Methods* [1]. Duer's 2004 book *Introduction to Solid-State NMR* focuses more on the solid state and a wide variety of more modern applications [2]. Duer's textbook contains an expanded discussion of the physics and theory of solid-state NMR.

Why do we specify that we are discussing the solid state? Liquid-phase (solution) NMR is another common technique that uses the same spectrometers but requires different probes. In solution NMR experiments, anisotropic NMR interactions are averaged to yield sharper line-shapes. This averaging occurs because the molecules move and tumble in the solution. In many ways, this process is mimicked in a solid-state magic-angle spinning experiment described in Section 14.4 (spinning is not required in solution NMR). The benefit of ssNMR is that these orientation-dependent interactions can give insights into the chemistry and structure of the solid-state material being studied.

The field of NMR has been rapidly evolving. The International Union of Pure and Applied Chemistry (IUPAC) sets standards for nomenclature, nuclear spin properties, and chemical shift referencing [3,4]. In 2004, the International Union of Crystallography established the Commission on NMR Crystallography and Related Methods to establish, implement, and maintain these and other standards [5]. They are in the process of establishing a common format for sharing NMR data and extracted information similar to what already exists for the X-ray diffraction community. An excellent textbook on the topic of NMR crystallography is *NMR Crystallography*, edited by Harris, Wasylischen, and Duer [6]. NMR crystallography uses solid-state NMR data to obtain crystallographic information to support the analysis of single-crystal and powder diffraction data. Overall, the *Encyclopedia of Nuclear Magnetic Resonance* is an excellent source of information about NMR experiments.

The expense of NMR spectrometers and probes is a hurdle challenging the widespread use of ssNMR characterization. User facilities exist as do labs that specialize in investigating hard-to-study nuclei. The National High Magnetic Field Laboratory (MagLab) associated with Florida State University in the United

States is a user facility that offers free magnet time allocated on the basis of scientific peer review. They can perform high-resolution NMR at very high fields, ultrafast magic-angle spinning (MAS), low gamma and quadrupolar NMR, and a variety of other techniques. They can also perform low-temperature MAS NMR (down to −140 °C). The National Ultrahigh-Field NMR facility for Solids, hosted by the Canada Foundation for Innovation in Ottawa, is another user facility that focuses on solid-state NMR research. Europe has two ssNMR user facilities. The Rhône-Alpes European Large Scale Facility for NMR (RALF) performs both solid-state and solution-NMR experiments and is a user facility for researchers from France, the EU, and associated countries. It is part of the European Photon and Neutron Campus in Grenoble, France. The Conditions Extrêmes et Matériaux: Haute Température et Irradiation Laboratory (CEMHTI) is part of the Université Orléans in Orleans, France, and focuses on materials and processes at high temperatures, defects, and optical properties. They developed Dmfit, a common program for interpreting NMR data [7]. At least one of their high-temperature probes is capable of reaching 1,500 °C, with several other probes capable of reaching temperatures above 500 °C. They can also use lasers to illuminate their samples during measurements. Similarly, users can apply for experimental time by writing a proposal. The National Facility for High-Field NMR at the Tata Institute of Fundamental Research in Mumbai, India, is another user facility that focuses on high-field NMR, with at least two spectrometers capable of solid-state measurements. Many other research and industrial groups have their own NMR spectrometers, so user facilities are not the only way to access an ssNMR spectrometer.

14.2.1 What is the physics of ssNMR?

In contrast to X-ray diffraction, which involves investigating the net structure and symmetry of a material from a global perspective, ssNMR involves investigating the net structure and symmetry of a material from the inside out, a local perspective. SsNMR offers this local perspective because localized nuclei are generating the signal. The physics of these interactions are important for understanding when ssNMR is most useful and understanding the data.

Every nucleus has a spin, I_i, that, if nonzero, will align with an applied magnetic field (derivation based on Duer *et al.* [2]). The nuclear magnetic moment of a given nucleus is related to the nuclear spin following (14.1), where γ is the gyromagnetic ratio of the nucleus:

$$\mu_i = \gamma I_i. \tag{14.1}$$

The net magnetization from nuclear contributions is given by (14.2), where J is the net nuclear spin angular momentum:

$$M = \sum_i \mu_i = \gamma \mathrm{J}. \tag{14.2}$$

A uniform magnetic field supplied by the spectrometer will exert a torque on these nuclear spins:

$$T = M \times B = \frac{d}{dt}J. \tag{14.3}$$

By substituting (14.2) into (14.3), one derives the differential equation shown in (14.4). One generally assumes that the static applied field, B_0 is along the $+z$-axis:

$$\begin{aligned}
\frac{dM}{dt} &= \gamma M \times B \\
\frac{dM_x}{dt} &= \gamma B_0 M_y \\
\frac{dM_y}{dt} &= -\gamma B_0 M_x \\
\frac{dM_z}{dt} &= 0.
\end{aligned} \tag{14.4}$$

To solve these differential equations, one must take the second derivative:

$$\begin{aligned}
\frac{dM_x^2}{dt^2} &= \gamma B_0 \frac{dM_y}{dt} = -\gamma^2 B_0^2 M_x \\
\frac{dM_y^2}{dt^2} &= \gamma B_0 \frac{dM_x}{dt} = -\gamma^2 B_0^2 M_y.
\end{aligned} \tag{14.5}$$

Solutions to these equations are shown in (14.6), assuming that M_{z0} is the initial magnetization in the z-direction, M_{xy} is the initial magnetization in the x–y plane, and M_{xy} is directed along the $+x$-axis at $t = 0$:

$$\begin{aligned}
M_x &= M_{xy0} \cos(\gamma B_0 t) = M_{xy0} \cos(\omega_0 t) \\
M_y &= M_{xy0} \sin(\gamma B_0 t) = M_{xy0} \sin(\omega_0 t) \\
M_z &= M_{z_0}.
\end{aligned} \tag{14.6}$$

The frequency $\omega_0 = \gamma B_0$ is known as the Larmor frequency and is characteristic of the nucleus being studied. If a radio-frequency (RF) pulse is applied, it supplies an additional magnetic field in the x–y plane. Only spins with Larmor frequencies near the pulse frequency will be affected by the RF pulse. This tilts the net magnetization away from the z-axis and, when the RF pulse stops, the magnetization is allowed to relax back to its equilibrium position. As the magnetization relaxes to its equilibrium position, it radiates energy. The nuclear spin states are quantized, so the energy radiated is due to transitions from one quantized energy to another.

To understand how a signal is collected, it is useful to go back to discussing single nuclear magnetic moments. When a magnetic moment is placed in a static

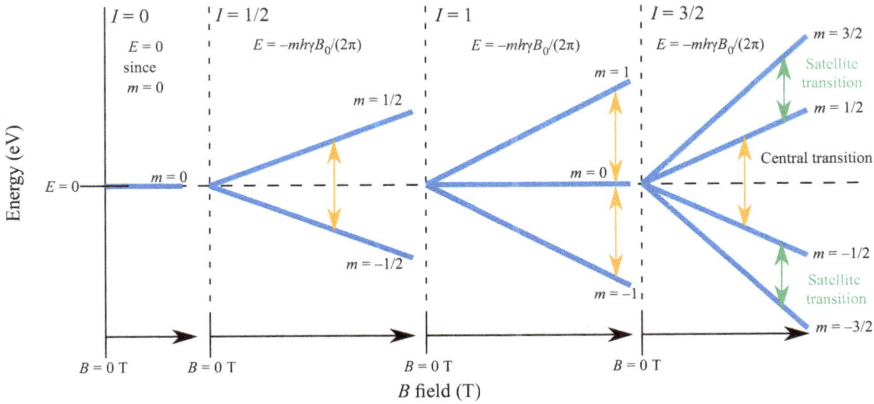

Figure 14.1 A magnetic field splits the nuclear spin states. Energy is released when spins transition between these quantized states

B field, the magnetic moment cannot perfectly align along the *z*-axis. The *z*-component of the magnetic moment is quantized such that $\mu_z = \frac{m\gamma h}{2\pi}$, where *m* is the magnetic quantum number in the set $(-I, -I+1, -I+2, \ldots, I-2, I-1, I)$. *I* is the spin quantum number, which can take on positive integer and half integer values including 0. $I = 0$ for nuclei with an even number of protons and neutrons. The energy of these states is given by (14.7):

$$E = -\mu_z \cdot B_0 = \frac{h}{2\pi} m\gamma B_0 = \hbar\omega_0 m. \tag{14.7}$$

An illustration of this is shown in Figure 14.1. It is clear that if *I* is zero, there is no separation at all. Whether or not *I* is a non-zero integer or half integer, the energy separation of the states surrounding $E = 0$ (called the central transition) is equal to the Larmor frequency, ω_0, times $\hbar$. This is related to why RF pulses near the Larmor frequency can change state occupancies. (The net magnetization axis is rotated by the pulse.)

14.2.1.1 T_1 and T_2 relaxation

In an NMR experiment, spins are perturbed and allowed to relax back to their equilibrium positions. A coil is used to detect signals along a single axis perpendicular to the static magnetic field. This is why the signal detected in an ssNMR experiment is called a free induction decay (fid). There are two time constants that are related to this relaxation: T_1, the spin-lattice relaxation constant, and T_2, the spin-spin relaxation constant. Both of these slightly modify (14.4), with T_2 acting on the $\frac{dM_x}{dt}$ and $\frac{dM_y}{dt}$ terms (introducing $-M_x/T_2$ and $-M_y/T_2$ terms, respectively) and T_1 acting on the $\frac{dM_z}{dt}$ term (introducing a $-(M_z - M_{\text{equilibrium}})/T_1$ term). The T_1 relaxation constant describes the rate at which a system gains or loses magnetization in the direction of the static magnetic field. This can be measured using the inversion recovery $(\pi - \tau - \frac{\pi}{2})$ pulse sequence. The first pulse inverts the magnetization ($+z$ to $-z$ direction), so it is along the axis of the static magnetic field but in the opposite direction to the field. The magnetization is allowed to relax for a time τ. Next, an RF pulse along the x-axis rotates the magnetization to the y-axis, bringing it into the plane where free induction decay can be measured. The intensity of the fid signal is plotted for a variety of τ relaxation intervals. The location of the null in this plot is generally proportional to the T_1 constant following $\tau_{\text{null}} = T_1\ln(2)$. The T_2 relaxation constant describes the dephasing of spins. This is related to different regions of the sample experiencing slightly different magnetic fields, which cause them to precess at slightly different rates. The variation in magnetic fields, in a well-shimmed spectrometer, is due to variations in the material being studied rather than variations in the external applied field. T_2 is less than T_1, necessarily. A Hahn (or spin)-echo pulse sequence ($\frac{\pi}{2} - \tau - \pi$) may be used to measure the T_2 relaxation constant. With this sequence, nuclear spins are first tilted into the x–y plane and then inverted after waiting a period of time. After the tilt into the x–y plane, the spins are allowed to dephase. The π pulse rotates all magnetizations by 180°, causing the magnetizations to come back in phase, producing an echo signal at a time of 2τ. The intensity of the echo signal is related to T_2 following (14.8), where D, the spin diffusion coefficient (G = magnetic field gradient) [8]:

$$\mathrm{A}(\text{echo at } 2\tau) \propto \exp\left[-(2\tau/\mathrm{T}_2) - \frac{2}{3}\gamma^2 G^2 D\tau^3\right]. \tag{14.8}$$

To measure T_2, the interval τ is varied. To account for the effect of diffusion, a modified pulse sequence called the Carr–Purcell technique ($\frac{\pi}{2} - \tau - \pi - 2\tau - \pi - 2\tau - \pi - 2\tau$...) may be used.

14.2.2 Interactions contributing to ssNMR signals

The local magnetic fields in a sample are not equal to the static applied magnetic field, allowing researchers to learn about bonding in the materials being studied. These effects are not always strong enough to be detected and not all interactions listed below apply for all nuclei (for example, spin-$\frac{1}{2}$ nuclei do not exhibit the quadrupolar interaction). The interaction described in Section 14.2.1 was the Zeeman interaction. This Zeeman interaction is generally much stronger than internal NMR interactions, allowing these other interactions to be treated as a perturbation of the Zeeman–Hamiltonian and, thus, added to the Zeeman–Hamiltonian to yield the total Hamiltonian of the spin system. A summary of these interactions is given in Table 14.1. The dipole–dipole interaction, chemical shift interaction, and quadrupolar interactions are discussed in detail in the following subsections. The simulated spectra were generated using WSolids1 software [9].

14.2.2.1 Dipole–dipole interaction

The nuclear spins in an NMR experiment do not only interact with the static magnetic field and RF pulses. They also interact with other nuclear spins from

Table 14.1 A summary of the major internal spin interactions that allow one to learn about local chemistry in an ssNMR experiment. $\widehat{I}$ is the spin operator of one nucleus. $\widehat{S}$ is the spin operator of the overall spin field, describing the nuclear spin interaction and its orientation dependence. σ is the chemical shielding tensor. D is the dipolar coupling tensor. V is the electric field gradient tensor. I is the nuclear spin quantum number. $\widehat{J}_{\text{molecule}}$ is the molecule's angular momentum operator. C is the spin-rotation tensor. J is the indirect spin coupling tensor

Interaction name	Abbreviation	Description
Dipolar coupling	$\widehat{H}_{DD} = -2\widehat{I} \cdot D \cdot \widehat{S}$	The nuclear spin of one nucleus interacts with (couples to) the nuclear spin of another, nearby nucleus. This can either be the same type of nucleus or a different one.
Quadrupolar interactions	$\widehat{H}_Q = \frac{eQ}{2I(2I-1)\hbar}\widehat{I} \cdot V \cdot \widehat{I}$	The electric field gradient (V) within the material tilts the equilibrium position of the nuclear magnetic moment slightly away from the static applied field direction. This is proportional to the quadrupolar moment, Q.
Chemical shielding	$\widehat{H}_{CS} = \gamma\widehat{I} \cdot \sigma \cdot B_0$	The B-field perturbs electrons surrounding the nucleus, changing the magnetic field experienced locally. Other interactions that tend to be small in solid-state powders (not small molecules)
Spin-rotation coupling	$\widehat{H}_{\text{CR}} = h\widehat{I} \cdot C \cdot \widehat{J}_{\text{molecule}}$	Magnetic fields are generated at a nucleus by the movement of the molecular magnetic moment associated with the electron distribution of the molecule.
Indirect spin (J) coupling	$\widehat{H}_J = \sum_{i \neq j} \widehat{I}_i \cdot J \cdot \widehat{I}_j$	The indirect nuclear spin–spin coupling through bonds

nearby nuclei. The direct interaction of these nuclear magnetic dipoles through space is called dipole–dipole coupling. Unlike J-coupling, there is no mediation by the bonds (electrons) for this interaction. The strength of the interaction is related to the distance, r, between nuclei (in this case, as $1/r^3$) and the molecular orientation, since the static B field selects an equilibrium axis for the otherwise unperturbed spins. This is why the interaction is important to the study of proteins, glasses, and other materials that are difficult to crystallize. In static ssNMR experiments, the dipole–dipole interaction can significantly broaden spectra depending on the nuclei being studied.

As shown in Table 14.1, the Hamiltonian for the dipole-dipole interaction is $\widehat{H}_{DD} = 2\widehat{I} \cdot D \cdot \widehat{S}$. The dipolar coupling tensor, D, describes the strength of the interaction. When this tensor is diagonalized, the principal values are $-d/2$, $-d/2$, and d where d, the dipolar coupling constant, is given in (14.9):

$$d = \hbar\left(\frac{\mu_0}{4\pi}\right)\frac{1}{r^3}\gamma_I\gamma_S. \tag{14.9}$$

As one can see from the principal values of D, the trace of this matrix is zero, so it is averaged out in solution NMR where molecules are tumbling. It can be shown that the homonuclear dipolar coupling Hamiltonian can be written following (14.10) and the heteronuclear dipolar coupling Hamiltonian can be written following (14.11) [2]:

$$\widehat{H}_{DD}^{\text{home}} = -\frac{d}{2}(3\cos^2\theta - 1)[3\widehat{I}_z\widehat{S}_z - \widehat{I}\cdot\widehat{S}] \tag{14.10}$$

$$\widehat{H}_{DD}^{\text{home}} = -d(3\cos^2\theta - 1)\widehat{I}_z\widehat{S}_z. \tag{14.11}$$

Here, it is clear that when θ equals 54.74°, known as the magic angle, the strength of this interaction goes to zero. Therefore, spinning the sample at the magic angle (see Section 14.4) will remove the effects of the dipolar interaction. The general effects of introducing homonuclear and heteronuclear dipolar coupling on static spectra are shown in Figure 14.2.

To determine whether dipolar coupling needs to be included in a model, one can calculate the dipolar coupling constant, d (14.9), in Hz for expected interaction lengths. The separation of the peaks in the spectrum is equal to 3d/2.

14.2.2.2 Chemical shielding

A material contains electrons as well as nuclei. These electrons can redistribute in a magnetic field and shield nuclei from it. There are two main components to chemical shielding: diamagnetic and paramagnetic. The diamagnetic portion comes from core electrons that shield the nucleus. The paramagnetic portion comes from the circulation of electrons between excited and ground states in the presence of a magnetic field close to the nucleus. Consequently, this term contains more information about bonding because of the near-Fermi level contributions. This effect tends to deshield the nucleus.

The absolute chemical shielding of a nucleus is not experimentally accessible because one cannot put bare nuclei in an ssNMR spectrometer. Nonetheless, one must define some frequency as 0. One generally compares the nucleus of interest to that of a standard reference compound, as defined by the IUPAC [3]. When one does a Fourier transform on a fid signal, the data are represented in terms of frequency. One is, however, generally interested in the difference in shielding between different nuclei. For this reason, one generally reports chemical shifts in terms of ppm. The chemical shift in ppm, δ, is related to the absolute chemical shielding of the sample, σ_s, and reference, σ_{ref}, following (14.12). These quantities can be related to the measured frequencies of reference, ν_{ref}, and sample, ν_s, chemical shifts:

$$\delta(\text{ppm}) = 10^6\frac{\sigma_{ref} - \sigma_s}{1 - \sigma_{ref}} = 10^6\frac{\nu_s - \nu_{ref}}{\nu_{ref}}. \tag{14.12}$$

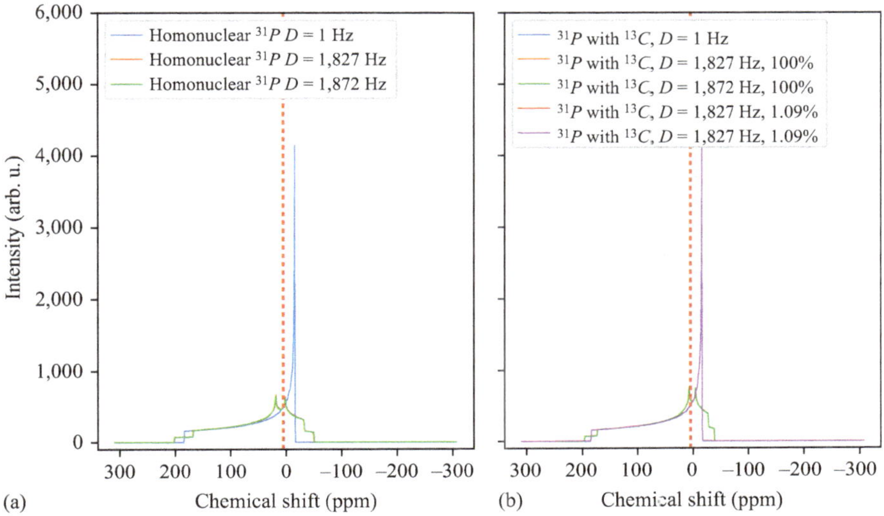

Figure 14.2 A common P–P bond is 2.21 Å, which corresponds to a dipolar coupling constant of 1,827 Hz. A common P–C bond is 1.87 Å, which corresponds to a dipolar coupling constant of 1,872 Hz. The left figure shows the effects of homonuclear coupling on ssNMR spectra. The right figure shows the effects of heteronuclear dipolar coupling on ssNMR spectra. In order to observe an effect, there must be enough nuclei to interact (the abundance of the coupled nucleus matters). Dipolar coupling tends to decrease the maximum intensity of lineshapes, splitting one peak into two and broadening them

Reference compounds were chosen to give sharp signals with well-defined chemical shifts. The use of a chemical shift rather than a frequency to describe the chemical shift allows one to compare data collected using spectrometers with different strengths of magnetic field. Some of the older literature may use different standards, but one can often convert between these standards. To do this conversion, assuming both experiments were Fourier-transform NMR experiments using the same nucleus, one must know the chemical shift of at least one compound represented using the IUPAC reference and referenced to the other standard that one is converting from.

There are three major conventions for describing the strength of chemical-shift interactions: Haeberlen, Herzfeld-Berger, and Standard (Table 14.2). One can readily convert between the three conventions. Since chemical shift is a second-rank tensor (a matrix), one can find a coordinate system such that the components of the tensor are directed along the x', y', and z' axes with no cross-terms (diagonalize the matrix). Consequently, only three values are important. Using the standard convention, these are denoted δ_{11}, δ_{22}, and δ_{33} and are ordered from largest to smallest. These appear in samples dominated by chemical shift as shown in

Table 14.2 Three major conventions for describing chemical shift

Convention name	Name	Description
Standard	Isotropic, δ_{iso}	$\delta_{iso} = \frac{\delta_{11}+\delta_{22}+\delta_{33}}{3}$
	Largest principal component, δ_{11}	δ_{11}
	Middle principal component, δ_{22}	δ_{22}
	Smallest principal component, δ_{33}	δ_{33}
Herzfeld-Berger	Isotropic, δ_{iso}	$\delta_{iso} = \frac{\delta_{11}+\delta_{22}+\delta_{33}}{3}$
	Span, Ω	$\Omega \approx \delta_{11} - \delta_{33}$ such that $\Omega \geq 0$
	Skew, κ	$\kappa = \frac{3(\delta_{22}-\delta_{iso})}{\Omega}$ such that $-1 \leq \kappa \leq 1$ For an axially symmetric tensor, the skew, κ, equals ± 1.
Haeberlen	Isotropic, δ_{iso}	$\delta_{iso} = \frac{\delta_{11}+\delta_{22}+\delta_{33}}{3}$ The principal components (δ_{11}, δ_{22}, and δ_{33}) are ordered such that: $\lvert\delta_{zz} - \delta_{iso}\rvert \geq \lvert\delta_{xx} - \delta_{iso}\rvert \geq \lvert\delta_{yy} - \delta_{iso}\rvert$ $\delta = \delta_{zz} - \delta_{iso}$
	Reduced anisotropy, δ	$\Delta = \delta_{zz} - \delta_{iso}$
	Anisotropy, $\Delta\delta$	$\Delta\delta = \frac{3\delta}{2} = \delta_{zz} - \frac{(\delta_{xx}+\delta_{yy})}{2}$
	Asymmetry, η	$\eta = \frac{\delta_{yy}-\delta_{xx}}{\delta}$ such that $0 \leq \eta \leq 1$ For an axially symmetric tensor, the asymmetry, η, equals 0.

Figure 14.3. The isotropic chemical shift, δ_{iso}, is defined as $\delta_{iso} = \frac{\delta_{11}+\delta_{22}+\delta_{33}}{3}$ for all the conventions. The other conventions are related to the relationships between δ_{11}, δ_{22}, and δ_{33} and provide more insights into the symmetry of the interaction. For an axially symmetric tensor, the skew in the Herzfeld-Berger convention, κ, equals ± 1 and the asymmetry, η, in the Haeberlen convention equals 0. The effects of these parameters on spectra are described in Section 14.2.1.

14.2.2.3 Quadrupolar interaction

The quadrupolar interaction is only present in the quadrupolar nuclei shown in Figure 14.4. Quadrupolar nuclei have nuclear spins greater than 1 ($I > 1/2$). Physically, these nuclei do not have a spherical distribution of positive charge in their nuclei, which makes them interact with two electric field gradients. This interaction tilts the equilibrium nuclear spin in the static *B*-field slightly away from the static *B*-field direction. The contribution to the Hamiltonian from the quadrupolar interaction is: $H_Q = eQ/(2I(2I-1)\hbar)\,\widehat{I} \cdot V \cdot \widehat{I}$, where $\widehat{I}$ is the spin operator of the nucleus, *eQ* is the quadrupole moment, *V* is the electric field gradient, and *I* is the nuclear spin quantum number. Luckily, the quadrupole moments (eQ) of all nuclei have been experimentally

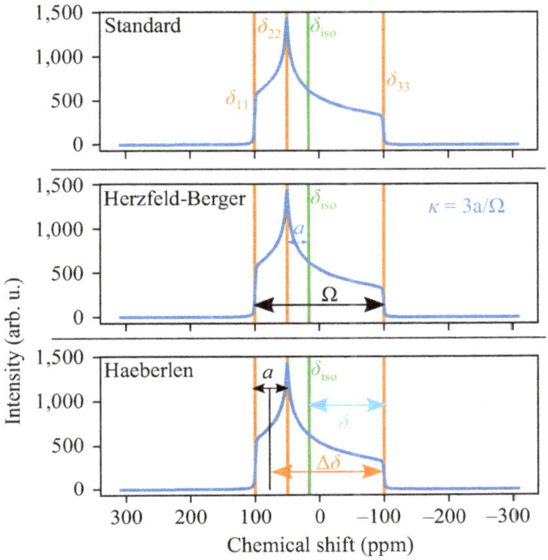

Figure 14.3 The principal components of the chemical shift can be easily read off a static ssNMR spectrum using the standard convention. The other two conventions provide more information about the symmetry of the interaction and can also be derived from the spectrum. NMR spectra are reported such that negative chemical shifts are to the right (the shielded side) and positive chemical shifts are to the left (deshielded side)

measured and tabulated so measurements of the quadrupolar interaction can be used to measure local electric field gradients in materials [10].

Two quantities, including the quadrupolar constant, C_Q, and the quadrupolar asymmetry parameter, η_Q, fully describe the strength of the quadrupolar interaction and can be measured (only the absolute value for C_Q) in an ssNMR experiment. The electric field gradient can be decomposed into three components, V_{zz}, V_{yy}, and V_{xx}, which are ordered from greatest to least. C_Q is proportional to the largest magnitude of the electric field gradient, V_{zz}:

$$C_Q = \frac{eQV_{zz}}{h}. \tag{14.13}$$

Quadrupolar constants are generally not stronger than the Zeeman interaction but can still be quite strong. They are generally less than 30 MHz in magnitude. η_Q describes the relationship between the other components:

$$\eta_Q = \frac{V_{yy} - V_{xx}}{V_{zz}} \tag{14.14}$$

If axial symmetry is present, η_Q is zero.

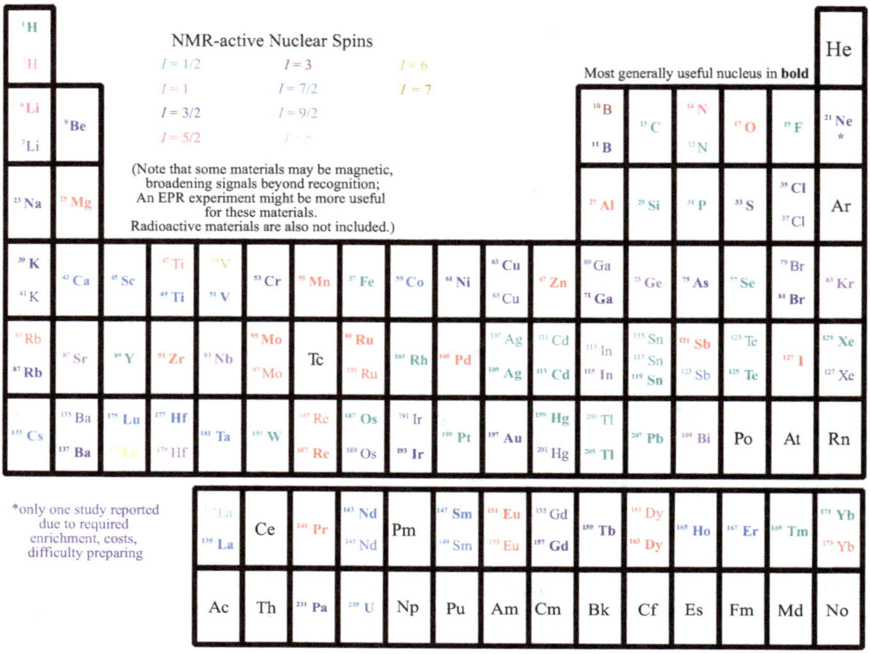

Figure 14.4 The nuclei listed here are NMR-active, with the most generally useful one for a given element written in bold. Please note that not all of these nuclei (and elements) are equally easy to study

One can see why only two quantities are needed despite there being three components to the electric field gradient by looking at the potential energy of a nuclear charge distribution of charge density ρ_c at some electrical potential φ. This is just a Taylor expansion around the nucleus:

$$U = \phi_0 \int \rho_c dx' dy' dz' + \sum_{i=1}^{3} \left(\frac{\partial \phi}{\partial r_i} \right)_0 \int r_i \rho_c dx' dy' dz' + \frac{1}{2} \sum_{i=1}^{3} \sum_{j=1}^{3} \left(\frac{\partial^2 \phi}{\partial r_i \partial r_j} \right)_0 \int r_i r_j \rho_c dx' dy' dz'. \tag{14.15}$$

The first integral that appears is the nuclear charge so the first term in the Taylor expansion can be represented by a constant, U_{pc}, that describes the potential energy of the nuclear point charge. The second term is necessarily zero because charge distributes itself evenly around a nucleus even if it lacks spherical symmetry (the charges do not congregate to one side of the nucleus). There is no electric dipole moment but there is a quadrupolar moment (the third term). A coordinate system can be chosen such that r_i and r_j are along the axes of integration:

$$\begin{aligned} U &= U_{pc} + \frac{\partial^2 \phi}{\partial x_0^2} \int x^2 \rho dxdydz + \frac{\partial^2 \phi}{\partial y_0^2} \int y^2 \rho dxdydz + \frac{\partial^2 \phi}{\partial z_0^2} \int z^2 dxdydz \\ &= U_{pc} + V_{xx} \int x^2 \rho dxdydz + V_{yy} \int y^2 \rho dxdydz + V_{zz} \int z^2 \rho dxdydz. \end{aligned} \tag{14.16}$$

At this point, one must use Gauss' law to obtain another constraint for the problem. Since this component does not include nuclear charge (that was accounted for in U_{pc}), the charge density at the nucleus associated with this component is zero.

$$\nabla \cdot E = \rho/\epsilon = 0 = \nabla^2 \phi = \left(\frac{\partial^2 \phi}{\partial x^2}\right) + \left(\frac{\partial^2 \phi}{\partial y^2}\right) + \left(\frac{\partial^2 \phi}{\partial z^2}\right) = V_{xx} + V_{yy} + V_{zz} \tag{14.17}$$

Equation (14.17) shows that, if two components of the electric field gradient are known, the third is already determined by (14.17). Subtracting (14.17) from (14.16) and defining the quadrupole moment along a given axis as $eQ_z = \rho(3z^2 - r^2)\, dz\, r\, dr\, d\theta$, one gets (14.18):

$$U = U_{pc} + V_{xx} \frac{eQ_x}{3} + V_{xx} \frac{eQ_y}{3} + V_{zz} \frac{eQ_z}{3}. \tag{14.18}$$

What is actually observed in an ssNMR experiment is the spectroscopic quadrupole moment due to quantum mechanics, but this derivation suggests why the quadrupolar moment and electric field gradients are important to the interaction.

14.2.2.4 Other effects

Although the interactions listed in Table 14.1 are the only ones involved in ssNMR, the coordinate systems used to describe the interactions may be different. The constants usually reported are in the principal coordinate system of the interaction (the describing matrix is diagonalized and all elements are along the *x*-, *y*-, and *z*-axes of that coordinate system). Nonetheless, it is common for the principal coordinate system of one interaction to be different from that of another interaction. For this reason, Euler angles (α, β, and γ) are necessary to relate the coordinate system of the other interactions to that of the chemical shift interaction. In addition, to model data, one must introduce certain site-independent convolution parameters that include both Gaussian and Lorentzian broadening.

14.2.3 Magic-angle spinning

In Section 14.2.2, one may have noticed that many of the interactions described have a $(1 - 3\cos^2\theta)$ term. Since θ describes the angle between the static magnetic field and the line connecting two different nuclei, we can make the terms containing this $(1 - 3\cos^2\theta)$ term go to zero by spinning the sample at this angle, called the magic angle. The $(1 - 3\cos^2\theta)$ term equals zero when $\theta = 54.74°$.

Consequently, spinning the sample at this "magic angle" is called magic angle spinning (MAS).

MAS allows one to remove a variety of interactions from the ssNMR signal. Specifically, it removes the effects of chemical shift anisotropy (except at integer multiples of the spinning frequency) and dipolar coupling (provided it is spinning fast enough). It also removes second-order quadrupolar interaction-related line broadening and, at extremely high speeds, can also remove homonuclear dipolar coupling. This is very useful when fitting and modeling spectra because the ability to fit multiple spectra collected from a sample under different conditions with the same parameters increases the likelihood that these parameters are actually representative of the material and not just a local fitting minimum of little physical significance. MAS also allows one to collect the spectra with good signal-to-noise ratios much faster since the signal is localized under the spinning sidebands (and isotropic chemical shift).

Since MAS eliminates the chemical shift anisotropy at all points of the spectrum except at δ_{iso} and integer multiples of the spinning frequency, one can use MAS to readily measure δ_{iso} by collecting spectra at different frequencies. These peaks at integer multiples of the spinning frequency are known as spinning sidebands. A detailed discussion of the effects of MAS on spectra is found in Section 14.4.2 and Figure 14.7.

14.2.4 Pulses and experiment types

The basic design of an ssNMR experiment involves sending in a pulse, acquiring data, and then waiting for the spins to fully relax to their equilibrium position. This final step is called the recycle delay and is crucial for obtaining accurate integrated intensities.

A variety of different pulse sequences can be used for ssNMR. The easiest of these is just the $\pi/2$ direct polarization sequence. This sequence tilts the magnetization away from the equilibrium position by 90° and then allows the spins to precess back to equilibrium. It is not always necessary to tilt spins fully to 90° if the signal is large enough and this may be preferred to speed up measurements or reduce differential relaxation effects.

A variety of pulse sequences involving echoes are also useful for investigating nuclei, particularly quadrupolar nuclei. The Hahn-echo sequence is the same sequence as that described for the T_2 measurement ($\frac{\pi}{2} - \tau - \pi$). This sequence can be used to improve the NMR baseline (less distortion, more ssNMR sensitivity to the signal) and can be very useful for quadrupolar nuclei. The first pulse generates the signal, tilting nuclear spins into the x–y plane, and the second one refocuses it. There are a variety of variants on this sequence that rely on the same concept.

The Carr–Purcell–Meiboom–Gill (QCPMG) sequence further increases the sensitivity of a spectrum to quadrupolar nuclei by concentrating intensity into spikelets that map out the spectrum that would have been made much more slowly using an echo excitation. For this spectrum, a train of pulses is used to concentrate the intensity into the spikelets. The separation of the spikelets is related to the reciprocal of the time separation of the pulses. Like the Hahn Echo sequence and the direct polarization, the first pulse generates the signal, tilting nuclear spins into the *x–y* plane, and the second one refocuses that signal. In the QCPMG sequence, subsequent identical refocusing pulses are then introduced and data are acquired between each of them.

Not all NMR-active nuclei are equally abundant. For this reason, it is sometimes necessary to use a different technique to transfer polarization from abundant nuclei to these rarer nuclei, speeding up the measurement. This is called a cross-polarization experiment and is often done in conjunction with MAS. It requires nuclei that are dipolar coupled to each other. To do this type of experiment, one must adjust the RF fields such that the Larmor frequency of the abundant nucleus matches that of the rare nucleus, allowing both to precess at the same rate and transferring some of the spin polarization.

A wide variety of other types of ssNMR experiments may be performed. Unfortunately, not all of them can be discussed here. A wide variety of 2-D experiments can also be performed that investigate the correlations between signals and, occasionally, atomic separations in molecules. The general sequence in these experiments starts with a preparation period (recycle delay). Then, a pulse perturbs the equilibrium and magnetic moments are allowed to evolve for a time, t_1. There may or may not be a mixing period after this (another pulse). Next, another pulse occurs and the signal is detected for an additional period of time, t_2. The evolution time, t_1, is varied in a set of experiments. A Fourier transform of the signal collected over t_2 at each t_1 yields one axis of the signal. A second Fourier transform of these spectra over t_1 yields the other axis. Some popular 2-D experiments include COSY, NOESY, HETCOR, and EXSY. A broader discussion is beyond the scope of this chapter, but more information on these techniques and their interpretation can be found in a variety of textbooks dedicated to NMR [11,12].

14.3 Sample types and preparation

When designing a new experiment, it is important to know what types of samples are necessary and the basic procedures and instrumentation involved. This section focuses on these practicalities. The basic design of an NMR spectrometer is discussed as are the initial steps in running an experiment. Next, the types of sample holders are discussed.

14.3.1 Spectrometer setup

A schematic of a typical NMR spectrometer configured for solids measurement is shown in Figure 14.5. The superconducting magnet supplies the static magnetic field. It is submerged in liquid helium. To reduce the evaporative loss of the liquid helium, the liquid helium is encased in a liquid nitrogen reservoir. The NMR probe is placed in the bore of the magnet. Typically, the strength of the magnetic field for an NMR spectrometer is listed in terms of the resonance frequency of hydrogen atoms in the spectrometer (in MHz) rather than *T*. Therefore, a 300 MHz

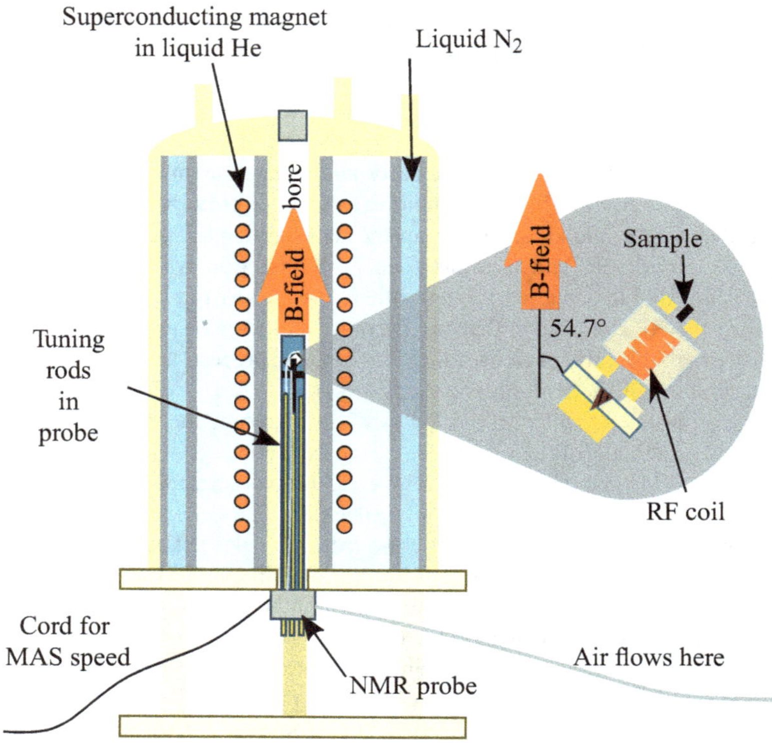

Figure 14.5 Diagram of a typical NMR spectrometer

spectrometer has a magnetic field strength of 7.04 T. The sample is placed in a rotor (described in Section 14.3.2, the next section), which is placed in the spinning module in the probe. Not all probes can do MAS. If the probe is capable of MAS, an air or nitrogen source must be connected to the probe. The probe is then raised into the spectrometer and, if MAS will be performed, the spinning speed is adjusted to the desired speed.

To set up the spectrometer, the probe must be tuned to the correct frequency. As discussed in Section 14.1, the frequency used for the experiment depends on the nucleus of interest. NMR frequency tables for a wide variety of nuclei are readily available. There may be some very slight variations in the tuning between static and spinning samples. Once the samples are loaded and the tuning of the spectrometer verified, pulses must be calibrated to ensure that the pulses sent into the material are performing the expected operations. After performing these calibrations, experiments may begin.

A wide variety of NMR probes are available. They are designed for a given rotor size and set the maximum spinning speed for MAS experiments. Smaller rotors may be spun faster than larger rotors and require less sample. In addition, probes can be designed to have different numbers of resonance configurations. For example, in a single tuning configuration, a single probe could measure hydrogen, tin, and copper spectra without needing to retune. In practice, finding the optimal configuration of tuning rods can be difficult, but the versatility can be useful. Cryogenic probes for solid-state NMR do exist, although temperatures can generally only go down to ~−40 °C at this time. Similarly, high-temperature probes exist, although ones that approach typical synthesis or annealing temperatures of PV materials are difficult to find [13].

14.3.2 Rotor types

To do an NMR experiment, one must place the sample into a rotor and raise this apparatus inside the magnet. The nature of this rotor becomes important for MAS experiments because samples must often be able to spin at rates higher than 1 kHz. Solid-state NMR rotors come in a variety of sizes ranging from ~0.7 mm up to ~7 mm. Smaller rotors can spin faster for MAS experiments but hold less material. This is a trade-off because faster spinning shortens the amount of time necessary to obtain a good signal (and increases the separation of spinning sidebands) whereas less material decreases the signal itself. The rotor diameter is set by the size of the RF solenoid that surrounds the rotor and collects the signal; one cannot, with a given spectrometer probe, arbitrarily change the rotor size. A diagram of a typical rotor design is shown in Figure 14.6. Rotors can be made from a wide variety of materials. For many experiments, ~200 mg of material is sufficient. In many cases, less is required. Samples that interact with the magnet will be difficult to use for MAS measurements, and paramagnetic samples can exhibit large shifts related to the Knight shift.

14.4 Understanding and analyzing NMR data

The physics of ssNMR may be relatively straightforward but the analysis of ssNMR data often is not. For this reason, this section is a basic primer that should help the reader understand how interaction parameters are extracted, the limitations and challenges of ssNMR data analysis, and the types of questions that ssNMR is capable of addressing.

A variety of software programs may be used to analyze NMR data. Dmfit is useful for modeling both solid and MAS spectra. It can be downloaded from the CEMHTI website (http://nmr.cemhti.cnrs-orleans.fr) for free and works on Windows, MacOS (with WINE), and Linux (with WINE) [7]. WSolids1 is another solid-state NMR analysis software. It has many of the same capabilities as Dmfit but does not model quadrupolar MAS spectra as well. It is, however, very user-friendly and helpful for intuitively understanding NMR spectra (and the sensitivity of the model). WSolids1 was used to generate the spectra in this chapter [9]. Bruker's TopSpin (Bruker Corporation) software package is a commercial software that is useful for data analysis and acquisition. Bruker

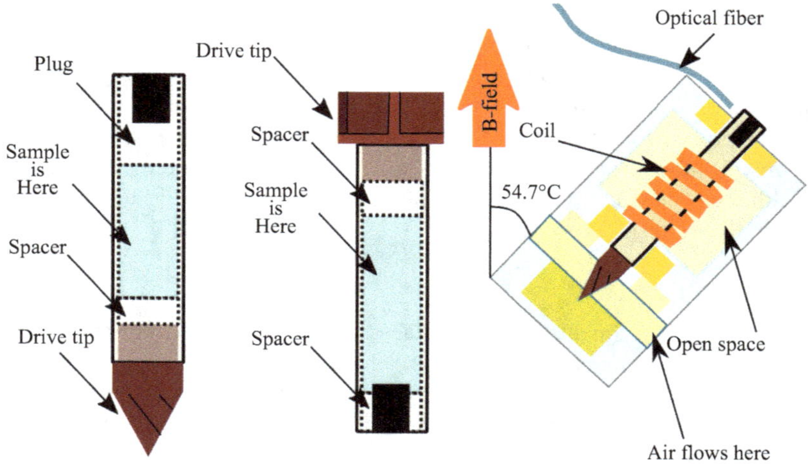

Figure 14.6 NMR rotors can come in a variety of sizes. For MAS experiments, rotors will either have fins on the tip or tail end. When compressed air or nitrogen gas is directed through the rotor chamber and past the fins, the rotor will spin. The spinning rate can be detected by changes in light reflectance associated with the black mark located at the top or bottom of the rotor. The spinning rate is measured by counting the number of light to dark (or dark to light) transitions. The image to the right shows how the rotor fits in the NMR probe coil

spectrometers use this software for data collection. Recently, Bruker began allowing academic users to use the software for free.

In Section 14.1.1, free induction decay was introduced. The raw data in a fid resemble a decaying sinusoid or series of sinusoids and are described by both real and imaginary components. To analyze these data in a more meaningful way, one performs a Fourier transform on it. Not all parts of the fid are equally important. Since the intensity of the packet decays to zero, the first portion of this packet is most important (and less sensitive to noise). For this reason, it is customary to multiply the spectrum by a weighting function, often $W(t) = \exp(-Rt)$, where R is a value that the user chooses. This improves the signal-to-noise ratio (SNR) by suppressing noise in the less important portion of the fid signal and slightly broadening the lineshape. Adding zeros to the end of the fid and then applying the Fourier transform can also make peaks better defined in the Fourier transformed spectrum.

Unfortunately, the signals in the fid are not always in phase with the instrument receiver. To correct for this, one must "phase" the spectrum by calculating the phase angles needed to shift the measured $t = 0$ to where $t = 0$ should actually be without any delays (see (14.6)). Generally, this means adjusting the phase so the real portion of the Fourier-transformed signal is entirely positive.

Assuming that the full signal can be collected, the integrated intensity of an NMR lineshape is proportional to the fraction of that type of nuclei contributing to it. This means that if the signal can be decomposed into two nuclear sites, one can determine the fraction of nuclei on each site. This is only valid when the nuclear spins are allowed to completely relax back to their equilibrium positions (there is a long-enough recycle delay).

Although the physics of ssNMR is well defined, it is important to understand how this manifests in actual data. The following sections demonstrate how the interaction parameters described in Section 14.1 manifest in ssNMR data.

14.4.1 Spin-1/2 nuclei

For spin-$\frac{1}{2}$ nuclei, one is often primarily concerned with the chemical shift interaction. One can relatively easily read off the three values, δ_{11}, δ_{22}, and δ_{33}, from a static ssNMR scan as shown in Figure 14.3. However, it generally takes significantly more time to collect static compared to MAS data. Overlapping sites can make interpretation less straightforward, for which MAS measurements are useful for clarification. Consequently, this section will focus on understanding MAS data, as shown in Figure 14.7. δ_{iso} can easily be identified by overlaying spectra collected at different spinning frequencies. δ_{iso} is the peak that does not move as the spinning frequency is changed and may not be the most intense peak in the spectrum. Figure 14.7 also shows how δ_{aniso} and η_{delta} affect a MAS spectrum. δ_{aniso} affects the extent of the spinning sidebands whereas η_{delta} affects the weighting of them. If the sample is spinning sufficiently fast, no spinning sidebands may be visible, making it impossible to measure η_{delta} and δ_{aniso} from the MAS spectrum. Often, δ_{iso} is the more important quantity.

14.4.2 Quadrupolar nuclei

The interpretation of quadrupolar spectra is more difficult compared to spin-$\frac{1}{2}$ nuclei. In addition to the interactions involved in spin-$\frac{1}{2}$ nuclei, the quadrupolar interaction introduces two additional parameters (C_Q and η_{Q}), and three Euler angles. Generally, the ssNMR spectra for quadrupolar nuclei are represented on a MHz scale rather than the chemical shift scale because the quadrupolar interaction often dominates spectra. The individual effects on spectra of changing the quadrupolar constant and η_{Q} are shown in Figure 14.8. Nuclei in more symmetric sites have smaller values of C_Q and more distorted sites have larger values of C_Q.

As Section 14.3.1 discussed, the coordinate systems of the chemical shift interaction and the quadrupolar interaction are not the same. The effects of changing the Euler angles are shown in Figure 14.9. Magic-angle spinning also changes the spectra of quadrupolar nuclei, introducing spinning sidebands and removing the first-order quadrupolar interaction. However, the second-order quadrupolar interaction cannot be removed by spinning at a single angle. The effects of MAS are shown in Figure 14.10.

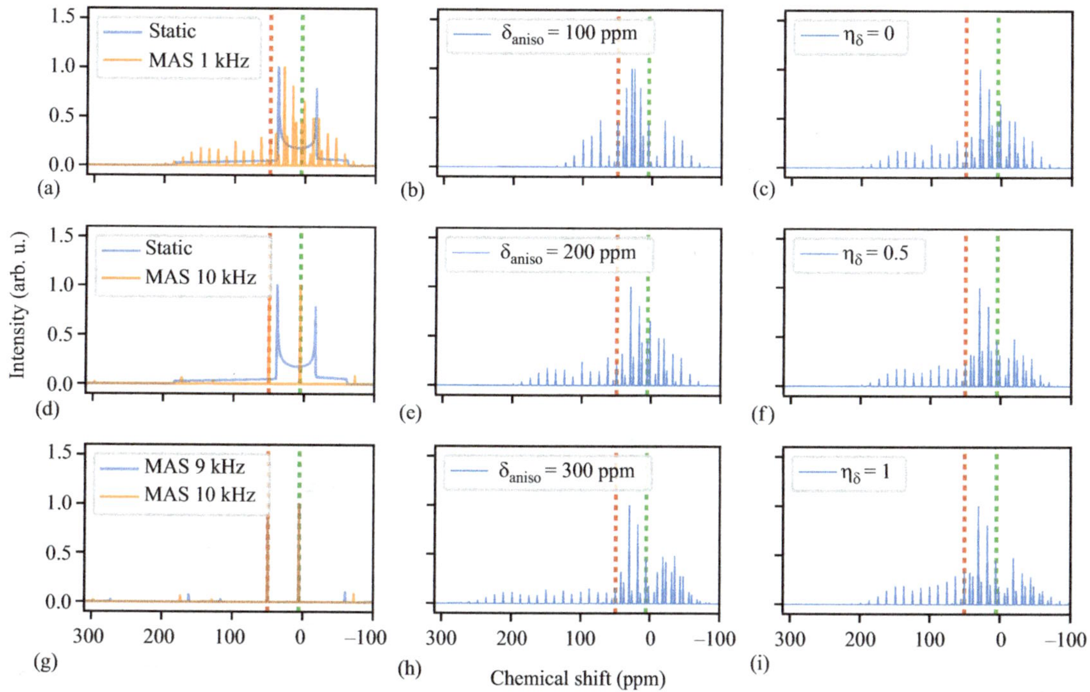

Figure 14.7 *(a) and (d) show a static scan with two overlapping sites. One site has $\delta_{iso} = 50$, $\delta_{aniso} = 200$, and $\eta_{delta} = 0$, and the other site has $\delta_{iso} = 5$, $\delta_{aniso} = 100$, and $\eta_{delta} = 0$. These were held constant in subsequent plots unless explicitly denoted. δ_{iso} for both sites is marked with a dotted line in each plot. The static scan is overlaid by a 1 kHz MAS scan (a) and a 10 kHz MAS scan (d). The spinning sidebands do not exactly trace the static scan and can extend slightly beyond it. δ_{iso} can be determined by spinning the sample at two frequencies and noting the peaks that do not move with spinning frequency (g). As seen in (a), δ_{iso} may not be the most intense peak. (b), (e), and (h) show how changing δ_{aniso} affects spectra. (c), (f), and (i) show how changing η_δ affects MAS spectra*

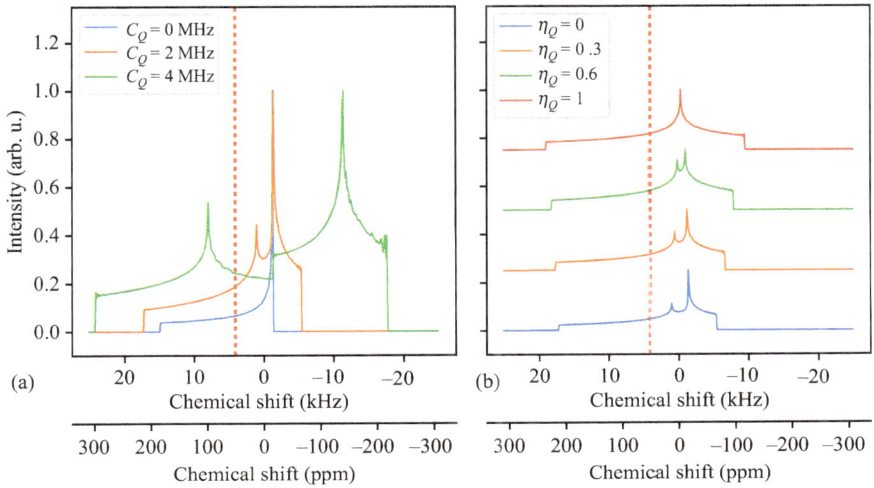

Figure 14.8 The effects of the quadrupolar interaction on ssNMR spectra. For all of these spectra, $\delta_{iso} = 50$ *ppm,* $\delta_{aniso} = 200$ *ppm, and* $\eta_\delta = 0$*. (a) The effects of changing the quadrupolar constant,* C_Q*, with* $\eta_Q = 0$*. (b) The effects of changing* η_Q *with* $C_Q = 0$

14.5 Limitations of ssNMR

SsNMR does not work well for some nuclei due to their nature. For example, from (14.2), the magnetization of a nucleus with a small γ will be small. Consequently, detecting a fid signal from these nuclei is difficult. Special low-γ probes are necessary (and scans generally take more time to get sufficient signal). Similarly, nuclei of low abundance will yield only small signals. For those types of experiment, enrichment is often required. In this section, we identify the nuclei that can be studied using ssNMR and some practical limitations to their study.

14.5.1 What nuclei are useful for ssNMR?

NMR can only probe nuclei with nuclear spins (*I*) greater than zero. This does not allow all nuclei to be studied. If the mass number is even and the atomic number is even, the nucleus is inactive and NMR cannot be used to study it. The classic examples are ^{12}C and ^{16}O. Fortunately, this means that all nuclei with odd mass numbers and all nuclei with even mass numbers and odd atomic numbers can be studied (theoretically). A summary of the useful elements that can be studied using ssNMR and the challenges one tends to face when studying various elements is shown in Figure 14.11.

Despite being NMR-active, some nuclei are not very prevalent. Only a small fraction of a sample contributes to the NMR signal so, for less abundant NMR-active nuclei, NMR can be infeasible without expensive enrichment procedures. To

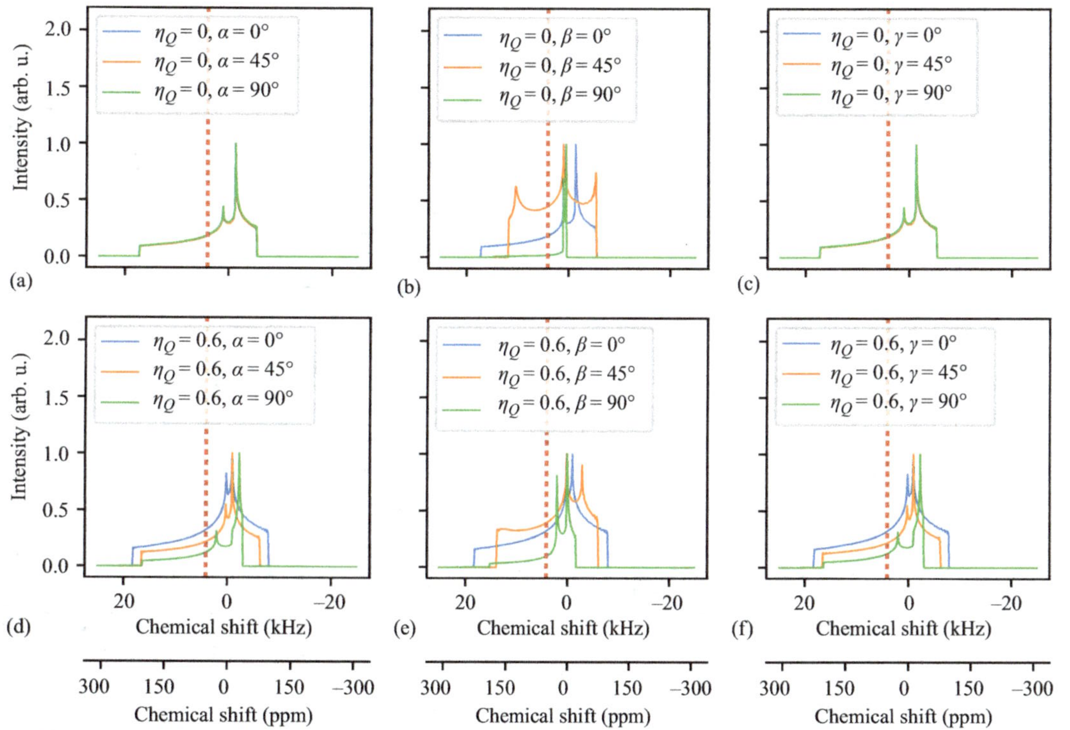

Figure 14.9 The Euler angles can significantly alter quadrupolar spectra if there is a difference between the principal coordinate system of chemical shift and the quadrupolar interaction

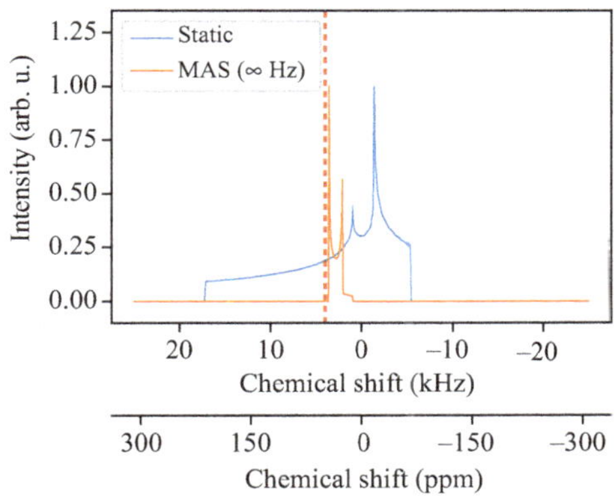

Figure 14.10 Magic-angle spinning results in a narrowing of the spectrum. Again, δ_{iso} is denoted by the dotted line. Spinning at infinite frequency cannot remove the second-order (and fourth) quadrupolar interaction, so the spectrum will always have some width. At lower spinning frequencies, spinning sidebands develop. Although higher-field spectrometers can improve resolution, they can make MAS data more difficult to interpret because sidebands are closer together and, for a given CQ, are closer together. For this reason, it can be useful to measure MAS spectra at a lower B-field if the excitation bandwidth is large enough to excite the entirety of a given site

get a sense of the proportion of ^{1}H nuclei that contribute to a signal, the calculation in (14.19) is informative where $N_{\text{upper}}/N_{\text{lower}}$ is the ratio of spins in the upper versus the lower nuclear spin states from Figure 14.1. :

$$N_{\text{upper}}/N_{\text{lower}} = \exp\left(\frac{-\delta \text{E}}{\text{kT}}\right) = \exp\left(\frac{-h\nu}{kT}\right)$$
$$\text{for protons at 18.8 T field, } \nu = 800 \text{ MHz, T} = 300 \text{ K} \tag{14.19}$$
$$N_{\text{upper}}/N_{\text{lower}} = 0.999871.$$

Consequently, for 1 million nuclei in the lower state, there would be 999,872 nuclei in the upper state, leaving a difference in occupancy between the two states of only 128 nuclei. Only a small portion of nuclei contribute to an ssNMR signal even for abundant nuclei. Nonetheless, spectra can be collected and, in the case of spin 1/2 nuclei like ^{13}C, this lack of abundance can be overcome using techniques like MAS to enhance the signal (and remove dipolar coupling terms). Some classic but commonly studied examples of comparatively low-abundance nuclei are ^{13}C, ^{15}N and ^{29}Si, ^{81}Br.

Unfortunately, low abundance is not the only limitation regarding which nuclei can be studied. In addition, some materials may have a small gyromagnetic ratio, γ (this determines the precession frequency), that leads to low sensitivity ($|\gamma_3| \cdot I\ (I+1)$). To study these nuclei, specialized "low-γ probes" must be used.

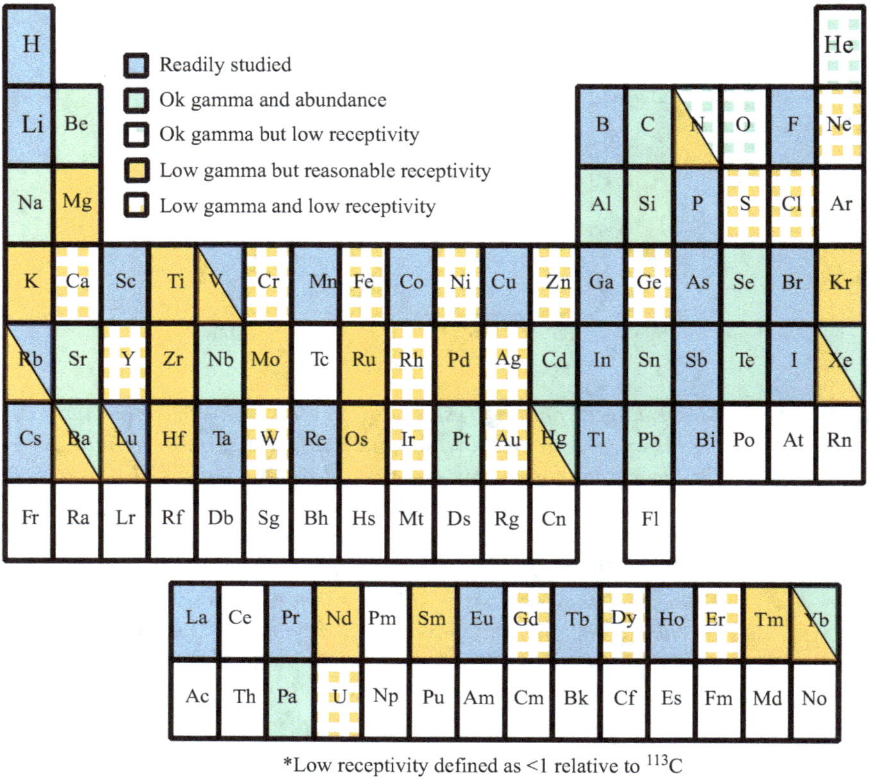

Figure 14.11 *Not all nuclei are readily studied using solid-state NMR. Both the gyromagnetic ratio, γ, and the receptivity (14.20) matter. Receptivity is related to both abundance and the gyromagnetic ratio*

To quantify how well ssNMR can be used on a given nucleus, a quantity called the receptivity can be defined. It is generally represented relative to another nucleus, commonly ^{1}H or ^{13}C, so R^Y are the actual values reported. R^Y is defined by (14.20):

$$R^Y = [\text{Natural abundance}(\%)] \cdot |\gamma^3| \cdot I(I+1). \tag{14.20}$$

The eventual SNR is related to the number of nuclear spins being observed, n, the magnetic field, B_0, the gyromagnetic ratio of the spin being excited, γ_e, the gyromagnetic ratio being detected, γ_d, and the measurement time, t. As discussed in Section 14.5, γ_e may not equal γ_d if one is doing a cross-polarization experiment:

$$S/N \propto n\gamma_e \sqrt{\gamma_d^3 B_0^3 t}. \tag{14.21}$$

Unpaired electronic spins can also broaden NMR signals, resulting in difficulty detecting the signal. For this reason (and difficulties doing magic-angle spinning measurements), it is difficult to do ssNMR on magnetic samples. Paramagnetic inclusions, similarly, can interfere with MAS measurements.

14.5.2 Other considerations

In addition to the previously identified limitations, broad signals can make ssNMR spectra difficult to obtain and interpret. This can happen in quadrupolar nuclei if the electric field gradients are too large, particularly when compared to the portion of the spectrum excited. The portion of the spectrum excited by a pulse is called the excitation bandwidth. Although one can collect several spectra centered around different frequencies and splice them together, this process can be time-consuming and introduce artifacts into the spliced data that make the spectra difficult to model and interpret. Alternate pulse shapes like the WURST pulse may be used to excite a larger bandwidth [14]. Whole-integer nuclei tend to give broader spectra compared to half integer nuclei [15]. In addition, a few nuclei like ^{113}In and ^{115}In or ^{47}Ti and ^{49}Ti have very similar resonance frequencies, so both (^{113}In and ^{115}In or ^{47}Ti and ^{49}Ti) nuclei can be observed simultaneously in a single spectrum. When you are considering a given nucleus to study, the website of the Hebrew University Institute of Chemistry NMR Lab is very helpful for determining the practicality of that effort (http://chem.ch.huji.ac.il/nmr/index.html) [15].

14.6 Applications: recent ssNMR characterization of semiconductors relevant to PV

SsNMR has recently been used to investigate a variety of materials in the Cu–Zn–Sn–S system, CdTe, polymers, and lead halide perovskites. This section focuses on these applications.

When considering using a characterization technique, it is important to know what questions the technique is capable of answering. Unfortunately, for ssNMR, some of this depends on the nuclei being studied, as discussed in Section 14.5. The questions that NMR is most useful for addressing involve structure and symmetry, especially when X-ray diffraction results are unclear. It can also detect some types of defect complexes if their concentration is high enough and, thereby, allow the study of defects, defect complexes, and disorder. In some cases, it can be used to monitor diffusion and transport.

14.6.1 Cu–Zn–Sn–S system

Solid-state NMR has recently been used to characterize Cu_2ZnSnS_4 and other materials in the Cu–Zn–Sn–S system in a variety of ways. As a quaternary alloy, Cu_2ZnSnS_4 generally deviates from perfect 2–1–1–4 stoichiometry. These deviations must be accommodated by point defects that may impact device performance. Unfortunately, the study of these point defects can be difficult using other techniques. Most luminescence-based techniques are not chemically specific. Solid-state NMR, however, allows chemically specific investigations of local order and symmetry. When the composition of a sample is known, it can be used to identify point defects. In 2013, Choubrac *et al.* carried out the first solid-state NMR investigations on Cu_2ZnSnS_4 [16]. In this investigation, ^{67}Zn, ^{119}Sn, and ^{65}Cu spectra were collected and compared to what would be expected for the kesterite structure based on X-ray diffraction. They performed calculations to approximate the expected C_Q values and compared them to values that they measured from their spectra. They also studied quenched samples, which would be expected to have more disorder than slowly cooled samples. A distinct broadening of spectra was observed in the quenched samples when compared to slowly cooled samples. Furthermore, since two Cu sites with equal integrated areas were observed, it was possible to confirm that Cu_2ZnSnS_4 adopts the kesterite structure (when slow-cooled). In quenched samples, the NMR lineshapes broadened, indicating that the Cu and Zn sites disordered, retaining a kesterite ordering as it does so and not reverting to stannite ordering.

Expanding on this study, in 2014, Paris *et al.* (the same research group) investigated Cu/Zn disorder and other defect complexes in Zn-rich Cu_2ZnSnS_4 [17]. Additional NMR small peaks were present in spectra that deviated from stoichiometry (three different deviations and a stoichiometric sample). By looking at spectra from samples with a variety of types and degrees of stoichiometry deviation, trends could be observed. Contributions from $[V_{Cu} + Zn_{Cu}]$ and $[2Zn_{Cu} + Zn_{Sn}]$ defect complexes were identified in the spectra at specific chemical shifts based on the appearance of additional peaks in the NMR spectra in the non-stoichiometric samples. The relative intensities of these peaks were used to suggest that $[V_{Cu} + Zn_{Cu}]$ complexes tend to cluster in well-defined ways. ZnS inclusions could also be identified. The second portion of this study was devoted to comparing quenched, disordered Cu_2ZnSnS_4 samples to slow-cooled samples, allowing one to investigate these defect complexes at different temperatures. By beginning with well-characterized samples like this, one might be able to more definitively investigate defects disorder using optical techniques. For example, Paris *et al.* used these samples to investigate how these defects impact Raman spectra, a technique that is much more amenable to quick, inexpensive, and easy thin-film characterization.

We (the author and collaborators) investigated the effect of oxygen on Cu_2ZnSnS_4 [18]. We found that oxygen, though it has negligible solubility in Cu_2ZnSnS_4, removes excess Sn from Cu_2ZnSnS_4, reducing disorder in Cu_2ZnSnS_4. Several groups had employed air-annealing steps to their Cu_2ZnSnS_4 solar cell synthesis processes and observed improved efficiencies and/or photoluminescence intensities. Since solid-state NMR is sensitive to local ordering (and, thereby, disorder), it was useful for demonstrating the effect of oxygen on disorder and attributing it to Sn loss as SnO_2. Room-temperature photoluminescence was then used to connect our structural observations to optical properties of the materials, connecting a shift and significant growth in the PL peak intensity to Sn loss (as SnO2).

Cu_2ZnSnS_4 is not the only phase in the Cu–Zn–Sn–S system of interest to the solar community; Cu_2SnS_3 is also potentially useful [19]. We (the author and collaborators) used solid-state NMR to investigate the controversy over whether

Cu_2SnS_3 is tetragonal or monoclinic. Even though the diffraction patterns of monoclinic and tetragonal Cu_2SnS_3 are very similar, ssNMR clearly distinguishes these phases. By investigating materials that deviated from stoichiometry in predictable ways (along the $Cu_4Sn_7S_{16}$–Cu_2SnS_3 line or along the Cu_2SnS_3–CuS line), we were able to show that the monoclinic phase forms only under Sn-rich conditions whereas the tetragonal phase, exhibiting much more disorder, forms under Cu-rich conditions. This composition difference had not been previously reported. We then put our Cu_2SnS_3 and Cu_2ZnSnS_4 results in the context of the broader Cu–Zn–Sn–S phase space [20]. We were able to show that both Cu_2ZnSnS_4 and Cu_2SnS_3 begin to disorder between 200 °C and 400 °C and produced phase diagrams at room-temperature, 200 °C, and 400 °C for the system.

14.6.2 CdTe

CdTe has been investigated using ssNMR since at least 1962 [21]. The most relevant isotopes for CdTe are ^{113}Cd and ^{125}Te. ^{111}Cd has similar abundance to ^{113}Cd but is slightly less sensitive. ^{123}Te is less sensitive and has a significantly lower abundance compared to ^{125}Te [22]. All four are spin-$\frac{1}{2}$ nuclei and spectra are dominated by the chemical shift interaction. Signals tend to be relatively narrow. In early ssNMR studies, instead of performing Fourier transforms of a fid signal, the strength of the static magnetic field itself was varied at a constant excitation frequency. Data were reported in terms of the magnetic field of resonance for the compound of interest and a standard reference compound. Consequently, it is necessary to convert to the newer standard to make use of these prior data. The initial ssNMR studies focused on the shielding of bulk CdTe [21,22]. Look and Moore measured the effective g factor of conduction electrons in degenerate CdTe, doping samples with In donors and Cl acceptors [23]. In the 1990s, researchers began to focus on defects in CdTe, primarily in n-doped material [24]. The focus on n-doped material occurred because n-doped materials have shorter T_1 relaxation constants (making the measurements take less time–still meaning many hours to days). Shroyer investigated the effects of doping CdTe with In and Ga. Ga-doping of CdTe is relatively ineffective, with 10^{19} atoms/cm^3 of Ga yielding a carrier concentration of only 10^{14} electrons/cm^3. Indium is more effective, yielding carrier concentrations of 3.5×10^{17} atoms/cm^{3}at the same 10^{19} atoms/cm^3 of In added [25]. By illuminating the samples while measuring ssNMR spectra, Shroyer was able to investigate the bistable defects that would explain the ineffective doping. For such materials with D-X bistable defects, carriers excited by the illumination do not recombine at low temperatures once the light is removed. D-X centers would distort the local bonding and localize charge, so spectra where carriers are trapped in the D-X center would look different from spectra for which carriers are free. These D-X centers were apparent in ssNMR spectra. A Knight shift caused by free carriers shifted the spectra relative to that of an undoped sample. Also, the T1 relaxation constant changed upon illumination at different temperatures as one would expect from bistable paramagnetic centers.

Recently, we investigated phosphorus and arsenic-doping of bulk CdTe [26]. Both are relatively ineffective dopants, but their ^{125}Te and ^{111}Cd spectra are different. Initially, the main peak of both doped samples was shifted by a similar amount relative to the undoped sample. The P-doped sample did not change over time but the As-doped sample did. For As-doped CdTe, an additional lattice site was observed in both ^{125}Te and ^{111}Cd spectra. Over time, the deshielded site became more shielded, suggesting that the number of free carriers decreased, and the more-shielded site grew in relative intensity and became comparatively less shielded. This suggests that As-doping is less stable in CdTe than P-doping and suggests that the limitation on As-doping is different from that for P-doping and the type of compensation in the As-doped sample introduces a new chemically distinct As site.

14.6.3 Polymer

There is a long history of using ssNMR to study polymeric and organic materials. For this reason, it was only natural to apply this technique to polymer solar cell materials. NMR (both solution and solid-state) is often used to confirm the structure of the polymers. Jacobsen *et al.* provides some guidance for interpreting this type of data [11]. SsNMR can also be used to confirm polymer morphology and investigate the miscibility of polymer blends [27]. ^{1}H and ^{13}C ssNMR are understandably most common for these types of materials. Both are spin-1/2 nuclei so chemical shift, dipolar coupling, and J-coupling tend to be important and there is no quadrupolar interaction. Mens *et al.* used differences in the T_1 decay times to compare the crystallinity of PCBM thin films to that of bulk materials and compare MDMO-PV/PCBM blended thin films to pure PCBM thin films [27]. They were also able to compare various processing and casting techniques in this fashion, estimating the minimum crystallite size in these materials. This study was the first quantitative estimation of the amount of phase separation in these polymeric solar materials.

14.6.4 Halide perovskites

SsNMR has been used to study halide perovskites relatively extensively. Researchers have used the technique to investigate the dynamics of cations in halide perovskite structures. They have also used ssNMR to investigate whether ions introduced to precursor solutions are incorporated into the structure, form secondary phases, or segregate [28]. SsNMR is useful for detecting randomly dispersed ion clusters, which can be difficult to detect using X-ray diffraction. In such cases, the local structure is not well-represented by the global average. It has also been used for understanding the diffusion of ions through the structure, which was particularly useful for understanding the observed *I–V* curve hysteresis and issues with chemical stability. Franssen *et al.* wrote an excellent review on the use of ssNMR to study halide perovskite materials [28].

A variety of groups have used ssNMR to investigate atom substitutions and solid solutions. The studies discussed here were chosen to highlight the range of capabilities of ssNMR. Rosales *et al.* used ssNMR to show that, in mixed halide

$CH_3NH_3PbX_{1.5}Y_{1.5}$ with X, Y = Cl, Br, or I, mixed halide compositions occur only in I/Br and Cl/Br mixtures and phase separate in I/Cl mixtures [29]. Roiland *et al.* evaluated the applicability of ^{207}Pb, ^{79}Br, ^{14}N, ^{13}C, and ^{2}H ssNMR to study $CH_3NH_3PbX_3$ where X = Cl, Br, or I and found that the ^{207}Pb chemical shift is very sensitive to the halogen atoms [30]. Kubicki *et al.* was the first to look at mixed A-atom (AA'BX_3) hybrid perovskites when they investigated $(CH_3)_x(CH(NH_2)_2)_{1-x}NH_3PbI_3$, showing that the formamidinium $((CH(NH_2)_2)^+)$ ions incorporated into the structure [31]. Van Gompel *et al.* then showed that formamidinium ions in $(CH(NH_2)_2)I$ can be distinguished from those in $(CH(NH_2)_2)NH_3PbI_3$. Similarly, ^{133}Cs and ^{87}Rb NMR have been used to probe Cs and Rb incorporation in a variety of halide perovskites [32]. ^{133}Cs ssNMR was also used to investigate copper-doping of $Cs_2SbAgCl_6$, europium doping of $CsPbI_2Br$, and manganese and cobalt doping of $CsPbBr_3$ and $CsPbCl_3$, all inorganic halide perovskites or double perovskite [33–35]. ^{121}Sb ssNMR was used to show that Cu can dope $Cs_2SbAgCl_6$ [34]. The local orientation dynamics of the organic cation can also be studied [28].

SsNMR has also been used to better understand the chemical stability of halide perovskite materials. This stability issue has been a continued issue for hybrid perovskite materials, necessitating encapsulation and further efforts to make them more robust [36,37]. Askar *et al.* observed $CH_3NH_3PbI_3$ ^{207}Pb ssNMR spectra that were collected at different times over a 22-day period. The samples were kept at room-temperature and 80% relative humidity. They were able to distinguish $CH_3NH_3PbI_3$ from its monohydrate and dihydrate forms and observe the transition to the monohydrate form. No decomposition to PbI_2 or dihydrate form was observed for the sample kept in a humid environment. However, when water was directly added to $CH_3NH_3PbI_3$, the sample would further transform to the dihydrate form and eventually decompose to PbI_2 and $CH_3NH_3PbI_3$ upon drying. The decomposition to PbI_2 was not seen when the sample was just kept in the humid environment (it dehydrated to $CH_3NH_3PbI_3$). At a lower relative humidity level of 45% and a temperature of 85 °C, decomposition to PbI_2 was observed with no hydrate formation.

I–V curve hysteresis is attributed to ion buildup at the electrodes, so the diffusion of species within the perovskite is also important [28]. Askar *et al.* and Karmakar *et al.* observed halide diffusion in $CH_3NH_3Pb(Cl/Br)_3$ and $(CH(NH_2)_2)Pb(Cl/Br)_3$ [38,39]. SsNMR cannot probe diffusion processes with timescales longer than the T_1 relaxation time. Nonetheless, if there is exchange between sites with different NMR signals, longer timescales up to milliseconds and seconds can often be probed [28].

14.7 Recent developments in solid-state NMR that may be relevant to PV in the future

New techniques and methods are expanding the capabilities of ssNMR, expanding the types of materials suitable for study and questions that ssNMR is capable of addressing. Many of these are related to improved sensitivities, the proliferation and greater accessibility of higher-field spectrometers, and advances in NMR crystallography. Of particular interest is an increased focus of the International Union of Crystallography and IUPAC on compiling and sharing NMR data among researchers. In 2014, the International Union of Crystallography created a Commission of NMR Crystallography to help standardize this process [5]. It is anticipated that the standardization of NMR data exchange and creation of databases containing NMR data will do for ssNMR what PDF and CIF reference databases did for X-ray diffraction.

Despite the limitations discussed in Section 14.5, a variant of NMR and atomic force microscopy called magnetic resonance force microscopy (MRFM) was recently (2013) developed for thin-film use [40]. Unlike the techniques mentioned previously, this is an imaging technique and is much more similar to medical MRIs than the methods mentioned previously. This exciting technique may become more prevalent in the future and should be surface sensitive.

14.8 Conclusions

Solid-state NMR is a useful tool for studying the structures of materials. It is capable of detecting defect complexes in a chemically specific manner and is highly sensitive to local order and symmetry. This local sensitivity makes it especially useful for understanding how these complexes manifest. In some cases, it can be used to study diffusion through materials and the sources of metastabilities. Local defects and disorder often inhibit solar cell device performance, so ssNMR is a useful characterization technique for many PV materials. Although the required amount of sample can make it difficult (but not impossible) to use on thin films, ssNMR is an excellent technique for probing the thermodynamics of defects and complexes in representative bulk materials, the results of which can be generalized to thin-film materials. ssNMR has been successfully used to understand CdTe, Cu_2ZnSnS_4, hybrid perovskite, and polymer-based solar materials with great success. Recent advances in ssNMR will only make the technique more useful as spectrometers improve and become more widely available.

References

[1] T. C. Farrar and E. D. Becker, *Pulse and Fourier Transform NMR Introduction to Theory and Methods.* New York, New York, United States: Academic Press, 1971.

[2] M. J. Duer, *Introduction to Solid-State NMR Spectroscopy.* Oxford, UK: Blackwell Publishing Ltd, 2004.

[3] R. K. Harris, E. D. Becker, S. M. Cabral De Menezes, R. Goodfellow, and P. Granger, "NMR Nomenclature. Nuclear Spin Properties and Conventions for Chemical Shifts (IUPAC Recommendations 2001)," *Pure and Applied Chemistry*, vol. 73, pp. 1795–1818, 2001.

[4] R. K. Harris, E. D. Becker, S. M. Cabral De Menezes, P. Granger, R. E. Hoffman, and K. W. Zilm, "Further Conventions for NMR Shielding and Chemical Shifts (IUPAC Recommendations 2008)," *Pure and Applied Chemistry*, vol. 80, pp. 59–84, 2008.

[5] M. Dacombe, "International Union of Crystallography Report of the Executive Committee for 2014," *Acta Crystallographica Section A*, vol. 72, pp. 412–455, 2016.

[6] R. K. Harris, R. E. Wasylishen, and M. J. Duer, editors, *NMR Crystallography*. West Sussex, UK: John Wiley & Sons, Ltd, 1 ed., 2009.

[7] D. Massiot, F. Fayon, M. Capron, *et al.*, "Modelling One- and Two-Dimensional Solid-State NMR Spectra," *Magnetic Resonance in Chemistry*, vol. 40, pp. 70–76, 2002.

[8] H. Carr and E. Purcell, "Effects of Diffusion on Free Precession in Nuclear Magnetic Resonance Experiments," *Physical Review*, vol. 94, p. 630, 1954.

[9] K. Eichele, "WSolids1: Solid-state NMR Spectrum Simulation Version 1.21.3," Universität Tübingen, 2015.

[10] G. Fuller, "Nuclear Spins and Moments," *Journal of Physical and Chemical Reference Data*, vol. 5, pp. 835–1092, 1976.

[11] N. E. Jacobsen, *NMR Data Interpretation Explained: Understanding 1D and 2D NMR Spectra of Organic Compounds and Natural Products*. Hoboken, NJ: John Wiley & Sons, Inc., 2017.

[12] J. H. Simpson, *Organic Structure Determination Using 2-D NMR Spectroscopy: A Problem-Based Approach*. Boston, MA: Academic Press, second edition, 2012.

[13] J. F. Stebbins, "NMR at High Temperature," in *Modern Methods in Solid-State NMR: A Practitioner's Guide* (P. Hodgkinson, ed.), Chapter 9, pp. 262–288, Royal Society of Chemistry, 2018.

[14] L. A. O'Dell, "The WURST Kind of Pulses in Solid-State NMR," *Solid State Nuclear Magnetic Resonance*, vol. 55–56, pp. 28–41, 2013.

[15] O. Gidron, Y. Ozery, and R. Hoffman, "Hebrew University NMR Lab."

[16] L. Choubrac, M. Paris, A. Lafond, C. Guillot-Deudon, X. Rocquefelte, and S. Jobic, "Multinuclear (^{67}Zn, ^{119}Sn and ^{65}Cu) NMR Spectroscopy–An Ideal Technique to Probe the Cationic Ordering in Cu_2ZnSnS_4 Photovoltaic Materials.," *Physical Chemistry Chemical Physics*, vol. 15, pp. 10722–10725, 2013.

[17] M. Paris, L. Choubrac, A. Lafond, C. Guillot-Deudon, and S. Jobic, "Solid-state NMR and Raman Spectroscopy to Address the Local Structure of Defects and the Tricky Issue of the Cu/Zn Disorder in Cu-Poor, Zn-Rich CZTS Materials," *Inorganic Chemistry*, vol. 53, pp. 8646–8653, 2014.

[18] E. A. Pogue, A. Sutrisno, N. E. Johnson, D. P. Shoemaker, and A. A. Rockett, "Oxygen-Induced Ordering in Bulk Polycrystalline Cu_2ZnSnS_4 by Sn Removal," *Inorganic Chemistry*, vol. 56, pp. 12328–12336, 2017.

[19] E. Pogue, M. Paris, A. Sutrisno, *et al.*, "Identifying Short-Range Disorder in Crystalline Bulk Cu_2SnS_3 Phases: A Solid-State NMR Investigation," *Chemistry of Materials*, vol. 30, pp. 6624–6635, 2018.

[20] E. A. Pogue, A. Sutrisno, N. E. Johnson, *et al.*, "Phase Stability and Structural Comparison of Phases in the Cu-Zn-Sn-S System Using Solid-State NMR," *Solar Energy Materials and Solar Cells*, vol. 190, pp. 37–48, 2019.

[21] I. Weinberg, "Shifts of the Nuclear Spin Resonance in Cadmium Telluride," *The Journal of Chemical Physics*, vol. 37, pp. 1571–1572, 1962.

[22] A. Nolle, "Direct and Indirect Dipole-Dipole Coupling Between ^{11}Cd, J ^{3}Cd and ^{12}STe in Solid CdTe," *Zeitschrift fur Physik B*, vol. 182, pp. 175–182, 1979.

[23] D. Look and D. Moore, "Nuclear-Magnetic-Resonance Measurement of the Conduction-Electron g Factor in CdTe," *Physical Review B*, vol. 5, pp. 3406–3412, 1972.

[24] M. Shroyer, J. Furdyna, A. Ryskin, and W. Warren, "NMR Study of Bistable Defects Under In Situ Illumination," *Physica B: Condensed Matter*, vol. 273–274, pp. 852–855, 1999.

[25] M. H. Shroyer, *An NMR Study of Bistable Defects in CdTe and* CdF_2. Doctor of Philosophy, Oregon State University, 1999.

[26] E. Pogue, A. Sutrisno, A. Nagaoka, M. Scarpulla, and A. Rockett, "Solid-state NMR for the detection of P and As doping-induced defects in CdTe," in preparation, 2019.

[27] R. Mens, P. Adriaensens, L. Lutsen, *et al.*, "NMR Study of the Nanomorphology in Thin Films of Polymer Blends Used in Organic PV Devices: MDMO-PPV/PCBM," *Journal of Polymer Science Part A: Polymer Chemistry*, vol. 46, pp. 138–145, 2008.

[28] W. M. Franssen and A. P. Kentgens, "Solid-State NMR of Hybrid Halide Perovskites," *Solid State Nuclear Magnetic Resonance*, vol. 100, pp. 36–44, 2019.

[29] B. A. Rosales, L. Men, S. D. Cady, M. P. Hanrahan, A. J. Rossini, and J. Vela, "Persistent Dopants and Phase Segregation in Organolead Mixed-Halide Perovskites," *Chemistry of Materials*, vol. 28, pp. 6848–6859, 2016.

[30] C. Roiland, G. Trippé-Allard, K. Jemli, *et al.*, "Multinuclear NMR as a Tool for Studying Local Order and Dynamics in $CH_3NH_3PbX_3$ (X = Cl, Br, I) Hybrid Perovskites," *Physical Chemistry Chemical Physics*, vol. 18, pp. 27133–27142, 2016.

[31] D. J. Kubicki, D. Prochowicz, A. Hofstetter, *et al.*, "Formation of Stable Mixed Guanidinium-Methylammonium Phases with Exceptionally Long Carrier Lifetimes for High-Efficiency Lead Iodide-Based Perovskite Photovoltaics," *Journal of the American Chemical Society*, vol. 140, pp. 3345–3351, 2018.

[32] D. J. Kubicki, D. Prochowicz, A. Hofstetter, S. M. Zakeeruddin, M. Grätzel, and L. Emsley, "Phase Segregation in Cs-, Rb- and K-Doped Mixed-Cation $(MA)_x (FA)_{1-x} PbII_3$ Hybrid Perovskites from Solid-State NMR," *Journal of the American Chemical Society*, vol. 139, pp. 14173–14180, 2017.

[33] W. Xiang, Z. Wang, D. J. Kubicki, *et al.*, "Europium-Doped $CsPbI_2$ Br for Stable and Highly Efficient Inorganic Perovskite Solar Cells," *Joule*, vol. 3, pp. 205–214, 2019.

[34] A. Karmakar, M. S. Dodd, S. Agnihotri, E. Ravera, and V. K. Michaelis, "Cu (II)-Doped $Cs_2SbAgCl_6$ Double Perovskite: A Lead-Free, Low-Bandgap Material," *Chemistry of Materials*, vol. 30, pp. 8280–8290, 2018.

[35] D. J. Kubicki, D. Prochowicz, A. Pinon, *et al.*, "Doping and Phase Segregation in Mn^{2+}- and Co^{2+}-Doped Lead Halide Perovskites from ^{133}Cs and ^{1}H NMR Relaxation Enhancement," *Journal of Materials Chemistry A*, vol. 7, pp. 2326–2333, 2019.

[36] D. Wang, M. Wright, N. K. Elumalai, and A. Uddin, "Stability of Perovskite Solar Cells," *Solar Energy Materials and Solar Cells*, vol. 147, pp. 255–275, 2016.

[37] S. Yang, W. Fu, Z. Zhang, H. Chen, and C.-Z. Li, "Recent Advances in Perovskite Solar Cells: Efficiency, Stability and Lead-Free Perovskite," *Journal of Materials Chemistry A*, vol. 5, pp. 11462–11482, 2017.

[38] A. Karmakar, A. M. Askar, G. M. Bernard, *et al.*, "Mechanochemical-Synthesis of Methylammonium Lead Mixed-Halide Perovskites: Unraveling the Solid-Solution Behavior Using Solid-State NMR," *Chemistry of Materials*, vol. 30, pp. 2309–2321, 2018.

[39] A. M. Askar, A. Karmakar, G. M. Bernard, *et al.*, "Composition-Tunable Formamidinium Lead Mixed Halide Perovskites via Solvent-Free Mechanochemical Synthesis: Decoding the Pb Environments Using Solid-State NMR Spectroscopy," *The Journal of Physical Chemistry Letters*, vol. 9, pp. 2671–2677, 2018.

[40] S. Won, S.-B. Saun, S. Lee, S. Lee, K. Kim, and Y. Han, "NMR Spectroscopy for Thin Films by Magnetic Resonance Force Microscopy," *Scientific Reports*, vol. 3, p. 3189, 2013.

Chapter 15
Summary and outlook

Daniel Abou-Ras[1] *and Steve Harvey*[2]

In the various chapters of this book, numerous characterization techniques are presented that can be applied to thin-film solar cells to determine (micro)structural, compositional, electrical, and optoelectronic properties. What has yet to be more fully elucidated is to what extent these characterization techniques can be combined, in a correlative way, to enhance the information gathered on materials and devices. Indeed, it is valuable to consider combining techniques to verify the relevance of the measured materials properties or to obtain them on different length scales—to compare surface with bulk properties, or to correlate structure and composition of materials with electrical and optoelectronic properties.

Point defects and their properties are often characterized by capacitance–voltage (*C–V*) measurements (Chapter 9), but they can also be analyzed by photoluminescence (PL) spectroscopy (Chapter 6). For example, Burgelman *et al.* [1], who studied $CuInSe_2$ thin films using both techniques, report that admittance and deep-level transition spectroscopy can give the density of deep-level states in the bandgap and their properties as electron and hole traps; likewise, PL can provide parallel insight into shallow defect levels. Spindler *et al.* [2] present an overview of defect spectroscopy on $Cu(In,Ga)Se_2$ and also compare the information on point defects in $Cu(In,Ga)Se_2$ obtained by PL and *C–V* measurements (see Appendix E of Ref. 2). Combining *C–V* and PL analyses while varying the materials composition as well as the measurement temperature during thin-film growth provides a powerful means for thoroughly analyzing defect states in a semiconductor.

As outlined in Chapter 11, electron energy-loss spectroscopy (EELS) in scanning transmission electron microscopy (STEM) gives insight into various types of materials properties on the (sub)nanometer level by detecting the energy loss of electrons scattered at different particles in the specimen under investigation; the magnitude of energy loss can range in scale from meV to keV. At loss energies on the order of meV, due to phonon scattering, vibrational properties of materials are probed, which can be compared directly with results from infrared or Raman

[1]Helmholtz-Zentrum Berlin für Materialien und Energie, Berlin, Germany
[2]National Renewable Energy Laboratory, Golden, CO, USA

spectroscopy (Chapter 7). Local variations of bandgap energies can be determined by evaluating the energy losses of around 1–5 eV, which should agree with corresponding analysis by PL spectroscopy, for example. Finally, the energy range from 50 eV to several keV features signals due to scattering at electrons bound to core levels. Thus, the information contained in this high-energy range of the EEL spectrum is equivalent to that of photoelectron spectroscopy (Chapter 12). Thus, elemental distributions can be investigated, but in addition, information can be gathered on chemical binding, density of electronic states, as well as structural ordering. EELS can access the materials properties discussed above on the (sub) nanometer scale, so it is interesting to check whether any of these materials properties—which can be confirmed by corresponding macroscopic measurements—vary on the much smaller length scale accessed by STEM.

Scanning probe microscopy (SPM) and its related techniques can provide insight on surface roughness, electrical properties, and density of states of semiconductor materials (Chapter 10). However, most SPM methods are strongly surface-sensitive (especially in the case of experimental setups in which the contacting needles are positioned on the surface and the measured current is running along it). To achieve a better understanding of the complete thin film being investigated, it is always recommended to perform additional analyses with methods that are more sensitive to the bulk of the thin film at a similar spatial resolution, such as scanning electron microscopy (SEM) and its related techniques. A good example of correlative SPM and SEM characterization is the investigation of grain boundaries in polycrystalline semiconductor thin films (as often applied in solar cells). Baier *et al.* [3] performed measurements using scanning Kelvin probe force microscopy (SKPFM) to determine differences in the contact potential between grain interiors and grain boundaries. These differences can be interpreted as upward or downward band bending at grain boundaries, and thus, they provide information on the charge densities and on barriers for charge carriers at the grain boundaries. Nevertheless, SKPFM signals are strongly influenced by the topographies of the thin-film surfaces, and hence, the locations of the grain boundaries often cannot be identified unambiguously. Therefore, Baier *et al.* also applied electron backscatter diffraction (EBSD) on the same areas as for the SKPFM measurements to be able to identify but also classify them according to their symmetries. Baier *et al.* showed that most highly symmetrical $\Sigma 3$ grain boundaries in $Cu(In,Ga)Se_2$ thin films do not exhibit substantial barriers for charge carriers.

EBSD measurements can also be combined with other SEM techniques on the identical positions to investigate structure-property relationships. An approach frequently applied for structural and chemical phase analysis on the submicrometer scale is combining EBSD with energy-dispersive X-ray spectroscopy on identical specimen positions [4]. Another example concerns recombination velocities at grain boundaries of polycrystalline thin films. In an effort to measure such quantities for polycrystalline $CuInSe_2$ layers in completed solar cells, Abou-Ras *et al.* [5] acquired EBSD maps as well as electron-beam-induced current (EBIC) and cathodoluminescence (CL) images on the identical areas. By using appropriate models to evaluate EBIC and CL profiles extracted across grain boundaries—which were identified and classified by EBSD—it was possible to estimate recombination velocities for more than 100 grain boundaries. In addition, the individual values were verified for each grain boundary by comparing EBIC and CL results.

When acquired on cross-sectional specimens of thin-film solar cells with *p–n* junctions at various bias voltages, EBIC images provide valuable insight into the lateral distributions of the widths of the space-charge regions (w_{SCR}) of the *p–n* junctions [6], from which the net-doping densities in the absorber layers can be estimated. These quantities can also be accessed by *C–V* analysis (Chapter 9), and thus, the EBIC results can be verified. In addition to the average w_{SCR} values, electrostatic potential fluctuations can be evaluated from EBIC images via local variations in w_{SCR}. At p–n junctions, these fluctuations can be attributed to inhomogeneously distributed, localized charge densities within the *p*-type and *n*-type parts (equivalent to the net-doping densities) as well as at the interfaces of the junctions [7,8]. The free electrons and holes redistribute to screen these charge densities, so the local electrostatic potentials exhibit fluctuations according to Poisson's equation. When electrostatic fluctuations become too large, they can lead to enhanced nonradiative recombination, and thus, to a deterioration of the device performance [9].

Time-of-flight secondary-ion mass spectrometry can be combined with optical imaging techniques (electroluminescence and thermography) to identify failure regions on the macroscopic (full-module) scale down to small-area defects (several millimeters field of view). A wealth of information about root-cause mechanisms of module failure can be obtained by combining these optical imaging methods with module coring and subsequent SEM- and TEM-based imaging and characterization with various marking methods to ensure analysis on the same exact area across techniques. In Section 13.2.4 (Chapter 13), a study was discussed that used these many methods on the same area of a cell. Results show that potential-induced degradation (PID) stems from sodium accumulation in structural defects, revealing at the several-angstrom level the mechanisms causing module power loss at the meter length scale. In addition, these methods were also used to show that PID recovery is due to out-diffusion of sodium from the structural defect [10].

More examples exist of the fruitful combinations of various techniques to achieve enhanced information from the characterization, but they go beyond the scope of this book. Of particular importance are reliable simulations of experimental work as well as the adjustment of the corresponding simulation parameters to fit to the experimental results. Naturally, correlative analyses applying several techniques on the same material or device always require greater effort and demands in terms of time and equipment. However, it is worthwhile to consider combined and correlative characterization wherever possible. In the future, in view of ever more complex materials systems and device structures, we can expect the correlative approach to become increasingly important.

References

[1] M. Burgelman, F. Engelhardt, J.F. Guillemoles, *et al.*, Defects in Cu(In, Ga) Se_2 semiconductors and their role in the device performance of thin-film solar cells. Prog. Photovolt: Res. Appl., 5 (1997) 121–130.

[2] C. Spindler, F. Babbe, M.H. Wolter, *et al.*, Electronic defects in Cu(In,Ga) Se_2: Towards a comprehensive model, Phys. Rev. Mater. 3 (2019) 090302.

[3] R. Baier, D. Abou-Ras, T. Rissom, M.C. Lux-Steiner, S. Sadewasser, Symmetry-dependence of electronic grain boundary properties in polycrystalline $CuInSe_2$ thin films, Appl. Phys. Lett. 99 (2011) 172102-1-3.

[4] D. Abou-Ras, M. Nichterwitz, R. Caballero, *et al.*, (Enhanced) insight in the microstructure and composition of chalcopyrite-type thin film solar cells, in: Proceedings of the 22nd European Photovoltaic Solar Energy Conference and Exhibition, Milan, Italy, September 3–7, 2007, WIP Munich, 2007, pp. 1911–1914.

[5] D. Abou-Ras, N. Schäfer, T. Rissom, *et al.*, Grain-boundary character distribution and correlations with electrical and optoelectronic properties of $CuInSe_2$ thin films, Acta Mater. 118 (2016) 244–252.

[6] D. Abou-Ras, N. Schäfer, N. Baldaz, S. Brunken, C. Boit, Electron-beam-induced current measurements with applied bias provide insight to locally resolved acceptor concentrations at p-n junctions, AIP Advances 5 (2015) 077191-1-7.

[7] D. Abou-Ras, N. Schäfer, C.J. Hages, S. Levcenko, J. Marquez, T. Unold, Inhomogeneities in Cu(In,Ga)Se_2 thin films for solar cells: The issue of length scales, RRL Solar 2 (2018) 1700199.

[8] R. Scheer, H.W. Schock, Chalcogenide Photovoltaics: Physics, Technologies, and Thin Film Devices, Weinheim, Germany, John Wiley & Sons (2011), pp. 19ff.

[9] J.H. Werner, J. Mattheis, U. Rau, Efficiency limitations of polycrystalline thin film solar cells: case of Cu(In,Ga)Se_2, Thin Solid Films 480–481 (2005) 399–409.

[10] S.P. Harvey, J. Moseley, A. Norman, *et al.*, Investigating PID shunting in polycrystalline silicon modules via multiscale, multi-technique characterization, Progress in Photovoltaics: Research and Applications 26(6) (2018) 377–384.

Index

www.ingramcontent.com/pod-product-compliance
Lightning Source LLC
LaVergne TN
LVHW011557200926
848484LV00001B/325